The Science of Well-Being

The Science of Well-Being

Edited by

Felicia A. Huppert
Department of Psychiatry
University of Cambridge

Nick Baylis
Department of Social and Developmental Psychology
University of Cambridge

and

Barry Keverne
Department of Zoology
University of Cambridge

Contributions originating from
Philosophical Transactions of
the Royal Society, Series B.

OXFORD
UNIVERSITY PRESS

Great Clarendon Street, Oxford OX2 6DP

Oxford University Press is a department of the University of Oxford.
It furthers the University's objective of excellence in research, scholarship,
and education by publishing worldwide in

Oxford New York

Auckland Cape Town Dar es Salaam Hong Kong Karachi
Kuala Lumpur Madrid Melbourne Mexico City Nairobi
New Delhi Shanghai Taipei Toronto

With offices in

Argentina Austria Brazil Chile Czech Republic France Greece
Guatemala Hungary Italy Japan Poland Portugal Singapore
South Korea Switzerland Thailand Turkey Ukraine Vietnam

Oxford is a registered trade mark of Oxford University Press
in the UK and in certain other countries

Published in the United States
by Oxford University Press Inc., New York

Chapters 6, 11, 13, 15, 16 © Oxford University Press 2005
Chapters 1–5, 8–10, 12, 14, 17, 19 © The Royal Society 2005
Chapter 7 © Bernard Gesch 2005
Chapter 18 © 2005 by the Massachusetts Institute of Technology and the
American Academy of Arts and Sciences
Chapter 20 © nef 2005

British Library Cataloguing in Publication Data
Data available

Library of Congress Cataloging in Publication Data
The science of well-being / edited by Felicia A. Huppert, Nick Baylis, and Barry Keverne.
Includes bibliographical references and index.
1. Success–Psychological aspects.   2. Happiness.   3. Well-being.
I. Huppert, Felicia A.   II. Baylis, N. (Nick)   III. Keverne, B. (Barry)
BF637.S8S382 2005   155.2′5–dc22   2005015599

Typeset by Newgen Imaging Systems (P) Ltd., Chennai, India
Printed in Great Britain
on acid-free paper by
Biddles Ltd, King's Lynn

ISBN 0–19–856752–9 (Pbk. : alk. paper)   978–0–19–856752–3 (Pbk. : alk. paper)
ISBN 0–19–856751–0 (Hardback : alk. paper)   978–0–19–856751–6 (Hardback : alk. paper)

10 9 8 7 6 5 4 3 2 1

Preface

Until recently, the focus of social and biomedical science has been on pathology and its remediation—on moving people from a condition of disorder towards one of adequate functioning. But most people want more out of life than just 'getting by'; they want a life that is enjoyable and satisfying. A new approach, the scientific study of well-being, is needed to advance our understanding of life going well—a life characterised by health and vitality, by happiness, creativity and fulfilment—along with an understanding of the social relationships and civic institutions that can foster these positive experiences.

Well-being is more than pleasant emotions; it is a positive and sustainable condition that allows individuals, groups or nations to thrive and flourish. Our concept also encompasses human resilience; there will always be setbacks inherent in the process of living, and the ability to develop and thrive in the face of such adversity is a key element of well-being.

Despite the substantial improvements in material circumstances that have been taking place in many parts of the world, there has not been a corresponding increase in individual or societal well-being. In many countries there is more wealth than ever before, better health care, improved access to education, and more opportunities for travel and leisure pursuits. Yet, in this era of apparent plenty, there are increasingly high rates of depression, crime, and social disintegration. Our conventional approaches seem to be failing.

Understanding how individuals and communities can be helped to thrive and flourish could be of great benefit to our citizens, our educators, and our leaders. It is promising that so many governments are beginning to identify National Indicators of Well-being as potentially a more pertinent goal than increasing Gross National Product. Numerous institutions and policy-making bodies are now actively considering ways to increase aspects of well-being, such as human capability, resilience, and social capital. Only a genuinely scientific exploration of this developing field can provide a source of trustworthy information to guide interventions.

Well-being, by its very nature, requires an integrated approach, one that embraces mind, body, society, and the environment. This multi-disciplinary volume unites a diverse range of leading researchers and practitioners who introduce the latest theoretical concepts and scientific evidence, some focusing on well-being at an individual level, others at a societal level.

This book arose out of the highly successful Royal Society Discussion Meeting held in London in November 2003, although the scope of the book goes well beyond the topics covered at that meeting. We are delighted to bring this important and innovative field to a wider audience, through the contributions of twenty of the world's leading authorities. We warmly welcome you to *The science of well-being*.

Felicia Huppert
Nick Baylis
Barry Keverne

Acknowledgements

We are very grateful to the Royal Society for hosting the Discussion Meeting on The Science of Well-being which provided the foundation and impetus for this book. In particular, we appreciate the encouragement given to us by the Society's President, Robert May.

The enthusiastic support for this volume shown by our editor, Martin Baum, enabled us to extend it well beyond the scope of the Royal Society meeting. The invaluable assistance provided by the many colleagues who reviewed the manuscripts has been greatly appreciated.

Our very special thanks go to Julie Aston for her untiring work in contacting authors, co-ordinating editors and publishers, her detailed correction and refinement of manuscripts, and above all, her cheerfulness and good humour throughout.

Contents

Part 4 **Cultural perspectives**

Part 5 **Social and economic considerations**

Abbreviations

5-HT	5-hydroxytryptamine (serotonin)		GDP	gross domestic product
5-HTT	serotonin transporter (gene)		GLA	gamma-linolenic acid
5-HTTP	5-hydroxy-L-trytophan		GNP	gross national product
AA	arachidonic acid		GP	general practitioner
ACC	anterior cingulate cortex		GWP	GoodWork Project
ADD	attention deficit disorder		HDI	human development index
ADHD	attention deficit hyperactivity disorder		HPA	hypothalamic pituitary adrenocortical (axis)
ALA	alpha-linolenic acid		HUFA	highly unsaturated fatty acids
ASP	antisocial personality disorder		IBIDS	International Bibliographic Information on Dietary Supplements (database)
BAS	behavioural activation scales			
BDI	Beck Depression Inventory		LC	locus ceruleus
BIS	behavioural inhibition scales		MIDUS	Midlife in the United States (study)
BMI	body mass index			
CBR	community-based rehabilitation (project)		MPQ	multidimensional personality questionnaire
CEO	chief executive officer		MRI	magnetic resonance imaging
CI	confidence interval		Myr	million years
CRH	corticotropin-releasing hormone		MZ	monozygotic
CSF	cerebrospinal fluid		NFE	non-formal education (courses)
CWIN	Child Workers in Nepal		NGO	non-governmental organization
DBD	disruptive behaviour disorders		NK	natural killer (cells)
DHA	docosahexaenoic acid		OCD	obsessive compulsive disorder
DIY	do-it-yourself		OECD	Organization of Economic Cooperation and Development
DLPFC	dorsolateral prefrontal cortex			
DRM	day reconstruction method		OFC	orbitofrontal cortex
DSM	*Diagnostic and statistical manual of mental disorders*		ONS	Office for National Statistics (UK)
			PACE	physician-based assessment and counselling for exercise (US project)
DZ	dizygotic			
EEG	electroencephalogram			
EMG	electromyography		PANAS	positive and negative affect scales
EPA	eicosapentoenic acid		PET	positron emission tomography
ESC	equality, security, and community		PFC	prefrontal cortex
ESM	experience-sampling method		PFCA	prefrontal cortical asymmetry
EVS	European Values Survey		PHC	primary health care
FDG	fluorodeoxyglucose		POMS	profile of mood states (scale)
fMRI	functional magnetic resonance imaging		PTSD	post-traumatic stress disorder
			QALYs	quality-adjusted life years

RDA	recommended dietary allowance (US)
ROI	region of interest
RwR	relationship with reality
SAI	sensual awareness inventory
SAM	sympathetic adrenomedullary (axis)
SD	standard deviation
SEM	standard error of the mean
SIPI	short imaginal processes inventory
SRHS	self-reported health status
SSRIs	serotonin-specific reuptake inhibitors
SWB	subjective well-being
TTM	transtheoretical model (of behaviour change)
vmPFC	ventromedial prefrontal cortex
WHO	World Health Organization
WVS	World Values Survey
Z-CAI	Zambia cognitive assessment instrument

Contributors

David J.P. Barker
University of Southampton and MRC
Epidemiology Centre,
Southampton General Hospital,
Southampton SO16 6YD, UK

Nick Baylis
Department of Social and
Developmental Psychology,
University of Cambridge,
Cambridge, UK

Stuart J.H. Biddle
British Heart Foundation National
Centre for Physical Activity and Health,
School of Sport and Exercise Sciences,
Loughborough University,
Loughborough,
Leicestershire LE11 3TU, UK

George W. Burns
Milton H. Erickson
Institute of Western Australia,
62 Churchill Avenue, Subiaco,
Western Australia 6008, Australia

Richard J. Davidson
Laboratory for Affective Neuroscience,
W.M. Keck Laboratory for Functional
Brain Imaging and Behavior,
University of Wisconsin–Madison,
1202 West Johnson Street,
Madison,
Wisconsin 53706, USA

Antonella Delle Fave
Dipartimento di Scienze Precliniche
LITA Vialba,
Università degli Studi di Milano,
Via G. B. Grassi, 74, Milan, Italy

Panteleimon Ekkekakis
Department of Health and Human
Performance,
Iowa State University, Ames,
Iowa, USA

Robert H. Frank
Johnson Graduate School of
Management,
Cornell University, Ithaca,
New York 14853, USA

Barbara L. Fredrickson
Department of Psychology,
University of Michigan,
525 East University Avenue,
Ann Arbor,
Michigan 48109–1109, USA

Johan Galtung
TRANSCEND: A Network for Peace
and Development and TRANSCEND
Peace University <www.transcend.org>

Howard Gardner
Harvard Graduate School of Education,
Harvard University,
Cambridge,
Massachusetts, USA

Bernard Gesch
University Laboratory of Physiology,
Parks Road,
Oxford OX1 3PT, UK

Elena L. Grigorenko
PACE Center, Department of
Psychology,
Yale University, Box 208358,
New Haven, CT 06520, USA

John F. Helliwell
Department of Economics,
University of British Columbia,
997–1873 East Mall, Vancouver,
British Columbia V6T 1Z1, Canada

Felicia A. Huppert
Professor of Psychology,
Department of Psychiatry,
University of Cambridge,
Box 189 Addenbrooke's Hospital,
Cambridge CB2 2QQ, UK

Daniel Kahneman
Department of Psychology,
Princeton University,
Princeton,
New Jersey 08544, USA

Eric B. Keverne
Subdepartment of Animal Behaviour,
University of Cambridge,
Madingley,
Cambridge CB3 8AA, UK

Sonia J. Lupien
Laboratory of Human Stress Research,
Douglas Hospital Research Center,
Department of Psychiatry,
McGill University,
6875 Boulevard Lasalle,
Montreal,
Quebec H4H 1R3, Canada

Nic Marks
The New Economics Foundation,
3 Jonathan Street,
London SE11 5NH, UK

Fausto Massimini
Dipartimento di Scienze Precliniche
LITA Vialba,
Università degli Studi di Milano,
Via G. B. Grassi, 74, Milan, Italy

Randolph M. Nesse
Department of Psychiatry,
Department of Psychology, and
Institute for Social Research,
The University of Michigan,
426 Thompson Street,
Room 5261, Ann Arbor,
Michigan 48104, USA

Acacia C. Parks
Positive Psychology Center,
University of Pennsylvania,
Philadelphia,
Pennsylvania 19104, USA

Robert D. Putnam
Kennedy School of Government,
Harvard University,
Cambridge,
Massachusetts 02138, USA

Jason Riis
Center for Health and Well-being,
Princeton University,
Princeton,
New Jersey 08544, USA

Martin E.P. Seligman
Positive Psychology Center and
Department of Psychology,
University of Pennsylvania,
Philadelphia, Pennsylvania, USA

Hetan Shah
The New Economics Foundation,
3 Jonathan Street,
London SE11 5NH, UK

Tracy Steen
Charles O'Brien Center for
Addiction Treatment,
University of Pennsylvania,
Philadelphia,
Pennsylvania, USA

Robert J. Sternberg
PACE Center,
Department of Psychology,
Yale University, Box 208358,
New Haven, CT 06520, USA

Susan Verducci
Stanford Center on Adolescence,
Stanford University, Stanford,
California, USA

Nathalie Wan
Laboratory of Human Stress
Research,
Douglas Hospital Research Center,
Department of Psychiatry,
McGill University,
6875 Boulevard Lasalle,
Montreal,
Quebec H4H 1R3, Canada

Part 1

Evolution and development

Randolph M. Nesse is professor of psychiatry and professor of psychology at the University of Michigan where he directs the Evolution and Human Adaptation Program. He is trying to discover the evolutionary origins of capacities for mood and for committed relationships as foundations for a general understanding of mental disorders.

Chapter 1*

Natural selection and the elusiveness of happiness

Randolph M. Nesse

The core dilemma of modernity

At the beginning of the third millennium, we humans in technological societies are not sure how to proceed. Ever since the Enlightenment people in the West have pursued happiness by trying to eliminate the causes of suffering, on the sensible belief that this would lead to a world that would be, if not utopia for all, at least much happier for most. Subsequent advances in technology have made life easier, safer, and more pleasurable and comfortable to a degree that could not have been imagined. The sufferings caused by hunger, pain, sickness, boredom, and even the early death of loved ones have been largely eliminated from the experiences of many people. The triumph of technology over most of the specific causes of human suffering is nothing short of miraculous. But the deeper hope that this would lead to general happiness is not only unfulfilled, it is almost a cruel joke. Even among those who have succeeded beyond measure in getting what people always wanted, vast numbers of people remain deeply unhappy, and many of the rest live lives that feel frantic, meaningless, or both. This is the core dilemma of modernity. What we have been doing to increase general happiness is no longer working, and there is no consensus about what we should try next.

This quest for happiness is, of course, much more ancient and diverse than just the Western attempt to use rationality to solve problems. Every religion offers its own answers, most of which emphasize the need to act according to duty instead of desire, and the power to change one's own attitudes in an unchangeable world. A related range of solutions offered by ancient philosophers in the Western world can be arrayed on a continuum from hedonistic pursuits to balanced living to ascetic denial (Nussbaum 1994). In more recent historical times, utilitarian social philosophers offered various schemes for organizing human groups, advocating variations on policies that provide the greater happiness for the greatest number (Mill 1848; Bentham and Mill 1973). The great experiments in communism have foundered spectacularly on the shoals of human nature, leaving the appearance that capitalistic democracies are the stable final condition of human social groups (Fukuyama 1992). In very recent years, attention has shifted from grand social theory to the effectiveness of mood-altering drugs and the beliefs they foster, namely, that severe states of unhappiness are usually the products of brain abnormalities (Andreasen 1984; Valenstein 1998).

This is not the place to review the attempts by priests, philosophers, politicians, and physicians to understand and relieve unhappiness. For our purposes, it is essential to note only the diversity, persistence, and passion of the quest over the centuries, and the tendency to focus on relieving unhappiness. The strategy at the social level has mainly been to give people more of what they want by promoting economic growth and solving specific problems. Although a reliable method to compare well-being across the centuries is lacking, it seems clear that the remarkable growth in wealth and the technological advances of recent decades have improved human well-being considerably. However, we have now reached a point where average life satisfaction is stable at best, and possibly declining rapidly (Kahneman *et al.* 1999). The time is ripe to consider new perspectives on the factors that influence well-being.

Social science tackles well-being

Human subjective well-being has long been a central focus of social science. From Durkheim's attempts to understand the social origins of anomie to Freud's interpretations of an individual's impulses and defences, the ancestors of modern social science have doggedly persisted in the task of finding out why people are unhappy. More modern social science has applied objective methods that provide detailed epidemiological pictures of the correlates not only of anomie, but also of depression, anxiety, and other aversive states (Tsuang and Tohen 2002). It examines how developmental experiences influence individual relationships and emotions and the effects of different kinds of therapies (Rutter and Rutter 1993). Increasingly, it explores not only the effects of genetic factors (Bouchard and Loehlin 2001), but also the effects of specific genes (Bouchard and Loehlin 2001; Caspi *et al.* 2002). The vast bulk of this research has been structured as attempts to explain pathology. Although effective interventions have only recently become available, a great deal of work to find the correlates of unhappiness has been accomplished. Because resources for intervention are so much more available than those for prevention, the effects of early abuse and neglect and the effects of dysfunctional social groups have been overshadowed by the dominant paradigm of biological psychiatry, with its emphasis on individual differences in brain mechanisms and drug treatments (Valenstein 1998).

In the past decade, the challenge of finding routes to human happiness has taken a new turn, with an explicit focus not only on the causes of suffering, but now also on the origins of positive states of well-being (Kahneman *et al.* 1999). This, in part, reflects a growing recognition that happiness and flourishing do not automatically emerge when the swamps of suffering are drained, and it results also from indications that positive and negative affect are not necessarily opposite ends of one continuum (Watson *et al.* 1988). Furthermore, some people experience states of flow and flourishing that are far more positive than ever appear in the lives of average people (Csikszentmihalyi 1977). If some proportion of their extraordinary well-being could be shared with others, the benefits would be truly wonderful. Other chapters in this volume highlight many of those advances.

Many findings from these research endeavours are increasingly solid. Although most people describe their lives as 'generally happy' (Diener and Diener 1996), 16% of individuals in the USA have experienced episodes of serious depression and, in any given year, approximately 6% of adults will experience 2 weeks or more during which depressive symptoms interfere with their ability to function effectively (Kessler *et al.* 2003). Rates of anxiety disorders are of the same order of magnitude, and the comorbidity of anxiety and depression is extraordinary (Maser and Cloninger 1990). The definitions of these severe disorders are quite restrictive, and considerably more people report milder states of low mood or anxiety. Another major finding is that the comorbidity among many mental disorders is so striking that most of the disorders occur in the small proportion of the population who have many diagnoses, while most people have none. That is not quite true. Almost half of the population has qualified for a lifetime psychiatric diagnosis using the criteria from the American Psychiatric Association's (1994) *Diagnostic and statistical manual of mental disorders* (DSM) criteria, and approximately 30% qualify for a diagnosis in a given year (Kessler *et al.* 1994). The problem of understanding how such a large proportion of the population can report that they are 'happy', despite so many suffering from a mental disorder, is a complex one whose resolution will require comparison of the quite different methods used to gather these two kinds of data.

Research that moves beyond pathology to measures of well-being has now been conducted with large samples in dozens of countries (Diener and Suh 1999; Prescott-Allen 2001; Helliwell 2002). There is some consistency in the shape of the distribution, with most people reporting general satisfaction with their lives and a fraction saying that they are dissatisfied. The correlates are also consistent and significant. The average well-being for a society increases as the average income increases up to the equivalent of approximately US $10 000 per year; above that level additional gross national product (GNP) adds little to the average happiness ratings of the populace (Kahnemann *et al.* 1999). Within societies, the picture is different; increasing income increases ratings of well-being, although the benefits taper off in high income brackets, perhaps in part because of ceiling effects for the scales used.

Dozens of studies have examined demographic correlates of well-being (Argyle 1999). Women, on average, have well-being levels similar to those of men, but men are more likely to report extreme levels. Married people have higher well-being than single or divorced individuals but the differences are small, last only a few years, and the direction of causation is often uncertain. A particularly large and careful study that provided a multivariate analysis, including both individual factors within cultures and differences among cultures, found the strongest effect from health, closely followed by unemployment, results that tend to support the old view that happiness is indeed mostly lack of unhappiness (Helliwell 2002). Other effects found in the study were the positive effects of being married, believing in God, having a personal sense that people can be trusted, living in a country where trust is high, and the tendency for negative affect in the prime adult years and for those living in the former Soviet Union. The grand sum of variance explained in this study is 25%. Adding variables for societal differences accounted for an additional 1% of the variance. These estimates are comparable to the estimated 20% of subjective well-being (SWB) variance explained by demographic factors in other studies (Diener and Lucas 1999).

Although accounting for only 25% of the variance may seem low, it is remarkably high given the profound and disturbing evidence that individual levels of well-being are remarkably stable over time almost irrespective of what happens (Loewenstein and Shane 1999). Even dramatic events such as becoming paralysed or winning the lottery have effects on SWB that are strikingly small and temporary (Brickman *et al.* 1978). While reconsideration of these findings suggests mood changes more congruent with everyday expectations (Easterlin 2003)—the lottery makes most people somewhat happier and serious medical problems leave most people less happy—the stability of SWB remains impressive. Furthermore, studies of twins and adopted children suggest that much of the variation in SWB among individuals, more than half in most estimates, can be attributed to their genetic differences (Tellegen *et al.* 1988; Lykken and Tellegen 1996). Perhaps even more surprising is that little, if any, of the correlations among siblings can be attributed to being raised in the same family, suggesting that the similarities arise from shared genes (Bouchard and Loehlin 2001). Personality traits, especially extraversion and neuroticism, predict SWB even over and above demographic factors (Diener and Lucas 1999). Variations in these traits arise, like those for SWB, mostly from genetic variation, with considerable similarity for sibs raised in different families, but little similarity among adopted children raised together in the same household (Bouchard and Loehlin 2001).

Shared genetic factors could, of course, account for the association of certain personality features, especially extraversion, with SWB. But it is increasingly clear that the observed high heritability estimates for such traits do not mean that they are in any way directly 'encoded in the genes'. Far from it. First of all, there may be many different genetic routes to a trait, some more direct than others. But equally important is recognizing that the crucial inherited factors may be preferences. For instance, an individual's likelihood of having experienced a severe life event in the past 12 months is highly heritable (Saudino *et al.* 1997). There are obviously no genes that directly cause life events, but there certainly are inherited tendencies such as risk-taking or having a short temper that are likely to lead to life events (Bouchard and Loehlin 2001). Consideration of such possibilities is even more important for conditions such as depression, where there is a pronounced tendency to assume that the influence of genetic factors implies brain defects. An evolutionary functional view of the involved systems suggests many other routes for genetic influences. One person might inherit a tendency for ambition that can never be satisfied, whereas another inherits a tendency to become so attached to a partner that he/she will not leave despite mistreatment. Perhaps most depressogenic of all, in the model developed here, would be tendencies to undertake huge difficult goals, and to be unwilling to give them up, even when all efforts are obviously in vain.

Another line of research is explicating the brain mechanisms that mediate states of well-being. Progress is coming rapidly in uncovering the neurochemistry of mood, with the role of dopamine in attention and motivation increasingly well understood, and the role of serotonin in anxiety and depression becoming more clear (Nathan *et al.* 1995; Gershon *et al.* 2003). The anatomic correlates of mood are becoming clear, with evidence showing an association of depression with increased activity in a region of the right forebrain (Davidson 1992). Even within-subject manipulations of mood influence activation of this brain region. Transcranial magnetic brain stimulation, which disrupts neuronal activity in this area, seems to relieve depression in some individuals, although the matter of finding adequate control conditions is difficult (Gershon *et al.* 2003).

On a social and political level, it is abundantly clear that certain policies can increase average SWB in a society (Layard 2005). More equitable income distribution is highly correlated with the average level of well-being in a society, and high taxes on high incomes and luxury goods would result in only infinitesimal decrements in the positional pleasures provided by luxury goods (Frank 1999). Most democratic societies seem unable, however, to enact laws based on this knowledge to increase the well-being in their societies. Or is it possible, as Layard (2005) and Felicia Huppert have separately suggested, that confirmation and dissemination of these findings may help political groups to transcend short-term self-interest and to enact policies that would improve well-being for whole societies?

In a parallel phenomenon at the individual level, people often do not act in ways that would bring them happiness. Addiction is obviously a huge cause of misery but, even in domains where people are imagined to make more considered and free choices, they still seem to pursue goals that bring pleasure more reliably than happiness. The human tendency to make decisions with a short-term time horizon has been noted for several millennia, but it is nonetheless a powerful explanation for human behavioural tendencies that limit SWB (Kimball 1993). At least in modern Western societies, people also tend to make disproportionate investments in the pursuit of money, status, and attractiveness, domains in which the value of resources is intrinsically relative to what others have, resulting in escalating arms races that

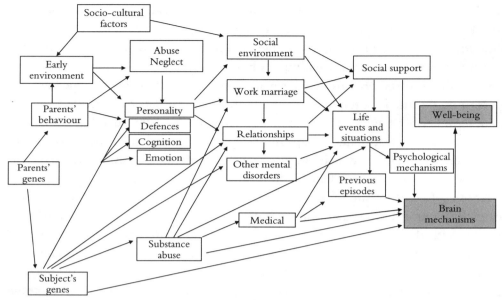

Fig. 1.1 Factors and pathways that influence SWB.

sap time and energy from friendships and social engagement that are highly correlated with SWB (Crocker and Wolfe 2001).

Many of the other chapters in this book will further explicate the extraordinary knowledge base we now have on factors that influence SWB. But what are we to do with all this information? We need a model. All the variables studied and their connections need to be incorporated into a path diagram so we can see their interrelationships. Figure 1.1 summarizes some major pathways.

Taken together, the studies that measure the coefficients for each of these relationships comprise a remarkable achievement. Lest it be thought that Fig. 1.1 is just an imaginary exercise, models nearly this complex have been published for depression (Kendler *et al.* 2002). Continuing research is investigating the details and the interaction effects in an attempt to make the model stronger. However, from another perspective, the model vividly illustrates the limits of this paradigm. How exactly are we to use this information? What should we do next? Carry out still more careful studies of these relationships? Add yet more variables? It seems that there is something missing from the picture. Perhaps there is some way to begin to make sense of all this information.

What evolution offers to the study of well-being

New questions

All of the research already described is about proximate mechanisms. It explicates the components and operation of the systems that influence well-being, from the level of neurochemistry to brain loci, life events, relationships, personality, cultural factors, and methodological factors that influence how an individual reports SWB. Two things are omitted.

First, the model assumes that different individuals will have the same pathways to SWB. We all know, of course, that a life event such as divorce can devastate one person while rejuvenating another. Recent studies are increasingly investigating how genetic differences interact with different life experiences to generate changes in SWB (Kendler *et al.* 1999; Caspi *et al.* 2003). However, few studies look at differences in individual life goals and strategies to provide an idiographic (individualized narrative) context in which innate and acquired differences can be analysed (Brown and Harris 1978). The divorce of idiographic and nomothetic methods poses a very serious problem for social science studies in general and studies of SWB in particular. Everyone knows someone who became depressed because his wife was having an affair, or someone who fell in love and got a new lease on life, but the effects of these powerful individualized factors are not captured in studies that rely on age, sex, social class, education, and other demographic variables. One study that has attempted to bridge the gap finds that SWB is influenced mainly by events and resources that influence a person's ability to reach his or her individual goals (Diener and Fujita 1995). This offers an explanation for why levels of resources so poorly predict SWB, and it offers the beginnings of a scientific basis for bringing data on individual situations to bear on the problems of psychology.

Many people assume that an evolutionary approach necessarily treats all humans crudely as if they were all the same. Instead, however, an evolutionary approach offers a theoretical bridge from the richness of idiographic data to the solidity of nomothetic understanding. Once you know the core goals that humans tend to pursue and the mechanisms that regulate their pursuit of these goals, you can understand how individuals deal with these desires, and with the exigencies that inevitably keep people from their goals. This section concludes by returning to this issue to argue that natural selection provides a nomothetic framework that can incorporate the effects of individual differences in goals and strategies.

The other gap is the absence, with a few exceptions (Buss 2000), of evolutionary explanations of happiness to complement existing proximate knowledge. Even if we knew every connection of every neuron and the effects of every life event on all levels of the system, we would still only have an explanation of how the mechanisms work. Also needed, and equally important, is knowledge about why those mechanisms exist at all and if they give a selective advantage (Tinbergen 1963). Such questions about behavioural mechanisms are structurally identical to questions about physical traits such as the shape of a finch's beak or a fish's fin. They are best addressed using the comparative method to determine if differences in the environments encountered match the characteristics of species in different environments. For instance, guppies vary considerably in their degree of bright coloration. The hypothesis that bright colour is a product of sexual selection is supported by strong and consistent female preferences for brightly coloured males across environments. However, in deep water, where predators are common, the bright colour is often fatal; there, natural selection has selected camouflage coloration instead (Endler 1986). Even when the comparative method is not applicable, however, there are other strategies for assessing the viability of evolutionary hypotheses about why behavioural regulation mechanisms are the way they are (Alcock 2001).

Behaviours and associated emotional mechanisms are not shaped directly, but the brain mechanisms that regulate them are (Symons 1990). No mechanism can give rise to the exact optimal action in every circumstance but, on average, ultimately, in the natural environment, we can safely assume that evolved behaviour regulation mechanisms will give rise to actions that tend to maximize reproductive success.

Six reasons why the body is not better designed

Most people like to imagine that normal life is happy and that other states are abnormalities that need explanation. This is a pre-Darwinian view of psychology. We were not designed for happiness. Neither were we designed for unhappiness. Happiness is not a goal left unaccomplished by some bungling designer, it is an aspect of a behavioural regulation mechanism shaped by natural selection. The utter mindlessness of natural selection is terribly hard to grasp and even harder to accept. Natural selection gradually sifts variations in DNA sequences. Sequences that create phenotypes with a less-than-average reproductive success are displaced in the gene pool by those that give increased success. This process results in organisms that tend to want to stay alive, get resources, have sex, and take care of children. But these are not the goals of natural selection. Natural selection has no goals: it just mindlessly shapes mechanisms, including our capacities for happiness and unhappiness, that tend to lead to behaviour that maximizes fitness. Happiness and unhappiness are not ends; they are means. They are aspects of mechanisms that influence us to act in the interests of our genes.

Although the capacities for unhappiness and other aversive states are normal, their expression may be normal or abnormal, useful or harmful. Even after recognizing the utility of affects, we still need to ask why unhappiness is so widespread. Is it a design flaw present in only some individuals? Is it an acquired disease resulting from some kind of damage to the brain? Is it a typical response to an abnormal environment? Or is most such suffering useful even now? The six categories of evolutionary explanations for vulnerability to disease offered by Darwinian medicine (Williams and Nesse 1991; Nesse and Williams 1994) offer a framework that is helpful in attempts to address the question of why so many people find well-being so elusive. They apply to mental conditions just as well as to physical ailments, and the link to physical disease provides a solid grounding for analysis of mental disorders.

One obvious possibility is that natural selection is too slow to adapt us to rapidly changing environments. Our modern world is vastly different from the environments in which we evolved. Much, even most, chronic disease results from this mismatch; atherosclerosis, diabetes, hypertension, and the complications of smoking and alcohol are rare in hunter–gatherers even at older ages. In discussing these findings Eaton *et al.* (1988) have described modern people as 'stone-agers in the fast lane'. Many before and after Rousseau have attributed human unhappiness to our separation from a state of nature (Stevens and Price 1996) and most generations imagine that life was more calm and happy in times past, even if not more healthy. One recent book uses growing knowledge about evolution and how our lives differ from those of our ancestors as a guide to authentic happiness (Grinde 2002).

A second reason is that no speed of selection can keep us ahead of other organisms that evolve faster than we can. We can never hope to evolve defences against all pathogens, and our defences and their ways to get past our defences give rise to arms races that yield expensive and dangerous bodily devices. Perhaps even more disturbing is recognition that competition with other humans, especially mate competition (Buss and Schmitt 1993), has shaped similar arms races, including products such as jealousy that help to explain the elusiveness of well-being (Buss *et al.* 1999).

Third and fourth, there are many things that no system can accomplish and some that are impossible for organisms shaped by selection. Design trade-offs leave every aspect of any machine, including the body, somewhat less than optimal. For instance, thicker bones would break less easily, but they would be heavy and unwieldy. Natural selection is subject to additional constraints, especially the requirement that changes can take place only by incremental alternations of existing designs, and the result must work well in every generation. There is some hope that the awkward qwerty keyboard will be replaced some day, but our eyes will always be an absurdly designed device, with the nerves and vessels running between the light and the retina where they cast shadows and cause an unnecessary blind spot.

Fifth, some traits shaped by natural selection are useful even though they seem like mistakes. A tendency to compete to the death for a mate, for instance, seems like the height of absurdity. However, selection shapes many such traits that benefit the genes at the expense of the individual and of community peace and solidarity. Organisms are designed for maximum reproduction, nothing else (Dawkins 1976). An allele that increases reproductive success will increase in prevalence over the generations, even at the cost of health, longevity, and happiness.

Finally, other apparent abnormalities are actually useful defences. Pain, fever, nausea, and cough arise when something is wrong, but they themselves are useful responses. Their very aversiveness is a product of selection, almost certainly a component of motivational systems that move organisms out of bad circumstances and that promote future avoidance of harm. Likewise, the aversiveness of anxiety and other negatives is also useful (Marks and Nesse 1994). We would rather live without them, but they

Table 1.1 Six reasons why the body is not better designed

1 The body is poorly adapted to modern environments
2 Arms races with other fast-evolving organisms
Selection cannot do everything:
3 Constraints
4 Trade-offs
What appears to be a defect is actually useful:
5 Trait increases reproduction at the expense of health
6 The trait is a defence that is useful even if aversive

are essential for our own welfare; people born without a capacity for pain are dead by the age of 30. Other aversive emotions motivate us to do things that advance the interests of our genes; most people would be better off without envy, jealousy, and sudden lust, but we experience them nonetheless.

The six evolutionary explanations in Table 1.1 offer a robust framework for examining the possible origins of lack of well-being. Although the focus on negative states may seem to be the old 'negative psychology' approach that positive psychology is trying to transcend, this evolutionary approach is actually quite novel in that it expands attention from the causes of individual pathologies to the evolutionary explanations for why emotional systems are designed the way they are, and to the benefits of negative as well as positive emotions. For this reason I address the evolutionary origins and functions of negative affects prior to considering positive affects.

The utility and prevalence of negative emotions and affects

In seeking evolutionary explanations for emotions and affects, the temptation is to jump directly to hypotheses about possible functions. This is a mistake. Although emotional states would not exist unless they had been useful, there is no one-to-one correspondence between an emotion and a function. One emotion can serve multiple functions, and one function may be served by several different emotions. Anxiety motivates escape and future avoidance, and it can serve as a warning to others. Disgust also motivates escape, prepares the body to make escape more likely, and motivates future avoidance. Instead of jumping directly to postulated functions, a more explicitly evolutionary approach attends to the situations that shaped each emotion (Nesse 1990; Tooby and Cosmides 1990). It is these situations, with their adaptive challenges recurring over evolutionary time, that shaped special states of the organism that facilitate coping with these challenges. If a situation contains neither threats nor opportunities it will have no influence on fitness. This is why there are few, if any, neutral emotions.

Aversive emotions arise in situations when a loss has occurred or when the risk of loss is high. From the primal capacity of one-celled organisms to move away from excess heat, dryness, acidity, or salinity, natural selection has gradually differentiated a host of responses to cope with different kinds of threats. This phylogenetic perspective emphasizes that the differentiation of defensive states will be only partial and that we should expect much messy overlap in both regulatory and effector mechanisms. Our human minds insist on parsing the world into nice neat categories such as fear, sadness, anger, and jealousy; our minds rebel at the untidiness of psychological reality. The history of emotions research may be interpreted as a reflection of human attempts to view the mind as a machine with neat components designed by some intelligence (Plutchik 1980). Many attempts have been made to locate the emotions in a two- or three-dimensional space, often with a positive–negative axis and another axis that reflects the level of arousal. Others have tried to pair each emotion with its opposite in circumplex models. Much has been made of the distinction between emotions, with their short duration and specific referents, and affects with their longer duration and relative disconnection from specific stimuli (Morris 1992). All of these descriptions are informative, but match only very roughly the evolutionary origins of emotional states in the specific situations that shaped them.

This is not the place for details about how best to parse the negative emotions but, as an example, consider the subtypes of anxiety disorders. Even the avowedly atheoretical DSM system (American Psychiatric Association 1994) differentiates general anxiety, phobic anxiety (with several subtypes), panic, and social anxiety, among others. The observation that excesses of one kind of anxiety predispose to others has spawned a small research industry to examine the origins of this 'comorbidity' (von Hecht *et al.* 1989; Maser and Cloninger 1990). From an evolutionary point of view, however, there is every reason to expect that these states will be only partly differentiated (Marks and Nesse 1994). Note the contrast between this position and the massive modularity sometimes espoused by evolutionary psychologists (Cosmides and Tooby 1994). An evolutionary view certainly does imply that the mind cannot be a blank slate, but it equally strongly implies that the mind's structure consists not of distinct modules, each shaped to carry out a particular task, but of jury-rigged and partly overlapping mechanisms that in one way or another tend to lead to adaptive behaviour most of the time (Nesse 2000*b*). The mechanisms do process information, but calling them algorithms incorrectly suggests that their mechanisms are like those of a digital computer. Nevertheless, the emotions are closer to modules than anything else is. Each was shaped to cope with the challenges associated with a particular kind of situation.

The prevalence of negative affect seems excessive. As already noted, 15% of the US population has had an episode of severe depression. Many other people just have many bad days when they are too worried, sad, or angry to function, or at least nowhere near being full of energy and loving well-being. Most attempts to understand this state of affairs seeks explanations in individual differences, often based on the assumption that there is something wrong with these suffering people. In some cases there is. But an entirely separate question is why we are all designed in ways that leave us likely to experience negative emotions, often for no apparent reason (Gilbert 1989).

Because many defences are inexpensive compared with the harm they protect against, false alarms are both normal and common for many defences. For instance, if successful panic flight costs 200 calories but being clawed by a tiger costs the equivalent of, say, 20 000 calories, then it will be worthwhile to flee in panic whenever the probability of a tiger being present is greater than 1%. This means that the normal system will express 99 false alarms for every time a tiger is actually present; the associated distress is unnecessary in almost all individual instances. Blocking the tendency to panic would be an unalloyed good. Except, that is, for that 1 time in 100. This has been called the 'smoke detector principle' after our willingness to accept false alarms from making toast because we want a smoke detector that will give early warning about any and every actual fire (Nesse 2005).

The same principle explains how it is possible for physicians to relieve so much suffering with so few complications. Much of general medicine consists of using drugs to block perfectly normal expressions of cough, fever, vomiting, pain, etc. This is generally safe because the regulation of these defences follows the smoke detector principle and the vast majority of the time the defence is not needed to the degree expressed and because the body has redundant defences that offer good substitutes if one defence is blocked (Nesse and Williams 1994). Attention to the risks of mindlessly blocking defences is, however, gradually growing.

Positive emotions and affects

Just as negative emotions give an advantage in situations that pose threats, positive emotions give advantages in situations that offer opportunities (Fredrickson 1998) or when progress towards a goal is faster than expected (Carver 2003). The common notion that positive and negative emotional states are simply opposites has been challenged. Even on questionnaires, scales of positive and negative affect can be remarkably unrelated to each other (Watson *et al.* 1988). Nonetheless, brain research has documented the substantially different pathways and mechanisms for the 'behavioural inhibition system' as contrasted with systems that regulate appetitive behaviour (Gray 1987). These findings are perfectly congruent with an evolutionary view of emotions. In bacteria, information that feeds in from cell surface receptors for over 30 different substances all gets funnelled into a binary output— continue swimming ahead, or tumble randomly and swim in some other direction. Bacteria are too small to be able to detect chemical gradients over the length of their bodies, but they can compare the concentration of a substance now with the concentration half a second previously. This rudimentary memory allows extraordinarily adaptive behaviour, movement towards food and away from dangers

(Dusenbery 1996). Our positive and negative emotions are mere elaborations, albeit vastly more complex after at least an additional 600 Myr of evolution.

It is, however, somewhat harder to specify the situations that arouse positive as compared with negative emotions. Although it is fairly easy to see the advantage that negative emotions offer by facilitating escape from threats, it is harder to see exactly why someone with a capacity for joy would have a selective advantage over someone who lacked the capacity. Why not just move towards opportunities and take advantage of them; why get all emotional about it? One answer involves the social functions of emotions. These we will save for consideration later. Even for an organism that is simply foraging, however, states of positive arousal can be useful and are quite distinct, even neurologically, from incentive motivation (Berridge 1996). In the simplest possible terms, an organism needs to do only three things well for its behaviour to maximize fitness. First, it must exert the right amount of energy in the current task, walking versus running, for instance. Second, it must stop its current activity at the point when some other activity offers a greater pay-off per minute. Third, it must choose what subsequent activity would most advance fitness. All of these may be usefully regulated by affective states.

If a major opportunity suddenly becomes temporarily available, say the appearance of a tree full of fruit that is also being harvested by birds, then an intense burst of effort will be worthwhile to take advantage of the opportunity. When nearly all the fruit is gone, continuing to spend hours to get the last small bits at the top will not be worthwhile; it is better to quit and do something else. Whether that something else involves looking for another tree, going back to camp, or something else depends on the detailed characteristics of the available alternatives and inner states.

Different kinds of positive experiences seem to be somewhat differentiated according to domain, from the generic differences between physical pleasure versus social joy, to the more distinct differences in positive feelings from being praised versus the pleasure from vanquishing an enemy (Ellsworth and Smith 1988b). Nonetheless, much about positive feelings fits nicely into a larger framework based on the situations that arise in the pursuit of generic goals. If an organism is pursuing a goal and encounters an obstacle, it becomes aroused, even aggressive (Klinger 1975). The increased effort and risk-taking soon determine if the obstacle is surmountable or not. If it is, or if other circumstances make approach to the goal more rapid than expected, the resulting positive emotions facilitate investing yet more effort. If, however, increasing efforts lead to positions further and further from the goal, then low mood tends to disengage effort from the goal, or at least from the specific strategy. This prevents wasted effort and tends to allocate effort efficiently among various possible enterprises and strategies.

This model has been elaborated in 2 decades of work by a score of psychologists (Klinger 1975; Janoff-Bulman and Brickman 1982; Pyszczynski and Greenberg 1987;

Cantor 1990; Carver and Scheier 1990; Emmons 1996; Martin and Tesser 1996; McGregor and Little 1998; Mackey and Immerman 2000). Together, they offer a sophisticated description of a domain general mechanism for regulating effort allocation. Surprisingly, it has hardly been recognized by psychiatry, and it is only now being mapped to the behavioural ecology equivalent, foraging theory (Charnov 1976). At the core of foraging theory is the principle that organisms should stay in the same feeding patch until the rate of return per unit time decreases to the average rate of return over all patches including the search time needed to find a new patch.

Organisms do not need to calculate differential equations to behave optimally: they just need to have a mechanism that disengages effort from the current patch whenever the rate of return falls below the average recent rate of return. They also need to monitor the magnitude and trajectory of the average rate of return. As the evening cools, flying costs a bumble-bee more and more calories per minute. When the rate of expenditure is greater than the rate of return, it is best to give up foraging for that day (Heinrich 1979).

Whether the bumble-bee experiences anything like joy or sadness is of little consequence here. What is important is that people experience positive affect when they are reaching their goals more quickly than expected, and they experience negative affect and decreased motivation when goals seem to be unavoidably slipping away (Carver and Scheier 1990). As many psychologists have noted, opposing this normal emotional blockade of motivation only makes the negative affect stronger, and an inability to disengage from a major unreachable life goal is a recipe for serious depression. Although this effect has been documented for subjects in experimental and cross-sectional designs (Martin and Tesser 1996; Carver and Scheier 1998; Wrosch *et al.* 2003), it has yet to be applied to clinical or community samples to see how well it can explain episodes of mild and more severe depression (Nesse 2000*a*).

This view suggests locating positive affect not as a response to a domain-specific situation but as a response to the various situations that arise in pursuit of individual goals. The simplest framework incorporates the distinctions between opportunities and threats, and between the time before and after the outcome to yield four core emotions that regulate motivation (see Table 1.2). The earliest reference to such a model is from Plato, but the basic model was elaborated by Cicero and the Stoics, then endorsed by Virgil and other leading Romans. Later it was slightly

Table 1.2 A simple model of emotions for goal pursuit

	Before	After
Opportunity	Desire	Pleasure
Threat	Fear	Pain

Table 1.3 Emotions for situations that arise in the pursuit of social and physical goals

Situation	Before	Usual progress	Fast progress	Specific obstacle	Slow/no progress	Success	Failure
Opportunity							
Physical	Desire	Productive effort	Flow	Frustration	Resignation	Pleasure	Disappointment
Social	Excitement	Friendship, engagement	Gratitude	Anger	Low mood	Happiness	Disappointment
Threat or loss							
Physical	Fear	Defensive behaviour	Confidence	Despair	Despair	Relief	Pain
Social	Anxiety	Defensive behaviour	Confidence	Anger or helplessness	Helplessness	Relief	Sadness

modified by the replacement of desire by hope in Hume's simplest formulation (Fiesler 1992).

The model can be elaborated further, however, by distinguishing physical from social domains, and by expanding the sequence of situations that are routinely encountered in the course of goal pursuit (see Table 1.3). The word 'goal' can be problematic both because some goals are things people want to avoid (e.g. the goal of avoiding sickness) and also because many goals are very personal or even spiritual quests (such as trying to be good, or seeking a state of transcendence) that are far removed from the competitive tone of the word 'goal' (Emmons 1999). Nonetheless, as things or states that people want, they can be described usefully as goals, and this construct makes important links with work on motivation (Gollwitzer and Moskowitz 1996), efficacy (Costa *et al.* 1985), possible selves (Oyserman and Markus 1993), and control theories (Wrosch and Heckhausen 2002), in addition to the burgeoning literature on goal pursuit and mood (Klinger 1975; Carver and Scheier 1990; Cantor 1994; Emmons 1996; Little 1999; Wrosch *et al.* 2003).

It appears that generic emotions for coping with the situations that arise in pursuit of all goals have been differentiated by natural selection and experience to deal with specific kinds of opportunity or threat. For instance, the threat that involves the possible loss of a mate's fidelity arouses emotions that are aspects of jealousy (Buss *et al.* 1999). If the threat involves a risk of loss of social position, the specific emotions are humiliation, pride, etc. (Gilbert and Andrews 1998). The important point is that emotional states were shaped to deal with the situations that arise in the pursuit of goals, and they were subsequently partially differentiated to deal more effectively with the situations that arise in the pursuit of certain specific goals. Social goals are of particular importance (Kenrick *et al.* 2002). There is no attempt here to suggest that all of the differentiation is a result of natural selection. Life experiences, including the richness of human language and forethought, elaborate and further differentiate variations on the basic themes to yield the complexities we observe and experience.

A phylogeny of emotions

The tendency in emotions research has been to attempt, in the best traditions of science, to describe as many data as possible with the simplest and smallest number of concepts and relationships. This has given rise to three main traditions. One posits two or more dimensions and tries to locate emotions in the space described (Larsen and Diener 1992). Usually one dimension is positive versus negative affect and another is aroused versus non-aroused. A related approach orients each emotion directly opposite from its paired converse emotion in a circumplex (Plutchik 1980). A second main approach tries to define a few basic emotions, usually starting with joy, sadness, fear, anger, surprise, and disgust (Izard 1992). Finally, appraisal theories explain emotions more explicitly in the framework of how events change an individual's perception of his or her ability to reach personal goals, and thus take us further towards a functional evolutionary view (Ellsworth and Smith 1988*a*; Ortony *et al.* 1988).

A further step in that direction is to consider the phylogeny of emotions. The behavioural repertoire of bacteria is limited to moving forward in the same direction, or tumbling randomly to some new direction (Dusenbery 1996). This allows them to move towards food and away from danger, a dichotomy that matches the neurologically instantiated differentiation between behavioural approach and behavioural avoidance systems in mammalian brains (Gray 1987; Davidson 1992). After the fundamental division of affect into positive and negative, selection seems to have gradually further differentiated responses to increase the ability of organisms to cope with the vicissitudes that arise in efforts to obtain the three main kinds of life resources recognized by behavioural ecologists: personal, reproductive, and social effort. Further differentiation would facilitate coping with opportunities and threats involving more specific kinds of resources in each domain. The illustration of these relationships in Fig. 1.2 is not intended as a definitive proposal for how selection actually shaped the emotions, but it offers a framework for thinking about the origins of emotions that is rooted in biology. It suggests that some long-standing debates about emotions may arise from our human cognitive tendency to try to organize reality into crisp simple categories that may not match what selection has shaped. Furthermore, the diagram intentionally shows overlap between emotions. The key concept is that emotions are partly differentiated from precursor states. They are neither fully distinct nor areas in a dimensional space; instead, they are states that are simultaneously both overlapping and somewhat separate. Note that observing this phylogenetic tree from above would give the appearance of separate emotions, possibly even in an arrangement something like a circumplex model.

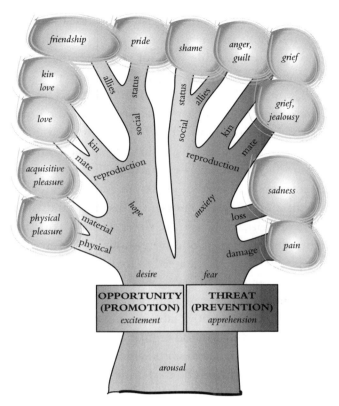

Fig. 1.2 A phylogeny of emotions. Resources are in upright font, emotions in italic font, and situations in capitals.

Diagonal psychology

The costs of negative emotions are obvious enough, but neglect of their benefits is so prevalent it has been called 'the clinician's illusion' (Nesse and Williams 1998). Clinicians are especially prone to being unable to recognize the utility of defences because negative states are aroused when something is wrong, because they are aversive, and because they can usually be blocked without adverse consequences. A parallel illusion, perhaps even stronger, is prevalent in considerations of positive emotions. Only rarely are the dangers of positive emotions considered, except in the context of mania or drug abuse. However, the utility of any emotion depends entirely on whether it is expressed in the situation in which it is useful. In other situations, it can be expected to be harmful. The case for the dangers of positive emotions is made most straightforwardly by individuals with mania. Their joy is infectious, their optimism and self-confidence unbounded. However, they usually lack the ability to stick with one thing long enough to accomplish anything, they often are incensed by the least criticism, and their respect for other people diminishes to the vanishing point. One manic may give away his life's savings on a whim, while another joyfully drives 100 m.p.h. to a sexual liaison with a potentially dangerous stranger.

Much milder is hypomania. This state of high mood without justification often leads to substantial accomplishments and its charisma may bring an expanding network of admirers. It is hard to see any downside without very carefully looking for life problems that are being neglected and dangers that are unprepared for. Enquiries of several chronically happy people revealed that two had stayed for years in untenable marriages that, in retrospect, should have ended much sooner. Others continued without much dissatisfaction in jobs that did not begin to use their potentials. Another admitted that she knew cognitively that her spontaneous and exuberant affairs were dangerous to her health and her marriage, but lack of anxiety left her to continue until catastrophe occurred. These anecdotes offer only a few examples of some costs of high mood. They do not mean that most individuals with hypomania would be better off with more average states of mood; that is an open empirical question. If higher than average mood were shown to be useful for most people, an evolutionary perspective suggests the need to look hard for explanations in aspects of the modern novel environment.

One of the main questions facing happiness research is whether most people would be better off if they experienced more positive affect and, if that proves to be the case, how it can be accounted for given that mood regulation mechanisms were shaped by natural selection. Much happiness research starts with the folk psychology notion that happiness is good for you and proceeds to demonstrate correlations between positive affect and a variety of other indicators of well-being including friendships, achievement, health, and longevity. The confounding factor here is, of course, that people who are in good circumstances or are otherwise doing well can be expected to experience positive affect as a result of these circumstances. The solution is to conduct longitudinal studies that either look at changes in well-being or that experimentally induce positive affect to examine the effects on other variables. Studies of both kinds have been conducted, and both show that changes towards positive affect are associated with other positive changes.

Aside from the quite plausible possibility that file drawers are full of unpublished negative studies, this conclusion combines with the documented heritability of positive affect to pose a serious question. If positive affect is strongly heritable and improves function, and presumably reproduction, then why did natural selection not long ago shape a higher average level of positive affect? More directly, why are there so many very successful people with many friends and resources who remain in states of chronically low mood? They may, of course, have a disorder. However, determining what is and is not a disorder will be secure only when we have an understanding, so far elusive, of the origins, functions, and normal regulation of the mood system.

Affect and goal pursuit

Our brains could have been wired so that good food, sex, being the object of admiration, and observing the success of one's children were all aversive experiences.

However, any ancestor whose brain was so wired would probably not have contributed much to the gene pool that makes human nature what it is now. Similarly, if there were someone who experienced no upset at failure, no anxiety in the face of danger, and no grief at the death of a child, his or her life might be free of suffering but also would probably be without much accomplishment, including having offspring. These evolved preferences for pursuing certain resources and avoiding their loss are at the very centre of human experience. It is not surprising that bad feelings are reliably aroused by losses, threats of losses, and inability to reach important goals (Emmons 1996).

This has led thinkers over several millennia to a common prescription for happiness. Just adjust your goals to what is possible, be satisfied with who you are and what you have, and happiness will be yours. It is wise advice with an increasingly strong scientific foundation. Occasionally, someone is able to follow it, usually with good effect. Some follow the established middle way of Buddhism, striving to transcend the tangles of desire that are seen as the origins of all suffering. However, most of us muddle on, trying to do things in life, some feasible, some grand, others mundane, some successful, others sources of constant frustration, and some that lead to abject failure after huge efforts. The effects of discrete successes or failures on mood are strong, but not as strong as efforts that are steadily productive or increasingly ineffective despite great effort. It is important to recognize that only some of the goals in question are tangible, such as getting a job or buying a house. Success for many other goals, such as winning the golf tournament, being chosen as the beauty queen or the valedictorian, or having higher social status than others in a group, depends on winning a zero-sum game with escalating competition. Other goals that influence our states of mind are more elusive yet. How may people spend their lives trying to get their mothers finally to love them, to get a spouse to want to have sex again, to stop a child from taking drugs, or trying to control their own habits? In such desperate enterprises that cannot be given up are the seeds of intense dissatisfaction that often precede serious depression.

The motivational structure of individual lives

The actual situation is often even more challenging. The task is not just to choose sensible goals; individuals must at every juncture make decisions that allocate effort to the pursuit of one goal and not another. This is not just a human problem. As noted in Fig. 1.2, behaviour ecologists routinely categorize effort into one of several domains. Somatic effort gets material resources and preserves health. Reproductive effort is spent finding and wooing mates, and caring for offspring. Social effort is divided between helping others and otherwise making alliances, and striving for status, recognition, and power in a group (Table 1.4).

Table 1.4 Categories of effort

Somatic effort
Effort to get external resources such as food and shelter
Effort to get and keep personal resources such as health and staying alive
Reproductive effort
Mating effort
Parenting effort
Social effort
Effort to make alliances
Effort to gain status and social power

Most animals must allocate their effort carefully between feeding and watching for predators (Krebs and Davies 1997). Without adequate monitoring, sudden death will come soon, but too much time looking about will result in starvation. The amount of time an animal spends feeding takes away from time guarding its nestlings. With too little guarding they will probably become victims, but with too little time feeding they will starve. For males especially, there is a major trade-off between effort displaying to attract mates, and the effort that can go into foraging and protecting offspring and the existing mate. In species that have hierarchies, such as most primates, yet another major trade-off requires allocating the optimal amount of effort and risk-taking to competing for position in the group, and every bit of this effort takes away from effort available for protection, foraging, mating, and parenting. Reproductive success depends on managing these trade-offs effectively. Many of the decisions made by the brain, usually at an intuitive and automatic level, involve allocation of resources. And many of the difficulties that we call 'stress' arise from conflicts among the demands for pursuing multiple resources. These conflicts are not pathological, or found only in extreme circumstances; they are the substance of everyday decisions.

When there are sufficient time, energy, and resources to successfully pursue current goals, life is good. Recent survey research has made it clear that a simple level of resources is not a strong predictor of well-being; both theory and laboratory studies suggest that rate of approach to goals is much more closely correlated with affect (Carver and Scheier 1990). However, an evolutionary assessment suggests considering a still broader view of what is important to the organism. What really counts is the viability of the overall motivational structure, that is, the degree to which all major goals can be pursued successfully without unduly compromising others. If this is correct, it means that survey studies of well-being will overlook most of what is important (Diener and Fujita 1995). Not only do individuals differ in their values, they also have different commitments to pursuing certain goals at different times in life (Heckhausen and Schultz 1995; McGregor and Little 1998). Furthermore, a person with substantial resources in every area may nonetheless be faced with impossible dilemmas. The ambitious executive who is climbing nicely, and who has a wonderful spouse and child at home, may be faced with demands from a spouse to participate more in caring for the children at the very time when promotion in the company requires accepting more assignments out of town. This is no different, in principle, from the dilemma of a hunter–gatherer who has to choose to forage far afield or to stay at home and protect his family.

The implications for methodology are severe. To better predict positive affect, not only must multiple efforts in different domains be considered, but also individual differences in commitment to idiosyncratic goals and how much progress can be made towards these goals in the current context, which includes the effects of all other efforts to reach other goals. This takes us close to the perspective of those who argue that only narrative includes information detailed and idiographic enough to allow a real understanding of an individual's life (Oatley 1999). However, instead of giving up on trying to find nomothetic generalities, this evolutionary approach to affect anticipates which goals people will typically pursue at each life stage. It also maps generic emotional responses on to the nomothetic framework of situations that arise during goal pursuit, and it provides a strong expectation that affects will reflect an individual's assessment of their ability to make progress towards important life goals. In this perspective, the distress of people who seem to be successful in every area need not be automatically attributed to autonomous brain processes or deep unconscious complexes, but instead may arise from the necessity of coping with impossible trade-offs and other threats to the stability of the whole system.

In the grip of our genes

The systems that regulate our emotions were shaped not to benefit individuals or the species, but only to maximize the transmission of the genes themselves. Thus, every species experiences a tension between efforts to maintain individual welfare and efforts to maximize reproductive success. An oyster allocates substantial effort to making the usual 10–20 million eggs; an oyster that makes 100 million eggs will probably compromise its ability to survive. We humans experience the situation much more acutely, feeling ourselves drawn into status competitions, driven to pursue sexual partners, and subject to envy despite our very best intentions. Many of these tendencies benefit our genes but not our individual selves. The dissatisfactions arising from these unending pursuits drain our capacities for happiness; however, even many people who know this still find themselves unable to enjoy what they have because of their efforts to get what their genes motivate them to want. The very same dynamic, however, plays out in taking care of children. Here, too, people spend effort and take risks that assist their reproductive success. The difference is that the investment is in quality and success of offspring, something that is often achievable, in contrast to the unending struggle for positional goods such as status and relative wealth and recognition. This said, many people spend years distressed because they cannot recover from an illness, help a child cope with drug abuse, or reform an abusive spouse.

Modern goals

The mechanisms that regulate behaviour were shaped for an environment considerably different from the one we live in now. The size and duration of personal goals may be a major difference that accounts for much pathology. Just a few thousand years ago, most individuals allocated their effort among a limited variety of tasks: gathering food; taking care of kin; participating in the group; etc. But as social groups enlarged and roles became more specialized, the requirements for success have escalated so that the big rewards now go to those who allocate a huge proportion of their life's effort to one domain, and sometimes even just to one goal. The Olympic athlete, for instance, has to devote him or herself so completely to the goal of gold that a balanced life is impossible. Likewise, someone who wants to be a music star, a top academic, or a chief executive officer (CEO) will almost always have to sacrifice much to strive for the goal. What is worse, the goal may take years of specialized effort, it may not be reached, and, if failure occurs, no satisfactory alternative may be available. Our brain regulation systems were never designed to cope with efforts so long in duration, towards goals so large, with all-or-none outcomes that offer few alternatives. It is relatively easy to give up on looking for nuts when several days of foraging have proved fruitless. Giving up on a PhD programme after 5 years, or a marriage after 10 years, are decisions orders of magnitude larger, decisions whose costs often lead individuals to persist in the pursuit of hopeless or unwanted goals, creating the exact situations that disengage motivation and cause depression.

The competition for scarce elite social roles requires not only extreme efforts, but also extravagant displays whose significance is proportional to their expense (Veblen 1899). Once Rolex watches are no longer a reliable indicator, status display moves on to extravagant cars and homes (Frank 1999). Once the pool of available spouses expands to thousands of people, simple attractiveness is overlooked in favour of exaggerated sexual characteristics. Many bemoan the human costs of such competitions, but when employers seek a worker, the Harvard graduate gets the edge, and when people with options decide who to marry, the drop-dead good-looking partner is rarely resisted.

In short, the connections between modern life and compromised well-being may be far more complex than simple increased exposure to situations that foster upwards social comparison (Gilbert *et al.* 1995). Instead, modern social structures often seem to induce motivational structures that leave individuals sacrificing much in life to pursue major efforts to reach huge goals whose attainment is uncertain and whose alternatives are few and unsatisfactory. Furthermore, these pursuits may foster personal relationships based on exchange, and even exploitation, instead of the commitment that characterizes small kinship groups (Fiske 1992).

Individual differences

Although some variation in happiness arises from life circumstances, most is a subject characteristic that remains remarkably stable despite changing life circumstances. Twin studies find that half of the variance in SWB in most modern populations can be attributed to genetic differences between individuals (Bouchard and Loehlin 2001; Lykken and Tellegen 1996). This poses an evolutionary question entirely separate from the questions addressed so far. Instead of asking why the capacities for mood exist and how they are regulated, it asks instead why natural selection would leave such wide variation in baseline mood, and in the responsiveness of the mood system. A full answer depends on a better understanding than we now have of the evolutionary origins and functions of mood and mood regulation systems. However, some general principles nonetheless apply.

There are good reasons to expect that selection should leave baseline mood and its responsiveness with parameters somewhere in the vicinity of those that maximized reproductive success in the ancestral environment. How selection acts on a trait such as mood depends, of course, on the relative fitness of individuals with different phenotypes. Consider baseline mood first. We have already addressed the illusion that steady levels of very positive mood are optimal. If they were, we would expect that selection would quickly shift mood for most people to a steady high positive level. Of course, the data on depressive realism do suggest that the average person is more optimistic than is justified, at least about his or her own abilities (Alloy and Abramson 1988; Taylor and Brown 1988). One can readily imagine how social factors could shape such a tendency. People prefer to associate with others who are happy and successful, so an advantage might well accrue to those with a tendency towards optimistic distortion. Whether natural selection, life experience, or both would shape such a proposed distortion is unanswered. Nonetheless, as with other traits, fitness should be highest at some intermediate level of baseline mood with selection acting against genes that foster extremes of high or low baseline mood. It might well be that the distribution is skewed, perhaps with fitness falling off much faster on the side of low baseline mood than on that of high baseline mood. More important is the width of the distribution. It might well be that the average fitness difference between someone with a chronic moderate low mood and a steady high mood would be small, especially since the fitness value of such moods depends mainly on the circumstance.

A previous section emphasized that the fitness value of emotions depends on whether their expression occurs in the situations in which they are adaptive. Baseline levels of mood might well be less important than whether mood changes in response to cues that accurately indicate situations in which high and low mood are useful, and if changes are appropriately intense, and for the right duration. Lowered fitness should be characteristic of individuals whose moods swing too readily to extremes, and of individuals whose mood does not change in situations where it should.

If it is correct that affect reflects rates of progress towards goals, then the baseline mood may matter little so long as its changes adjust future behaviour in ways that allocate effort efficiently. While we attend to, and care very much about, steady levels of mood, they may have little effect on fitness, compared with the importance of changes in mood. Furthermore, what seem to us to be substantial variations in mood and mood regulation may have such small and erratic effects on fitness that the resulting systems are left quite variable. We keep looking for the exact cues and mechanisms that regulate mood. However, because the mind was not designed by an engineer, but was instead shaped by natural selection, the mood system may be far more irregular than we can imagine.

Implications of evolutionary insights about happiness

An evolutionary view of well-being shifts most people's perspectives substantially. It is disturbing to recognize that negative emotions exist because they have been useful in certain situations, and that positive emotions can be maladaptive. It is harder yet to recognize that the apparent superiority of positive emotions arises mainly because they feel good and they are associated with beneficial situations. More difficult still is recognition that selection does not shape emotion regulation systems for our benefit, and that the motives we experience often benefit our genes at the expense of the quality of our lives. Much easier to accept is the recognition that our brains were designed to function in a vastly different environment, with the resulting mismatch accounting for many malfunctions. In addition, there are the many unnecessary expressions of aversive states whose normality is accounted for by the smoke detector principle.

These insights should promote neither despair nor resignation. Just because a negative emotion is normal and useful in the natural environment, or even in our environment, does not mean that we should just accept it. Instead, an understanding of its normal regulation and functions provides crucial information for deciding if it is or is not safe to block a negative emotion in a particular instance. The pursuit of strategies to increase happiness will get farther faster if we understand the origins and functions of positive emotions (Johnson 2003). An evolutionary view suggests, however, that an excessively direct pursuit of happiness is likely to lead to frustration and, paradoxically, unhappiness, because happiness is not a reachable goal, but a state that emerges when an individual is making good progress towards his or her individual life goals.

An evolutionary perspective does not come down definitively on the side of one of the traditional prescriptions for pursuing well-being. It recognizes the pleasures of hedonism and the reasons for satiation and adaptation that limit that approach to happiness. It acknowledges the success of ascetic avoidance of temptations in preventing unhappiness, and the reasons why so few people are able to give up, or even control, their desires. It appreciates the good sense in the golden mean, and the reasons why so many who pursue it nonetheless lurch from unbridled desire to restraint and back again. Although it is disappointing to recognize that there is no formula for happiness, at least not one that applies to everyone, it is reassuring to understand the origins of the emotions in whose loops we dance. It may be so hard to control our emotions, so difficult to foster happiness, and to damp down sadness and envy, because those who could do that were deprived of crucial tools for adaptation, while those whose emotional experiences mapped accurately on to the situations in which they were useful had a selective advantage.

We are entering the period of history when technology will allow us to control our emotions, if not at will, at least far more easily. For now, the techniques are mostly chemical manipulations in the form of psychotropic drugs. Electrical brain influences via transcranial magnetic stimulations are developing quickly and implanted brain electrodes are already being used to influence mood. Within the lifetime of most readers, we will identify the genetic differences that leave some people morose throughout life and provide a fortunate few others with a steady sunny disposition. These advances will lead to profitable technologies for manipulating mood that will be widely used. Governments will no doubt worry initially that such interventions will sap the strength of their citizens, but they will also recognize the benefits of a happier populace who may well demand such interventions. The rich will certainly take advantage of these advances first, but the less fortunate, who have less happiness and disproportionate suffering, may get even greater benefits. The effects of such technologies on social and political structures will be substantial. If we are to make sensible decisions about these issues, we need major efforts now to reach deeper understandings about how our capacities for happiness were shaped by natural selection.

Our growing knowledge about the evolutionary functions of happiness and the situations in which it gives advantages also provides a scientific foundation for individual and social interventions. People imagine that, because acquiring things causes joy, having things will cause happiness. Although most people will be unable to use such insights, it is possible that some people will make good use of them, in the same way that some people now use information to modify their diets in ways that decrease immediate pleasure but increase long-term health and vigour. Perhaps most hopeful of all is the possibility that these findings can be used to structure human organizations, or perhaps even societies, in ways that offer people achievable goals that satisfy our deep human needs. We have used science and technology to protect ourselves from want and many diseases. It seems quite possible that we will be able to use our growing knowledge to make similar advances in promoting human happiness.

References

Alcock, J. (2001). *The triumph of sociobiology*. Oxford University Press, New York.

Alloy, L.B. and Abramson, L. (1988). Depressive realism: four theoretical perspectives. In *Cognitive processes in depression* (ed. L. Alloy), pp. 223–65. Guilford, New York.

American Psychiatric Association (1994). *Diagnostic and statistical manual of mental disorders (-IV)*, 4th edn. APA Press, Washington, DC.

Andreasen, N.C. (1984). *The broken brain: the biological revolution in psychiatry*. Harper and Row, New York.

Argyle, M. (1999). Causes and correlates of happiness. In *Wellbeing* (ed. D. Kahneman, E. Diener, and N. Schwartz), pp. 353–73. Russell Sage Foundation, New York.

Bentham, J. and Mill, J.S. (1973). *The utilitarians: an introduction to the principles of morals and legislation*. Anchor Press/Doubleday, Garden City, New York.

Berridge, K.C. (1996). Food reward: brain substrates of wanting and liking. *Neurosci. Biobehav. Rev.* **20**, 1–25.

Bouchard, J.T.J. and Loehlin, J.C. (2001). Genes, evolution, and personality. *Behav. Genet.* **31**, 243–73.

Brickman, P., Coates, D., and Janoff-Bulman, R. (1978). Lottery winners and accident victims: is happiness relative? *J. Pers. Soc. Psychol.* **36**, 917–27.

Brown, G.W. and Harris, T. (1978). *Social origins of depression*. Free Press, New York.

Buss, D.M. (2000). The evolution of happiness. *Am. Psychol.* **55**, 15–23.

Buss, D.M. and Schmitt, D.P. (1993). Sexual strategies theory: an evolutionary perspective on human mating. *Psychol. Rev.* **100**, 204–32.

Buss, D.M., Shackelford, T.K., Kirkpatrick, L.A., Choe, J., Hasegawa, M., Hasegawa, T., and Bennett, K. (1999). Jealousy and beliefs about infidelity: tests of competing hypotheses in the United States, Korea, and Japan. *Person. Rel.* **6**, 125–50.

Cantor, N. (1990). From thought to behavior: 'having' and 'doing' in the study of personality and cognition. *Am. Psychol.* **45**, 735–50.

Cantor, N. (1994). Life task problem solving: situational affordances and personal needs. *Person. Soc. Psychol. Bull.* **20**, 235–43.

Carver, C.S. (2003). Pleasure as a sign you can attend to something else: placing positive feeling within a general model of affect. *Cogn. Emotion* **17**, 241–61.

Carver, C.S. and Scheier, M.F. (1990). Origins and functions of positive and negative affect: a control-process view. *Psychol. Rev.* **97**, 19–35.

Carver, C.S. and Scheier, M. (1998). *On the self-regulation of behavior*. Cambridge University Press, Cambridge.

Caspi, A., McClay, J., Moffitt, T.E., Mill, J., Martin, J., Craig, I.W., Taylor, A., and Poulton, R. (2002). Role of genotype in the cycle of violence in maltreated children. *Science* **297**, 851–4.

Caspi, A. (and 11 others) (2003). Influence of life stress on depression: moderation by a polymorphism in the 5-HTT gene. *Science* **301**, 386–9.

Charnov, E.L. (1976). Optimal foraging: the marginal value theorem. *Theor. Popul. Biol.* **9**, 129–36.

Cosmides, L. and Tooby, J. (1994). Origins of domain specificity: the evolution of functional organization. In *Mapping the mind: domain specificity in cognition and culture* (ed. L.A. Hirschfeld and S.A. Gelman), pp. 85–116. Cambridge University Press, Cambridge.

Costa, D., Mogos, I., and Toma, T. (1985). Efficacy and safety of mianserin in the treatment of depression of women with cancer. *Acta Psychiat. Scand.* **320** (suppl.), 85–92.

Crocker, J. and Wolfe, C.T. (2001). Contingencies of self-worth. *Psychol. Rev.* **108**, 593–623.

Csikszentmihalyi, M. (1977). *Finding flow: the psychology of engagement with everyday life*. Basic Books, New York.

Davidson, R. (1992). Emotion and affective style: hemispheric substrates. *Psychol. Sci.* **3**, 39–43.

Dawkins, R. (1976). *The selfish gene*. Oxford University Press, Oxford.

Diener, E. and Diener, C. (1996). Most people are happy. *Psychol. Res.* **7**, 181–5.

Diener, E. and Fujita, F. (1995). Resources, personal strivings, and subjective well-being: a nomothetic idiographic approach. *J. Pers. Soc. Psychol.* **68**, 926–35.

Diener, E. and Lucas, R.E. (1999). Personality and subjective well-being. In *Well-being: the foundations of hedonic psychology* (ed. D. Kahneman, E. Diener, and N. Schwartz), pp. 213–29. Russell Sage Foundation, New York.

Diener, E. and Suh, E.M. (1999). National differences in subjective well-being. In *Well-being* (ed. D. Kahneman, E. Diener, and N. Schwartz), pp. 434–50. Russell Sage Foundation, New York.

Dusenbery, D.B. (1996). *Life at small scale: the behavior of microbes*. Scientific American Library, New York.

Easterlin, R.A. (2003). Inaugural article: explaining happiness. *Proc. Natl Acad. Sci. USA* **100**, 11176–83.

Eaton, S. B., Shostak, M., and Konner, M. (1988). *The paleolithic prescription*. Harper and Row, New York.

Ellsworth, P.C. and Smith, C.A. (1988*a*). From appraisal to emotion: differences among unpleasant feelings. *Motiv. Emotion* **12**, 271–302.

Ellsworth, P.C. and Smith, C.A. (1988*b*). Shades of joy: patterns of appraisal differentiating pleasant emotions. *Cogn. Emotion* **2**, 301–31.

Emmons, R.A. (1996). Striving and feeling: personal goals and subjective well-being. In *The psychology of action: linking cognition and motivation to behavior* (ed. P. M. Gollwitzer), pp. 313–37. Guilford Press, New York.

Emmons, R. A. (1999). *The psychology of ultimate concerns: motivation and spirituality in personality*. Guilford Press, New York.

Endler, J.A. (1986). *Natural selection in the wild*, Monographs in Population Biology. Princeton University Press, Princeton.

Fiesler, J. (1992). Hume's classification of the passions and its precursors. *Hume Stud.* **18**, 1–17.

Fiske, A.P. (1992). The four elementary forms of sociality: framework for a unified theory of social relations. *Psychol. Rev.* **99**, 689–723.

Frank, R.H. (1999). *Luxury fever: why money fails to satisfy in an era of excess*. Free Press, New York.

Fredrickson, B.L. (1998). What good are positive emotions? *Rev. Gen. Psychol.* **2**, 300–19.

Fukuyama, F. (1992). *The end of history and the last man*. Free Press, New York.

Gershon, A., Dannon, P., and Grunhaus, L. (2003). Transcranial magnetic stimulation in the treatment of depression. *Am. J. Psychiatry* **160**, 835–45.

Gilbert, P. (1989). *Human nature and suffering*. Lawrence Erlbaum, Hove, UK.

Gilbert, P. and Andrews, B. (1998). *Shame: interpersonal behavior, psychopathology, and culture*, Series in Affective Science. Oxford University Press, New York.

Gilbert, P., Price, J., and Allen, S. (1995). Social comparison, social attractiveness and evolution: how might they be related? *New Ideas Psychol.* **13**, 149–65.

Gollwitzer, P.M. and Moskowitz, G.B. (1996). Goal effects on action and cognition. In *Social psychology: handbook of basic principles* (ed. E.T. Higgins and A.W. Kruglanski), pp. 361–99. Guilford Press, New York.

Gray, J.A. (1987). *Fear and stress*. Cambridge University Press, Cambridge.

Grinde, B. (2002). *Darwinian happiness: evolution as a guide for living and understanding human behavior*. Darwin Press, Princeton, New Jersey.

Heckhausen, J. and Schultz, R. (1995). A life-span theory of control. *Psychiatry Rev.* **102**, 284–304.

Heinrich, B. (1979). *Bumblebee economics*. Harvard University Press, Cambridge, Massachusetts.

Helliwell, J.F. (2002). How's life? Combining individual and national variables to explain subjective well-being. National Bureau of Economic Research working papers, paper 9065.

Izard, C.E. (1992). Basic emotions, relations among emotions, and emotion–cognition relations. *Psychol. Rev.* **99**, 561–5.

Janoff-Bulman, R.and Brickman, P. (1982). Expectations and what people learn from failure. In *Expectations and action* (ed. N.T. Feather), pp. 207–37. Lawrence Erlbaum, Hillsdale, New Jersey.

Johnson, V.S. (2003). The origins and functions of pleasure. *Cogn. Emotion* **17**, 167–79.

Kahneman, D., Diener, E., and Schwarz, N. (eds.) (1999). *Well-being: the foundations of hedonic psychology*. Russell Sage Foundation, New York.

Kendler, K.S., Karkowski, L.M., and Prescott, C.A. (1999). Causal relationship between stressful life events and the onset of major depression. *Am. J. Psychiatry* **156**, 837–41.

Kendler, K.S., Gardner, C.O., and Prescott, C.A. (2002). Toward a comprehensive developmental model for major depression in women. *Am. J. Psychiat.* **159**, 1133–45.

Kenrick, D.T., Maner, J.K., Butner, J., Li, N.P., and Becker, D.V. (2002). Dynamical evolutionary psychology: mapping the domains of the new interactionist paradigm. *Pers.* Soc. *Psychol. Rev.* **6**, 347–56.

Kessler, R.C., McGonagle, K.A., Zhao, S., Nelson, C.B., Hughes, M., Eshelman, S., Willchen, H.U., and Kendler, K.S. (1994). Lifetime and 12-month prevalence of DSM-III-R psychiatric disorders in the United States: results from the national comorbidity survey. *Arch. Gen. Psychiatry* **51**, 8–19.

Kessler, R.C., Berglund, P., Demler, O., Jin, R., Koretz, D., Merikangas, K.R., Rush, A.J., Walters, E.E., and Wang, P.S. (2003). The epidemiology of major depressive disorder results from the National Comorbidity Survey Replication (NCS-R). *J. Am. Med. Assoc.* **289**, 3095–105.

Kimball, M.S. (1993). Standard risk aversion. *Econometrica* **61**, 589–611.

Klinger, E. (1975). Consequences of commitment to and disengagement from incentives. *Psychol. Rev.* **82**, 1–25.

Krebs, J.R. and Davies, N.B. (1997). *Behavioral ecology: an evolutionary approach*. Blackwell Science, Oxford.

Larsen, R.J. and Diener, E. (1992). Problems and promises with the circumplex model of emotion. *Rev. Pers. Soc. Psychol.* **13**, 25–59.

Layard, R. (2005). *Happiness—lessons from a new science*. Penguin, London.

Little, B.R. (1999). Personal projects and social ecology: themes and variation across the life span. In *Action and self development: theory and research through the life span* (ed. J. Brandtstadter and R.M. Lerner), pp. 197–221. Sage, Thousand Oaks, California.

Loewenstein, G. and Shane, F. (1999). Hedonic adaptation. In *Well-being: the foundations of hedonic psychology* (ed. E. Diener, D. Kahnemann, and N. Schwartz), pp. 302–29. Russell Sage Foundation Press, New York.

Lykken, D. and Tellegen, A. (1996). Happiness is a stochastic phenomenon. *Psychol. Sci.* **7**, 186–9.

Mackey, W.C. and Immerman, R.S. (2000). Depression as a counter for women against men who renege on the sex contract. *Psychol. Evol. Gender* **2**, 47–71.

Marks, I.M. and Nesse, R.M. (1994). Fear and fitness: an evolutionary analysis of anxiety disorders. *Ethol. Sociobiol.* **15**, 247–61.

Martin, L.L. and Tesser, A. (1996). *Striving and feeling: interactions among goals, affect, and self-regulation.* Lawrence Erlbaum, Hillsdale, New Jersey.

Maser, J. and Cloninger, R. (eds) (1990). *Comorbidity of mood and anxiety disorders.* American Psychiatric Press, Washington, DC.

McGregor, I. and Little, B.R. (1998). Personal projects, happiness, and meaning: on doing well and being yourself. *J. Pers. Soc. Psychol.* **74**, 494–512.

Mill, J.S. (1848). *Principles of political economy.* John W. Parker, London.

Morris, W.N. (1992). A functional analysis of the role of mood in affective systems. *Rev. Pers. Soc. Psychol.* **21**, 736–46.

Nathan, K.I., Musselman, D.L., Schatzberg, A.F., and Nemeroff, C.B. (1995). Biology of mood disorders. In *Textbook of psychopharmacology* (ed. A.F. Schatzberg and C.B. Nemeroff), pp. 439–77. APA Press, Washington, DC.

Nesse, R.M. (1990). Evolutionary explanations of emotions. *Hum. Nature* **1**, 261–89.

Nesse, R.M. (2000*a*). Is depression an adaptation? *Arch. Gen. Psychiatry* **57**, 14–20.

Nesse, R.M. (2000*b*). Natural selection, mental modules and intelligence. In *The nature of intelligence* (ed. G.R. Bock and J.A. Goode), pp. 96–115. Wiley, Chichester.

Nesse, R.M. (2005). Natural selection and the regulation of defensive responses. *Evol. Hum. Behav.* **26**(1), 88–105.

Nesse, R.M. and Williams, G.C. (1994). *Why we get sick: the new science of Darwinian medicine.* Vintage, New York.

Nesse, R.M. and Williams, G.C. (1998). Evolution and the origins of disease. *Sci. Am.* **289** (5), 86–93.

Nussbaum, M.C. (1994). *The therapy of desire: theory and practice in Hellenistic ethics.* Princeton University Press, Princeton.

Oatley, K. (1999). Why fiction may be twice as true as fact: fiction as cognitive and emotional simulation. *Rev. Gen. Psychol.* **3**, 101–17.

Ortony, A., Clore, G.L., and Collins, A. (1988). *The cognitive structure of emotions.* Cambridge University Press, New York.

Oyserman, D. and Markus, H. (1993). The sociocultural self. In *Psychological perspectives on the self,* Vol. 4. (ed. J. Suls and A.G. Greenwald), pp. 187–220. Lawrence Erlbaum, Hillsdale, New Jersey.

Plutchik, R. (1980). *Emotion: a psychoevolutionary synthesis.* Harper and Row, New York.

Prescott-Allen, R. (2001). *The wellbeing of nations: a country-by-country index of quality of life and the environment.* Island Press, Washington, DC.

Pyszczynski, T. and Greenberg, J. (1987). Self-regulatory perseveration and the depressive self-focusing style: a self-awareness theory of reactive depression. *Psychol. Bull.* **102**, 122–38.

Rutter, M. and Rutter, M. (1993). *Developing minds: challenge and continuity across the life span.* Basic Books, New York.

Saudino, K. J., Pederson, N.L., Lichentenstein, P., McClearn, G.E., and Plomin, R. (1997). Can personality explain genetic influences on life events? *J. Pers. Soc. Psychol.* **72**, 196–206.

Stevens, A. and Price, J. (1996). *Evolutionary psychiatry: a new beginning.* Routledge, London.

Symons, D. (1990). Adaptiveness and adaptation. *Ethol. Sociobiol.* **44**, 427–44.

Taylor, S.E. and Brown, J.D. (1988). Illusion and well-being: a social psychological perspective on mental health. *Psychol. Bull.* **103**, 193–210.

Tellegen, A., Lykken, D.T., Bouchard, T.J., Wilcox, K.J., Segal, N.L., and Rich, S. (1988). Personality similarity in twins reared apart and together. *J. Pers. Soc. Psychol.* **54**, 1031–9.

Tinbergen, N. (1963). On the aims and methods of ethology. *Z. Tierpsychol.* **20**, 410–63.

Tooby, J. and Cosmides, L. (1990). The past explains the present: emotional adaptations and the structure of ancestral environments. *Ethol. Sociobiol.* **11**, 375–424.

Tsuang, M.T. and Tohen, M. (2002). *Textbook in psychiatric epidemiology.* Wiley-Liss, New York.

Valenstein, E.S. (1998). *Blaming the brain: the truth about drugs and mental health.* Free Press, New York.

Veblen, T. (1899). *The theory of the leisure class: an economic study in the evolution of institutions.* Macmillan, New York.

von Hecht, H.Z.D., Krieg, C., Possl, J., and Wittchen, H.U. (1989). Anxiety and depression: comorbidity, psychopathology, and social functioning. *Comp. Psychiatry* **30**, 420–33.

Watson, D., Clark, L.A., and Carey, G. (1988). Positive and negative affect and their relation to anxiety and depressive dis orders. *J. Abnorm. Psychol.* **97**, 346–53.

Williams, G.W. and Nesse, R.M. (1991). The dawn of Darwinian medicine. *Q. Rev. Biol.* **66**, 1–22.

Wrosch, C. and Heckhausen, J. (2002). Perceived control of life regrets: good for young and bad for old adults. *Psychol. Aging* **17**, 340–50.

Wrosch, C., Scheier, M.F., and Carver, C.S. (2003). The importance of goal disengagement in adaptive self-regulation: when giving up is beneficial. *Self Ident.* **2**, 1–20.

Eric B. Keverne DSc, FRS, is professor of behavioural neuroscience and director of the Subdepartment of Animal Behaviour, University of Cambridge. His current interests are brain evolution and behaviour and the ways in which these have been influenced by genomic imprinting. In the past, he has worked on brain and behaviour in a number of species, particularly living primates.

Chapter 2*

Understanding well-being in the evolutionary context of brain development

Eric B. Keverne

Introduction

Much of the work on well-being and positive emotions has tended to focus on the adult, partly because this is when problems are manifest and well-being often becomes an issue by its absence. However, it is pertinent to ask if early life events might engender certain predispositions that have consequences for adult well-being. The human brain undergoes much of its growth and development postnatally until the age of seven and continues to extend its synaptic connections well into the second decade. Indeed, the prefrontal association cortex, areas of the brain concerned with forward planning and regulatory control of emotional behaviour, continue to develop until the age of 20. In this review I want to consider the significance of this extended postnatal developmental period for brain maturation and how brain evolution has encompassed certain biological changes and predispositions that, with our modern lifestyle, represent risk factors for well-being. Awareness of these sensitive phases in brain development is important to understanding how we might facilitate secure relationships and high self-esteem in our children. This will provide the firm foundations on which to develop meaningful lifestyles and relationships that are crucial to adult well-being.

Brain evolution

Allometric scaling and a large comparative database of mammalian brains have enabled assessments of brain evolution based on the widely accepted assumption that modern day phylogenies represent evolutionary progression (Harvey and Pagel 1991). Most hypotheses of brain evolution have suggested that larger brain size correlates with greater cognitive ability and that selection pressures for such abilities have come from ecological variables where the knowledge of available food sources in time and space requires complex cognitive maps. Equally important has been the complex nature of social life, especially in primate mammals, and the cognitive skills required to process and store this social information. While the size of the brain as a whole has been linked to different mammalian lifestyles or ecological features, this approach fails to take into account the different functions served by its component parts. For example, the hippocampus is concerned with processing spatial information and is remarkably enlarged in species that store and hide food, while the neocortex is enlarged in social living primates, with those living in larger groups having the larger neocortex (Dunbar 1992). Certain areas of the mammalian limbic brain (hypothalamus, amygdala, medial pre-optic area) are important for primary motivated behaviour such as maternal care, feeding, and sexual and aggressive behaviour (Keverne 1985). These areas of the brain are under strong hormonal and visceral influences and in small-brained mammals are primarily activated by olfactory cues. Most mammals only show maternal care after exposure to the hormones of pregnancy and parturition while sexual behaviour responds to the demands of the gonadal hormones that determine sexual motivation. However, large-brained human and non-human primates are spontaneously maternal while most sexual activity is non-reproductive and gonadal hormones are facilitating rather than determining human sexual behaviour. How then have these differing behavioural control mechanisms been influenced by evolutionary changes in the brain and how do they impinge on subjective well-being?

While certain regions of the primate 'executive' brain have expanded relative to the rest of the brain, regions of the brain that regulate primary motivated behaviour (hunger, sex, aggression, maternal care) have become reduced in size (Keverne et al. 1996). Areas of the cortex that are concerned with forward planning have increased exponentially, while those regions that respond to gonadal and visceral hormones have decreased in size. This does not mean that motivated behaviour has also declined in large-brained primates, but the controlling mechanisms for the behaviour have shifted away from tight linkage with physiological determinants in favour of deployment of cognitive behavioural strategies (Keverne et al. 1996).

The evolutionary remodelling of the mammalian brain has provided increased executive control of behaviour while concomitant decreases in certain limbic areas have produced a degree of emancipation from hormonal determinants in human behaviour. Human sexual activity is not contingent on oestrus, maternal care readily occurs without the hormonal priming of pregnancy, and feeding behaviour in affluent societies is driven by habit more than by hunger. This evolution from biological to cognitive regulation of behaviour has required considerable increases in executive brain power that, because of the limiting capacity of the uterine placenta and birth canal, has been achieved by the postponing of substantial brain growth to the postnatal period. This, in turn, has required extended maternal care that goes well beyond the weaning period. Such executive brain control over behaviour not only permits a mother to extend her care-giving beyond lactation but also enables fathers, older siblings, and grandparents to participate in this behaviour. However, only the mother has components of both the biology and cognition, thereby facilitating her care-giving and bonding for the infant.

Child development

As young children develop, their knowledge base expands especially in regard to the changing ways in which their mother responds to them and, what is more important, how each is likely to respond to the other (e.g. Stevenson-Hinde and Verschueren 2002). Not surprisingly, the infant's executive brain and its connections are undergoing active growth during this same period. Indeed, evolution of a large neocortex that develops extensively in the postnatal period has required an extended interdependence of mother and child and would not have been possible if parental care had ceased at weaning. The child's knowledge base steadily becomes organized in the form of internal working models of self and mother, encompassing an understanding of both her moods and intentions. Building on this early knowledge provides the infant with an ability to simulate happenings in an expanding world of relationships. The advantages gained from these experiences enable forward planning and, because these working models are in constant daily use, their influence on thought, emotions, and behaviour becomes routine and all-pervasive (Fonagy *et al.* 2002).

In the realm of socio-emotional development, the measurement of the security of infant–parent attachment has proved a powerful predictive source of future competence in later life (Denham *et al.* 2002). From the perspective of developmental psychopathology, there are several ways in which to view the relationship between attachment and subsequent clinical disturbance. Anxious attachments may be conceived as a risk factor for subsequent socio-emotional problems. Attachment theory predicts that by middle childhood the internal working models relate to coping strategies evoked to deal with situations where distress and insecurities are aroused. Children classified as secure are more likely to seek help from others than are children classed as insecure. Anxious attachments predict more difficult, aggressive peer relationships and few good, close friends. While inadequate peer relationships in middle childhood have been identified as a sturdy predictor of adult maladjustment (Sroufe *et al.* 1999), the expansion of the child's social world has undoubtedly been made easier by having a secure base from which to explore this world (Bretherton and Munholland 1999).

It is also important to note that, in these early years of child development when the neocortex is forming and making connections and associations with other parts of the brain, the limbic emotional brain is already well developed. The overt expression of a large emotional repertoire in young infants bears testimony to this. Understanding emotions, curbing these emotions, and channelling these emotions for beneficial purposes must represent an important phase in brain maturation.

Equally important has been the interplay of emotions with cognitive development, which has undoubtedly prospered from the exaggerated expression of emotions between mother and child. The child's need for an attachment figure and the mother's predisposition to bond provide an optimal social environment in which the human executive brain can develop.

The nature of adaptive social functioning changes throughout development and these changes are accompanied by parallel reorganizations of ways to deal with emotional issues (Denham *et al.* 2002). Young children must learn to control the disorganization that stems from a tantrum and to think effectively about a distressing situation, while development of a wider social network introduces the need to meet the social expectations of persons other than their parents. Managing how and when to show emotion becomes paramount as does knowing with whom emotional experiences can be shared. Conversation assumes particular importance, and replaces the more overt behavioural emotionality. Older and wiser parents provide the foresight whereby immature brains can learn and develop strategies for regulating emotions. Children model their emotions from parental interactions and are exquisitely attentive to parental reactions. This may, of course, contribute to weaknesses as well as strengths in emotional development.

Mechanisms underpinning behavioural development

What do we know about the brain at a mechanistic level that might help us to understand how certain early life events or stressors may shape its developmental organization? There is considerable support for the concept of long-term effects of developmental stress from experimental animal studies. Even in the small-brained rat, single 24 hour periods of maternal deprivation in neonates, or recurrent periods of 3 hour maternal deprivation during the first 10 days of life can result in long-lasting alteration in stress hormones (corticosteroids) and increased anxiety-like behaviour in a range of test situations (Caldji *et al.* 1998; Ladd *et al.* 1996). In addition to the neuro-endocrine and behavioural alterations, neonatal rats experiencing the 3 hour repeated maternal deprivation are more likely to adopt cocaine and alcohol self-administration later in life (Ploj *et al.* 2003). Adult rats reared in this way can be treated with serotonin-selective antidepressants. However, discontinuation of the antidepressants results in the re-emergence of addictive behaviour and endocrine alterations, illustrating the long-lasting and persistent nature of these early separations.

Naturally occurring variations in maternal behaviour by rats such as licking, grooming, and nursing are associated with individual differences in both behavioural and hypothalamic–pituitary responses to stress in their infants (Francis *et al.* 1999). Mothers that score high on these maternal behaviours have offspring that are less fearful and are less susceptible to stress compared with infants whose mothers perform these maternal behaviours at low levels. Cross-fostering studies show that it is the maternal phenotype that influences the stress responsiveness in offspring, suggesting that the style of maternal behaviour can serve as a mechanism for differences in stress reactivity across generations (Meaney 2001).

The early development (first 10 weeks) of monkey social behaviour occurs predominantly in the context of interactions with the mother. These early social interactions are almost totally under the mother's control in terms of both the amount and kinds of interaction permitted. By 40 weeks of age, infants are considerably more independent from their mother, and much of their behaviour is oriented toward peers. Nevertheless, mothers continue to monitor their infants and quickly intervene in response to risks arising during play (Simpson *et al.* 1989). The mother serves as a secure base from which the infant can obtain contact and grooming while developing and strengthening its social bonds with peers and other kin.

Administration of opioids has been shown to reduce the distress shown by infants of various species when separated from their mothers. For example, the opiate agonist morphine reduces distress vocalization rates in chicks, puppies, and rhesus monkeys (Panksepp *et al.* 1997). Processes involving endogenous opioid reward may therefore underlie infant attachment and this has been investigated in young rhesus monkeys given acute treatment with the opioid receptor blocker, naloxone, and observed in their natal group (Martel *et al.* 1995). Naloxone increases the duration of affiliative infant–mother contact and the amount of time the infant spends on the nipple. This occurs even at 1 year of age when the mothers are no longer lactating. Indeed, feeding is unaffected by naloxone treatment of infants, but play activity decreases and their distress vocalizations increase. Whereas the opiate agonist, morphine, reduced distress in the absence of mother, the opiate antagonist, naloxone, promotes distress, but in the presence of mother the infant seeks solace by tactile social contact with mother. These results may be interpreted in terms of opiate receptor blockade reducing the 'positive affect' that accrues from having developed an attachment relationship with mother, as a result of which the young infant returns to mother as an established secure base. Moreover, the opioid system of both infant and mother coordinates intimate contact during reunion (Kalin *et al.* 1995).

Maternal bonding

Bowlby himself was interested in evolutionary biology and was struck by the imprinting work on birds by Conrad Lorenz, which influenced his own thinking about 'sensitive periods' for human infant development. If, as Bowlby suggested, an infant's attachment behaviour evolved jointly with maternal care-giving behaviour (George and Solomon 1999), then there is a possibility for these sensitive periods during infant attachment and maternal bonding to have common underlying mechanisms.

In most mammals, the female is committed to the major share of parental care. By virtue of internal fertilization and viviparity, the female mammal commits considerable time and resources to the developing offspring. The placenta, a lifeline for the developing fetus, produces hormones that increase maternal food intake, shut down sexual motivation, prime the brain for maternal care, and prime the mammary gland for milk production contingent on the timing of birth. These aspects of maternalism are an integral part of physiology that not only provide for the developmental energetic requirements of the offspring, but have the capacity to physiologically plan ahead for post-partum needs of lactation. The onset, sustain-ability, and termination of maternal care are, therefore, tightly locked into maternal physiology and synchronized with the mother's ability to provide nutrient until the offspring is able to fend for itself.

This well-established biology has undoubtedly formed the foundations for mam-malian reproductive success, but raises the question as to what has happened in the brain when the maternal behaviour of human and certain non-human primates is unlocked from this biology? The hormones of pregnancy and parturition are, at best, only produced until the termination of lactation, while human parental as well as alloparental care may extend well beyond this period. Moreover, female human and non-human primates are very motherly to infants without even being pregnant. What makes this behaviour so rewarding and fulfilling when it is not determined by the hormones of pregnancy, and are the subserving neural changes that are so important for well-being also available for subversion with the changing lifestyle that we now experience?

Opioids and maternal bonding

It has been suggested that the activation of the endogenous opioid system at partu-rition and during suckling promotes the positive affect arising from maternal behaviour. In the early post-partum period, a mother's social interactions are almost exclusively with her infant, and opiate receptor blockade in mother has marked effects on this relationship.

Studies on naloxone treatment of post-partum rhesus monkey mothers living in social groups have addressed the importance of opioids in maternal bonding. Naloxone treatment reduces the mothers' care-giving and protective behaviour shown toward their infants. In the first weeks of life when infant retrieval is nor-mally very high, naloxone-treated mothers neglect their infants and show little retrieval even when the infant moves a distance away. As the infants approach 8 weeks of age, when a strong grooming relationship normally develops between mother and infant, mothers treated with naloxone fail to develop a close grooming relationship. Moreover, they permit other females to groom their infants, while saline-treated control females are very possessive and protective of their infants (Martel et al. 1993).

The infant is not rejected from suckling, but a mother's usual possessive preoccu-pation with the infant declines with opioid receptor blockade. The mother is not the normal attentive care-giver, and mother–infant interactions are invariably infant-initiated. Primates and other mammals both have in common opioid involvement in maternal care, but the consequences of opioid blockade in small-brained mammals are much greater for the biological aspects of maternal behaviour. In rodents and sheep, for example, interference with the endogenous opioid system severely impairs maternal behaviour, including suckling, whereas monkeys neglect their focused preoccupation with infant care but still permit suckling. These differences may reflect the degree of emancipation from endocrine determinants that maternal behaviour has undergone in primates, and the increased importance of 'emotional reward' for the bonding mechanism. If the endogenous opioid system in the monkey is positively linked to mother–infant bonding then heroin addiction, which acts as an agonist to downregulate the functioning of the same opioid receptor, would be predicted to have severe consequences for human maternal bonding. Women who are heroin-addicted have many aspects of their social and economic life disrupted making the data difficult to disentangle. Nevertheless, the facts are that by 1 year of age nearly 50% of children are living away from their biological mothers, and by school age only 12% remain with their biological mothers (Mayes 1995). These infants have been abandoned for adoption or are taken into the care of their grandparents and other female kin. Moreover, in a follow-up of 57 methadone-maintained mothers compared with controls matched for ethnicity, socio-economic status, infant birthweight, and gestational age, opiate-addicted mothers were far less likely to have remained the child's primary parent and the children were significantly more likely to have been referred to child protective care or special service agencies for neglect, abandonment, or abuse.

Integral to the bonding process in large-brained primates is the endogenous opioid system, which has been shown to act on receptors in the ventral striatum (Koob and Le Moal 1997). This area of the brain is involved in 'reward' and involves the mesolimbic dopamine projection, which detects rewarding stimuli and also detects ways in which they occur differently from prediction to enable 'updating' of the stimulus (Schultz and Dickinson 2000). The mother–infant bonding process entails obsessive grooming especially to hands, face, and genitalia by mothers, and these are the phenotypic traits of infant monkeys that show the greatest changes during development. Because primates show extended post-partum care, offspring recognition requires the continual updating of any changes in these morphological features and in behavioural development. This updating of infant recognition probably involves prefrontal–ventral striatal pathways that are also intimately linked to the emotional brain via the amygdala. The positive emotional responses that infants generate in females enable parental care to occur without the continual priming by pregnancy and parturition.

Human mothers also experience preoccupations and rituals in the context of maternal care, and, even before the birth of their baby, they are obsessive with cleaning and creating a safe environment. After birth, safety is a major concern and mothers frequently check on their baby even at times when they know the baby is fine (Leckman *et al.* 1999). The evolution of these obsessive psychological and behavioural states can be seen as a developmental extension of the preoccupations that all primates show for their infants. Interestingly, areas of the human brain that, using magnetic resonance imaging (MRI), have been shown to be responsive to babies crying include the brain's reward structures (mesolimbic dopamine from the ventral tegmental area, ventral striatum, and amygdala; Lorberbaum *et al.* 2002).

The evolutionary progression away from hormonal centric determinants of maternal behaviour to emotional, reward-fulfilling activation involves dopaminergic and opioidergic activity in the ventral striatum. The enhanced role of this circuitry for regulating behaviour in humans may also provide vulnerability to various forms of psychopathology such as obsessive compulsive disorder and substance abuse. Mild forms of addictive behaviour (gambling, video games, internet use, and consumption of caffeine and chocolate) are such indicators of this neurological predisposition for obsessive behaviour seen in humans (Greenberg *et al.* 1999) that are not conducive to well-being. Obsessive compulsive disorder (OCD) itself is characterized by intrusive thoughts and preoccupations, rituals, and compulsions. *In vivo* neuroimaging studies identify the orbital frontal cortex, head of the caudate, and closely associated ventral striatum and anterior cingulate as being involved in OCD (Rauch 2000), while acquired OCD occurs later in life in patients with striatal lesions (Chacko *et al.* 2000). There is also evidence that cerebrospinal fluid (CSF) levels of the neuropeptide, oxytocin, are elevated in OCD, and this peptide also plays a fundamental role in many obsessive aspects of maternalism (Leckman *et al.* 1994). Not surprisingly, therefore, OCD is more common in women, and female OCD probands have a higher rate of relatives with OC spectrum disorders. The influence of gonadal hormones on periodicity of OCD (Weiss *et al.* 1995) and the post-partum exacerbation of OCD symptoms in women suggest that the course of this disorder may be influenced by the hormones of pregnancy (Williams and Koran 1997). Moreover, psychological challenges directed at the prefrontal cortex are predictive of the severity of OCD symptoms in female patients and this predictive value discriminates between males and females (Zohar *et al.* 1999). OCD is less common in males although the obsessional behaviour of males, especially in the postpubertal period when first experiencing romantic love, shows remarkable similarities with mother–infant love (Leckman and Mayes 1999). Hence there are components of behaviour that occur in the postpubertal and post-partum period that are influenced by hormones but, interestingly, they relate to areas of the brain concerned with reward and not motivation.

Puberty and behavioural development

Puberty is a vulnerable period for the development of numerous behavioural problems including eating disorders, obsessive compulsive disorders, addictive disorders, onset of depression, and, in some cases, suicide. What is happening in the brain at puberty that makes this a special developmental period for consideration, and why do these problems appear to be occurring both more frequently and at a progressively younger age (Fombonne 1998a) than in previous generations?

In early and middle adolescence there is a marked increase in symptoms of depression, and this is more prevalent for girls than boys (Grant and Compass 1995). Increased instability and higher demands, both in society and within families, have been proposed to play important aetiological roles explaining increased prevalence rates of depressive conditions among young people in recent decades (Goodyer 1995). Chronic daily stress and daily hassles have been strongly associated with depressive symptoms among adolescents, with girls reporting more interpersonal stressful events than boys. In an extensive study of psychosocial correlates of depressive symptoms in 12–14 year old adolescents, daily hassles (e.g. someone criticized me), stressful life events especially at school, gender, and lack of friends correlated very strongly with depressive symptoms (Sund et al. 2003). In adolescents, identity formation and self-esteem are central issues in psychological development, and having more than four close friends seems to be protective against the development of depressive symptoms.

Several studies in recent years have revealed an earlier age of onset for psychosocial disorders with most of these disorders starting in early adolescent years (Joyce et al. 1990). Findings over the last 3 decades have also revealed a progressive increase in disorders and a progressively earlier age of onset. It is paradoxical that, at a time where economic conditions and physical health are improving, psychosocial disorders of youth are also increasing in most developed countries (Fombonne 1998a). Of even more concern is the finding that many of the psychiatric disorders that have an onset in adolescence have a strong continuity with adult disorders (Fombonne et al. 2001b).

Considerable clinical literature has evaluated the possible impact of early and ongoing psychosocial stress in the onset and recurrence of unipolar and bipolar affective disorders. In a population of 258 patients who were followed prospectively with clinical assessments, those with a positive history of early adversity (physical or sexual abuse) not only experienced an earlier onset of bipolar illness and more suicide attempts, but these patients were relatively more treatment-resistant in the prospective follow-up (Leverich et al. 2002). Puberty has always been a life event that produces a challenge for well-being, but why is it becoming more of a problem? Can we put these problems down to failures in early development, or does the protracted developmental period produce problems in its own right that are exacerbated by modern lifestyles.

Predisposing factors for early puberty onset

One major difference in human reproduction and that of other mammals is the extended developmental period between birth and puberty. In the mouse this is about 40 days, in the monkey it is 2–3 years, and in humans at the turn of the century it was about 16–17 years. This long interval between birth and reproductive age enables the brain to mature in its ability to establish and manage peer relationships, to regulate emotions, and, through knowledge of self gained by experience, to develop high self-esteem. On the basis of secure attachment, the extended period from infancy to puberty, longer for humans than for any other primate, has enabled the expansion of social relationships that provide a buffer for regulating the emotional turmoil of puberty.

In the last 100 years there has been a significant trend towards early puberty in boys and early menarche in girls. The age at which girls first menstruate has decreased from 16–17 years according to records for Norway, Germany, Finland, and Sweden (Tanner 1966), to between 12 and 13 years in modern-day populations. Health care and socio-economic living standards have improved during this time and of particular importance has been nutrition. Body weight, and notably body fat signalled to the brain via the hormone leptin, is a critical factor in the onset of puberty in many mammalian species (Apter 2003). Evidence in support of this viewpoint in humans is abundant (Klentrou and Plyley 2003; Biro *et al.* 2003). Malnutrition is associated with delayed menarche, while moderately obese girls experience earlier menarche than lean girls. Amenorrhoea is common in ballerinas who tend to be at the low end of fatness, and also in female athletes who may have greater than normal lean body mass but almost undetectable levels of fat. Hence it would appear that for humans, like other mammals, body size and fat reserves that are sufficient to deal with the energetic demands of pregnancy are a signal for the brain to commence the neuroendocrine cascade that initiates puberty. The trend towards obese phenotypes in recent decades is likely to have contributed to this shortening of childhood and early puberty onset (Shalitin and Phillips 2003). It is certainly the case that body fat at 5 years of age predicts earlier pubertal development among girls aged 9 (Davison *et al.* 2003).

Brain development at puberty

Longitudinal studies of brain development from childhood through adolescence have shown that the volume of white matter (myelinated nerve fibres) increases linearly between 4 and 22 years of life (Sowell *et al.* 1999, 2001*b*). Changes in the volume of cortical grey matter (neural cell bodies) are nonlinear and regionally specific. In the frontal lobes, grey matter increases to maximum at puberty (males, age 12.1; females, age 11) and is followed by a decline during post-adolescence (Fig. 2.1). Temporal lobe grey matter development is also nonlinear and peaks at 16.5 years. In contrast to non-human primates, development of the human neocortex is non-synchronous with changes occurring in visual and auditory cortex long before those in the frontal and temporal cortex. The pre-adolescent increase and post-adolescent decrease in prefrontal cortical grey matter are also shown in positron emission tomography (PET) studies of glucose metabolism and in electroencephalography (EEG) studies. Electrocortical recordings of the brain indicate

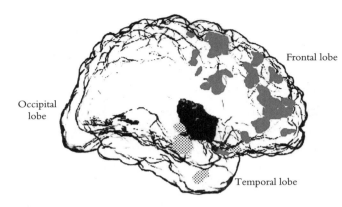

Frontal lobe

Occipital
lobe

Temporal lobe

Fig. 2.1 Areas of the frontal cortex (dorsal medial and orbital frontal—shaded grey), temporal lobe (speckled), and subcortical striatum (black) that show reductions in grey matter postpuberty (modified after Sowell *et al.* 1999).

a wave of synaptic proliferation in the frontal lobes around the age of puberty. If this increased frontal cortex activity is related to a wave of synaptic overproduction, it may herald a stage of pubertal development when psychosocial and emotional activities guide selective synapse elimination during adolescence. *In vivo* structural imaging studies (high resolution MRI) from childhood to adolescence (age 12–16 years) have shown an inverse relationship between the reduction of grey matter and increased brain growth, particularly in areas of the frontal cortex that control executive cognitive functioning (Giedd *et al.* 1999). These events are thought to represent loss of neurons that fail to make appropriate connections (synaptic pruning) and progressive myelination of neural connections that do make appropriate connections. This frontal cortex remodelling is also associated with improved memory functioning (Sowell *et al.* 2001*a*).

Frontal cortex grey matter peaks 1 year earlier in females than in males, corresponding to the earlier age of puberty onset. Studies have revealed significant sex by age interactions in cerebral grey and white matter volumes and suggest there are age-related sex differences in brain maturational processes (De Bellis *et al.* 2001). In tests of emotional cognition where subjects were required to make a decision about emotion expressed in a face or a word or both, reaction times slowed significantly at the onset of puberty and stabilized by 15 years of age. Girls started to react more slowly a year before boys, and had longer reaction times from age 15–17, but this sex difference was transient and not present in 18–22 year olds (Bremner *et al.* 2001). It is also noteworthy that, while adults relative to adolescents are better able to engage the orbitofrontal cortex (executive brain) when required to attend to different facial expressions, adolescents relative to adults exhibit greater modulation of emotional circuitry (amygdala, ventral striatum, anterior cingulate, and frontal cortex) based on these same tasks (Monk *et al.* 2003). This emotional circuitry is rich in opioid receptors.

Studies show the effects of opioid receptor blockade on attachment in infant monkeys (as already described) to be the same in males and females, such that both increase their contact time with their mother. At puberty, however, feral males tend to leave their natal social group while females stay with the group. This type of social organization in Old World monkeys, where females remain with the group, is referred to as being 'female-bonded'. Why do we find this sex difference in social bonding and, if the mechanisms serving infant attachment extend to social bonding, why should it differ between males and females? In males entering puberty, opioid receptor blockade results in significant increases in the time these males spend with the mother and decreases the time they spend with others. At the same age, females tend to spend more time alone than males but, when subjected to opioid receptor blockade, they spend more time with other females, not with their mother (Keverne et al. 1997). This suggests that female monkeys have a predisposition to develop other socially meaningful relations and have expanded their secure base beyond mother, although these social relationships are invariably with matrilineal kin. Postpubertal males running to mother would not be favourably tolerated by dominant males, would invite aggression, and hence the peripheralization and subsequent mobility of young males from the natal social group. In this context it is interesting that synthetic androgenic steroids increase opioid activity in the ventral striatum, hypothalamus, and periaqueductal grey, brain regions that regulate opiate dependence, defensive reactions, and aggression (Johansson et al. 2000), The attachment that rhesus monkey infants develop with the matriline is especially enduring in females, lasting a lifetime, whereas in males it rarely lasts beyond puberty.

Puberty is undeniably a period of great emotional turmoil when changes in physical phenotype synchronize with reorganization of the prefrontal cortex, an area of the brain intimately concerned with emotional regulation and forward planning. The fact that girls are likely to experience psychological problems earlier than boys may be related to the consistent findings that they enter puberty earlier, and also undergo the developmental changes in the brain earlier. A major change in the last few decades has been an even earlier onset of puberty in boys and girls as a result of better nutrition. Whether this earlier onset of puberty is responsible for the progressively earlier symptoms of psychological disturbance needs more rigorous study. It is certainly the case that girls with more body fat at 5 years of age are predictive of early puberty at 9 years (Davison et al. 2003), and it is also the case that we are seeing unprecedented increases in childhood obesity. The human body and brain have evolved in synchrony for millions of years. Is the maturation of the body phenotype becoming out of phase with brain development or prematurely precipitating brain maturation? If so, does this underpin vulnerability to adolescent disorders of the psyche?

Functional brain development in an evolutionary framework

Our mammalian ancestors gained mobility and the possibility of exploring land environments away from water by virtue of ovoviviparity which entailed internal fertilization, placentation, and the ability to protect offspring from predators by transporting them internally. There was an inherent need to plan ahead in the sense that increased food intake early in pregnancy enabled reserves to be laid down to meet the extra demands of the exponential growth of their offspring later in pregnancy, and for the production of milk needed for post-partum lactation. This investment of time and energetic resources throughout pregnancy ensures that the female is biologically committed to maternal care and to provision her offspring with food and water (milk) and warmth until they became independently able to regulate thermogenesis and forage for themselves. Most female mammals spend many weeks or months engaged in maternal care but only a very few hours of their lives engaged in sexual behaviour.

Both pregnancy and post-partum care are regulated by neuropeptides and hormones, but an important difference between these hormones and those that regulate other aspects of behaviour has been established for the hormones of pregnancy that prime the brain for maternal behaviour. Pregnancy hormones are produced or regulated by the fetal placenta, while the infants' post-partum suckling sustains maternal behaviour by activating neural pathways in those brain areas primed by hormones during pregnancy. The sequence of events that represent maternalism is an exquisite piece of biology determined by hormones acting in the mother that are under the control of the fetal genome.

A major progression across mammalian evolution has been the massive increase in brain size relative to body size. This has required some important adjustments in the biology, not least of which are those involved with the mother–infant relationship. The brain is limited in its growth within the mother because of size constraints of the uterus and birth canal. Hence, in those primate mammals that have the largest brains, much of the brain and body growth is postponed to the postnatal period. This extensive postnatal growth and development of the brain, especially in humans, requires extended maternal, paternal, and alloparental care. Breaking the link that binds maternal care to hormonal mechanisms has required extensive modification in the mother's brain, while the infant's brain, by virtue of postponing much of its development to the post-partum period, has constructed a niche for itself that transforms the evolutionary dynamic. Much of the brain's connectivities that are made during the postnatal period are under epigenetic control. The multiplicity of axonal projections and synaptic connections made in the brain are not determined by any specific genetic programme but by activity in other neurons, and many of the misconnections that are hence inevitably made are eliminated by programmed cell death. In this way, human brain development comes under the prolonged influence of the postnatal social environment in which the infant is reared.

Mother and infant behaviours have evolved as a unit. While the mother's brain has been maternalized by the hormones of pregnancy generated by the fetus, the infant's developing brain requires social stimulation from a mother committed to provide the emotional rewards of warmth and suckling. Not only do infants have to learn to sit, stand, and eventually walk and talk, but they have to learn social rules and emotional control. It is clear that the process of infant socialization bene-fits from this close relationship but whether this occurs during a critical period in brain development is open to question. We know, for example, that the ability to learn language arises from a synergy between early brain development and lan-guage experience and is seriously compromised when language is not experienced early in life (Mayberry *et al.* 2002). It is, therefore, probably safer to think of this whole early developmental period as critical, but modifiable while the brain remains in a plastic state. The synergy between early brain development and the learning of social rules is also likely to be compromised when the developing infant–mother relationship is compromised early in life. Certainly, mother–infant separations in monkeys are known to have long-term consequences (Hinde *et al.* 1978) and extreme consequences for infant abuse when infants are separated from mother and reared with peers (Harlow *et al.* 1971; Kraemer *et al.* 1991). However, so long as we are able to recognize developmental problems and providing they are not too traumatic, the plasticity of the developing brain may provide for remedial action. Humans tend to worry about the uterine environment and toxic agents or drugs that may damage the fingers and toes of babies, but perhaps we should pay more attention to the post-partum period when the social environment exercises its effects on the developing brain and lays down the foundations for future well-being.

Prefrontal cortex

A second major evolutionary event unique to the hominid brain has been the extended development of the prefrontal cortex and its connections with the striatum (dorsal and ventral) and the cingulate cortex. The pattern of brain maturation in the post-adolescent period is distinct from earlier development and is localized to large areas of the dorsal, medial, and orbital frontal cortex together with striatum (Sowell *et al.* 1999). This patterning of development, elucidated from using MRI, is supported by post-mortem studies of the brain and the development of cognitive functions attributed to these structures. Neuropsychological studies show that these regions of frontal cortex are essential for such functions as forward planning and organization, emotional regulation, and the ability to block a habitual behaviour and execute a less familiar behaviour. The latter function depends on the lateral prefrontal cortex and the anterior cingulate cortex (Fig. 2.2). The anterior cingulate is thought to detect conflicts between plans of action and, in response to these conflicts, the anterior cingulate recruits greater cognitive control in the lateral prefrontal cortex (Botvinick *et al.* 2001). An important question about the nature of cognitive control is how do the processes involved in implementing control become engaged. Recent studies, using MRI, have demonstrated that conflict-related activity engages greater

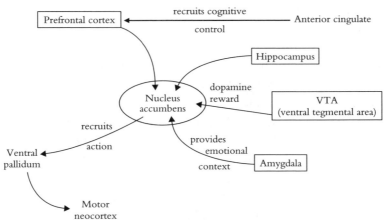

Fig. 2.2 The ventral striatum (nucleus accumbens) is a focal area for brain reward, with the amygdaloid input providing an emotional bias.

prefrontal cortex activity preceded by activity in the anterior cingulate, which consequently provides for adjustments in behaviour. These findings support a role for the anterior cingulate in conflict monitoring and the engagement of prefrontal cognitive control (Kerns *et al.* 2004). Forward planning inevitably depends on learning experiences, especially those experiences that have proved rewarding in the past, while decision-making sometimes depends on 'gut feelings' that are linked with visceral brain connections to the amygdala (Fig. 2.2). The amygdala may be viewed as providing an emotional bias to frontostriatal projections that represent the neural substrates that serve for brain reward (Damasio 1998). Patients with bilateral amygdala damage fail to show the normal emotional response to reward (Bechara *et al.* 1999), while the ventral striatal region responds to actual and anticipation of reward (Schultz *et al.* 1997), and the frontal cortex codes for the value of reward (Tremblay and Schultz 2000).

Neurophysiological studies in patients with orbital frontal cortex lesions suggest that the deficits in decision-making shown by these patients are attributable to impaired ability to evaluate the consequences of their actions (Bechara *et al.* 2000). Responses to reward in humans have been assessed in recent functional imaging studies and it is clear that brain areas involved in secondary rewards (e.g. financial reward) overlap with regions responding to primary rewards like food. The amygdala, dorsal and ventral striatum, and the dopaminergic ventral tegmental area respond to the presence of reward regardless of its value. However the orbitofrontal cortex responds nonlinearly, being particularly responsive to the lowest and highest reward values (Elliott *et al.* 2003). This is consistent with a role for orbitofrontal cortex in coding the complex relative values of reward.

It is clear, therefore, that different regions of the prefrontal cortex that are engaged in decision-making, forward planning, and emotional control undergo a surge of development modifications at puberty that continue throughout adolescence, are complete around the age of 20–22 years, and are important for adult well-being. The prefrontal cortex projects heavily to the striatum, both the dorsal and ventral areas, which are important for calling into action motor patterns of behaviour aided by dopaminergic inputs from the midbrain (substantia nigra → dorsal striatum; ventral tegmental area → ventral striatum; see Fig. 2.2). The amygdala and anterior cingulate also interrelate with different components of the prefrontal cortex (PFC; amygdala → medial PFC; anterior cingulate → lateral PFC). I have already described how parts of these neural interconnections get 'locked' into creating fixed or obsessive action patterns in the context of OCD. Moreover, this same circuitry is integral to disorders such as psychosis and depression, and is vulnerable to addictive drugs. Hence the human brain has evolved a circuitry that develops in the context of social well-being and is itself crucial to social well-being. Not surprisingly, this circuitry takes years to mature in the human brain, extending its vulnerability into the adolescent period.

Subversion of the neurochemistry of attachment and bonding

I have already mentioned the role of endogenous opioids for bonding in monkeys and the negative consequences of opiate addiction on maternal care in humans. This raises the possibility that the same neurochemical systems that form the glue of social relationships become vulnerable to drugs of abuse.

Drug addiction can be understood in terms of normal learning and memory systems of the brain that, through the action of chronically administered drugs, are pathologically subverted. Drug use becomes compulsive and is characterized by a pattern of drug seeking and consumption that become habitual and progressively less modified (Everitt *et al.* 2001; Robbins and Everitt 1999). Many drugs of abuse (heroine, cocaine, amphetamine, nicotine, and alcohol) have the common action of increasing dopamine transmission in the ventral striatum. These common findings of drug action have led to a widely held view that the mesolimbic dopamine system has a general reinforcing effect on drugs that act in mediating aspects of natural reward. For example, rats will self-administer opiates to the ventral tegmental area from which dopamine neurons project, and opiates also seem to have reinforcing effects mediated by dopamine-independent mechanisms in the ventral striatum (Koob and Le Moal 1997). Dopaminergic neurons from the ventral tegmental area project to the ventral striatum, which responds to unpredicted rewards that with training transfer to stimuli predictive of reward. Thus, by signalling reward prediction errors, dopamine may serve as an instruction for striatal learning in the context of hedonic reward (Schultz and Dickinson 2000).

In the real world, drugs are not freely available and the drug abuser has to attend to cues that predict their availability. Drug-seeking behaviour can become powerfully associated with environmental cues. This process of associative learning by which particular places become connected with the euphoric drug-induced state involves brain structures that have strong anatomical links to the ventral striatum. When addicts are allowed to see the paraphernalia associated with self-administration, several interconnected areas of the ventral striatum are activated including the amygdala and frontal cortex (Everitt *et al.* 2001). Drug abusers often undertake ritualized preparation with the paraphernalia for drug administration that in itself becomes rewarding. This may lead to the people developing a drug-seeking habit even when the euphoric effects that initially encouraged it are reduced.

Hence the neural circuitry, neurochemistry, and obsessive behaviour associated with numerous forms of addiction (alcohol, drugs) appear to subvert those parts of the human brain that have evolved to subserve hedonic and social reward. Experimental rats can also be addicted to drugs such as heroin, but I guarantee we will never find an addicted rat in the wild. Clearly, they have the same neurochemical/neuroanatomical reward system, but missing from the rat brain is the prefrontal association cortex that is important for decision-making and forward planning and provides the means with which to make drugs available. Since these same parts of the brain enable planning ahead and rational decision-making, then why do individuals, especially teenagers, make such destructive decisions over drug abuse? Is addiction more of a problem in Western adolescents because they are reaching the age of puberty well in advance of reaching the age of reasoning?

In the last decade, the significance of self-esteem has been identified in maintaining wellness among adolescents, and adolescents treated in clinical settings for drug use problems are often observed to have low self-esteem (Modrein-Talbott *et al.* 1998). Susceptibility to peer pressure and lack of family caring have also been associated with teenage drug abuse, while lack of social support and low self-esteem measures strongly correlate with remission of adolescent addiction (Richter *et al.* 1991). In other words, how teenagers rate themselves with respect to others (self-esteem) and how they want to be seen by others are strong predisposing factors for addiction at a time when their brain is immature and insecure in these social domains.

Conclusions

For thousands of years, our ancestors had to contend with finding food, a task that required cooperation, and rearing children, which also required cooperation. This cooperation was invariably within part of an extended family grouping, and any small community encompassed relatively few relationships that needed to be handled and understood. Our brains matured at a leisurely pace, while the collective knowledge of the community for achieving life's comforts and well-being was not greater than any one person might understand in their lifetime. Indeed, life was short and the outcome of decision-making was relatively short term. Decisions as to how we obtain the next meal, how to optimize child survival, how to secure warmth and comfort required a degree of forward planning for which the rewards were life-enhancing and failures were life-threatening. Living was made easier by the reward values of correct decisions because these kinds of decision sustained life. In modern societies, individuals are required to make decisions about careers, marriages, and financial security for old age and pensions. These are some of the most important decisions we have to make, they are exceptionally long term, and there is no guaranteed way of getting them right. The outcome depends on the decisions of others and events that are far removed from the control of the decision maker. No one person, no one community, or even one nation can understand the collective accumulated knowledge of mankind, let alone predict life's well-being when expectations have expanded way beyond the basic necessities of life. Living with such long-term decisions can produce anxiety and frustration that are not conducive to well-being.

Even though we know eating makes us fat, technology has enhanced the reward value to the brain of food by additions of hedonic flavourings, and the hormonal signals to inhibit overeating (leptin) are no longer effective in humans. Maternal care is dissociated from the hormones of pregnancy and parturition, and the emotional rewards of the developing parent–infant relationship have many competing new relationships that tap into this very same neural circuitry. Relationships are frequently changing due to moving home, moving school, and family break-up, not to mention the thousands of fantasy relationships with which we become involved through television soap dramas. This saturation with both real and fantasy relationships must be potentially confusing to the adolescent brain that is desperately trying to come to terms with a whole new dimension to relationships, namely sexuality. Long gone are the ancestral rituals associated with puberty and the collective sharing and participation of the community in these rituals. It is not surprising, therefore, that the infatuated romantic love of today's teenagers takes on such obsessional qualities, often to the cost of other meaningful relationships.

Evolution of the mammalian brain from one that acts on hormonal instructions for integrating behaviour with bodily needs to one that determines behaviour based primarily on cognitive reasoning has represented a phase shift for the evolutionary dynamic. The removal of these biological restrictions to behaviour has enabled the cognitive brain not only to emancipate our lifestyle from the external environment but to create an artificial environment that caters for and satisfies its own hedonistic needs. The brain has become a victim of its own evolutionary success. Our future well-being depends upon understanding these vulnerabilities and changing lifestyles so that the care of our children and the nurturing of relationships take into account the predispositions that are imposed by our protracted brain development and longevity.

Acknowledgements

Leverhulme Trust Award. Evolution, post genomics, and contextual biology.

References

Apter, D. (2003). The role of leptin in female adolescence. *Ann. N.Y. Acad. Sci.* **997**, 64–76.

Bechara, A., Damasio, H., Damasio, A.R., and Lee, G.P. (1999). Different contribution of the human amygdala and ventromedial prefrontal cortex to decision making. *J. Neurosci.* **19**, 5473–81.

Bechara, A., Damasio, H., Damasio, A.R., and Lee, G.P. (2000). Emotion, decision making and the orbitofrontal cortex. *Cereb. Cortex* **10**, 295–307.

Biro, F.M., Lucky, A.W., Simbartl, L.A., Barton, B.A., Daniels, S.R., Striegel-Moore, R., Kronsberg, S.S., and Morrison, J.A. (2003). Pubertal maturation in girls and the relationship to anthropometric changes: pathways through puberty. *J. Pediatr.* **142**, 643–6.

Botvinick, M.M., Braver, T.S., Barch, D.M., Carter, C.S., and Cohen, J.D. (2001). Conflict monitoring and cognitive control. *Psychol. Rev.* **108**, 624–52.

Bremner, J.D., Soufer, R., McCarthy, G., Delaney, R., Staib, L.H., Duncan, J.S., and Charney, D.S. (2001). Gender differences in cognitive and neural correlates of remembrance of emotional words. *Psychopharmacol. Bull.* **35**, 55–78.

Bretherton, I. and Munholland, D.A. (1999). Internal working models in attachment relationships: a construct revisited. In *Handbook of attachment: theory, research, and clinical application. The caregiving behavioural system* (ed. J. Cassidy and P.R. Shaver), pp. 89–111. Guilford Press, New York.

Caldji, C., Tannenbaum, B., Sharma, S., Francis, D., Plotsky, P.M., and Meaney, M.J. (1998). Maternal care during infancy regulates the development of neural systems mediating the expression of feafulness in the rat. *Proc. Natl Acad. Sci. USA* **95**, 5335–40.

Chacko, R.C., Corbin, M.A., and Harper, R.B. (2000). Acquired obsessive-compulsive disorder associated with basal ganglia lesions. *J. Neuropsychiatry Clin. Neurosci.* **12**, 269–72.

Damasio, A.R. (1998). Emotion in the perspective of an integrated nervous system. *Brain Res. Rev.* **26**, 83–6.

Davison, K.K., Susman, E.J., and Birch, L.L. (2003). Percent body fat at age 5 predicts earlier pubertal development among girls at age 9. *Pediatrics* **111**, 815–21.

De Bellis, M.D., Keshaven, M.S., Beers, S.R., *et al.* (2001). Sex differences in brain maturation during childhood and adolescence. *Cereb. Cortex* **11**, 552–7.

Denham, S., von Salisch, M., Olthof, T., Kochanoff, A., and Caverly, S. (2002). Emotional and social development in childhood. In *Handbook of child social development* (ed. P.K. Smith and C.H. Hart), pp. 307–28. Blackwell, Oxford.

Dunbar, R.I.M. (1992). Neocortex size as a constraint on group size in primates. *J. Hum. Evol.* **20**, 469–93.

Elliott, R., Newman, J.L., Longe, O.A., and Deakin, J.F.W. (2003). Differential response patterns in the striatum and orbitofrontal cortex to financial reward in humans: a parametric functional magnetic resonance imaging study. *J. Neurosci.* **23**, 303–7.

Everitt, B.J., Dickinson, A., and Robbins, T.W. (2001). The neuropsychological basis of addictive behaviour. *Brain Res. Rev.* **36**, 129–38.

Fombonne, E. (1998*a*). Increased rates of psychosocial disorders in youth. *Eur. Arch. Psychiatry Clin. Neurosci.* **248**, 14–21.

Fombonne, E., Wostear, G., Cooper. V., Harrington, R. and Rutler, M. (2001). The Mandsley long-term follow-up of adolescent depression. Suicidality, Criminality, and Social dysfunction in adulthood. *Brit. J. Psychiat.* **179**, 218–24.

Fonagy, P., Gergely, G., Jurist, E.L., *et al.* (2002). *Affect regulation, mentalization, and the development of the self*. Other Press, New York.

Francis, D.J., Diorio, J., Liu, D., and Meaney, M.J. (1999). Non-genomic transmission across generations of maternal behavior and stress responses in the rat. *Science* **286**, 1155–8.

George, C. and Solomon, J. (1999). Attachment and caregiving. In *Handbook of attachment theory, research, and clinical applications. The caregiving behavioural system* (ed. J. Cassidy and P.R. Shaver), pp. 649–70. Guilford Press, New York.

Giedd, J.N., Blumenthal, J., Jeffries, N.O., Castellanos, F.X., Liu, H., Zijdenbos, A., Paus, T., Evans, A.C., and Rapoport, J.L. (1999). Brain development during childhood and adolescence: a longitudinal MRI study. *Nat. Neurosci.* **2**, 861–3.

Goodyer, I.M. (1995). Life events and difficulties: their nature and effects. In *The depressed child and adolescent* (ed. I.M. Goodyer), pp. 171–93. Cambridge University Press, Cambridge.

Grant, K.E. and Compass, B.E. (1995). Stress and anxious-depressed symptoms among adolescents: searching for mechanisms of risk. *J. Consult. Clin. Psychol.* **63**, 1015–21.

Greenberg, J.L., Lewis, S.E., and Dodd, D.K. (1999). Overlapping addictions and self-esteem among college men and women. *Addict. Behav.* **24**, 565–71.

Harlow, H.F., Harlow, M.K., and Suomi, S.J. (1971). From thought to therapy: lessons form a primate laboratory. *Am. Sci.* **59**, 538–49.

Harvey, P.H. and Pagel, M.D. (1991). *The comparative method in evolutionary biology*. Oxford University Press, Oxford.

Hinde, R.A., Leighton-Shapiro, M.E., and McGinnis, L. (1978). Effects of various types of separation experience on rhesus monkeys 5 months later. *J. Child Psychol. Psychiatry* **19**, 199–211.

Johansson, P., Lindqvist, A., Nyberg, F., and Fahlke, C. (2000). Anabolic androgenic steroid affects alcohol intake, defensive behaviors and brain opioid peptides in the rat. *Pharmacol. Biochem. Behav.* **67** (2), 271–9.

Joyce, P.R. Oakley-Brown, M.A., Wells, J.E., Bushnell, J.A. and Hornblow, A.R. (1990). Birth cohort trends in major depression: increasing rats and earlier onset in New Zealand. *J. Affect. Disord.* **18**, 83–89.

Kalin, N.H., Sheldon, S.E., and. Lynn, D.E. (1995). Opiate systems in mother and infant primates coordinate intimate contact during reunion. *Psychoneuroendocrinology* **7**, 735–42.

Kerns, J.G., Cohen, J.D., MacDonald III, A.W., Cho, R.Y., Stenger, A., and Carter, C.S. (2004). Anterior cingulate conflict monitoring and adjustments in control. *Science* **303**, 1023–6.

Keverne, E.B. (1985). Reproductive behaviour. In *Reproduction in mammals: 4. Reproductive fitness*, 2nd edn (ed. C.R. Austin and R.V. Short), pp. 133–75. Cambridge University Press, Cambridge.

Keverne, E.B., Martel, F.L., and Nevison, C.M. (1996). Primate brain evolution: genetic and functional considerations. *Proc. R. Soc. Lond. B* **262**, 689–96.

Keverne, E.B., Nevison, C.M., and Martel, F.L. (1997). Early learning and the social bond. In *The integrative neurobiology of affiliation* (ed. C.S. Carter, I.I. Lederhendler, and B. Kirkpatrick), pp. 329–39. New York Academy of Sciences, New York.

Klentrou, P. and Pyley, M. (2003). Onset of puberty, menstrual frequency, and body fat in elite rhythmic gymnasts compared with normal controls. *Br. J. Med.* **37**, 490–4.

Koob, G.F. and Le Moal, M. (1997). Drug abuse: hedonic homeostatic dysregulation. *Science* **278**, 52–8.

Kraemer, G.W., Ebert, M.H., Schmidt, D.E., and McKinney, W.T. (1991). Strangers in a strange land: a psychobiological study of infant monkeys before and after separation from real or inanimate mothers. *Child Dev.* **62**, 548–66.

Ladd, C.O., Owens, M.J., and Nemeroff, C.B. (1996). Persistent changes in corticotropin-releasing factor neuronal systems induced by maternal deprivation. *Endocrinology* **137**, 1212–18.

Leckman, J.F. and Mayes, L.C. (1999). Preoccupations and behaviors associated with romantic and parental love: Perspectives on the origin of obsessive–compulsive disorder. *Child Adolesc. Psychiatr. Clin. N. Am.* **8**, 635–65.

Leckman, J.F., Goodman, W.K., North, W.G., Chappell, P.B., Price, L.H., Pauls, D.L., Anderson, G.M., Riddle, M.A., McDougle, C.J., Barr, L.C., and Cohen, D.J. (1994). The role of central oxytocin in obsessive compulsive disorder and related normal behavior. *Psychoneuroendocrinology* **19**, 723–49.

Leckman, J.F., Mayes, L.C., Feldman, R., Evans, D.W., King, R.A., and Cohen, D.J. (1999). Early parental preoccupations and behaviors and the possible relationship to the symptoms of obsessive–compulsive disorder. *Acta Psychiatr. Scand. Suppl.* **396**, 1–26.

Leverich, G.S., Perez, S., Luckenbaugh, D.A., and Post, R.B. (2002). Early psychosocial stressors: relationship to suicidality and course of bipolar illness. *Clin. Neurosci. Res.* **2**, 161–70.

Lorberbaum, J.P., Newman, J.D., Horwitz, A.R. Dubno, J.R., Lydiard, R.B., Hamner, M.B., Bohning, D.E, and George, M.S. (2002). A potential role for thalamocingulate circuitry in human maternal behavior. *Biol. Psychiatry* **51**, 431–45.

Martel, F.L., Nevison., C.M., Rayment, F.D., Simpson, M.D.A., and Keverne, E.B. (1993). Opioid receptor blockade reduces maternal affect and social grooming in rhesus monkeys. *Psychoneuroendocrinology* **18**, 307–21.

Martel, F.L., Nevison., C.M., Simpson, M.D.A., and Keverne, E.B. (1995). Effects of opioid receptor blockade on the social behavior of rhesus monkeys living in large family groups. *Psychobiology* **28**, 71–84.

Mayberry, T.I., Lock, E., and Kazami, H. (2002). Linguistic ability and early language exposure. *Nature* **417**, 38.

Mayes, L.C. (1995). Substance abuse and parenting. In *Handbook of parenting: applied and practical parenting*, Vol. 4 (ed. M.H. Bornstein), pp. 101–26. Lawrence Erlbaum Associates, New Jersey.

Meaney, M.J. (2001). Maternal care, gene expression, and the transmission of individual differences in stress reactivity across generations. *Annu. Rev. Neurosci.* **21**, 1161–92.

Modrein-Talbot, M.A. Pullen, L., Zandstra, K., Ehrenberger, H., and Muenchen, B. (1998). A study of self-esteem among well adolescents: seeking a new direction. *Issues Compr. Pediatr. Nurs.* **21**, 229–41.

Monk, C.S., McClure, E.B., Nelson, E.E. Zarahn, E., Bilder, R.M., Leibenluft, E., Charney, D.S., Ernst, M., and Pine, D.S. (2003). Adolescent immaturity in attention-related brain engagement to emotional facial expression. *Neuroimage* **20**, 420–8.

Panksepp, J., Nelson, E., and Bekkedal, M. (1997). Brain systems for the mediation of social separation-distress and social reward. *Ann. NY Acad. Sci.* **807**, 78–100.

Ploj, K., Roman, E., and Nylander, I. (2003). Long-term effects of maternal separation on ethanol intake and brain opioid and dopamine receptors in male Wistar rats. *Neuroscience* **121**, 787–99.

Rauch, S.L. (2000). Neuroimaging research and the neurobiology of obsessive–compulsive disorder: where do we go from here? *Biol. Psychiatry* **47**, 168–70.

Richter, SS., Brown, S.A., and Mott, M.A. (1991). The impact of social support and self-esteem on adolescent substance abuse treatment outcome. *Subst. Abuse* **3**, 371–85.

Robbins, T.W. and Everitt, B.J. (1999). Drug addiction: bad habits add up. *Nature* **398**, 567–70.

Schultz, W. and Dickinson, A. (2000). Neuronal coding of prediction errors. *Annu. Rev. Neurosci.* **23**, 473–500.

Schultz, W., Dayan, P., and Montague, P.R. (1997). A neural substrate of prediction and reward. *Science* **275**, 1593–9.

Shalitin, S. and Phillips, M. (2003). Role of obesity and leptin in the pubertal process and pubertal growth—a review. *Int. J. Obes. Relat. Metab. Disord.* **27**, 869–74.

Simpson, M.J.A., Gore, M.A., Janus, M., and Rayment, F.D. (1989). Prior experience of risk and individual differences in enterprise shown by rhesus monkey infants in the second half of their first year. *Primates* **30**, 493–509.

Sowell, E.R., Thompson, P.M., Holmes, C.J., Jernigan, T.L., and Toga, A.W. (1999). *In vivo* evidence for post-adolescent brain maturation in frontal and striatal regions. *Nat. Neurosci.* **2**, 859–61.

Sowell, E.R., Delis, D., Stiles, J., and Jerrigan, T.L. (2001*a*). Improved memory functioning and frontal lobe maturation between childhood and adolescence: a structural MRI study. *J. Int. Neuropsychol. Soc.* **7**, 312–22.

Sowell, E.R., Thompson, P.M., Tessner, K.D., and Toga, A.W. (2001*b*). Mapping continued brain growth and gray matter density reduction in dorsal frontal cortex: inverse relationships during postadolescent brain maturation. *J. Neurosci.* **21**, 8819–29.

Sroufe, L.A., Carlson, E.A., Levy, A.K., and Egeland, B. (1999). Implications of attachment theory for developmental psychopathology. *Dev. Psychopathol.* **11**, 1–13.

Stevenson-Hinde, J. and Verscheuren, K. (2002). Attachment in childhood. In *Handbook of child social development* (ed. P.K. Smith and C.H. Hart), pp. 182–204. Blackwell, Oxford.

Sund, A-M., Larsson, B., and Wichstrom, L. (2003). Psychosocial correlates of depressive symptoms among 12–14-year-old Norwegian adolescents. *J. Child Psychol. Psychiatry* **44**, 588–97.

Tanner, J.M. (1966). *Growth at adolescence*. Blackwell Science Publications, Oxford.

Tremblay, L. and Schultz, W. (2000). Modifications of reward expectation-related neuronal activity during learning in primate orbitofrontal cortex. *J. Neurophysiol.* **83**, 1877–85.

Weiss, M., Baerg, E., Wiseboard, S., and Temple, J. (1995). The influence of gonadal hormones on periodicity of obsessive–compulsive disorder. *Can. J. Psychiatry* **40**, 205–7.

Williams, K.E. and Koran, L.M. (1997). Obsessive–compulsive disorder in pregnancy, the puerperium, and the premenstruum. *J. Clin. Psychiatry* **58**, 330–4; quiz 335–6.

Zohar, J., Hermesh, H., Weizman, A., Voet, H., and Gross-Isseroff, R. (1999). Orbitofrontal cortex dysfunction in obsessive–compulsive disorder? I. Alternation learning in obsessive–compulsive disorder: male–female comparison. *Eur. Neuropsychopharmacol.* **9**, 407–13.

David J.P. Barker FRS is professor of clinical epidemiology at the University of Southampton, UK, and professor in the Department of Medicine at Oregon Health and Science University, USA. His research shows that chronic diseases, including cardiovascular disease and type 2 diabetes, originate through poor nutrition and other adverse influences during early development.

Chapter 3*

The developmental origins of well-being

David J.P. Barker

Introduction

The recent discovery that people who develop chronic disease grew differently to other people during fetal life and childhood has led to a new 'developmental' model for a group of diseases, including coronary heart disease, stroke, high blood pressure, and type 2 (adult onset) diabetes (Barker *et al.* 1989). To explore the developmental origins of chronic disease required studies of a kind that had not hitherto been carried out. It was necessary to identify groups of men and women now in middle–late life, whose size at birth had been recorded at the time. Their birthweights could thereby be related to the later occurrence of coronary heart disease and other disorders. In Hertfordshire, UK, from 1911 onwards, when women had their babies they were attended by a midwife, who recorded the birthweight. A health visitor went to the baby's home at intervals throughout infancy, and the weight at 1 year was recorded. Table 3.1 shows the findings in 10 636 men born during 1911–30. Standardized mortality ratios for coronary heart disease fell with increasing birthweight. There were stronger trends with weight at 1 year. The association between low birthweight and coronary heart disease has now been replicated among men and women in Europe, the USA, and India (Osmond *et al.* 1993; Frankel *et al.* 1996; Rich-Edwards *et al.* 1997; Stein *et al.* 1996; Leon *et al.* 1998; Eriksson *et al.* 2001). Low birthweight has been shown to predict type 2 diabetes in studies around the world (Hales *et al.* 1991; McCance *et al.* 1994; Lithell *et al.* 1996; Rich-Edwards *et al.* 1999; Forsén *et al.* 2000).

These findings suggest that influences linked to early growth have an important effect on the risks of coronary heart disease and type 2 diabetes. It has been argued, however, that people whose growth was impaired *in utero* and during infancy may continue to be exposed to an adverse environment in childhood and adult life, and it is this later environment that produces the effects attributed to intrauterine influences. There is now strong evidence that this argument cannot be sustained (Frankel *et al.*

Table 3.1 Hazard ratios (95% confidence intervals) for death from coronary heart disease according to birthweight and weight at 1 year in 10 636 men in Hertfordshire

Weight (pounds)	Hazard ratio (95% confidence interval)	
	Before 65 years	All ages
Birthweight		
≤5.5 (n = 486)	1.50 (0.98–2.31)	1.37 (1.00 to 1.86)
5.6–6.5 (n = 1385)	1.27 (0.89–1.83)	1.29 (1.01 to 1.66)
6.6–7.5 (n = 3162)	1.17 (0.84–1.63)	1.14 (0.91 to 1.44)
7.6–8.5 (n = 3308)	1.07 (0.77–1.49)	1.12 (0.89 to 1.40)
8.6–9.5 (n = 1564)	0.96 (0.66–1.39)	0.97 (0.75 to 1.25)
≥10 (n = 731)	1.00	1.00
p for trend	0.001	0.005
Weight at age 1 year		
≤18 (n = 715)	2.22 (1.33–3.73)	1.89 (1.34–2.66)
18.1–20 (n = 1806)	1.80 (1.11–2.93)	1.58 (1.15–2.16)
20.1–22 (n = 3404)	1.96 (1.23–3.12)	1.66 (1.23–2.25)
22.1–24 (n = 2824)	1.52 (0.95–2.45)	1.36 (1.00–1.85)
24.1–26 (n = 1391)	1.36 (0.82–2.26)	1.29 (0.93–1.78)
≥27 (n = 496)	1.00	1.00
p for trend	<0.001	<0.001

1996; Rich-Edwards *et al.* 1997; Eriksson *et al.* 2001). The associations between low birthweight and later disease have been shown to be independent of influences such as socio-economic status and cigarette smoking, though adult lifestyle adds to the effects of early life: for example, the prevalence of type 2 diabetes is highest in people who had low birthweight and became obese as adults (Hales *et al.* 1991; McCance *et al.* 1994; Lithell *et al.* 1996; Rich-Edwards *et al.* 1999; Forsén *et al.* 2000).

Biological basis

During development the organs and systems of the body go through sensitive periods when they are plastic and sensitive to the environment. For most organs and systems the sensitive period occurs *in utero*. Developmental plasticity enables the production of phenotypes that are better matched to their environment than would be possible if the same phenotype was produced in all environments (West-Eberhard 1989; Bateson and Martin 1999). If a mother is poorly nourished, the baby responds by adaptations, such as reduced body size and altered metabolism, that help it to survive. Since, as Mellanby (1933) noted long ago, the ability of a human mother to nourish her baby is partly determined when she herself is *in utero* and by her childhood growth; the human fetus responds not only to conditions at the time of the pregnancy, but to conditions a number of decades before. Until recently, we have overlooked a growing body of evidence that systems of the body that are closely related to adult disease, such as the regulation of blood pressure, are also plastic during early development. In animals it is surprisingly easy to produce lifelong changes in the blood pressure and metabolism of offspring by minor modifications to the diet of the mother before and during pregnancy (Widdowson and McCance 1963; Kwong *et al.* 2000).

The different sizes of newborn human babies exemplify plasticity. The growth of babies has to be constrained by the size of the mother; otherwise normal birth could not occur. Small women have small babies: in pregnancies after ovum donation they have small babies even if the woman donating the egg is large (Brooks *et al.* 1995). Babies may be small because their growth is constrained in this way or because they lack the nutrients for growth. As McCance (1962) wrote, 'The size attained in utero depends on the services which the mother is able to supply. These are mainly food and accommodation.' The mother's height or long pelvic dimensions are generally not found to be important predictors of the baby's long-term health, and research into the developmental origins of disease has focused on the nutrient supply to the baby, while recognizing that other influences such as hypoxia and stress also influence fetal growth. This focus on fetal nutrition was endorsed in a recent review (Harding 2001). Around the world, size at birth in relation to gestational age is a marker of fetal nutrition. In developing countries many babies are undernourished because their mothers are chronically malnourished. Despite current levels of nutrition in Western countries, the nutrition of many fetuses and infants remains suboptimal, because the nutrients available are unbalanced or because their delivery is constrained by the long and vulnerable fetal supply line.

Developmental origins hypothesis

The developmental (or fetal) origins hypothesis proposes that coronary heart disease, type 2 diabetes, stroke, and hypertension originate through developmental plasticity, in response to undernutrition during fetal life and infancy (Barker 1995; Barker *et al.* 2002*a*). Why should fetal responses to undernutrition lead to disease in later life? The general answer is clear. According to 'life history theory', increased allocation of energy to the development of one trait such as brain growth necessarily reduces allocation to one or more other traits such as tissue repair processes. Smaller babies, who have had a lesser allocation of energy, must incur higher costs, and these it seems include disease in later life. A more specific answer to the question is that people who were small at birth are vulnerable to later disease through at least three kinds of process. First, they have fewer cells in key organs, such as the kidney. This may be part of a general reduction in cell numbers or a selective one: undernourished babies may, for example, divert blood flow away from the trunk to protect the brain. One theory holds that high blood pressure is initiated by the reduced number of glomeruli in the kidneys of people who were small at birth (Brenner and Chertow 1993). A reduced number necessarily leads to increased blood flow through each glomerulus. Over time this hyperfiltration may lead to the development of glomerulosclerosis, which, combined with the loss of glomeruli that accompanies normal ageing, leads to accelerated age-related loss of glomeruli, and a self-perpetuating cycle of rising blood pressure and glomerular loss. Direct evidence in support of this hypothesis has come from a study of the kidneys of people killed in road accidents. Those being treated for hypertension had fewer but larger glomeruli (Keller *et al.* 2003).

Another process by which slow fetal growth may be linked to later disease is in the setting of hormones and metabolism. An undernourished baby may establish a 'thrifty' way of handling food. Tissue resistance to the effects of insulin, which underlies type 2 diabetes and is associated with low birthweight, may be viewed as persistence of a fetal response by which glucose concentrations are maintained in the blood for the benefit of the brain, but at the expense of glucose transport into the muscles and muscle growth (Phillips 1996).

A third link between low birthweight and later disease is that people who were small at birth are more vulnerable to adverse environmental influences in later life. Observations on animals show that the environment during development can permanently change not only the body's structure and function but also its responses to environmental influences encountered in later life (Bateson and Martin 1999). Table 3.2 shows the effect of low income in adult life on coronary heart disease among men in Helsinki (Barker *et al.* 2001). As expected, men who had a low taxable income had higher rates of the disease. There is no agreed explanation for this, but the higher rate of coronary heart disease among poorer people is a feature of the disease in Western countries and a major component of the social inequalities in health. Among the men in Helsinki this association was confined to those who had had slow fetal growth and were thin at birth, defined by a ponderal index (birthweight/length^3) of less than 26 kg/m^3 (Table 3.2). Around one-quarter of all baby boys born in Britain today are thin under this definition. Men who were not thin at birth were resilient to the effects of low income on coronary heart disease.

One explanation of these findings emphasizes the psychosocial consequences of a low position in the social hierarchy, as indicated by low income and social class, and suggests that perceptions of low social status and lack of success lead to changes in neuroendocrine pathways and hence to disease (Marmot and Wilkinson 2001). The

Table 3.2 Hazard ratios (95% confidence intervals) for coronary heart disease in 3676 men according to ponderal index at birth (birthweight/length^3) and taxable income in adult life

Household income (UK £/year)	Hazard ratio (95% confidence interval) for ponderal index at birth	
	≤26 kg/m^3 (*n* = 1475)	>26 kg/m^3 (*n* = 2154)
>£15 700	1.00	1.19 (0.65–2.19)
£15 700	1.54 (0.83–2.87)	1.42 (0.78–2.57)
£12 400	1.07 (0.51–2.22)	1.66 (0.90–3.07)
£10 700	2.07 (1.13–3.79)	1.44 (0.79–2.62)
≤£8 400	2.58 (1.45–4.60)	1.37 (0.75–2.51)
p for trend	<0.001	0.75

findings in Helsinki seem consistent with this. People who are small at birth are known to have persisting alterations in responses to stress, including raised serum cortisol concentrations (Phillips *et al.* 2000). It has been suggested that persisting small elevations of cortisol concentrations over many years may have effects similar to those seen when tumours lead to more sudden, large increases in glucocorticoid concentrations. People with Cushing's syndrome are insulin-resistant and have raised blood pressure, both of which predispose to coronary heart disease.

Childhood growth and chronic disease

Figure 3.1 shows the growth pattern of 357 men who were either admitted to hospital with coronary heart disease or died from it (Eriksson *et al.* 2001). They belong to a cohort of 4630 men who were born in Helsinki, and their growth is expressed as standardized scores, termed Z-scores. The Z-score for the cohort is set at zero, and a boy maintaining a steady position as large or small in relation to other boys would follow a horizontal path on the figure. Boys who later developed coronary heart disease, however, were small at birth and remained small in infancy, but had accelerated gain in weight and body mass index thereafter. In contrast, their heights remained below average. As in Hertfordshire, the hazard ratios for coronary heart disease fell with increasing weight at 1 year, and also with increasing length and, more strongly, with body mass index (weight/height^2). Small size at this age predicted coronary heart disease independently of size at birth. Hence there appear to be at least two pathways of development that lead to coronary heart disease among men. One begins with slow growth *in utero*, and low birthweight and thinness at birth, thought to be a consequence of fetal undernutrition. The other begins with poor weight gain during infancy, which occurs in babies born into poor living

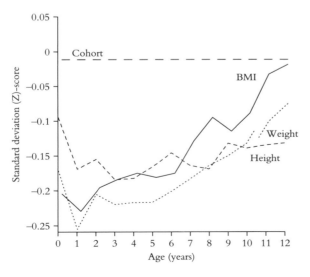

Fig. 3.1 Standardized mean scores (Z-scores) for height, weight, and body mass index (BMI) during childhood in 357 boys who later developed coronary heart disease within a cohort of 4630 boys. At any age the mean Z-score for the cohort is set at 0, while the standard deviation is set at 1.

conditions with overcrowding and consequent recurrent minor illness. Among the 4130 girls in the same birth cohort, the 87 who later developed coronary heart disease showed a broadly similar pattern of growth to that of the boys (Forsén *et al.* 2004). They were, however, short at birth rather than thin but became thin during infancy. This persisted up to the age of 4 years, after which they gained weight rapidly. In both sexes disease risk is related to the tempo of weight gain during childhood rather than to body size at any particular age. As D'Arcy Thompson wrote in 1917, 'To say that children of a given age vary in the rate at which they are growing would seem to be a more fundamental statement than that they vary in the size to which they have grown' (Thompson 1942).

High blood pressure and type 2 diabetes are associated with the same general pattern of growth as coronary heart disease. In both sexes risk of disease falls with increasing birthweight and rises with rapid weight gain in early childhood (Hales *et al.* 1991; McCance *et al.* 1994; Lithell *et al.* 1996; Rich-Edwards *et al.* 1999; Forsén *et al.* 2000; Curhan *et al.* 1996). Table 3.3 shows hazard ratios for hypertension, defined as high blood pressure requiring treatment, according to birthweight and fourths of body mass index at age 11 years among men and women born in Helsinki during 1924–44 (Barker *et al.* 2002a). At any birthweight, hazard ratios tend to be highest in people who had a high childhood body mass index. Conversely, at any childhood body mass the hazard ratios are highest among those who had low birthweight.

There is a substantial literature showing that birthweight is associated with differences in blood pressure within the normal range (Huxley *et al.* 2000). These differences are found in children and adults but they tend to be small. A one-kilogram (2.2 pound) difference in birthweight is associated with around 3 mmHg difference in systolic pressure. The contrast between this small effect and the large effect on hypertension (Table 3.3) suggests that lesions that accompany poor fetal growth and tend to elevate blood pressure, which may include a reduced number of glomeruli, have a small influence on blood pressure within the normal range because counterregulatory mechanisms maintain normal blood pressure levels.

Table 3.3 Odds ratios* (95% confidence intervals) for high blood pressure (2997 cases) according to birthweight and body mass index (BMI) at 11 years among 13 517 men and women

| Birthweight (kg) | Odds ratio (95% confidence intervals for BMI (kg/m²) at 11 years | | | |
	<15.7	15.8–16.6	16.7–17.6	>17.6
<3.0	2.0 (1.3–3.2)	1.9 (1.2–3.1)	1.9 (1.2–3.0)	2.3 (1.5–3.8)
3.0–3.5	1.7 (1.1–2.6)	1.9 (1.2–2.9)	1.9 (1.2–3.0)	2.2 (1.4–3.4)
3.6–4.0	1.7 (1.0–2.6)	1.7 (1.1–2.6)	1.5 (1.0–2.4)	1.9 (1.2–2.9)
>4.0	1.0	1.9 (1.1–3.1)	1.0 (0.6–1.7)	1.7 (1.1–2.8)

* Odds ratios adjusted for sex and year of birth.

As described already, a reduced number of glomeruli leads to increased blood flow. Experience in renal tranplantation surgery shows that such hyperfiltration is greatest in people with large body size, and these people are most vulnerable to the development of glomerulosclerosis and consequent glomerular destruction. This may explain why high childhood body mass index is deleterious (Table 3.3). Ultimately, homeostasis can no longer be maintained and blood pressure rises, which accelerates glomerular destruction. A cycle of rising blood pressure, further glomerular destruction, and further rise in blood pressure may be initiated (Ingelfinger 2003). Evidence to support the development of self-perpetuating cycles comes from a study of elderly people in Helsinki among whom the effect of birthweight on blood pressure was confined to those being treated for high blood pressure (Ylihärsilä *et al.* 2003). Despite their treatment, the blood pressures of those who had low birthweight were markedly higher. There was a more than 20 mmHg difference in systolic pressure between those who weighed 2500 g (5.5 pounds) or less at birth and those who weighed 4000 g (8.8 pounds) or more. Among the normotensive subjects, birthweight was unrelated to blood pressure. An inference is that by the time they reached old age most of the people with lesions acquired *in utero* had developed clinical hypertension.

Table 3.4 shows the relation between age at 'adiposity rebound' and later type 2 diabetes (Eriksson *et al.* 2003*a*). After the age of 2 years the degree of obesity of young children, as measured by body mass index, decreases to a minimum at around 6 years of age before increasing again—the so-called adiposity rebound. The age at adiposity rebound ranges from around 3 years to 8 years or more. Table 3.4 shows that early adiposity rebound is strongly related to a high body mass index in later childhood, as has previously been shown (Rolland-Cachera *et al.* 1987). It also predicts an increased incidence of type 2 diabetes in later life. The incidence of later type 2 diabetes was 8.6% in children whose adiposity rebound occurred at 4 years of age or less, falling to 1.8% in children whose body mass index did not start to increase until 8 years or more. This large difference emphasizes the importance of

Table 3.4 Body mass index (BMI) at 11 years of age and cumulative incidence of type 2 diabetes according to age at adiposity rebound in 8760 men and women

Age at adiposity rebound (years)	Mean BMI (kg/m^2) at age 11 (all subjects)	Cumulative incidence of diabetes [% (n)]		
		Men	Women	All
≤4	19.7	8.1 (86)	8.9 (112)	8.6 (198)
5	17.6	6.2 (904)	2.5 (864)	4.4 (1768)
6	17.0	3.7 (1861)	2.5 (1456)	3.2 (3317)
7	16.8	2.4 (249)	2.1 (243)	2.2 (492)
≥8	16.7	3.0 (135)	0.7 (150)	1.8 (285)
p for trend	<0.001	<0.001	0.002	<0.001

the tempo of childhood weight gain. The association between early adiposity rebound and impaired glucose tolerance has recently been replicated in a longitudinal study in Delhi, India (Bhargava *et al.* 2004). In both studies an early adiposity rebound was found to be associated with thinness at birth and at 1 year. It is not therefore the young child who is overweight who is at greatest risk of type 2 diabetes but the one who is thin but subsequently gains weight rapidly. Little is known about the physiological or environmental influences that determine the age at adiposity rebound.

Compensatory growth

When undernutrition during early development is followed by improved nutrition, many animals and plants stage accelerated or 'compensatory' growth. Compensatory growth has costs, however, that in animals include reduced life-span (Metcalfe and Monaghan 2001). There are a number of processes by which, in humans, undernutrition and small size at birth followed by rapid childhood growth could lead to cardiovascular disease and type 2 diabetes in later life (Eriksson *et al.* 2001; Forsén *et al.* 2000; Barker *et al.* 2002*a*). Rapid growth may be associated with persisting hormonal and metabolic changes. Larger body size may increase the functional demand on functional capacity that has been reduced by slow early growth—fewer glomeruli, for example. Rapid weight gain may lead to an unfavourable body composition. Babies that are small and thin at birth lack muscle, a deficiency that will persist as the critical period for muscle growth occurs *in utero* and there is little cell replication after birth (Widdowson *et al.* 1972). If they develop a high body mass during later childhood they may have a disproportionately high fat mass in relation to lean body mass, which will lead to insulin resistance (Eriksson *et al.* 2002*a*).

Pathways to disease

New studies, especially those in Helsinki where there is exceptionally detailed information on child growth and socio-economic circumstances, suggest increasingly that the pathogenesis of coronary heart disease and the disorders related to it depend on a series of interactions occurring at different stages of development. To begin with, the effects of the genes acquired at conception may be conditioned by the early environment (Dennison *et al.* 2001; Eriksson *et al.* 2002*b*). For example, the Pro12Pro polymorphism of the PPAR-Y gene is known to be associated with insulin resistance. In a study of 476 elderly people in Helsinki, however, this effect was found only among men and women who had low birthweight. Such interactions between the effects of genetic polymorphisms and those of birthweight, a marker of fetal nutrition (Harding 2001), would be predicted if chronic disease originates through developmental plasticity (West-Eberhard 1989; Bateson and Martin 1999).

Table 3.5 Cumulative incidence (%) of hypertension according to birthweight and father's social class in 8760 men and women

Birthweight (g)	Cumulative incidence (%) for father's social class			
	Labourer	**Lower middle class**	**Upper middle class**	**p for trend**
<3000	22.2	20.2	10.5	0.002
3000–<3500	18.8	15.2	10.6	<0.001
3500–4000	14.5	12.5	10.3	0.04
>4000	11.1	15.6	15.7	0.11
p for trend	<0.001	0.05	0.79	

The effects of the intrauterine environment on later disease are conditioned not only by events at conception, but by events after birth. Table 3.3 showed how the effects are conditioned by childhood weight gain. Table 3.2 showed that the effects of low ponderal index at birth are conditioned by living conditions in adult life. Table 3.5 shows how the effects of low birthweight on later hypertension are conditioned by living conditions in childhood, indicated by the occupational status of the father (Barker *et al.* 2000*b*). Among all the men and women, low birthweight was associated with an increased incidence of hypertension, as expected from previous findings (Curhan *et al.* 1996). This association, however, was present only among those who were born into families where the father was a labourer or of lower middle class. Again, this is a statistically strong interaction with the effects of birthweight.

It seems that the pathogenesis of cardiovascular disease and type 2 diabetes cannot be understood within a model in which risks associated with adverse influences at different stages of life add to each other. Rather, disease is the product of branching paths of development. The branchings are triggered by the environment. The pathways determine the vulnerability of each individual to what lies ahead. The pathway to coronary heart disease can originate either in slow fetal growth as a consequence of undernutrition, or in poor infant growth as a consequence of poor living conditions. Thereafter the pathway is determined by rates of weight gain in childhood, by socio-economic conditions, and, presumably, by other influences as yet undiscovered.

The effects of slow fetal growth and low birthweight, and the effects of postnatal development, depend on environmental influences and paths of development that precede and follow them. Low birthweight, or any other single influence, does not have 'an' effect that is best estimated by a pooled estimate from all published studies. As René Dubos (1960) wrote, 'the effects of the physical and social environments cannot be understood without knowledge of individual history.' Unravelling the causation of chronic disease, and hence the way to prevent it, will therefore require an understanding of heterogeneity.

Strength of effects

Low birthweight, though a convenient marker in epidemiological studies, is an inadequate description of the phenotypic characteristics of a baby that determine its long-term health. The wartime famine in Holland produced lifelong insulin resistance in babies who were *in utero* at the time, with little alteration in birthweight (Ravelli *et al.* 1998). In babies, as in children, slowing of growth is a response to a poor environment, especially undernutrition, but body size at birth does not adequately describe the long-term morphological and physiological consequences of undernutrition. The same birthweight can be attained by many different paths of fetal growth and each is likely to be accompanied by different gene–environment interactions (Harding 2001). Nevertheless, birthweight provides a basis for estimating the magnitude of the effects of the fetal phase of development on later disease, though it is likely to underestimate them.

Because the risk of cardiovascular disease is influenced both by small body size at birth and during infancy and by rapid weight gain in childhood, estimation of the risk of disease attributable to early development requires data on fetal, infant, and childhood growth. Currently, the Helsinki studies are the main source of information (Barker *et al.* 2002*a*). If each man in the cohort had been in the highest third of ponderal index at birth, and each woman in the highest third of birth length, and if each man or woman had lowered their body mass index between ages 3 and 11 years, the incidence of coronary heart disease would have been reduced by 25% in men and 63% in women (Barker *et al.* 2002*a*).

Sensitive periods

The famine in Holland began suddenly in November 1944 and ended 6 months later (Ravelli *et al.* 1998). Official rations fell to around 700 calories. The subsequent effect on people who were *in utero* at the time depended on the stage of gestation at which they were exposed to the famine. People who were conceived during the famine now have raised serum cholesterol and raised rates of coronary heart disease. In contrast, those conceived before the famine but exposed to it in mid–late gestation are insulin-resistant and have impaired glucose tolerance (Ravelli *et al.* 1998; Roseboom *et al.* 2001). These differences presumably reflect sensitive periods of organs such as the liver, heart, and muscle occurring at different stages of fetal development. The effects of growth failure at different phases of infancy also depend on its timing. Type 2 diabetes is associated with failure of linear growth between birth and 3 months, which may coincide with a sensitive period of development for insulin production by the endocrine pancreas (Eriksson *et al.* 2003*b*). Coronary heart disease is associated with growth failure between 3 months and 1 year, which may be one of the sensitive periods for cholesterol regulation by the liver.

Well-being

Most of the data that link development with later health, and hence well-being, come from studies of disease. Underlying the associations between early growth and later disease, however, are processes that are likely to affect the well-being of people who are, from a medical viewpoint, healthy. People who were small at birth have heightened stress responses (Phillips *et al.* 2000), one manifestation of lifelong settings of hormones and metabolism that are established before birth. These settings affect reproductive physiology, as is well known in animals (Bateson and Martin 1999; Harding 2001). In Hertfordshire, men who were small at birth are less likely to marry, an observation confirmed in Sweden (Phillips *et al.* 2001). Around 20% of men who weighed 5.5 pounds (2.5 kg) or less at birth remained unmarried in middle age compared with only 5% of men weighing 7.75 pounds (3.5 kg) or more. This is not an effect of adult height: at any height, those who were smaller at birth were less likely to marry. It seems that restriction of growth before birth alters some aspect of partner selection—sexuality, socialization, personality, or emotional responses.

Women born after term tend to have altered gonadotrophin production and polycystic ovaries (Cresswell *et al.* 1997). A possible explanation is that the human fetus produces large amounts of androgens that are converted to oestrogen by the placenta and pass to the maternal circulation. Placental failure associated with post-maturity could expose the fetal hypothalamus to increased concentrations of androgens or oestrogens and alter hypothalamic–pituitary set points for release of luteinizing hormone. Further evidence that a woman's reproductive fitness may be influenced *in utero* comes from studies of age at menarche. In a national sample of British girls, those who reached menarche at the youngest age had low birthweight but put on weight rapidly in childhood (Cooper *et al.* 1996). Similar findings came from a study of Filipino girls (Adair 2001). Again, the suggested link between fetal growth and age at menarche is the setting of the pattern of gonadotrophin release. Pulsatile release of luteinizing and follicle-stimulating hormone is initiated *in utero*, continues through infancy, and thereafter ceases until it resumes at puberty.

Mood and cognitive function

For many years depression in adult life has been thought to originate through parental indifference, abuse, and other adverse influences in childhood. In Hertfordshire men and women who committed suicide, which is commonly the result of depression, had low weight gain in infancy (Barker *et al.* 1995). This could be due to adverse psychosocial influences in infancy, but there are no data on this. Another possibility is that depression is initiated by *in utero* programming of hormonal axes that influence growth in infancy and mood in later life. Patients with depression have been found to have abnormal secretion of growth hormone and abnormalities in the hypothalamic–adrenal and hypothalamic–thyroid axes (Checkley 1992). There is evidence that each of these axes are programmed *in utero*. In a subsequent study, 867 elderly men and women completed the Geriatric Depression Scale and Mental State examinations (Thompson *et al.* 2001). Among the men, but not the women, the odds ratio for depression fell progressively with increasing birthweight. This was independent of other correlates of depression, which include low social class, recent bereavement, social isolation, and physical illness.

Early growth also has lasting effects on cognitive function, but brain growth during infancy and early childhood seems to be more important than growth during fetal life. In the well known study of children born in Newcastle-upon-Tyne during 1947, the 'Thousand Families Study', one of the strongest predictors of intelligence at age 11 years was height at 3 years (Miller *et al.* 1974). In a recent study of 9-year-old children, intelligence quotients were unrelated to head circumference at 18 weeks gestation or at birth, but strongly related to the circumference at 9 months and 9 years (Gale *et al.* 2004). These associations with postnatal brain growth persisted after adjustment for maternal intelligence, education, and history of low mood in the post-partum period.

The lifelong importance of early postnatal brain growth is indicated by findings in the Helsinki cohort. Men who had had slow linear growth between birth and 1 year had worse educational achievements and lower incomes at 50 years, than those who grew more rapidly (Barker *et al.* in press). Each 2 cm increase in length between birth and 1 year was associated with a 3.5% increase in income. The findings were similar within each social class at birth. Slow head growth during infancy is also associated with reduced cognitive function and an increased risk of dementia and Alzheimer's disease in later life (Gale *et al.* 2003; Schofield *et al.* 1997). While improved brain growth during infancy may help to protect against cognitive decline in later life, remarkably little is known about the genetic and environmental influences that control this.

Conclusions

The list of chronic diseases whose origins lie in early development extends beyond cardiovascular disease and type 2 diabetes. Depression, which is more common among people with these diseases, may also originate in prenatal life. There is now strong evidence that osteoporosis, failure of bone mineralization, is another of the body's 'memories' of undernutrition at a sensitive early stage of development (Cooper *et al.* 2002). The demonstration that normal variations in fetal size at birth have implications for health throughout life has prompted a re-evaluation of fetal development. There is increasing evidence that a woman's own fetal and childhood growth, and her diet and body composition at the time of conception as well as during pregnancy, play an important role in determining the lifelong well-being of her children.

References

Adair, L.S. (2001). Size at birth predicts age at menarche. *Pediatrics* **107**, e59.

Barker, D.J.P. (1995). Fetal origins of coronary heart disease. *Br. Med. J.* **311**, 171–4.

Barker, D.J.P., Osmond, C., Winter, P.D., Margetts, B., and Simmonds, S.J. (1989). Weight in infancy and death from ischaemic heart disease. *Lancet* **2**, 577–80.

Barker, D.J.P., Osmond, C., Rodin, I., Fall, C.H.D., and Winter, P.D. (1995). Low weight gain in infancy and suicide in adult life. *Br. Med. J.* **311**, 1203.

Barker, D.J.P., Forsén, T., Uutela, A., Osmond, C., and Eriksson, J.G. (2001). Size at birth and resilience to the effects of poor living conditions in adult life: longitudinal study. *Br. Med. J.* **323**, 1273–6.

Barker, D.J.P., Eriksson, J.G., Forsén, T., and Osmond, C. (2002*a*). Fetal origins of adult disease: strength of effects and biological basis. *Int. J. Epidemiol.* **31**, 1235–9.

Barker, D.J.P., Forsén, T., Eriksson, J.G., and Osmond, C. (2002*b*). Growth and living conditions in childhood and hypertension in adult life: longitudinal study. *J. Hypertens.* **20**, 1951–6.

Barker, D.J.P., Eriksson, J.G., Forsén, T., and Osmond, C. (in press). Infant growth and income fifty years later. *Arch. Dis. Child.*

Bateson, P. and Martin, P. (1999). *Design for a life: how behaviour develops.* Jonathan Cape, London.

Bhargava, S.K., Sachdev, H.P.S., Fall, C.H.D., Osmond, C., Lakshmy, R., Barker, D.J.P., Biswas, S.K.D., Ramji, S., Prabhakaran, D., and Reddy, K.S. (2004). Relation of serial changes in childhood body-mass index to impaired glucose tolerance in young adulthood. *New Engl. J. Med.* **350**, 865–75.

Brenner, B.M. and Chertow, G.M. (1993). Congenital oligonephropathy: an inborn cause of adult hypertension and progressive renal injury? *Curr. Opin. Nephrol. Hypertens.* **2**, 691–5.

Brooks, A.A., Johnson, M.R., Steer, P.J., Pawson, M.E., and Abdalla, H.I. (1995). Birth weight: nature or nurture? *Early Hum. Dev.* **42**, 29–35.

Checkley, S. (1992.) Neuroendocrinology. In *Handbook of affective disorders* (ed. E.S. Paykel), Churchill Livingstone, Edinburgh.

Cooper, C., Kuh, D., Egger, P., Wadsworth, M., and Barker, D. (1996). Childhood growth and age at menarche. *Br. J. Obstet. Gynaecol.* **103**, 814–17.

Cooper, C., Javaid, M.K., Taylor, P., Walker-Bone, K., Dennison, E., and Arden, N. (2002). The fetal origins of osteoporotic fracture. *Calcif. Tissue Int.* **70**, 391–4.

Cresswell, J.L., Barker, D.J.P., Osmond, C., Egger, P., Phillips, D.I.W., and Fraser, R.B. (1997). Fetal growth, length of gestation and polycystic ovaries in adult life. *Lancet* **350**, 1131–5.

Curhan, G.C., Chertow, G.M., Willett, W.C., *et al.* (1996). Birth weight and adult hypertension and obesity in women. *Circulation* **94**, 1310–15.

Dennison, E.M., Arden, N.K., Keen, R.W., Syddall, H., Day, I.N.M., Spector, T.D., and Cooper, C. (2001). Birthweight, vitamin D receptor genotype and the programming of osteoporosis. *Paediatr. Perinat. Epidemiol.* **15**, 211–19.

Dubos, R. (1960). *Mirage of health.* Allen and Unwin, London.

Eriksson, J.G., Forsén, T., Tuomilehto, J., Osmond, C., and Barker D.J.P. (2001). Early growth and coronary heart disease in later life: longitudinal study. *Br. Med. J.* **322**, 949–53.

Eriksson, JG., Forsén, T., Jaddoe, V.W.V., Osmond, C., and Barker, D.J.P. (2002*a*). Effects of size at birth and childhood growth on the insulin resistance syndrome in elderly individuals. *Diabetologia* **45**, 342–8.

Eriksson, J.G., Lindi, V., Uusitupa, M., *et al.* (2002*b*). The effects of the Pro12Ala polymorphism of the peroxisome proliferator-activated receptor-y2 gene on insulin sensitivity and insulin metabolism interact with size at birth. *Diabetes* **51**, 2321–2324.

Eriksson, J.G., Forsén, T., Tuomilehto, J., Osmond, C., and Barker, D.J.P. (2003*a*). Early adiposity rebound in childhood and risk of type 2 diabetes in adult life. *Diabetologia* **46**, 190–4.

Eriksson, J.G., Forsén, T.J., Osmond, C., and Barker, D.J.P. (2003*b*). Pathways of infant and childhood growth that lead to type 2 diabetes. *Diabetes Care* **26**, 3006–10.

Forsén, T., Eriksson, J., Tuomilehto, J., Reunanen, A., Osmond, C., and Barker, D. (2000). The fetal and childhood growth of persons who develop type 2 diabetes. *Ann. Intern. Med.* **133**, 176–82.

Forsén, T., Osmond, C., Eriksson, J.G., and Barker, D.J.P. (2004). The growth of girls who later develop coronary heart disease. *Heart* **90**, 20–4.

Frankel, S., Elwood, P., Sweetnam, P., Yarnell, J., and Davey Smith, G. (1996). Birthweight, body mass index in middle age, and incident coronary heart disease. *Lancet* **348**, 1478–80.

Gale, C.R., Walton S., and Martyn, C.N. (2003). Foetal and postnatal brain growth and risk of cognitive decline in old age. *Brain* **126**, 2273–8.

Gale, C.R., O'Callaghan, F.J., Godfrey, K.M., Law, C.M., and Martyn, C.N. (2004). Critical periods of brain growth and cognitive function in children. *Brain* **127**, 321–9.

Hales, C.N., Barker, D.J.P., Clark, P.M.S., *et al.* (1991). Fetal and infant growth and impaired glucose tolerance at age 64. *Br. Med. J.* **303**, 1019–22.

Harding, J. (2001). The nutritional basis of the fetal origins of adult disease. *Int. J. Epidemiol.* **30**, 15.

Huxley, R.R., Shiell, A.W., and Law, C.M. (2000). The role of size at birth and postnatal catch-up growth in determining systolic blood pressure: a systematic review of the literature. *J. Hypertens.* **18**, 815–31.

Ingelfinger, J.R. (2003). Is microanatomy destiny? *New Engl. J. Med.* **348**, 99–100.

Keller, G., Zimmer, G., Mall, G., Ritz, E., and Amann, K. (2003). Nephron number in patients with primary hypertension. *New Engl. J. Med.* **348**, 101–108.

Kwong, W.Y., Wild, A., Roberts, P., Willis, A.C., and Fleming, T.P. (2000). Maternal undernutrition during the pre-implantation period of rat development causes blastocyst abnormalities and programming of postnatal hypertension. *Development* **127**, 4195–202.

Leon, D.A., Lithell, H.O., Vagero, D., *et al.* (1998). Reduced fetal growth rate and increased risk of death from ischaemic heart disease: cohort study of 15 000 Swedish men and women born 1915–29. *Br. Med. J.* **317**, 241–5.

Lithell, H.O., McKeigue, P.M., Berglund, L., Mohsen, R., Lithell, U.B., and Leon, D.A. (1996). Relation of size at birth to non-insulin dependent diabetes and insulin concentrations in men aged 50–60 years. *Br. Med. J.* **312**, 406–10.

Marmot, M. and Wilkinson, R.G. (2001). Psychosocial and material pathways in the relation between income and health: a response to Lynch *et al. Br. Med. J.* **322**, 1233–6.

McCance, D.R., Pettitt, D.J., Hanson, R.L., Jacobsson, L.T.H., Knowler, W.C., and Bennett, P.H. (1994). Birth weight and non-insulin dependent diabetes: thrifty genotype, thrifty phenotype, or surviving small baby genotype? *Br. Med. J.* **308**, 942–5.

McCance, R.A. (1962). Food, growth and time. *Lancet* **2**, 621–6.

Mellanby, E. (1933). Nutrition and child-bearing. *Lancet* **2**, 1131–7.

Metcalfe, N.B. and Monaghan, P. (2001). Compensation for a bad start: grow now, pay later? *Trends Ecol. Evol.* **16**, 254–60.

Miller, F.J.W., Court, S.D.M., Knox, E.G., and Brandon, S. (1974). The school years in Newcastle-upon-Tyne, 1952–1962. Oxford University Press, London.

Osmond, C., Barker, D.J.P., Winter, P.D., Fall, C.H.D, and Simmonds, S.J. (1993). Early growth and death from cardiovascular disease in women. *Br. Med. J.* **307**, 1519–24.

Phillips, D.I.W. (1996). Insulin resistance as a programmed response to fetal undernutrition. *Diabetologia* **39**, 1119–22.

Phillips, D.I.W., Walker, B.R., Reynolds, R.M., *et al.* (2000). Low birth weight predicts elevated plasma cortisol concentrations in adults from 3 populations. *Hypertension* **35**, 1301–6.

Phillips, D.I.W., Handelsman, D.J., Eriksson, J.G., Forsén, T., Osmond, C., and Barker, D.J.P. (2001). Prenatal growth and subsequent marital status: longitudinal study. *Br. Med. J.* **322**, 771.

Ravelli, A.C.J, van der Meulen, J.H.P., Michels, R.P.J., Osmond, C., Barker, D.J.P., Hales, C.N., and Bleker, O.P. (1998). Glucose tolerance in adults after exposure to the Dutch famine. *Lancet* **351**, 173–7.

Rich-Edwards, J.W., Stampfer, M.J., Manson, J.E., *et al.* (1997). Birth weight and risk of cardiovascular disease in a cohort of women followed up since 1976. *Br. Med. J.* **315**, 396–400.

Rich-Edwards, J.W., Colditz, G.A., Stampfer, M.J., *et al.* (1999). Birthweight and the risk for type 2 diabetes mellitus in adult women. *Ann. Intern. Med.* **130**, 278–84.

Rolland-Cachera, M.F., Deheeger, M., Guilloud-Bataille, M., Avons, P., Patois, E., and Sempe, M. (1987). Tracking the development of adiposity from one month of age to adulthood. *Ann. Hum. Biol.* **14**, 219–29.

Roseboom, T., van der Meulen, J.H.P, Ravelli, A.C., Osmond, C., Barker, D.J.P., and Bleker, O.P. (2001). Effects of prenatal exposure to the Dutch famine on adult disease in later life: an overview. *Mol. Cell Endocrinol.* **185**, 93–8.

Schofield, P.W., Lagroscino, G., Andrews, H.F., Albert, S., and Stern, Y. (1997). An association between head circumference and Alzheimer's disease in a population-based study of aging and dementia. *Neurology* **49**, 30–7.

Stein, C.E., Fall, C.H.D., Kumaran, K., Osmond, C., Cox, V., and Barker, D.J.P. (1996). Fetal growth and coronary heart disease in South India. *Lancet* **348**, 1269–73.

Thompson, C., Syddall, H., Rodin, I., Osmond, C., and Barker, D.J.P. (2001). Birthweight and the risk of depressive disorders in late life. *Br. J. Psychiatry* **179**, 430–55.

Thompson, D.W. (1942). *On growth and form.* Cambridge University Press, Cambridge

West-Eberhard, M.J. (1989). Phenotypic plasticity and the origins of diversity. *Annu. Rev. Ecol. Syst.* **20**, 249.

Widdowson, E.M. and McCance, R.A. (1963). The effect of finite periods of undernutrition at different ages on the composition and subsequent development of the rat. *Proc. R. Soc. Lond. B* **158**, 329–42.

Widdowson, E.M., Crabb, D.E., and Milner, R.D.G. (1972). Cellular development of some human organs before birth. *Arch. Dis. Child.* **47**, 652–5.

Ylihärsilä, H., Eriksson, J.G., Forsén, T., Kajantie, E., Osmond, C., and Barker, D.J.P. (2003). Self-perpetuating effects of birth size on blood pressure levels in elderly people. *Hypertension* **41**, 446–50.

Sonia J. Lupien, Ph.D. is the director of the Laboratory of Human Stress Research at Douglas Hospital/McGill University, Canada, and the associate director of the McGill Center for Studies on Aging.

Nathalie Wan, MA is a sociologist and a research associate at the Laboratory of Human Stress Research at Douglas Hospital/McGill University.

Chapter 4*

Successful ageing: from cell to self

Sonia J. Lupien and Nathalie Wan

Introduction

When groups of young and old participants were asked to describe the first image of old age that came to mind, both groups spontaneously mentioned the following words: 'wise, slow, senile, ill, infirm, forgetful, frail and decrepit' (Levy 1996, p. 1104). For the past 30 years, negative age stereotypes have been widely accepted as the norm by most scientists and general public alike. The elderly have been viewed and labelled as, 'ill and/or disabled', 'impotent', 'ugly', 'mentally declining', 'mentally ill', 'useless', 'isolated', 'poor', and 'depressed' (Palmore 1990). Even gerontology has been defined as the science of drawing downwardly sloping lines (Minkler 1990). Negative stereotypes of ageing tended to marginalize perceptions of elderly individuals within a predominantly young–adult society, by placing them on opposite sides along the lifespan spectrum. Such stereotyping and discrimination is known as ageism.

In 1968, Pulitzer Prize winning author Robert N. Butler coined the word 'ageism' using this definition: 'ageism can be seen as a process of systematic stereotyping of and discrimination against people because they are old . . .' (Butler 1969, p. 243). A few years later at a symposium on geriatric medicine, he condemned health professionals and academics for bringing about these stereotypes: 'medicine and the behavioural sciences have mirrored social attitudes by presenting old age as a grim litany of physical and emotional ills' (Butler 1977, p. 14). Like all prejudices, ageism influences the self-view and behaviour of its victims. Elderly individuals have a tendency to take on negative definitions of themselves and to perpetuate the very stereotypes directed against them, hence reinforcing society's beliefs. Such emphasis on senescence without acknowledging the growth, recuperation, and improvement that occur in ageing is a form of ageism.

Fortunately, this negative view of the elderly is now challenged. As the number of elderly people in the developed world has grown, the public has become increasingly aware of the problems that the elderly face. The increased salience of the ageing population may have changed people's beliefs and prejudices about elderly individuals, with older individuals being recently seen in a more positive light. The term 'successful ageing' has thus made its entry into the popular and scientific literature. Although the study of successful ageing has certainly shed a positive light on the ageing population, many theoretical and methodological problems remain with regard to what we define as 'successful ageing'.

In the following section, we present a historical perspective of research on ageing, from the negative to the positive view of the ageing process. We then discuss the various definitions of successful ageing, from the biological, psychological, and psychosocial perspectives, and we conclude by presenting and commenting on the multicriterion models of successful ageing.

The mismeasure of ageing

The foundation of gerontology can be traced back to the early decades of the twentieth century, when the concept of ageing as a personal as well as a social problem dominated the images of researchers studying this domain (Ferraro 1997; Hirshbein 2001). Along with being considered a 'problem', there was an early acknowledgement that the study of ageing occupied several disciplines, in particular, medicine, psychology, and sociology. One of the professional groups that displayed increasing interest in the subject of ageing as a topic for professional intervention was psychology. As early as the 1920s, psychologists G. Stanley Hall and Lillian Martin advocated shifting the study of psychology from the early stages of life towards the later stages of life (Hall 1922; Martin and de Gruchy 1930). In the 1930s, several psychologists began to expand their work by incorporating old age into their work on developmental psychology. In their report on cross-sectional studies on the life trajectory of measurable abilities, both Miles (1933) and Bird (1940) showed that individuals peaked in childhood and declined in old age. However, in defining 'normal ability of performance' in the elderly, they compared the ability of older people with the average performance of children and young adults, thus generating a biased definition. Thus, despite psychologists' genuine efforts to expand scientific knowledge on old age, their bias towards children may have added to the growing idea that old age meant degeneration and declining ability.

During the 1940s, theoretical models of the biology of ageing concerning homeostatic mechanisms were produced (Griffiths 1997). As with psychology, the expanding biomedical interest in old age unintentionally extended the idea that old age was accompanied by inevitable pathology. Despite being supportive of the elderly, the medical model emphasized disease and disability as inevitable products of old age. In the 1950s the study of ageing once again began to focus heavily on a psychological perspective. Post–World War II was the launching period of the psychology of ageing in which myriad studies focused on linkages between early and later stages of human development, with particular emphasis on cognitive ageing. At the same time, there existed an increasing awareness of issues facing the older population that led to the expansion of research in social gerontology.

Early sociological models of ageing emphasized the negativity of the ageing process. In the 1960s, the disengagement theory emerged and proposed that ageing entailed a gradual social withdrawal or disengagement from personal relationships or society in general (Lynott and Lynott 1996). However, because of the negative connotation and conclusions of this theory, many sociologists became outraged and, as a result, made countless efforts to debunk and refute the main principles of disengagement theory. It may be argued that the theory was ageist, and hence attacked by sociologists and social psychologists in an attempt to make gerontology the study of normal ageing.

The 1970s witnessed a tremendous increase in the number of articles dealing with the changes in cognitive and physical functions observed during ageing, particularly those affecting memory and linguistic functions (for a complete review see Birren and Schaie 1985). As with the first developmental studies performed by psychologists in the 1930s, it is through cross-sectional studies comparing the performance of young individuals with that of older individuals that scientists were able to detect these modifications. However, such comparisons between different age groups revealed another important difference between young and old individuals: old individuals present a greater heterogeneity in cognitive performance when compared with young individuals.

Interestingly, however, most researchers neglected the increased variability in the cognitive manifestations of ageing for almost 2 decades (see Lupien and Lecours 1993). It was concluded that a general decline in cognitive performance with age was 'normal' (Dannefer 1988), leading again to the negative view of ageing. However, as Rowe and Kahn (1987) pointed out in their seminal *Science* paper on usual and successful ageing, if the group variance of elderly subjects is so high, it is in part a result of the fact that some individuals show very poor performance on the measure taken (pathological ageing), whereas others show very high levels of performance (successful ageing), thus making any concept of a 'norm' for ageing totally meaningless.

This view was confirmed in the 1980s and 1990s, when several major longitudinal studies of ageing were developed (reviewed in Birren and Schaie 1985; Baltes and Baltes 1990). These studies revealed that movement, speed, visual acuity, and several types of memory steadily and inexorably decline, but that this decline starts as early as the age of 20 years. Furthermore, several of these longitudinal studies have shown that many cognitive skills actually improve with age (Baltes and Lindenberger 1988; Schaie and Willis 1998). More recent data also revealed that the adult brain does not lose as many neurons as was once thought but, rather, it continues to sprout new neurons (Shingo *et al.* 2003). Finally, psychosocial studies show that, with age, individuals tend to become happier, they have better mental health, they are better at managing interpersonal relationships, and they present fewer negative emotions (Helmuth 2003). Given the plethora of studies showing that the young and teenagers present significant problems of well-being, often leading to depression and suicide, one could eventually conclude that 'we may do very well to study older people to see how to help younger people' (L. Carstensen; cited in Helmuth 2003, p. 1300). Statements such as this reveal the extremely important changes in our view of ageing that developed from the 1920s into the new century.

The meaning of successful ageing

Despite the varying perspectives of the biology, psychology, and sociology of ageing, it is evident that both public and professional attention to old age has increased dramatically over the latter half of the twentieth century. This growing interest may be explained by the increasing number of elderly people in the USA and other developed countries over the past century. The numbers of elderly people have been growing substantially: in 1900 the population aged 65 and older in the USA was 3.1 million (4.1% of the total population), in 1950, this number had increased to 12.2 million (8.1% of the total population), and in 2000 it grew to 35.0 million (12.4% of the population) (Department of Health and Human Services 2001). Undoubtedly, these numbers are staggering, but what is of considerable interest is the way in which the study of ageing has shifted, from portraying old age in terms of its losses to the notion of 'successful ageing'.

The concept of successful ageing has been a subject of escalating interest since its initiation as the main theme of the 1986 annual meeting of the Gerontological Society of America (Fisher 1995). The following year, Rowe and Kahn's (1987) seminal study, 'Human aging: usual and successful aging', in *Science* ignited another wave of interest in successful ageing by recommending that research on successful ageing should concentrate on people with above average physiological and psychosocial characteristics in later life, or 'successful agers' as opposed to 'usual agers'. This certainly had a beneficial effect on research on ageing as it drew the attention of scientists to the positive aspects of ageing. However, at this point, there is still a lack of consensus on what defines a 'successful ager'. In the following sections we give the various definitions of successful ageing from these different scientific domains.

Models of successful biological ageing: born to live long

Most of the models of successful biological ageing consider that successful ageing is represented by two main factors: compression of morbidity and longevity. To understand the importance of these factors for the definition of biological successful ageing, it is important to distinguish between 'maximum lifespan' and 'average lifespan' or 'life expectancy'. The maximum lifespan represents the longest-lived member(s) of the population or species. In humans, the oldest individual ever recorded was Jeanne Calment, who died in 1997 in France at the age of 122 years. By contrast, the average lifespan is represented by the age at which 50% of a given population are still alive. Interestingly, the average lifespan of humans has increased over time, while the maximum lifespan has remained constant, around the age of 90 to 100 years (Cutler 1990).

The average lifespan has been shown to depend on various environmental factors such as socio-economic status and nutritional status (Cutler 1990). It is mainly a result of a significant improvement in sanitary conditions over the past century that the average life expectancy at birth is now approximately 74.6 years in males and 79.8 years in females in the USA (World Health Organization 2003a, b). In 2000, the World Health Organization recognized that quality of life in old age is as important as increased longevity, so it created an index of health expectancy at birth that calculates the number of years an individual is expected to live without major diseases. Using this index, it has been calculated that the average healthy active life expectancy is approximately 67 years of age in males and 71 years of age in females in the USA (World Health Organization 2000).

Compression of morbidity

Although the average life expectancy at birth has increased in the past century, there are still many diseases that can decrease life expectancy. Sheldon (1948) identified diseases that were thought to be attributable to old age, which led to the theory that the development of these diseases was part of 'normal' ageing (age-associated diseases). The age-associated diseases that have been mostly studied in this context are cardiovascular disease, cancer, stroke, diabetes, and dementia. Although most of these age-associated diseases do not occur exclusively in older people and/or in all older people, it is interesting to note that the high prevalence of these diseases in the elderly has created the impression that disease is a necessary part of ageing. Given this later definition of ageing as being a time of disease, it is not surprising to see that one part of the biological definition of 'successful ageing' is compression of morbidity (Fries 1980). Hence, to increase successful ageing in humans, one should increase the number of people living into old age, while at the same time limiting the age-associated diseases to a shorter period before death (Fries 1980). The idea behind the notion of compression of morbidity is that, if the period from onset of chronic infirmity to death can be shortened, this would benefit both the individual and society (Fries 1993). Interventions based on primary preventions (smoking cessation, exercise, cholesterol reduction, etc.) have had significant effects on the number of age-associated diseases, which reveals that the notion of compression of morbidity has a positive impact on successful living (reviewed in Fries 1993).

The role of genetic factors

In contrast to average lifespan, which is sensitive to major diseases and environmental challenges, it has been shown that maximum lifespan, which represents the longest-lived member(s) of the population or species, is species-specific and very stable. This stability suggests that genetic factors might make a major contribution to the maximum lifespan of an individual. In this context, the centenarian phenotype has been used as a model of biological and genetic successful ageing. Centenarians are a model of successful ageing because they have escaped or survived the common age-associated diseases (see Perls *et al.* 2002*a*), and they are living close to the maximum lifespan. Many studies have examined the centenarian phenotype to identify factors associated with maximum lifespan. These include body fat and metabolism (Paolisso *et al.* 1995), cardiovascular risk factors (Barbagallo *et al.* 1995; Baggio *et al.* 1998), immune function (Effros *et al.* 1994; Franceschi *et al.* 1995), and cognitive function (Silver *et al.* 1998, 2001; Andersen-Ranberg *et al.* 2001; Hagberg *et al.* 2001). Interestingly, none of these extrinsic and intrinsic factors has been shown to correlate with the ability to survive into extreme old age (Karasawa 1979; Beregi 1990; Beard 1991; Poon 1992; Perls *et al.* 1999).

The factors that predict maximum lifespan are believed to be mainly genetic factors. A familial study comparing siblings of centenarians to siblings of non-centenarians revealed that siblings of centenarians have four times the probability of surviving to the age of 91 years when compared with the other group of siblings (Perls *et al.* 1998). However, this result does not necessarily confirm the major contribution of genes in maximum longevity, because other factors such as environmental, attitudinal, or lifestyle factors could explain these findings. Still, the results obtained from this and other familial studies (Perls and Fretts 1998; Rybicki and Elston 2000; Kerber *et al.* 2001; Perls *et al.* 2002*b*) led scientists to search for longevity-modifying genes. Here, the approach is to look for lifespan-altering genes in organisms in which the lifespan is short enough that it allows multigenerational experiments within a reasonable time frame (Guarente and Kenyon 2000). The organisms that have been mostly studied in this context are yeast, the worm *Caenorhabditis elegans*, *Drosophila*, and mice (Jazwinski 1999; Imai *et al.* 2000). Studies in these lower organisms, and particularly the yeast, revealed that lifespan is determined mainly by 19 genes, whose functions implicate four basic determinants of lifespan, namely, metabolic control, resistance to stress, gene dysregulation, and genetic stability (see Perls *et al.* 2002*a* for a comprehensive review).

Selective pressure on genes

If longevity is genetically determined, then it implies that selective pressure on specific genes has been applied throughout evolution. Here, it is interesting to note that species with a short lifespan have generally a higher fertility rate (Williams 1966). This trade-off between life expectancy and reproductive capacity has led to the disposable soma theory (Kirkwood 1977; Kirkwood and Holliday 1979), which comprises three major premises. The first is that in natural populations (with the exception of humans) most deaths generally occur accidentally. This means that the probability of reaching maximum life expectancy is low. The second premise is that the long-term survival of an individual depends on maintenance of the organism, a process that is energetically costly. The third premise is that it is not advantageous for an individual to invest a large fraction of metabolic resources in long-term survival if only a small fraction of these resources is necessary to survive in reasonably good condition. Basically, this theory recognizes the importance of the allocation of resources between growth, maintenance, and reproduction. If one increases maintenance of the organism, this will promote survival. However, this will be done at the expense of resources that could be used for growth and reproduction. In summary, the price to pay for reproduction of a species is finite survival. In these conditions, getting old with the minimum load of age-associated diseases is the biological definition of successful ageing.

Models of successful cognitive ageing: the power of comparisons

The measurement of individual differences is of central importance for the models of successful cognitive ageing. In this approach, the characteristics of individuals that are deviant from the age normals are contrasted with those of persons ageing more in accordance with the normative expectations. In this sense, the cognitive models split the ageing process into pathological, usual, and successful. Here, usual ageing implies the normative pattern, and individuals ageing successfully and those with premature or excessive cognitive frailty are ageing atypically (reviewed in Stones *et al.* 1990).

The entire approach of models of successful cognitive ageing stands on three types of comparison of an aged individual to other groups. In the first approach, the cognitive performance of an older person is compared with normative data obtained in individuals of the same chronological age (the normative approach). Here, a successful ager will be defined as someone being above normative values when controlling for age, education level, and socio-economic status. In the second approach, the cognitive performance of the individual is compared with the mean performance of a group of individuals within the same chronological age range (the age-related approach). Here, a successful ager is defined as someone showing a higher performance (by two or three standard deviations) than the mean of the group. In the third approach, the cognitive performance of the individual is compared with that of a group of young individuals. In this case, a successful ager is defined as someone having a cognitive performance as good as that of young individuals (the age-difference approach).

These three approaches are similar in the use they make of the increased interindividual variability with ageing. Indeed, all approaches index successful ageing by functions that are characterized by relative homogeneity in young adulthood but heterogeneity thereafter. Because it is known that aged individuals exhibit more heterogeneity in their cognitive performance, one can thus infer that inferior performance in an older individual represents an acquired decrement, whereas a performance similar to that of young individuals will mean successful ageing, because it implies that cognitive function was maintained throughout life. These three approaches carry with them very important differences for the ways in which they will categorize someone as being a successful ager. Indeed, all three approaches are independent in that a given individual could easily be defined as a successful ager using one type of comparison, and as a normal or pathological ager using another type of comparison.

Figures 4.1 and 4.2 give a schematic representation of these differences. Figure 4.1 depicts age-related inconsistencies in the categorization of a successful ager, whereas Fig. 4.2 presents age-difference inconsistencies in the categorization of a successful ager.

Age-related inconsistencies in the categorization of a successful ager

In Fig. 4.1 the distribution curve A represents memory data from a representative population sample in which the age is exactly 80 years (normative data). Here, let us assume that the normative mean for this group is 100. If an 80-year-old man

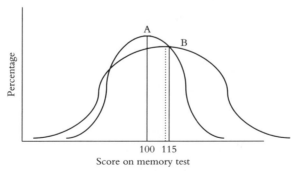

Fig. 4.1 Inconsistencies in the categorization of successful ageing using age-related data. The data represent scores on a memory test obtained by men with a mean education level of 10 years, from medium socio-economic status. The distribution curve A represents the scores of a normative sample of men aged exactly 80 years. The solid line represents the mean score (100) of this group on the test. The distribution curve B represents the scores of a group of men aged between 75 and 85 years selected for a particular study. The solid line represents the mean score (115) of this group on the test. The dotted line represents the score (113) of one individual on the test. This individual would be categorized as a successful ager if compared with the distribution curve A, but would be categorized as a normal ager if compared with the distribution curve B.

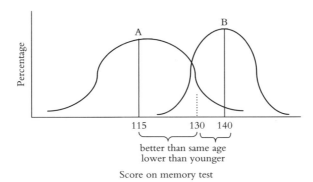

Fig. 4.2 Inconsistencies in the categorization of successful ageing using data on age differences. The distribution curve A represents the scores on a memory test for a group of men and women aged between 75 and 85 years, with mean education levels of 10 years, from medium socio-economic status. The mean score of this group was 115 (solid line), with an SD of 5 (dotted line within the same distribution curve). The distribution curve B represents the scores on a memory test for a group of young men and women ranging from 25 to 35 years, with a mean education levels of 12 years, from medium socio-economic status. The mean score of this group was 140 (solid line), with an SD of 2 (dotted line within the same distribution curve). The dotted line represents the score (130) of one aged individual on the test. This individual would be categorized as a successful ager if compared with the distribution curve A, but would be categorized as a normal ager if compared with the distribution curve B.

scores 113 on the memory test, one could say that this individual is a 'successful ager' because his score is higher than the norm. The distribution curve B represents memory data from a non–representative aged sample (with a mean age of 80 years) recruited for a particular study with a 10 year age range. One might think that the score distribution of this population should be the same as the normative database. However, the mean of this population is higher than the mean of the representative sample (distribution curve A), and the distribution curve is flatter than the distribution curve A.

Many factors can explain these differences. First, in population B, the individuals below the age of 80 could contribute in increasing the mean, and the age range could contribute in increasing the standard deviation of the group. Second, it is well known that aged individuals who tend to volunteer for a particular study are generally more educated, have a higher income, and are in better physical shape than those who do not readily volunteer (reviewed in Lupien and Lecours 1993). All of these factors have been shown to be related to better cognitive function in old age (Schaie 1993). This fact would thus tend to push the mean towards an upper limit (in this case, 115). The individual with a score of 113, who was previously categorized as a successful ager when compared with normative data, is now considered to be a usual ager, because his performance is not different than that of the mean of group B.

From this example, one could conclude that comparing the performance of an aged individual with normative data is the best way to categorize successful cognitive ageing. This might, unfortunately, not be the case because it has been shown that existing age-based norms for cognitive tests are almost certainly contaminated by cases of unsuspected incipient Alzheimer's disease and other age-related cognitive pathologies (Morris *et al.* 1991). The presence of unsuspected incipient cases of age-related cognitive impairments in normative databases will thus decrease the mean, allowing for a larger proportion of aged individuals to meet the criterion for successful cognitive ageing when one compares an individual with normative data.

Age-difference inconsistencies in the categorization of a successful ager

In Fig. 4.2, age-difference inconsistencies in the categorization of a successful ager are depicted. In this figure, the distribution curve A represents memory data from a non-representative aged sample (with a mean age of 80 years) recruited for a particular study with a 10 year age range. The distribution curve B represents memory data from a non-representative young sample (ranging from 25 to 35 years) recruited for the same study with a 10 year age range. The mean performance obtained in the aged population is 115 with a standard deviation of five, whereas the mean performance obtained in the young population is 140 with a standard deviation (SD) of two. If an 80-year-old man with a score of 130 on the test is compared with the aged population, he could be categorized as a successful ager, given that his score falls outside two standard deviations of the mean of the aged group ($115 + 5 + 5 = 125$; man's score is 130). However, if the score of the same man is compared with that of the young population, the man would not be categorized as a successful ager because his score falls more than two standard deviations below the mean of the young population ($140 - 2 - 2 = 136$; man's score is 130). Given that the young population shows less variability in performance (mean with 2 SD) when compared with the aged population (mean with 5 SD), this difference in categorization is an artefact that is induced by the large differences in variance in the two different populations.

This example illustrates the important difficulties that one faces when trying to define successful cognitive ageing. Because of the increased variability in cognitive performance known to exist in aged human populations, it is extremely hazardous to define a successful ager based on comparisons to either a norm or a mean from aged or young populations. Moreover, each approach defined above has its inherent methodological and theoretical pitfalls. First, as we have previously discussed, comparison with a norm can be difficult given the fact that these norms can be pushed downwards by the presence of incipient cases of age-related cognitive pathologies in the normative group tested. Second, comparison with the mean of an aged population leads directly to an elitist definition of successful ageing (Masoro 2001; Minkler and Fadem 2002). Indeed, it is well known that education level is positively related to cognitive function. Therefore, by using this method, there is a high probability that those individuals who will be defined as successful agers will be those who were successful during their entire life (high education, high income, etc.). Comparison with the mean of a young population decreases the probability of categorizing someone as being a successful ager given the presence of large differences in variances between the young and aged populations (see Lupien and Lecours 1993) and, more importantly, it increases the probability of having a cohort effect.

Gender balance of samples

There are other problems that arise when comparing young and aged populations. These include the gender balance of samples, and the functional equivalence of the tests used in these populations. Many scientists are aware of the inherent difficulties in establishing an equal proportion of aged women and men when creating an experimental sample, because the former generally volunteer more often than the latter. This may be explained by intrinsic characteristics of women, and/or by the fact that there is a smaller proportion of men than women particularly at older ages (Cutler and Harootyan 1975; Hendricks and Hendricks 1977; Botwinick 1978). Consequently, an experimental protocol incorporating an equal number of men and women in both the young and aged samples is not representative of the population. The disproportion between the men and women is largely a result of the relatively poor health and premature death of elderly men (Cutler and Harootyan 1975). Studying the performance of an equal number of men and women within an aged population means that, on average, the men are likely to be in poorer physical health, which would hardly be the case in a young population.

Education balance of samples

In addition to the problem of gender imbalance, the average level of education is lower in older compared with younger populations. In fact, there is no existing criterion that allows the assignment of equivalent education levels to individuals having different educational histories (Krauss 1980). We also know that it is not prudent to rely on number of years of education because many trades practised by older adults that were acquired through experience are now taught in specialized schools. An even more complex problem is related to the fact that a smaller proportion of aged people are highly educated, which does not reflect a difference in aptitude but in availability and social norms (Cutler and Harootyan 1975). Hence, the idea of selecting subjects on the basis of an equal number of years of education would lead the researcher to study an extreme group of elderly people if the education criterion is high, or a limited group of young subjects if the education criterion is low. The use of an apparently objective criterion (number of years of education) in establishing a control for the education levels of groups being studied will thus produce misleading conclusions about age differences, because the selected group would not be representative of the general population.

Functional equivalence of cognitive tests

The problem of the functional equivalence of cognitive tests arises mainly with experimental designs in which the emphasis is placed on obtaining an interaction between the 'age' variable and one or several tasks (Baltes and Goulet 1971; Labouvie 1980). The degree of validity that we can attribute to such designs depends on the assurance that the tasks evaluate the same processes in all age groups. It is possible that certain measures that are formally identical are linked to totally different types of information processing in two or more different age groups. For example, word meanings and word associations may be different for different generations, and the widespread use of rote learning as the predominant teaching method for the older generation contrasts with an emphasis on understanding meaning rather than rote learning among more recent generations. Such generational differences in the meaning or salience of stimuli, or in styles of learning and remembering, can be misinterpreted as genuine age differences when they are, in fact, generational or cohort differences.

Recent functional brain imageing data have revealed important differences in the pattern of brain activation in young and aged subjects, despite no differences in test performance. Cabeza *et al.* (2004) scanned younger and older adults with functional magnetic resonance imaging (fMRI) while performing three different tasks tapping into different cognitive processing: working memory; visual attention; and episodic retrieval. The results showed that older adults performed as well as younger individuals, although they presented important differences in the pattern of brain activation induced by each task. In all three tasks, older adults showed weaker occipital and hippocampal activity and stronger prefrontal and parietal activity than younger adults. In the working memory and visual attention tasks, the older adults showed more bilateral patterns of prefrontal activity than younger adults. Finally, in the episodic memory task, the older adults showed stronger parahippocampal activity than the younger participants. These results have been taken to suggest that both common and specific factors play an important role in cognitive ageing.

The importance of cognitive reserves

Finally, it is important to note that comparing an elderly individual to other people, to define successful ageing, does not take into account the fact that most aged individuals still show the capacity to learn new information, which suggests that one could become a successful ager later in life. Indeed, substantial reserve capacity exists in the domain of cognitive functioning. Studies have shown that healthy elderly people between the ages of 60 and 80 benefit from practice and demonstrate an increase in performance in the specific abilities that are trained. Further, this augmented performance is similar in its magnitude to the ageing decline found in untrained individuals in longitudinal studies (Baltes and Lindenberger 1988; Schaie and Willis 1998). In addition, it has been shown that the healthy elderly are able to learn and acquire new cognitive skills, for instance, by becoming memory experts (Kliegl *et al*. 1989). These findings suggest that, for many older individuals, cognitive functioning during old age continues to operate in generally the same way as it did in the earlier stages of life. Therefore, older adults are able to use their cognitive mechanisms to acquire new information.

As older adults have the cognitive reserves to acquire new forms of knowledge, cognitive ageing is not only made up of the preservation of past functioning. Rather, the data suggest, contrary to negative ageing stereotypes, that 'new learning' is possible and that older adults continue to produce new forms of adaptive capacity (Baltes and Baltes 1990).

Models of successful psychosocial ageing: the power of well-being

The models of successful psychosocial ageing emphasize social interaction, life satisfaction, and well-being as major determinants of successful ageing. One of the first views to emerge in the psychosocial models was the activity theory, which stated that, because growing older involves the loss of roles, for example, through retirement and widowhood, older adults who wish to retain a positive sense of self need to find satisfaction in other, often newly substituted roles (Ferraro 1997). In the 1960s, the disengagement theory posited that ageing involves a gradual withdrawal or disengagement from interpersonal relationships or society in general, thus resulting in the marginalization of other people in society. Given this theoretical perspective, social withdrawal or decreased interaction is a usual response to the expectation of death, perception of a decreasing lifespan, and decreasing energy experienced by ageing individuals (Ferraro 1997). However, this theory has been strongly criticized by researchers who found a lack of evidence that older people disengage from their surroundings. Whereas older people's social relationships change and they possibly have fewer social ties, it was found that their relationships are often deeper and more meaningful (Helmuth 2003).

The importance of social support

More recently, psychosocial research and approaches to defining successful ageing have focused on topics including subjective well-being (George 1979, 1981), social support, and interactions, as well as life satisfaction (Herzog and Rodgers, 1981). A substantial body of research shows that social support is a key determinant of successful ageing (Rowe and Kahn 1998; Leviatan 1999; Unger *et al*. 1999; Seeman *et al.* 2001*b*). Research has indicated the health benefits of social integration and social support. Older adults reporting a greater number of social ties have been shown to have lower mortality risks, and both social integration and social support have been associated with better mental and physical health outcomes (reviewed in Berkman 1995; Cohen and Herbert 1996; Seeman 1996). Studies have also suggested that social environment may have an impact on patterns of cognitive functioning. In a study examining the relationship of social ties and support to patterns of cognitive ageing in the MacArthur studies of successful ageing, Seeman *et al.* (2001*b*) showed that participants receiving more emotional support had higher baseline cognitive performance.

Furthermore, growing bodies of research have shown that social integration and support are linked to protection against physical and mental health conditions such as cardiovascular diseases, hypertension, and depression, which are each related to cognitive decline. It has been postulated that, because social interactions have essential intrinsic cognitive components and hence sustain ongoing cognitive engagement, they may also contribute to enhanced cognitive function: the 'use it or lose it' paradigm (Baltes and Baltes 1990; Rowe and Kahn 1998). Bassuk *et al.* (1999) conducted a longitudinal, population-based cohort study of older adults and, over a 12 year follow-up, a greater baseline of social engagement (indexed by more reported contact with friends and family and more engagement in group activities) was shown to be protective against cognitive impairment. Fratiglioni *et al.* (2000) had similar findings in that greater reported social networking was protective against onset of dementia.

Subjective well-being

Studies examining the notion of successful ageing and life satisfaction have shown that, although the two concepts are different, they both contributed to subjective well-being in the aged population. Fisher (1995) explored the meanings that older people attached to successful ageing and life satisfaction among 40 individuals between the ages of 61 and 92 years. It was shown that the term successful ageing reflected an attitudinal or coping orientation nearly twice as often as those for life satisfaction. Life satisfaction seemed to represent basic needs, whereas successful ageing was linked more closely to higher-order needs, such as self-understanding, helping others, and feeling like one has made a difference. Another key difference that emerged from the study was that life satisfaction was described in terms of past expectations and present circumstances, whereas successful ageing was more oriented to strategies for coping in later life and maintaining a positive outlook (Fisher 1992).

The importance of personality traits

It is important to note that most of the psychosocial models of successful ageing do not take into account the potential impact of personality traits on life satisfaction and well-being in old age. Scholars have long debated the malleability of human personality. Self-concept and attitudes, in addition to other factors, contribute to stability and change in personality, which in turn influence the prospect of successful ageing. Personality traits have been assessed in many studies to examine whether personality is stable or not during late adulthood. Indeed, it appears that as much of an argument can be made for stability as for change, depending on one's interpretation, and depending on the level at which personality is examined (McAdams 1994). Some longitudinal studies have shown that personality is very stable over time (Costa and McCrae 1988), whereas others have found changes in personality in the later years of life (Shanan 1991).

Recent research has shown that five relatively stable independent factors or dimensions may encapsulate the domain of personality (McCrae and Costa 1984; Costa and McCrae 1986; McCrae and John 1992; Goldberg 1993). These factors are referred to, by both clinicians and researchers, as the 'big five' model. The five dimensions include neuroticism, extraversion, openness to experience, agreeableness, and conscientiousness (McCrae and Costa 1984; Costa and McCrae 1986; McCrae and John 1992; Goldberg 1993). A substantial amount of research related to the big five model indicates the consistency and stability of personality across lifespan (Costa and McCrae 1986).

Many studies have shown that personality traits are among the most potent predictors of psychological well-being (reviewed in McCrae 2002). A study by Costa and McCrae (1980) revealed that personality traits early in life predicted well-being years in advance. For example, negative affect was associated with neuroticism and low levels of well-being, whereas positive affect was associated with extraversion and high well-being. These results suggest that well-being, as a marker of successful ageing, might be highly dependent on certain personality traits that develop very early in life. This would support the idea that the pathways to successful ageing are determined very early in life.

Multicriteria models of successful ageing

The analysis of the biological, cognitive, and psychosocial models of successful ageing reveals that unique markers of successful ageing often present weaknesses because they cannot predict or explain other aspects of the ageing process. For this reason, some scientists have used multicriteria models of successful ageing, that is, models that take into account the biological, cognitive, and psychosocial aspects of ageing.

Two main perspectives on successful ageing exist that use a multicriterion approach to successful ageing. The first looks at successful ageing as a state of being, a condition that can be objectively measured at a certain moment. The second views successful ageing as a process of continuous adaptation. Both perspectives are based on the recognition of heterogeneity among older adults. Rowe and Kahn (1987) hold the first view and have defined successful ageing in terms of multiple physio-logical and psychosocial variables, whereby successful ageing is the positive extreme of what is considered normal ageing.

Multicriterion model of Rowe and Kahn (1987)

In 1987, Rowe and Kahn published a seminal paper in *Science* in which they described successful ageing as having three main components: avoidance of disease and disability; maintenance of cognitive capacity; and active engagement in life. To examine the validity of this concept, various scientists collaborated on a study of successful ageing funded by the MacArthur Foundation in the USA. They tested 4030 participants as part of a larger study, and 1931 met the multicriterion of successful ageing, which included the three main components of Rowe and Kahn's definition, i.e. avoidance of disease and disability, maintenance of cognitive capacity, and active engagement in life (Berkman *et al.* 1993). After defining these individuals as successful agers based on cross-sectional data, scientists then followed this par-ticular group of aged individuals over time.

Main findings of the MacArthur studies on successful ageing

Studies performed on this group first revealed that, at the cross-sectional level, low levels of education were associated with poorer psychological function, poorer health behaviours and biological conditions, and larger social networks (Kubzansky *et al.* 1998). At the longitudinal level, it was found that some psychological, psychosocial, and physiological variables predicted cognitive decline later in life. At the psychological level, it was found that certain measures of learning and memory (Li *et al.* 2001; Chodosh *et al.* 2002; Tabbarah *et al.* 2002), as well as low emotional support at baseline (Seeman *et al.* 2001*b*), were strong predictors of cognitive decline measured 7 years later. At the psychosocial level, results showed that a low number of social ties at the time of entry into the study predicted cognitive decline 7 years later (Unger *et al.* 1999; Kubzansky *et al.* 2000). Finally, at the physiological level, it was reported that elevated baseline plasma levels of interleukin-6 are a significant predictor of cognitive decline 2.5 years later (Weaver *et al.* 2002).

Stress and ageing

The study also revealed the impact of stress on cognitive function in this population. First, it was found that older women are more reactive to stress in terms of secre-tion of cortisol (a stress hormone, than men (Seeman *et al.* 1995*a*) and that level of self-esteem is a potent predictor of cortisol reactivity to stress (Seeman *et al.* 1995*b*). Second, the study showed that aged women who presented a significant increase in cortisol levels (a stress hormone) over a period of 2.5 years were more likely to show declines in memory performance over the same period of time (Seeman *et al.* 1997*a*).

This result extended a previous study performed by our group showing that, in a population of aged men and women, a significant increase of stress hormone levels over a period of 4 years is significantly related to both memory impairments (Lupien *et al.* 1994) and atrophy of the hippocampus, a brain structure that is involved in learning and memory (Lupien *et al.* 1998).

The bulk of data obtained for the effects of stress hormones and other biological measures led to the evaluation of the model of allostatic load, proposed by McEwen and Stellar (1993). The term 'allostatic load' refers to a cumulative, multisystemic view of the physiological toll that may be exacted on the body through attempts at adaptation. In this sense, allostatic load is the price that the body may ultimately pay for its adaptational efforts (see Seeman *et al.* 1997b). Recent studies performed by Seeman and collaborators (Seeman *et al.* 1997a, b, 2001a; Karlamangla *et al.* 2002) examined the hypothesis that risks for declining cognitive and physical functioning in elderly individuals are related to differences in allostatic load. In their study, they measured allostatic load by 10 parameters reflecting levels of physiological activity across a range of important regulatory systems (e.g. blood pressure, cholesterol, cortisol levels, etc.). They reported that higher allostatic load scores were associated with poorer cognitive and physical functioning. More importantly, they also showed that allostatic load scores predicted larger decrements in cognitive and physical factors later on. These results showed that allostatic load measures could provide a basis for the description of major risk factors in the development of the pathological ageing processes.

Successful ageing: an elitist definition?

Clearly, the multiple studies that were performed within the MacArthur study of successful ageing led to very important findings that related to all aspects (biological, cognitive, and psychosocial) of ageing. However, although this study yielded significant results on the factors that can contribute to successful ageing, there is an important problem concerning how the researchers defined and studied successful ageing in the first cross-sectional study and in subsequent studies (Berkman *et al.* 1993). The study performed by the MacArthur Foundation was a cross-sectional study in which the authors chose, as successful agers, the top 30% of the population in terms of cognitive factors. However, and as we have shown in our discussion of the models of successful cognitive ageing, it is well known that education level and socio-economic income are both positively related to cognitive function. As we have discussed, splitting a cross-sectional population at the top 30% means that those individuals who will be defined as successful agers will be those individuals who were successful during their entire life (high education, high income, etc.), thus leading to an elitist definition of successful ageing. In fact, in the first published report from the MacArthur group, it was shown that, within the largest population from which the successful agers were selected, the participants who were included in the low-functioning group using the lowest tertile were almost three times as likely to have a low income compared with the high-functioning group (Berkman *et al.* 1993). This result confirms that those individuals who were categorized as successful agers in the MacArthur study were the same individuals who had been successful during their entire life.

Consequently, the results obtained by the MacArthur group are not necessarily generalized over the entire population. Moreover, this approach does not take into account the factors that prevent an individual from showing age-related decline in biological, cognitive, or psychosocial factors when this individual is not part of the upper tertile of the population in the first place, that is, when this individual is part of the norms early and/or late in life. Also, this approach prevents anyone with functional disabilities, such as the physicist Stephen Hawking, to be defined as a successful ager later in life (Minkler 1990; Minkler and Fadem 2002). Clearly, many individuals who are not part of the upper tertile of a given population have the capacity to become successful agers, but, unfortunately, the working definition of successful ageing used by the MacArthur study prevents the assessment of these important factors. In summary, although the MacArthur study of successful ageing has generated a wealth of very important data, the method that was used to define successful ageing may have created the impression that success in old age is limited to only a small percentage of the population.

Multicriterion model of Baltes and Baltes (1990)

Baltes and colleagues have developed a meta-model of selective 'optimization with compensation' for the assessment of successful ageing (Baltes and Baltes 1990). Their goal is to explain what people in fact do when they age successfully. Recognizing that the meaning of success is highly individualized in nature, their aim has been to widen the focus beyond the 'theoretically normative psychological outcome that has seriously limited our understanding of successful aging' (Baltes and Carstensen 1996, p. 398).

The model of selective optimization with compensation

The model of selective optimization with compensation describes a general process of adaptation and attempts to explain the dynamic interchange of gains and losses throughout life, and how age-related and self-produced changes in oneself can be seen as an example of the plasticity of the ageing mind. Selective optimization with compensation is a meta-model of successful development that explains how individuals make adaptations when faced with changes brought about by the ageing process. The model conceptualizes three main processes that demonstrate people's ability to be resilient in their thinking, feelings, behaviours, or environments to attain desired goals throughout development and ageing.

First, there is the element of selection that refers to the restricted options of available functional domains because of an ageing loss in the range of adaptive potential. Selection implies that individuals adjust their expectations to allow the subjective experience of satisfaction and personal control (Baltes and Baltes 1990). The second element, optimization, reflects a view that people engage in behaviours that assist them in reaching higher and more desirable levels of functioning. Examples of optimization include training, practice, or education. Studies on plasticity have shown that older people continue to be able to perform this optimizing process (Baltes and Baltes 1990). Optimization holds the potential for growth that comes about when we master the uncertainties and changes of old age. In this way, Baltes and Baltes view successful ageing as the process by which the elderly achieve their individual goals in the face of simultaneous losses. The third element, compensation, results from restrictions in the range of plasticity or adaptive potential, similar to selection. The process of compensation occurs when an individual's behavioural capacities are lost or reduced below a level necessary for adequate functioning. This limited capacity is experienced particularly in situations that require a wide range of activity and a high level of performance, such as competitive sports or situations that require quick thinking and memorization. Compensatory efforts include, for instance, use of mnemonic strategies when internal memory strategies are inadequate (Baltes and Baltes 1990). This meta-model of ageing provides strategies that individuals can follow to contribute to their own successful ageing.

Clearly, the multicriterion model of selective optimization with compensation proposed by Baltes and Baltes (1990) holds its premises in the psychosocial models of successful ageing that view successful ageing as the successful adaptation of an individual to the changes inherent to the ageing process. It is thus a model of resilience and coping, and it has the advantage of not using an elitist definition of successful ageing. However, there are two potential weaknesses of the model of selective optimization with compensation. First, although the model uses a multicriterion approach to define successful ageing, these criteria are psychosocial in nature and do not include any biological or cognitive criteria. However, individuals who are in excellent health up into their old age might never be confronted with the need to select, optimize, and compensate to deal with the changes that occur with ageing. The model would suggest, however, that, even in the case of these individuals, they would have to compensate for other age-related losses such as retirement, loss of income, loss of spouse, etc. Still, the capacity for resilience and coping might significantly differ as a function of various factors such as health and cognitive capacity. Therefore, it is unclear at this point how the model can include these variables. The second problematic aspect of the model is its reliance on the individual as the sole instigator of successful ageing. In this respect, the studies of the kibbutz communities teach us something interesting.

Studies on the Israeli kibbutz and the importance of social arrangements

Although several studies on successful ageing have shown that social interactions and support impact cognitive functioning, research on the ageing members of the Israeli kibbutz communities has shown that the impact of social arrangements and social support are strong predictors of successful ageing, expressed in high life expectancy and positive well-being. Twenty years of research on the elderly population of kibbutzim show that their demonstration of successful ageing is mostly a result of the social arrangements and policies assumed by their communities in the areas of work, social relationships, stability in social roles, and surroundings (Leviatan 1999). The kibbutz uniqueness in its way of life allows its aged members to demonstrate successful ageing. The continuation of the worker role after retirement, shared responsibility by both individual and community in health preservation, formal community institutions as a means of social support to replace non-existent close family members, and constant adjustment of jobs to changing abilities are a few of the social arrangements that allow for lower mortality rates and potential longevity among the kibbutz communities (Leviatan 1989, 1999).

Support for the importance of social arrangements is also shown by the negative effects of well-being that result from structural changes, experienced by some kibbutzim, in the course of becoming more similar to industrialized societies. As their population is very homogeneous compared with other societies, a study has shown that, when comparing the occurrence of structural changes, those elderly members from kibbutzim that underwent the structural changes, as compared with members from kibbutzim without them, experienced more alienation and 'feeling of powerlessness', less satisfaction with life, and less satisfaction with kibbutz life (Leviatan 1999). Therefore, in the case of the experience of the Israeli kibbutz, any structural or social change in the life domains of its elderly members would seem to result in a disruption of the stability in social roles, environmental conditions, social arrangements, and life experiences that has been shown to lead to successful ageing.

The study of the kibbutz shows that social stability is a key determinant of successful ageing among members of its society. However, when faced with structural changes, the question of adaptation arises. The uniqueness of the kibbutz is that its society adapts to the needs of its members, which, in turn, leads them to age successfully. By contrast, the model proposed by Baltes and Baltes (1990) suggests that successful ageing is the successful adaptation of the individual to changes during the course of ageing. Hence, the attainment of successful ageing can arguably be recognized as taking two approaches: the societal approach, in which the concern is with what society should or should not do to raise elderly individuals to the functioning level of successful ageing; and the individual approach, in which the concern is with what individuals should or should not do to experience successful ageing. Such divergence in the approaches to attaining successful ageing exemplifies how the notion of successful ageing continues to lend itself to more than one interpretation or perspective.

Successful ageing and the power of positive attitudes towards ageing

Research on the perception of age and ageing stereotypes has shown that well-being and a positive view of ageing are major protective factors against the effects of age on the organism (Linn and Hunter 1979; Levy 1996, 2001; Levy *et al.* 2000*a*, 2002). The effects of attitudes and beliefs about ageing have been observed in various areas of science, from biological to sociological. In the biological domain of ageing research, studies have shown that older individuals' beliefs about ageing can have a direct impact on their health and longevity (Levy *et al.* 2000*b*, 2002). In the first study of its kind, Levy *et al.* (2000*b*) showed that negative attitudes towards ageing heightened cardiovascular response to stress, whereas positive attitudes towards ageing exerted a protective effect, bringing the physiological changes back to baseline levels. The study suggests that negative attitudes towards ageing may contribute to health problems in the elderly without their being aware. This could lead to elderly individuals mistakenly attributing their deteriorating health to the inevitability of ageing, which may then reinforce the negative age stereotypes and prevent successful ageing (Levy *et al.* 2000*a*). Introducing positive views of ageing reduced cardiovascular stress in the sample being studied and therefore suggests that interventions designed to improve cardiovascular health in the elderly should emphasize the importance of including the promotion of positive ageing attitudes.

A further indication that positive attitudes play an important role in an elderly individual's physiological state is that positive self-perceptions of older individuals can influence longevity (Levy *et al.* 2002). Findings in this observational study showed a 7.5 year mean survival advantage for those who expressed a more positive self-perception of ageing compared with those holding more negative perceptions. This finding provides further support for the idea that the internalization of negative stereotypes can be seen as a significant health hazard, although it has to be noted here that these observational data may have numerous other interpretations related to the characteristics of the participants.

Research on the psychology of ageing has similarly contributed to studies on the impact of stereotypes on ageing (Linn and Hunter 1979; Levy 1996). In a recent study examining the impact of negative and positive attitudes towards ageing and their influence on memory performance, it was found that older participants exposed to negative age stereotypes tend to worsen their memory performance, self-efficacy, and judgements of other elderly people (Levy 1996). In this study, general memory performance was compared before and after subliminal exposure to positive or negative words related to ageing (e.g. wise versus senile). Results showed that memory performance was significantly lower after exposure to negative words related to ageing, whereas it was higher when participants had been exposed to positive words related to ageing. Although these results raise the possibility that age stereotypes may have an important effect on memory performance in the aged human population, it has to be noted that the emotional valence of the words (positive versus negative) may be a more important factor in determining memory performance (see Fredrickson and Levenson 1998; Fredrickson 2001) than the fact that the words are related to ageing. However, in the study of Levy (1996), young participants exposed to the same positive and negative words related to ageing did not exhibit any of the significant interactions observed in the old participants. These results suggest that the ageing component of the words may indeed be an important modulatory factor of memory performance in the aged population.

Other psychology-based studies have found that perception of age in the elderly has an impact on psychological functioning (Linn and Hunter 1979). Specifically, it was found that older individuals who perceived their age as younger than that of others of the same age had more internal control, which has been related to more positive functioning. Internal control refers to the extent to which an individual sees their outcomes as being dependent on their own efforts and abilities as opposed to external circumstances or chance or fate (external control). In a similar vein, Palmore and Luikart (1972) reported that beliefs in personal control were related to enhanced life satisfaction. Also, Wolk and Kurtz (1975) and Wolk (1976) showed the relation between internality and life satisfaction among non-institutionalized elderly.

Across various spheres of science, positive attitudes towards ageing have been shown to positively influence memory performance, longevity, health, well-being, life satisfaction, will-to-live, and other physiological and psychological functioning (Linn and Hunter 1979; Levy 1996; Levy *et al.* 2000a, 2002). However, debates concerning the determinants of successful ageing persist. Societal stereotypes as well as elderly individuals' self-perception play a key role in determining and affecting how an individual ages. From heightened cardiovascular responses to stress, longevity, will-to-live, and memory performance, these particular variables have been shown to affect older individuals' ageing. Although these studies explored only one dependent variable at a time, in the real world, it is likely that self-perception and self-stereotyping occur in various spheres at once, whereby the effects are mutually reinforcing. For instance, an elderly person exposed to negative age stereotypes may show impaired memory and heightened stress levels. The impaired recall may worsen stress, and the elevated stress could impede memory performance even further (Levy 2001). Given that the effects of age stereotypes are mutually reinforcing and infiltrate into our everyday thinking and behaviour, the consequences of negative age stereotypes can have detrimental effects on older individuals' self-image, abilities, and health, thereby preventing successful ageing.

Conclusions

In this chapter we have provided a historical and methodological analysis of the concept of successful ageing. Although the ageing process was originally viewed as a time of decline of physical and cognitive functions, the concept of successful ageing, put forward by Rowe and Kahn in 1987, triggered a large interest in the study of the biological, psychological, and psychosocial determinants of successful ageing. We have exposed each perspective independently and described the factors that are thought to determine and/or define successful ageing within the biological, psychological, and psychosocial perspectives.

This analysis revealed the presence of weaknesses intrinsic to each unique approach because the notion of successful ageing implies more than success at the biological, psychological or psychosocial level. Consequently, we have reviewed the multicriteria models of successful ageing as proposed by Rowe and Kahn (1987) and Baltes and Baltes (1990). Although these new multicriteria approaches have certainly contributed to the development of a multidisciplinary study of successful ageing, each of them also presents some problems in the definition of what constitutes a 'successful ager' and/or what one can do to become a 'successful ager'.

This analysis led us to review the impact of age stereotypes on the biological and cognitive determinants of ageing. Here, we see that the views that one holds about the process of ageing can have an important impact on physical health and cognitive performance in old age. It is, therefore, possible to propose that the determinants of successful ageing stem in part from the societal influences of age stereotypes and the older individual's self-perception of ageing. To strive towards a more optimal view of the ageing process, research on successful ageing needs to focus not only on the models of biological and cellular ageing that consider mainly the gradual deterioration of the organism, but also on psychological and sociologically related factors that are related to improvements or maintenance of function. With increasing research, there is strong evidence that positive self-concepts can have a reversing effect on what was once believed to be an inevitable declining process of ageing. With the knowledge that positive attitudes function as protective factors against the effects of age on the organism, viewing successful ageing from these perspectives provides us with an integrative view that blends the various realms of ageing, and goes beyond the cell to reach the self.

Acknowledgements

S.J.L. is funded by an Institute of Aging Investigator Award from the Canadian Institutes of Health Research.

References

Andersen-Ranberg, K., Vasegaard, L., and Jeune, B. (2001). Dementia is not inevitable: a population-based study of Danish centenarians. *J. Gerontol. B Psychol. Sci. Soc. Sci.* **56**, 152–9.

Baggio, G. (and 12 others) (1998). Lipoprotein(a) and lipoprotein profile in healthy centenarians: a reappraisal of vascular risk factors. *FASEB J.* **12**, 433–7.

Baltes, M.M. and Carstensen, L.L. (1996). The process of successful aging. *Aging Soc.* **16**, 397–422.

Baltes, P.B. and Baltes, M. (1990). Psychological perspectives on successful aging: the model of selective optimisation with compensation. In *Successful aging: perspectives from the behavioural sciences* (ed. P.B. Baltes and M.M. Baltes), pp. 1–36. Cambridge University Press, Cambridge.

Baltes, P.B. and Goulet, L.R. (1971). Explorations of developmental variables by simulation and manipulation of age differences in behavior. *Hum. Dev.* **14**, 149–70.

Baltes, P.B. and Lindenberger, U. (1988). On the range of cognitive plasticity in old age as a function of experience: 15 years of intervention research. *Behav. Ther.* **19**, 283–300.

Barbagallo, C.M., Averna, M.R., Frada, G., Barbagallo, C.M., and Averna, M.R. (1995). Plasma lipid apolipoprotein and Lp(a) levels in elderly normolipidemic women: relationships with coronary heart disease and longevity. *Gerontology* **41**, 260–6.

Bassuk, S.S., Glass, T.A., and Berkman, L.F. (1999). Social disengagement and incident cognitive decline in communitydwelling elderly persons. *Ann. Intern. Med.* **131**, 165–73.

Beard, B.B. (1991). *Centenarians, the new generation.* Greenwood Press, New York.

Beregi, E. (1990). Centenarians in Hungary. a social and demographic study. *Interdiscipl. Topics Gerontol.* **27**, 31–9.

Berkman, L.F. (1995). The role of social relations in health promotion. *Psychosom. Med.* **57**, 245–54.

Berkman, L.F. (and 16 others) (1993). High, usual and impaired functioning in community-dwelling older men and women: finding from the MacArthur Foundation Research Network on Successful Aging. *J. Clin. Epidemiol.* **46**, 1129–40.

Bird, C. (1940). As we grow old. *Sci. Digest* **8**, 23–7.

Birren, J.E. and Schaie, K.W. (1985). *Handbook of the psychology of aging: principles and experimentation.* Van Nostrand Reinhold, New York.

Botwinick, J. (1978). *Aging and behavior,* 2nd edn. Springer, New York.

Butler, R.A. (1969). Ageism: another form of bigotry. *Gerontologist* **9**, 212–52.

Butler, R.A. (1977). Successful aging and the role of the life review. In *Readings in aging and death: contemporary perspectives,* 2nd edn (ed. S.N. Zarit), pp. 13–19. Harper and Row, New York.

Cabeza, R., Daselaar, S.M., Dolcos, F., Prince, S.E., Budde, M., and Nyberg, L. (2004). Task-independent and task-specific age effects on brain activity during working memory, visual attention and episodic retrieval. *Cereb. Cortex* **14**, 364–75.

Chodosh, J., Reuben, D.B., Albert, M.S., and Seeman, T.E. (2002). Predicting cognitive impairment in high-functioning community-dwelling older persons: MacArthur studies of successful aging. *J. Am. Geriatr. Soc.* **50**, 1051–60.

Cohen, S. and Herbert, T.B. (1996). Health psychology: psychological factors and physical disease from the perspective of human psychoneuroimmunology. *Annu. Rev. Psychol.* **47**, 113–42.

Costa Jr, P.T. and McCrae, R.R. (1980). Influence of extraversion and neuroticism on subjective well-being: happy and unhappy people. *J. Pers. Soc. Psychol.* **38**, 668–78.

Costa Jr, P.T. and McCrae, R.R. (1986). Personality stability and its implications for clinical psychology. *Clin. Psychol. Rev.* **6**, 407–23.

Costa Jr, P.T. and McCrae, R.R. (1988). Personality in adulthood. *J. Pers. Soc. Psychol.* **54**, 853–63.

Cutler, N.E. and Harootyan, R.A. (1975). Demography of the aged. In *Aging: scientific perspectives and social issues* (ed. D.S. Woodruff and J.E. Birren), pp. 45–55. Van Nostrand, New York.

Cutler, R.G. (1990). Evolutionary perspective of human longevity. In *Principles of geriatric medicine and gerontology* (ed. W.R. Hazzard, R. Andres, E.L. Bierman, and J.P. Blass), pp. 15–21. McGraw-Hill, New York.

Dannefer, D. (1988). What's in a name? An account of the neglect of variability in the study of aging. In *Emergent theories of aging* (ed. J.E. Birren and V.L. Bengtson), pp. 432–56. Springer, New York.

Department of Health and Human Services (2000). Older Americans 2000: key indicators of well-being. In *Federal interagency forum on aging related statistics*, p. 1–123. [See <http://www.agingstats.gov/chartbook2000/default.htm>.]

Effros, R.B., Boucher, N., Porter, V., Zhu, X., Spaulding, C., Walford, R.L., Kronenberg, M., Cohen, D., and Schachter, F. (1994). Decline in CD28 T cells in centenarians and in long-term T cell cultures: a possible cause for both *in vivo* and *in vitro* immunosenescence. *Exp. Gerontol.* **29**, 601–9.

Ferraro, K.F. (1997). The gerontological imagination. In *Gerontology: perspectives and issues*, 2nd edn (ed. K.F. Ferraro), pp. 3–18. Springer, New York.

Fisher, B.J. (1992). Successful aging and life satisfaction: a pilot study for conceptual clarification. *J. Aging Stud.* **6**, 191–202.

Fisher, B.J. (1995). Successful aging, life satisfaction and generativity in later life. *Int. J. Aging Hum. Dev.* **41**, 239–50.

Franceschi, C., Monti, D., Sansoni, P., and Cossarizza, A. (1995). The immunology of exceptional individuals: the lesson of centenarians. *Immunol. Today* **16**, 12–16.

Fratiglioni, L., Wang, H.X., Ericsoon, K., Maytan, M., and Winblad, B. (2000). Influence of social network on occurrence of dementia: a community-based longitudinal study. *Lancet* **355**, 1315–19.

Fredrickson, B.L. (2001). The role of positive emotions in positive psychology: the broaden-and-build theory of positive emotions. *Am. Psychol.* **56**, 218–26.

Fredrickson, B.L. and Levenson, R.W. (1998). Positive emotions speed recovery from the cardiovascular sequelae of negative emotions. *Cogn. Emotion* **12**, 191–220.

Fries, J.F. (1980). Aging, natural death, and the compression of morbidity. *New Engl. J. Med.* **303**, 130–5.

Fries, J.F. (1993). Medical perspectives upon successful aging. In *Successful aging: perspective from the behavioural sciences* (ed. P.B. Baltes and M.M. Baltes), pp. 35–49. Cambridge University Press, Cambridge.

George, L. (1979). The happiness syndrome: methodological and substantive issues in the study of social-psychological well-being in adulthood. *Gerontologist* **19**, 210–16.

George, L.K. (1981). Subjective well-being: conceptual and methodological issues. In *Annual review of gerontology and geriatrics*, Vol. 2 (ed. C. Eisdorfer), pp. 33–45. Springer, New York.

George, L.K. (1986). Life satisfaction in later life. *Generations* **10** (Spring), 5–8.

Goldberg, L.R. (1993). The structure of phenotypic personality traits. *Am. Psychol.* **48**, 26–34.

Griffiths, T.D. (1997). Biology of aging. In *Gerontology: perspectives and issues*, 2nd edn (ed. K.F. Ferraro), pp. 53–67. Springer, New York.

Guarente, L. and Kenyon, C. (2000). Genetic pathways that regulate ageing in model organisms. *Nature* **408**, 255–62.

Hagberg, B., Bauer, A., Alfredson, B., Poon, L.W., and Homma, A. (2001). Cognitive functioning in centenarians: a coordinated analysis of results from three countries. *J. Gerontol. B Psychol. Sci. Soc. Sci.* **56**, 141–51.

Hall, G.S. (1922). *Senescence: the last half of life.* Appleton and Company, New York.

Helmuth, L. (2003). The wisdom of the wizened. *Science* **299**, 1300–2.

Hendricks, J. and Hendricks, C.D. (1977). *Aging in mass society: myths and realities.* Winthrop, Cambridge, Massachusetts.

Herzog, A.R. and Rodgers, W.L. (1981). Age and satisfaction: data from several large surveys. *Res. Aging* **7**, 209–33.

Hirshbein, L.D. (2001). Popular view of old age in America, 1900–1950. *Am. Geriatr. Soc.* **49**, 1555–60.

Imai, S.I., Armstrong, C.M., Kaeberlein, M., and Guarente, L. (2000). Transcriptional silencing and longevity protein Sir2 is an NAD-dependent histone deacetylase. *Nature* **403**, 795–800.

Jazwinski, S.M. (1999). Molecular mechanisms of yeast longevity. *Trends Microbiol.* **7**, 247–52.

Karasawa, A. (1979). Mental aging and its medico-social background in the very old Japanese. *J. Gerontol.* **34**, 680–6.

Karlamangla, A.S., Singer, B.H., McEwen, B.S., Rowe, J.W., and Seeman, T E. (2002). Allostatic load as a predictor of functional decline: MacArthur studies of successful aging. *J. Clin. Epidemiol.* **55**, 696–710.

Kerber, R.A., O'Brien, E., Smith, K.R., and Cawthon, R.M. (2001). Familial excess longevity in Utah genealogies. *J. Gerontol. A Biol. Sci. Med. Sci.* **56**, B130–9.

Kirkwood, T.B.L. (1977). Evolution of ageing. *Nature* **270**, 301–4.

Kirkwood, T.B.L. and Holliday, R. (1979). The evolution of aging and longevity. *Proc. R. Soc. Lond. B* **205**, 531–46.

Kliegl, R., Smith, J., and Baltes, P.B. (1989). Testing the limits and the study of adult age differences in cognitive plasticity of a mnemonic skill. *Dev. Psychol.* **25**, 247–56.

Krauss, I.K. (1980). Between- and within-group comparisons in aging research. In *Aging in the 1980s* (ed. L.W. Poon), pp. 132–45. American Psychological Association, Washington, DC.

Kubzansky, L.D., Berkman, L.F., Glass, T.A., and Seeman, T.E. (1998). Is educational attainment associated with shared determinants of health in the elderly? Findings from the MacArthur studies of successful aging. *Psychosom. Med.* **60**, 578–85.

Kubzansky, L.D., Berkman, L.F., and Seeman, T.E. (2000). Social conditions and distress in elderly persons: findings from the MacArthur studies of successful aging. *J. Gerontol. B Psychol. Sci. Soc. Sci.* **55**, 238–46.

Labouvie, E.W. (1980). Identity versus equivalence of psychological measures and constructs. In *Aging in the 1980s* (ed. L.W. Poon), pp. 31–47. American Psychological Association, Washington, DC.

Leviatan, U. (1989). Successful aging: the Kibbutz experience. *J. Aging Judaism* **4**, 71–90.

Leviatan, U. (1999). Contributions of social arrangements to the attainment of successful aging: the experience of the Israeli Kibbutz. *J. Gerontol.* **54**, 205–13.

Levy, B. (1996). Improving memory in old age through implicit self-stereotyping. *J. Pers. Soc. Psychol.* **71**, 1092–107.

Levy, B. (2001). Eradication of ageism requires addressing the enemy within. *Gerontologist* **41**, 578–9.

Levy, B., Ashman, O., and Dror, I. (2000a). To be or not to be: the effects of aging stereotypes on the will to live. *Omega* **40**, 409–20.

Levy, B., Hausdorff, J., Hencke, R., and Wie, J. (2000*b*). Reducing cardiovascular stress with positive self-stereotypes of aging. *J. Gerontol. Psychol. Sci.* **55B**, 205–13.

Levy, B., Slade, M., Kundel, S., and Kasl, S. (2002). Longevity increased by positive self-perceptions of aging. *J. Pers. Soc. Psychol.* **83**, 261–70.

Li, S.C., Aggen, S.H., Nesselroade, J.R., and Baltes, P.B. (2001). Short-term fluctuations in elderly people's sensorimotor functioning predict text and spatial memory performance: the MacArthur successful aging studies. *Gerontology* **47**, 100–16.

Linn, M.W. and Hunter, K. (1979). Perceptions of age in the elderly. *J. Gerontol.* **34**, 46–52.

Lupien, S. and Lecours, A.R. (1993). All things being otherwise unequal: reflection upon increased inter-individual differences with aging. *Rev. Neuropsychol. (Paris)* **3**, 3–35.

Lupien, S., Lecours, A.R., Lussier, I., Schwartz, G., Nair, N.P.V., and Meaney, M.J. (1994). Basal cortisol levels and cognitive deficits in human aging. *J. Neurosci.* **14**, 2893–903.

Lupien, S., DeLeon, M., DeSanti, S., Convit, A., Tarshish, C., Nair, N.P.V., Thakur, M., McEwen, B.S., Hauger, R.L., and Meaney, M.J. (1998). Longitudinal increase in cortisol during human aging predicts hippocampal atrophy and memory deficits. *Nat. Neurosci.* **1**, 69–73.

Lynott, R.J. and Lynott, P.P. (1996). Tracing the course of theoretical development in the sociology of aging. *Gerontologist* **36**, 749–60.

McAdams, D.P. (1994). Can personality change? Levels of stability and growth in personality across the lifespan. In *Can personality change?* (ed. T.F. Heatherton and J.L. Weingerger), pp. 299–313. American Psychological Association, Washington, DC.

McCrae, R.R. (2002). The maturation of personality psychology: adult personality development and psychological well-being. *J. Res. Pers.* **36**, 307–17.

McCrae, R.R. and Costa, P.T. (1984). *Emerging lives, enduring dispositions: personality in adulthood.* Little Brown, Boston, Massachusetts.

McCrae, R.R. and John, O.P. (1992). An introduction to the five-factor model and its applications. *J. Pers.* **60**, 175–215.

McEwen, B.S. and Stellar, E. (1993). Stress and the individual: mechanisms leading to disease. *Arch. Intern. Med.* **153**, 2093–101.

Martin, L.J. and de Gruchy, C. (1930). *Salvaging old age.* Macmillan, New York.

Masoro, E.J. (2001). Longevity: to the limits and beyond. *Gerontologist* **41**, 414–18.

Miles, W.R. (1933). Age and human ability. *Psychol. Rev.* **40**, 114–15.

Minkler, M. (1990). Aging and disability: behind and beyond the stereotypes. *J. Aging Stud.* **4**, 245–60.

Minkler, M. and Fadem, P. (2002). Successful aging: a disability perspective. *J. Dis. Policy Stud.* **12**, 229–36.

Morris, J.C., McKeel Jr, D.W., Storandt, M., Rubin, E.H., Price, J.L., Grant, E.A., Ball, M.J., and Berg, L. (1991). Very mild Alzheimer's disease: informant-based clinical psychometric and pathologic distinction from normal aging. *Neurology* **41**, 469–78.

Palmore, E. (1990). *Ageism: negative and positive.* Springer, New York.

Palmore, E. and Luikart, C. (1972). Health and social factors related to life satisfaction. *J. Health Soc. Behav.* **13**, 68–80.

Paolisso, G., Gambardella, A., Balbi, V., Ammendola, S., D'Amore, A., and Varricchio, M. (1995). Body composition, body fat distribution and resting metabolic rate in healthy centenarians. *Am. J. Clin. Nutr.* **62**, 746–50.

Perls, T. and Fretts, R. (1998). Why women live longer than men. *Sci. Am. Presents* **9**, 100–3.

Perls, T.T., Bubrick, E., Wager, C.G., Vijg, J., and Kruglyak, L. (1998). Siblings of centenarians live longer. *Lancet* **351**, 1560.

Perls, T.T., Bochen, K., Freeman, M., Alpert, L., and Silver, M.H. (1999). Validity of reported age and centenarian prevalence in New England. *Age Ageing* **28**, 193–7.

Perls, T., Kunkel, L.M., and Puca, A.A. (2002*a*). The genetics of exceptional human longevity. *J. Am. Geriatr. Soc.* **50**, 359–69.

Perls, T., Levenson, R., Regan, M., and Puca, A. (2002*b*). What does it take to live to 100? *Mech. Aging Dev.* **123**, 231–42.

Poon, L.W. (1992). *The Georgian centenarian study*. Baywook, Amityville, New York.

Rowe, J.W. and Kahn, R.L. (1987). Human aging: usual and successful aging. *Science* **237**, 143–9.

Rowe, J.W. and Kahn, R.L. (1998). *Successful aging*. Pantheon Books, New York.

Rybicki, B.A. and Elston, R.C. (2000). The relationship between the sibling recurrence–risk ratio and genotype relative risk. *Am. J. Hum. Genet.* **66**, 593–604.

Schaie, K.W. (1993). The optimization of cognitive functioning in old age: predictions based on cohort-sequential and longitudinal data. In *Successful aging: perspectives from the behavioural sciences* (ed. P.B. Baltes and M.M. Baltes), pp. 94–117. Cambridge University Press, Cambridge.

Schaie, K.W. and Willis, S.L. (1998). Can adult intellectual decline be reversed? *Dev. Psychol.* **22**, 223–32.

Seeman, T.E. (1996). Social ties and health. *Ann. Epidemiol.* **6**, 442–51.

Seeman, T.E., Singer, B., and Charpentier, P. (1995*a*). Gender differences in patterns of HPA axis response to challenge: MacArthur studies of successful aging. *Psychoneuroendocrinology* **20**, 711–25.

Seeman, T.E., Berkman, L.F., Gulanski, B.I., Robbins, R.J., Greenspan, S.L., Charpentier, P.A., and Rowe, J.W. (1995*b*). Self-esteem and neuroendocrine response to challenge: MacArthur studies of successful aging. *J. Psychosom. Res.* **39**, 69–84.

Seeman, T.E., McEwen, B.S., Singer, B.H., Albert, M.S., and Rowe, J.W. (1997*a*). Increase in urinary cortisol excretion and memory declines: MacArthur studies of successful aging. *J. Clin. Endocrinol. Metab.* **82**, 2458–65.

Seeman, T.E., Singer, B.H., Rowe, J.W., Horwitz, R.I., and McEwen, B.S. (1997*b*). Price of adaptation: allostatic load and its health consequence. *Arch. Intern. Med.* **157**, 2259–68.

Seeman, T.E., McEwen, B.S., Rowe, J.W., and Singer, B.H. (2001*a*). Allostatic load as a marker of cumulative biological risk: MacArthur studies of successful aging. *Proc. Natl Acad. Sci. USA* **98**, 4770–5.

Seeman, T.E., Lusignolo, T.M., Albert, M., and Berkman, L. (2001*b*). Social relationships, social support, and patterns of cognitive aging in healthy, high-functioning older adults: MacArthur studies of successful aging. *Health Psychol.* **20**, 1–13.

Shanan, J. (1991). Who and how: some unanswered questions in adult development. *J. Gerontol. Psychol. Sci.* **46**, 309–16.

Sheldon, J.H. (1948). *The social medicine of old age*. Oxford University Press, London.

Shingo, T., Gregg, C., Enwere, E., Fujikawa, H., Hassam, R., Geary, C., Cross, J.C., and Weiss, S. (2003). Pregnancy-stimulated neurogenesis in the adult female forebrain mediated by prolactin. *Science* **299**, 117–20.

Silver, M., Newell, K., Hyman, B., Growdon, J., Hedley-Whyte, E.T., and Perls, T. (1998). Unravelling the mystery of cognitive changes in old age: correlation of neuropsychological evaluation with neuropathological findings in the extreme old. *Int. Psychogeriatr.* **10**, 25–41.

Silver, M.H., Jilinskaia, E., and Perls, T.T. (2001). Cognitive functional status of age-confirmed centenarians in a population-based study. *J. Gerontol. B Psychol. Sci. Soc. Sci.* **56**, 134–40.

Stones, M.J., Kozma, A., and Hannah, T.E. (1990). The measurement of individual differences in aging: the distinction between usual and successful aging. In *Cognitive and behavioural performance factors in atypical aging* (ed. M.L. Have, M.J. Stones, and C.J. Brainerd), pp. 181–218. Springer, New York.

Tabbarah, M., Crimmins, E.M., and Seeman, T.E. (2002). The relationship between cognitive and physical performance: MacArthur studies of successful aging. *J. Gerontol. A Biol. Sci. Med. Sci.* **57**, 228–35.

Unger, J.B., McAvay, G., Bruce, L.M., Berkman, L., and Seeman, T. (1999). Variation in the impact of social network characteristics on physical functioning in elderly persons: MacArthur studies of successful aging. *J. Gerontol. B Psychol. Sci. Soc. Sci.* **54**, 245–51.

Weaver, J.D., Huang, M.H., Albert, M., Harris, T., Rowe, J.W., and Seeman, T.E. (2002). Interleukin-6 and risk of cognitive decline: MacArthur studies of successful aging. *Neurology* **59**, 371–8.

Williams, G.C. (1966). *Adaptation and natural selection*. Princeton University Press, Princeton.

Wolk, S. (1976). Situational constraints as a moderator of locus of control–adjustment relationship. *J. Consult. Clin. Psychol.* **44**, 420–7.

Wolk, S. and Kurtz, J. (1975). Positive adjustment and involvement during aging and expectancy for internal control. *J. Consult. Clin. Psychol.* **43**, 173–8.

World Health Organization (2000). *WHO Issues New Healthy Life Expectancy Rankings*. Press Releases 2000. [See <http://www.who.int/inf-pr-2000/en/pr2000-life.html>].

World Health Organization (2003a). *The World Health Report 2003: shaping the future*, pp. 1–131. World Health Organization, Geneva. [See <http://www.who.int/whr/2003/en/>].

World Health Organization (2003b). *The World Health Report 2003*, Annex 4, *Healthy life expectancy (HALE) in all member states: Estimates for 2002*, p. 166. World Health Organization, Geneva. [See <http://www.who.int/whr/2003/en/Annex4-en.pdf>.].

Part 2

Physiology and neuroscience

Richard J. Davidson is the William James and Vilas Research Professor of Psychology and Psychiatry and director of the W.M. Keck Laboratory for Functional Brain Imaging and Behavior at the University of Wisconsin–Madison. His numerous awards include the prestigious American Psychological Association Distinguished Scientific Contribution Award (2000). His innovative research explores the neural substrates of emotion.

Chapter 5*

Well-being and affective style: neural substrates and biobehavioural correlates

Richard J. Davidson

Introduction

One of the most salient characteristics of emotion is the extraordinary heterogeneity in how different individuals respond to the same emotionally provocative challenge. Such differences in patterns of emotional reactivity play a crucial role in shaping variations in well-being. Although individual differences in emotion processing can be found at many levels of phylogeny, they are particularly pronounced in primates and probably are most extreme in humans. A number of evolutionary theorists have speculated on the adaptive significance of such individual differences (Wilson 1994). Although these arguments have never been applied to the domain of emotion and affective style, it is not difficult to develop hypotheses about how such differences might provide advantages to individuals living in groups. However, rather than focus on the distal causes of such individual differences, which are so difficult to subject to rigorous test, I wish only to call attention to the possibility that variability in characteristics such as 'fearfulness' or 'cheerfulness' might provide some adaptive benefit to individuals living together in groups. Instead, this chapter examines the proximal mechanisms that underlie such individual differences, with a focus on well-being. The central substrates of individual differences in components of well-being will be described. The possible influence of the central circuitry of emotion on peripheral biological indices that are relevant to physical health and illness will also be considered. It is helpful to contrast well-being with specific types of psychopathology that involve dysfunctions in the circuitry of adaptive emotional responding. Accordingly, some mention of recent work on the neurobiology of mood and anxiety disorders will be given. Finally, plasticity in the underlying brain circuitry that instantiates affective style will be described and its role in promoting resilience will be considered.

Affective style refers to consistent individual differences in emotional reactivity and regulation (see Davidson 1998a; Davidson *et al.* 2000a, b). It is a phrase that is meant to capture a broad array of processes that, either singly or in combination, modulate an individual's response to emotional challenges, dispositional mood, and affect-relevant cognitive processes. Affective style can refer to valence-specific features of emotional reactivity or mood, or it can refer to discrete emotion-specific features. Both levels of analysis are equally valid and the choice of level should be dictated by the question posed.

Rapid developments in our understanding of emotion, mood, and affective style have come from the study of the neural substrates of these phenomena. The identification of the brain circuitry responsible for different aspects of affective processing has helped to parse the domain of emotion into more elementary constituents in a manner similar to that found in cognitive neuroscience, where an appeal to the brain has facilitated the rapid development of theory and data on the subcomponents of various cognitive processes (e.g. Kosslyn and Koenig 1992).

This chapter will highlight some of the advances that have been made in our understanding of the brain mechanisms that underlie affective style. These advances have emerged from three major sources: studies of patients with discrete lesions of the brain; neuroimaging studies of normal individuals; and studies of pathologies of brain function in patients with various psychiatric and neurological disorders that involve abnormalities in emotion. I will use the material on pathology to help to identify the neural circuitry crucial to certain forms of positive affect so that we can begin to place well-being squarely within a neurobiological framework.

Both lesion and neuroimaging studies provide information primarily on the 'where' question, that is, where in the brain are computations related to specific aspects of affective processing occurring. It is important at the outset to consider both the utility of knowing 'where' and how such information can provide insight into the 'how' question, that is, how might a particular part of the brain instantiate a specific process that is essential to affective style. The brain sciences are now replete with information on the essential nature of specific types of information processing in different regions of the brain. For example, there is evidence to suggest that the dorsolateral prefrontal cortex (DLPFC) is important for maintaining a representation of information online in the absence of immediate cues. The neurophysiological basis of this type of information processing has been actively studied in the animal laboratory (e.g. Goldman-Rakic 1996, 2000). If this region of the brain is activated at certain times in the stream of affective information processing, we can develop hypotheses on the basis of extant work about what this territory of prefrontal cortex (PFC) might be doing during the affective behaviour and how it might be doing it. A related consideration is the network of anatomical connectivity to and from a particular brain region. From a consideration of connectivity, insights may be gleaned as to how a particular brain region might react during a particular form of emotional processing. For example, we know that regions of the amygdala have extensive connectivity with cortical territories that can become activated following activation of the amygdala. In this way, the amygdala can issue a cortical call for further processing in response to potentially threatening stimuli that must be processed further to assess danger. Other regions of the amygdala have extensive connections to limbic and brainstem circuits that can modulate behavioural and autonomic outflow. Adjustments in autonomic responses and action tendencies are typical components of emotion.

Conceptual and methodological considerations in the study of affective style

Current research on well-being is based largely on the use of self-report measures to make inferences about variation among individuals in type and magnitude of well-being. One important component of neurobiological research on well-being is to begin to dissect well-being into more specific constituents that may underlie the coarse phenomenological descriptions provided by subjects. In addition, research on the neural correlates of well-being may provide an independent biological measure sensitive to variations in well-being that are not subject to the kinds of reporting and judgemental biases commonly found in the self-report measures. For example, researchers have found that questions that precede items asking about well-being can influence a subject's report of well-being. Variations in the weather can similarly affect such reports. These examples illustrate the fact that, when subjects are queried about global well-being, they frequently use convenient heuristics to answer such questions and typically do not engage in a systematic integration of utility values over time. It may be that certain parameters of brain function are better repositories of the cumulative experiences that inevitably shape well-being. At the present point in the development of this science, these are mere speculations in search of evidence but the time is ripe for such evidence to be gathered.

The status of research on well-being is now at a point occupied about a decade ago or more by research on mood and anxiety disorders, though it continues to suffer from some of the same problems. Mood and anxiety disorders are generally conceptualized as being caused, or at least accompanied by, dysfunctions of emotion. However, which specific affective process is dysfunctional is rarely, if ever, delineated, and nosological schemes for categorizing these disorders do not rely upon the specific nature of the affective dysfunction in question, but rather are based upon phenomenological description. Research in my laboratory over the past 15 years has been predicated on the view that more meaningful and rapid progress in understanding the brain bases of mood and anxiety disorders can be achieved if we move to an intermediate level of description that penetrates below the categorical, phenomenologically based classifications of the *Diagnostic and statistical manual of mental disorders* (DSM; American Psychiatric Association 1994) and seeks to characterize the specific nature of the affective styles that are associated with vulnerability to these forms of psychopathology.

Many of the parameters of affective style, such as the threshold to respond, magnitude of response, latency to peak of response, and recovery function, are features that are often opaque to conscious report, though they may influence the subjective experience of emotion. These parameters of responding can be measured in many different response systems including both central and peripheral systems. For example, magnitude of response can be measured in a peripheral measure such as the emotion-modulated startle (Lang 1995) or in a central measure such as activation in the amygdala assessed with functional magnetic resonance imaging (fMRI). The extent to which coherence across response systems in these parameters is present has not yet been systematically addressed. In previous work, we have argued that variations in some of these parameters in particular response systems are especially relevant to vulnerability to mood, anxiety, and other disorders and also to resilience (e.g. Davidson *et al.* 2000*a, b*). One of the important developments in emotion research in general, and in affective neuroscience in particular, is the capacity to objectively measure these parameters of responding. For example, in several studies we have used the emotion-modulated startle to capture the time-course of valence-specific emotion responding (Larson *et al.* 1998; Jackson *et al.* 2000). The startle reflex is controlled by a brainstem circuit that is influenced by activity in forebrain structures. Davis (1992) elegantly dissected the circuitry through which the magnitude of this reflex is modulated during the arousal of fear in rodents. He demonstrated that it is via a descending pathway from the central nucleus of the amygdala to the nucleus pontine reticularis in the brainstem that the magnitude of startle is enhanced in response to a conditioned fear cue. Lesions of the central nucleus of the amygdala abolish the fear potentiation of the startle but do not affect the magnitude of the baseline startle. Lang and his colleagues (Vrana *et al.* 1988) were the first to show systematically that the same basic phenomenon can be produced in humans. They took advantage of the fact that brief acoustic noise bursts produce the eyeblink component of the startle and little else, thus enabling their presentation as innocuous stimuli in the background. By measuring electromyographic activity from the orbicularis oculi muscle with two miniature electrodes under one eye, they were able to quantify the strength of the blink response and show that the magnitude of the blink was greater when subjects were presented with unpleasant pictures in the foreground, compared with the presentation of neutral pictures. Moreover, when subjects were exposed to positive stimuli, the magnitude of startle was actually attenuated relative to a neutral condition (Vrana *et al.* 1988). This same basic effect has now been reported with many different types of foreground stimuli in several modalities (see Lang 1995 for a review).

We have exploited the emotion-modulated startle to begin to characterize the time-course of affective responding, or what I have referred to as affective chronometry (Davidson 1998*a*). By inserting acoustic noise probes at different latencies before and after a critical emotional stimulus is presented, both the anticipatory limb and the recovery limb of the response can be measured. By using paradigms in the magnetic resonance imaging (MRI) scanner that were first studied in the psychophysiology laboratory, the neural circuitry underlying the different phases of affective processing can be interrogated with fMRI. Our current work in this area has emphasized the importance of the recovery function following negative events for vulnerability to certain forms of psychopathology as well as for resilience. We have argued that the failure to recover rapidly following a negative event can be a crucial ingredient of vulnerability to both anxiety and mood disorders, particularly when such a style is combined with frequent exposure to negative events over a sustained period of time. The failure to recover adequately would result in sustained elevations in multiple systems that are activated in response to negative events. By contrast, the capacity for rapid recovery following negative events may define an important ingredient of resilience. We have defined resilience as the maintenance of high levels of positive affect and well-being in the face of significant adversity. It is not that resilient individuals never experience negative affect, but rather that the negative affect does not persist. Such individuals are able to profit from the information provided by the negative affect and their capacity for 'meaning making' in response to such events may be part and parcel of their ability to show rapid decrements in various biological systems following exposure to a negative or stressful event (see Giese-Davis and Spiegel 2003).

Neural substrates of emotion and affective style

In the following three sections, a brief overview is provided of core components of the circuitry that instantiates some important aspects of emotion and affective style, with an emphasis on the PFC and the amygdala. It is not meant to be an exhaustive review, but rather will present selected highlights to illustrate some of the key advances that have been made in the recent past.

Emotion and affective style are governed by a circuit that includes the following structures, and probably also others: DLPFC; ventromedial prefrontal cortex (vmPFC); orbitofrontal cortex (OFC); amygdala; hippocampus; anterior cingulate cortex (ACC); and insular cortex. It is argued that different subprocesses are instantiated in each of these structures, and that they normally work together to process, generate, and regulate emotional information and emotional behaviour.

Prefrontal cortex

A large corpus of data at both the animal and human levels implicates various sectors of the PFC in emotion. The PFC is not a homogeneous zone of tissue but, rather, has been differentiated on the basis of both cytoarchitectonic and functional considerations. The three subdivisions of the primate PFC that have been consistently distinguished include the DLPFC, vmPFC, and OFC. In addition, there appear to be important functional differences between the left and right sides within some of these sectors.

The case for the differential importance of left and right PFC sectors for emotional processing was first made systematically in a series of studies on patients with unilateral cortical damage (Gainotti 1972; Sackeim *et al.* 1982; Robinson *et al.* 1984). Each of these studies compared the mood of patients with unilateral left- or right-sided brain damage and found a greater incidence of depressive symptoms following left-sided damage. In most cases, the damage was fairly substantial and probably included more than one sector of PFC and often also included other brain regions. The general interpretation that has been placed upon these studies is that depressive symptoms are increased following left-sided anterior PFC damage because this brain territory participates in certain forms of positive affect and, when damaged, leads to deficits in the capacity to experience positive affect, a hallmark feature of depression (Watson *et al.* 1995). It should be noted that not all studies support this conclusion. In a recent meta-analysis of lesion studies, Carson *et al.* (2000) failed to find support for this hypothesis. Davidson (1993) has previously reviewed many of these studies and has addressed a number of critical methodological and conceptual concerns in this literature. The most important of these issues is that, according to the diathesis–stress model of anterior activation asymmetry proposed by Davidson (1995, 1998*b*) and colleagues (Henriques and Davidson 1991), individual differences in anterior activation asymmetry, whether lesion-induced or functional, represent a diathesis. As such, they alter the probability that specific forms of emotional reactions will occur in response to the requisite environmental challenge. In the absence of such a challenge, the pattern of asymmetric activation will simply reflect a propensity but will not necessarily culminate in differences in mood or symptoms. In a study with the largest sample size to date ($n = 193$) for a study of mood sequelae in patients with unilateral lesions, Morris *et al.* (1996) found that, among stroke patients, it was only in those with small lesions that the relation between left PFC damage and depressive symptoms was observed. It is likely that larger lesions intrude on other brain territories and mask the relation between left PFC damage and depression.

A growing corpus of evidence in normal intact humans is consistent with the findings derived from the lesion evidence. Davidson and his colleagues have reported that induced positive and negative affective states shift the asymmetry in prefrontal brain electrical activity in lawful ways. For example, film-induced negative affect increases relative right-sided prefrontal and anterior temporal activation (Davidson *et al.* 1990), whereas induced positive affect elicits an opposite pattern of asymmetric activation. Similar findings have been obtained by others (e.g. Ahern and Schwartz 1985; Jones and Fox 1992).

Using a cued reaction time paradigm with monetary incentives, Sobotka *et al.* (1992) first reported that, in the anticipatory interval between the cue and the response, electroencephalogram (EEG) differences were observed between reward and punishment trials with greater left-sided frontal activation observed in response to the former compared with the latter trial type. In a more recent study, Miller and Tomarken (2001) replicated and extended this basic effect and, very recently, we (Shackman et al. 2003) replicated the Miller and Tomarken effect, showing that reward trials produced significantly greater left prefrontal activation in the anticipatory interval compared with no-incentive trials. Moreover, subjects in this study also participated in an fMRI study using the identical paradigm and we found that those subjects who showed a robust EEG difference between reward and no-incentive trials also showed a significant difference in asymmetric prefrontal signal change in response to these conditions, with greater left-sided PFC activation in the reward compared with the no-incentive condition. In addition to these studies that manipulated phasic emotion, we will review in the section, 'What are individual differences in PFC and amygdala activations associated with?', a body of evidence that supports the conclusion that individual differences in baseline levels of asymmetric activation in these brain regions are lawfully related to variations in dispositional affective style. Using an extended picture presentation paradigm designed to evoke longer-duration changes in mood (Sutton et al. 1997a), we measured regional glucose metabolism with positron emission tomography (PET) to ascertain whether similar patterns of anterior asymmetry would be present using this very different and more precise method to assess regional brain activity (Sutton et al. 1997b). During the production of negative affect, we observed right-sided increases in metabolic rate in anterior orbital, inferior frontal, middle, and superior frontal gyri, whereas the production of positive affect was associated with a pattern of predominantly left-sided metabolic increases in the pre- and postcentral gyri. Using PET to measure regional cerebral blood flow, Hugdahl and his colleagues (Hugdahl et al. 1995; Hugdahl 1998) reported a widespread zone of increased blood flow in the right PFC, including the orbitofrontal and dorsolateral cortices and inferior and superior cortices, during the extinction phase after aversive learning had occurred compared with the habituation phase, before the presentation of the experimental contingencies.

Other investigators have used clinical groups to induce a stronger form of negative affect in the laboratory than is possible with normal controls. One common strategy for evoking anxiety among anxious patients in the laboratory is to present them with specific types of stimuli that are known to provoke their anxiety (e.g. pictures of spiders for spider phobics; making a public speech for social phobics). Davidson et al. (2000c), in a study using brain electrical activity measures, have recently found that, when social phobics anticipate making a public speech, they show large increases in right-sided anterior activation. Pooling across data from three separate anxiety-disordered groups that were studied with PET, Rauch et al. (1997) found two regions of the PFC that were consistently activated across groups: the right inferior PFC and right medial orbital PFC.

The vmPFC has been implicated in the anticipation of future positive and negative affective consequences. Bechara *et al.* (1994) have reported that patients with bilateral lesions of the vmPFC have difficulty in anticipating future positive or negative consequences, although immediately available rewards and punishments do influence their behaviour. Such patients show decreased levels of electrodermal activity in anticipation of a risky choice compared with controls, while controls exhibit such autonomic change before they explicitly know that it is a risky choice (Bechara *et al.* 1996, 1997, 1999).

The findings from the lesion method when effects of small unilateral lesions are examined and from neuroimaging studies in normal subjects and patients with anxiety disorders converge on the conclusion that increases in right-sided activation in various sectors of the PFC are associated with increased negative affect. Less evidence is available for the domain of positive affect, in part because positive affect is much harder to elicit in the laboratory and because of the negativity bias (see Taylor 1991; Cacioppo and Gardner 1999). This latter phenomenon refers to the general tendency of organisms to react more strongly to negative compared with positive stimuli, perhaps as a consequence of evolutionary pressures to avoid harm. The findings from Bechara *et al.* (1996, 1997) on the effects of vmPFC lesions on the anticipation of future positive and negative affective consequences are based upon studies of patients with bilateral lesions. It will be of great interest in the future to examine patients with unilateral ventromedial lesions to ascertain whether valence-dependent asymmetric effects are also present for this sector of PFC.

Systematic studies designed to disentangle the specific role played by various sectors of the PFC in emotion are lacking. Many theoretical accounts of emotion assign it an important role in guiding action and organizing behaviour towards the acquisition of motivationally significant goals (e.g. Frijda 1994; Levenson 1994). This process requires that the organism have some means of representing affect in the absence of immediately present rewards and punishments and other affective incentives. Such a process may be conceptualized as a form of affective working memory. It is probable that the PFC plays a key role in this process (see Watanabe 1996). Damage to certain sectors of the PFC impairs an individual's capacity to anticipate future affective outcomes and consequently results in an inability to guide behaviour in an adaptive fashion. Such damage is unlikely to disrupt an individual's response to immediate cues for reward and punishment; it disrupts only the anticipation before and maintenance after an affective cue is presented. This proposal can be tested using current neuroimaging methods (e.g. fMRI) but has not yet been rigorously evaluated. With regard to the different functional roles of the dorsolateral, orbitofrontal, and ventromedial sectors of the PFC, Davidson and Irwin (1999) suggested, on the basis of considering both human and animal

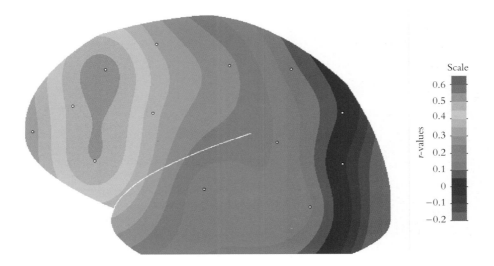

Plate 1 Relations between electrophysiological measures of asymmetry and the difference between the standardized score on the BIS/BAS (Carver and White 1994); $n = 46$. Electrophysiological data were recorded from each subject on two separate occasions separated by 6 weeks. The BIS/BAS were also administered on these two occasions. Data were averaged across the two time periods before performing correlations. The topographic map displays the correlations between alpha power asymmetry (log-right minus log-left alpha power; higher values denote greater relative left-sided activation) and the difference score between the standardized BAS minus BIS. After correlations were performed for each homologous region, a spline-interpolated map was created. The orange and red values of the scale denote positive correlations. The figure indicates that the correlation between the BAS–BIS difference score and the electrophysiology asymmetry score is highly positive in prefrontal scalp regions, denoting that subjects with greater relative left-sided activation report more relative behavioural activation compared with behavioural inhibition tendencies. The relation between asymmetric activation and the BAS–BIS difference is highly specific to the anterior scalp regions, as the correlation decreases rapidly more posteriorly. The correlation in the prefrontal region is significantly larger than the correlation in the parieto-occipital region. (From Sutton and Davidson (1997).) See Figure 5.2 in main text.

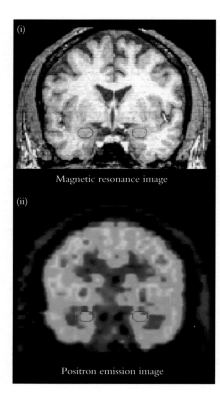

(i)

Magnetic resonance image

(ii)

Positron emission image

Plate 2 Images indicate (i) the MR and (ii) the corresponding PET image from one subject to illustrate our method of MRI co-registered ROIs around the amygdala. ROIs were individually drawn for each subject around the amygdala, and glucose metabolism was then extracted from the PET image in (ii). See Figure 5.7 (a) in main text.

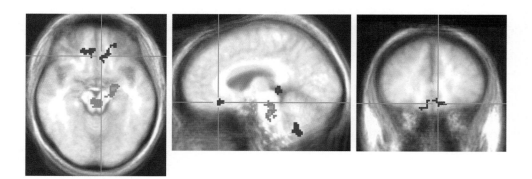

Plate 3 Image display of the associations between activation in the amygdala bilaterally and other regions of the brain. Blue areas denote inverse correlations whereas red areas denote positive correlations. The data reveal an inverse association between activation in vmPFC and signal change in the amygdala following aversive pictures in response to instructions to suppress (downregulate) negative affect, compared with a condition during which subjects were instructed to attend to the stimuli ($n = 17$). (From Urry *et al.* (2003).) See Figure 5.8 (a) in main text.

studies, that the ventromedial sector is most probably involved in the representation of elementary positive and negative affective states in the absence of immediately present incentives. The orbitofrontal sector has most firmly been linked to rapid learning and unlearning of stimulus–incentive associations and has been particularly implicated in reversal learning (Rolls 1999). As such, the orbitofrontal sector is probably key to understanding aspects of emotion regulation (see Davidson *et al.* 2000*d*). One critical component of emotion regulation is the relearning of stimulus–incentive associations that might have been previously maladaptive, a process probably requiring the OFC. The dorsolateral sector is most directly involved in the representation of goal states toward which more elementary positive and negative states are directed.

Amygdala

A large corpus of research at both the animal and human levels has established the importance of the amygdala for emotional processes (Aggleton 1993; LeDoux 1996; Cahill and McGaugh 1998; Davis and Whalen 2001). Since many reviews of the animal literature have appeared recently, a detailed description of these studies will not be presented here. LeDoux and his colleagues have marshalled a large corpus of compelling evidence to suggest that the amygdala is necessary for the establishment of conditioned fear. Whether the amygdala is necessary for the expression of that fear following learning and whether the amygdala is the actual locus of where the learned information is stored are still matters of some controversy (see Cahill *et al.* 1999; Fanselow and LeDoux 1999). The classic view of amygdala damage in non-human primates resulting in major affective disturbances, as expressed in the Kluver–Bucy syndrome where the animal exhibits abnormal approach, hyperorality, and sexuality and little fear, is now thought to be a function of damage elsewhere in the medial temporal lobe. When very selective excitotoxic lesions of the amygdala are made that preserve fibres of passage, nothing resembling the Kluver–Bucy syndrome is observed (Kalin *et al.* 2001). The upshot of this diverse array of findings is to suggest a more limited role for the amygdala in certain forms of emotional learning, though the human data imply a more heterogeneous contribution.

Although the number of patients with discrete lesions of the amygdala is small, they have provided unique information on the role of this structure in emotional processing. Several studies have now reported specific impairments in the recognition of facial expressions of fear in patients with restricted amygdala damage (Adolphs *et al.* 1995, 1996; Calder *et al.* 1996; Broks *et al.* 1998). Recognition of facial signs of other emotions was found to be intact. In a study that required subjects to make judgements of the trustworthiness and approachability of unfamiliar adults from facial photographs, patients with bilateral amygdala damage judged the unfamiliar individuals to be more approachable and trustworthy than did control subjects

(Adolphs *et al.* 1998). Recognition of vocal signs of fear and anger was found to be impaired in a patient with bilateral amygdala damage (Scott *et al.* 1997), suggesting that this deficit is not restricted to facial expressions. Other researchers (Bechara *et al.* 1995) have demonstrated that aversive autonomic conditioning is impaired in a patient with amygdala damage despite the fact that the patient showed normal declarative knowledge of the conditioning contingencies. Collectively, these findings from patients with selective bilateral destruction of the amygdala suggest specific impairments on tasks that tap aspects of negative emotion processing. Most of the studies have focused on the perceptual side, where the data clearly show the amygdala to be important for the recognition of cues of threat or danger. The conditioning data also indicate that the amygdala may be necessary for acquiring new implicit autonomic learning of stimulus–punishment contingencies. In one of the few studies to examine the role of the amygdala in the expression of already learned emotional responses, Angrilli *et al.* (1996) reported on a patient with a benign tumour of the right amygdala in a study that used startle magnitude in response to an acoustic probe measured from orbicularis oculi. Among control subjects, they observed the well-known effect of startle potentiation during the presentation of aversive stimuli. In the patient with right amygdala damage, no startle potentiation was observed in response to aversive versus neutral stimuli. These findings suggest that the amygdala might be necessary for the expression of already learned negative affect.

Since 1995, a growing number of studies using PET and fMRI to investigate the role of the amygdala in emotional processes have begun to appear. Many studies have reported activation of the amygdala detected with either PET or fMRI when anxiety-disordered patients have been exposed to their specific anxiety-provoking stimuli compared with control stimuli (e.g. Breiter *et al.* 1996*b*; Rauch *et al.* 1996). When social phobics were exposed to neutral faces, they showed activation of the amygdala comparable to what was observed in both the phobics and controls in response to aversive compared with neutral odours (Birbaumer *et al.* 1998). Consistent with the human lesion data, several studies have now reported activation of the amygdala in response to facial expressions of fear compared with neutral, happy, or disgust control faces (Morris *et al.* 1996; Phillips *et al.* 1997). In the Breiter *et al.* (1996*a*) fMRI study, they observed rapid habituation of the amygdala response, which may provide an important clue to the time-limited function of the amygdala in the stream of affective information processing. Whalen *et al.* (1998) observed activation of the amygdala in response to masked fear faces that were not consciously perceived. Unpleasant compared with neutral and pleasant pictures have also been found to activate the amygdala (Irwin *et al.* 1996). Finally, several studies have reported activation of the amygdala during the early phases of aversive conditioning (Buchel *et al.* 1998; LaBar *et al.* 1998). Amygdala activation in response to several other experimental procedures for inducing negative affect has been reported, including unsolvable anagrams of the sort used to induce learned helplessness (Schneider *et al.* 1996), aversive olfactory cues (Zald and Pardo 1997), and aversive gustatory stimuli (Zald *et al.* 1998). Other data on individual differences in amygdala activation and their relation to affective style will be treated in the next section. The issues of whether the amygdala responds preferentially to aversive versus appetitive stimuli, is functionally asymmetric, and is required for both the initial learning and subsequent expression of negative emotional associations have not yet been adequately resolved and are considered in detail elsewhere (Davidson and Irwin 1999), though some data clearly suggest that the amygdala does activate in response to appetitive stimuli (Hamann *et al.* 2002). It should be noted that one recent fMRI study (Zalla *et al.* 2000) found differential activation of the left and right amygdala in response to winning and losing money, with the left amygdala showing increased activation in response to winning more money, while the right amygdala showed increased activation in response to the parametric manipulation of losing money. Systematic examination of asymmetries in amygdala activation and function in appetitive and aversive contexts should be performed in light of these data. In several recent reviews, Whalen (e.g. Davis and Whalen 2001) has argued that a major function of the amygdala is the detection of ambiguity and the issuing of a call for further processing when ambiguous information is presented. I will return to this claim later in the chapter when the issue of individual differences is addressed.

These findings raise the question concerning the 'optimal' pattern of amygdala function for well-being. Based upon evidence reviewed in the section, 'What are individual differences in PFC and amygdala activations associated with?', in the context of individual differences, we will argue that low basal levels of amygdala activation, in conjunction with situationally appropriate responding, effective top–down regulation, and rapid recovery, characterize a pattern that is consistent with high levels of well-being.

Hippocampus and anterior cingulate cortex

In this section, brief mention will be made of the contributions of hippocampus and ACC to emotion. More extensive discussion of the contributions of this circuit to emotional processing is contained in several recent reviews (Davidson and Irwin 1999; Bush *et al.* 2000; Davidson *et al.* 2002).

The hippocampus has been implicated in various aspects of memory (see Zola and Squire 2000), particularly declarative memory of the sort we experience when we consciously recall an earlier occurring episode. Its role in emotion and affective style has only recently begun to be gleaned from the available corpus of animal studies on the role of the hippocampus in context-dependent memory (Fanselow 2000). This literature has generally supported a role for the hippocampus in the learning of context. For example, when an animal is exposed to a cue-conditioning procedure where a discrete cue is paired with an aversive outcome, in addition to learning the specific cue–punishment contingency, the animal also learns to associate the context in which the learning occurs with the aversive outcome. Lesions to the hippocampus will abolish this context-dependent form of memory but will have no effect on the learning of the cue–punishment contingency. The fact that the hippocampus is a site in the brain with a very high density of glucocorticoid receptors and participates in the feedback regulation of the hypothalamic–pituitary adrenal axis is particularly germane to the importance of this structure for emotion regulation. Basic research at the animal level has demonstrated the powerful impact of glucocorticoids on hippocampal neurons (Cahill and McGaugh 1998; McEwen 1998). There are data that indicate that exogenous administration of hydro cortisone to humans impairs explicit memory that is presumably hippocampally dependent (e.g. Kirschbaum *et al.* 1996), though there are other data that suggest that, in more moderate amounts, cortisol may facilitate memory (e.g. Abercrombie 2000). Several investigators have reported, using MRI-based measures, that hippo-campal volume is significantly decreased in patients with several stress-related disorders including post-traumatic stress disorder (PTSD; e.g. Bremner 1999) and depression (e.g. Sheline *et al.* 1996; Bremner *et al.* 2000), though there have also been several failures to replicate (e.g. Vakili *et al.* 2000; Rusch *et al.* 2001). In the studies where hippocampal atrophy has been found, the implication is that excess-ively high levels of cortisol associated with the stress-related disorder cause hippocampal cell death and result in hippocampal atrophy as seen on MRI. Although virtually all of these studies have focused on the implications of hippo-campal changes for cognitive function, particularly declarative memory, we (Davidson *et al.* 2000*a*) have proposed that the hippocampus plays a key role in the context-modulation of emotional behaviour. Moreover, we have suggested that it is in the affective realm where the impact of hippocampal involvement in psychopathology may be most apparent. We suggested that, in individuals with compromised hippocampal function, the normal context-regulatory role of this brain region would be impaired and individuals would consequently display emotional behaviour in inappropriate contexts. This argument holds that what may be particularly abnormal in disorders such as PTSD and depression is not the display of 'abnormal emotion' but rather the display of perfectly normal emotion in inappropriate contexts. For example, in the case of the PTSD, the extreme fear and anxiety is probably very adaptive in the original traumatic context. This extreme emotional response probably plays an important role in facilitating the organism's withdrawal from a threatening situation. However, in PTSD, this response is elicited in inappropriate situations. The patient with PTSD behaves similarly to the animal with a hippocampal lesion in failing to modulate emotional responses in a context-appropriate manner. These suggestions are only inferential at the present time. Neuroimaging studies are needed to document the role of the hippocampus in this process in normal and disordered populations. In addition, more attention is needed to understand how and why the hippocampus may preferentially extract and process information about context. Finally, some research (e.g. Davis and Lee 1998) indicates that other structures with direct connections to the hippocampus (for example, the bed nucleus of the stria terminalis) also play a role similar to the hippocampus. More work is needed to understand the differential contributions of the different components of this circuitry.

The findings on the role of hippocampal pathology in disease provide us with insights into the role of this structure for adaptive function and well-being. We suggest that effective context-modulation of emotional behaviour is a hallmark sign of well-being and promotes adaptive emotion regulation.

Many studies that have used neuroimaging methods to probe patterns of brain activation during the arousal of emotion have reported that the ACC activates in response to emotion. Several investigators (Whalen et al. 1998; Bush et al. 2000) have distinguished between cognitive and affective subdivisions of the ACC based upon where activations lie in response to tasks that are purely cognitive versus those that include aspects of emotion. For example, in response to the classical colour–word Stroop task (subjects are required to name the colour of colour words that are inconsistent with the colour in which they are printed, for example, the word 'red' printed in blue), ACC activation is found consistently more dorsal to the locus of activation observed in response to an emotional Stroop task with emotional words. However, the question of just what role the more ventral portions of the ACC might be playing in emotion has not been systematically addressed. On the basis of Cohen's model of the role of ACC in conflict monitoring in the cognitive domain (Carter et al. 1999), we have proposed that the affective subdivision of the ACC might play a similar role in emotion. When emotion is elicited in the laboratory, this itself presents something of a conflict since social norms dictate certain rules for participant behaviour that do not usually include the display of strong emotion. Thus, the very process of activating emotion in the unfamiliar context of a laboratory environment might activate the ACC. Cohen has suggested that ACC activation results in a call for further processing by other brain circuits to address the conflict that has been detected. In most individuals, automatic mechanisms of emotion regulation are probably invoked to dampen strong emotion that may be activated in the laboratory. The initial call for the processes of emotional regulation may result from ACC activation.

Again, considering the possible role of ACC function in well-being, we will argue on the basis of data we present in the section, 'Emotion regulation: a key component of affective style', that high levels of ACC activation in situations requiring emotional regulation will be associated with more effective regulatory skill and thus facilitate well-being. In the next section, attention is turned to individual differences in the key components of the circuitry we describe, with a focus on the pattern of individual differences that form the basis for well-being.

What are individual differences in PFC and amygdala activations associated with?

In both infants (Davidson and Fox 1989) and adults (Davidson and Tomarken 1989) there are large individual differences in baseline electrophysiological measures of prefrontal activation, and such individual variation is associated with differences in aspects of affective reactivity. In infants, Davidson and Fox (1989) reported that 10-month-old babies who cried in response to maternal separation were more likely to have less left- and greater right-sided prefrontal activation during a preceding resting baseline compared with those infants who did not cry in response to this challenge. In adults, we first noted that the phasic influence of positive and negative emotion elicitors (e.g. film clips) on measures of prefrontal activation asymmetry appeared to be superimposed upon more tonic individual differences in the direction and absolute magnitude of asymmetry (Davidson and Tomarken 1989).

During our initial explorations of this phenomenon we needed to determine if baseline electrophysiological measures of prefrontal asymmetry were reliable and stable over time and thus could be used as a trait-like measure. Tomarken *et al.* (1992) recorded baseline brain electrical activity from 90 normal subjects on two occasions separated by approximately 3 weeks. At each testing session, brain activity was recorded during eight 1 minute trials, four eyes open and four eyes closed, presented in counterbalanced order. The data were visually scored to remove artefacts and then Fourier-transformed. Our focus was on power in the alpha band (8–13 Hz), though we extracted power in all frequency bands (see Davidson *et al.* 1990, 2000*b* for methodological discussion). We computed coefficient alpha as a measure of internal consistency reliability from the data within each session. The coefficient alphas were quite high, with all values exceeding 0.85, indicating that the electrophysiological measures of asymmetric activation indeed showed excellent internal consistency reliability. The test–retest reliability was adequate, with intraclass correlations ranging from 0.65 to 0.75 depending upon the specific sites and methods of analysis. The major conclusion from this study was the demonstration that measures of activation asymmetry based upon power in the alpha band from prefrontal scalp electrodes showed both high internal consistency reliability and acceptable test–retest reliability sufficient to be considered a trait-like index.

On the basis of our prior data and theory, we reasoned that extreme left- and extreme right-frontally activated subjects would show systematic differences in dispositional positive and negative affect. We administered the trait version of positive and negative affect scales (PANAS; Watson *et al.* 1988) to examine this question and found that the left-frontally activated subjects reported more positive and less negative affect than their right-frontally activated counterparts (Tomarken *et al.* 1992; Fig. 5.1). More recently (Sutton and Davidson 1997), we showed that scores on a self-report measure designed to operationalize Gray's concepts of the behavioural

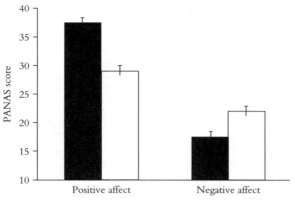

Fig. 5.1 Dispositional positive affect (from scores on the PANAS-General Positive Affect Scale) in subjects who were classified as extreme and stable left-frontally active (black bars; $n = 14$) and extreme and stable right-frontally active (grey bars; $n = 13$) on the basis of electrophysiological measures of baseline activation asymmetries on two occasions separated by 3 weeks. Error bars denote standard error of the mean. (From Tomarken *et al.* (1992).)

inhibition scales (BIS) and behavioural activation scales (BAS; Carver and White 1994) were even more strongly predicted by electrophysiological measures of prefrontal asymmetry than were scores on the PANAS (see Fig. 5.2). Subjects with greater left-sided prefrontal activation reported more relative BAS to BIS activity compared with subjects exhibiting more right-sided prefrontal activation.

In a recent study, we extended these early findings and found that baseline measures of asymmetric prefrontal activation predicted reports of well-being among individuals in their late 50s (Urry *et al.* 2004; Fig. 5.3). Moreover, this association was present even when the association between prefrontal activation asymmetry and dispositional positive affect was statistically removed. These findings indicate that prefrontal activation asymmetry accounts for variance in well-being over and above that accounted for by positive affect.

We also suggested that our measures of prefrontal asymmetry would predict reactivity to experimental elicitors of emotion. The model that we have developed over recent years (see Davidson 1992, 1994, 1995, 1998*a, b* for background) features individual differences in prefrontal activation asymmetry as a reflection of a diathesis that modulates reactivity to emotionally significant events. According to this model, individuals who differ in prefrontal asymmetry should respond differently to an elicitor of positive or negative emotion, even when baseline mood is partialled out. We (Wheeler *et al.* 1993; see also Tomarken *et al.* 1990) performed an experiment to examine this question. We presented short film clips designed to elicit positive or negative emotion. Brain electrical activity was recorded before the presentation of the film clips. Immediately after the clips were presented, subjects were asked to rate their emotional experience during the preceding film clip.

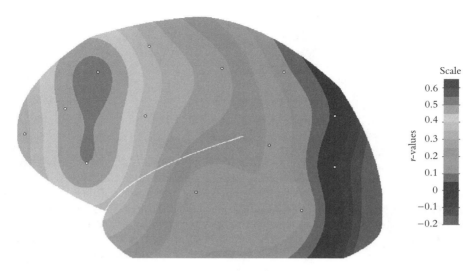

Fig. 5.2 Relations between electrophysiological measures of asymmetry and the difference between the standardized score on the BIS/BAS (Carver and White 1994); $n = 46$. Electrophysiological data were recorded from each subject on two separate occasions separated by 6 weeks. The BIS/BAS were also administered on these two occasions. Data were averaged across the two time periods before performing correlations. The topographic map displays the correlations between alpha power asymmetry (log-right minus log-left alpha power; higher values denote greater relative left-sided activation) and the difference score between the standardized BAS minus BIS. After correlations were performed for each homologous region, a spline-interpolated map was created. The orange and red values of the scale denote positive correlations. The figure indicates that the correlation between the BAS–BIS difference score and the electrophysiology asymmetry score is highly positive in prefrontal scalp regions, denoting that subjects with greater relative left-sided activation report more relative behavioural activation compared with behavioural inhibition tendencies. The relation between asymmetric activation and the BAS–BIS difference is highly specific to the anterior scalp regions, as the correlation decreases rapidly more posteriorly. The correlation in the prefrontal region is significantly larger than the correlation in the parieto-occipital region. (From Sutton and Davidson (1997).) See Plate 1 for full colour version.

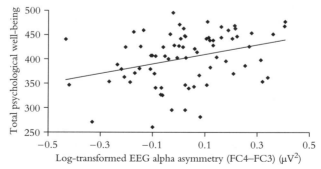

Fig. 5.3 Scatter plot depicting the correlation between frontal EEG asymmetry (FC4–FC3) and total psychological well-being. Relative left-frontal asymmetry (denoted by positive values on the abscissa) is associated with higher levels of well-being; $r(74) = 0.33$; $p = 0.002$. (From Urry *et al.* (2004).)

In addition, subjects completed scales that were designed to reflect their mood at baseline. We found that individual differences in prefrontal asymmetry predicted the emotional response to the films even after measures of baseline mood were statistically removed. Those individuals with more left-sided prefrontal activation at baseline reported more positive affect in response to the positive film clips and those with more right-sided prefrontal activation reported more negative affect in response to the negative film clips. These findings support the idea that individual differences in electrophysiological measures of prefrontal activation asymmetry mark some aspect of vulnerability to positive and negative emotion elicitors. The fact that such relations were obtained following the statistical removal of baseline mood indicates that any difference between left- and right-frontally activated PFC in baseline mood cannot account for the prediction of the film-elicited emotion effects that were observed. What has not yet been answered by these studies that use baseline measures of brain electrical activity to predict emotional reactivity is what other components of affective circuitry are upregulated and downregulated in response to affective challenges in subjects who differ on measures of baseline prefrontal activation asymmetry. This is a question that must be pursued using a combination of electrophysiological and neuroimaging measures.

Depression is clearly a heterogeneous disorder. In a review of the depression literature from the perspective of affective neuroscience (Davidson *et al.* 2002), we suggested that there was a subtype that was associated with deficits in approach-related positive affect, whose proximal cause was predicted to be hypoactivation in certain left prefrontal regions that we have previously implicated in approach-related positive affect. The relation between individual differences in brain electrical measures of prefrontal activation asymmetry and depression is a topic that has received extensive treatment in several recent articles. There has been a failure to replicate (Reid *et al.* 1998) our initial findings of decreased left prefrontal activation in depression (Schaffer *et al.* 1983; Henriques and Davidson 1990, 1991), though there have also been several published independent replications or conceptual replications (e.g. Allen *et al.* 1993; Field *et al.* 1995). Moreover, using PET, Drevets *et al.* (1997) have reported decreased activation in the left subgenual PFC in patients with depression. We interpreted the decrease in left-sided prefrontal activation as a diathesis related to deficits in the approach system and in reward-related responding (Henriques *et al.* 1994; Henriques and Davidson 2000). We also argued that this pattern of left prefrontal hypoactivation would be found only in certain subgroups of mood-disordered patients in light of the heterogeneity of the disorder (see Davidson 1998*b* for an extended discussion). Most importantly, we have suggested that it is crucial to move beyond descriptive phenomenology and to examine with objective laboratory methods variations in reactivity to emotion elicitors in individuals with this proposed diathesis. We have suggested that individuals who display left prefrontal hypoactivation will show specific deficits in reactivity to reward, though the need to consider other components of the circuitry with which the PFC is interconnected must be underscored in any effort to understand the neural bases of emotion and its disorders.

In addition to the studies described above using self-report and psychophysiological measures of emotion, we have examined relations between individual differences in electrophysiological measures of prefrontal asymmetry and other biological indices that, in turn, have been related to differential reactivity to stressful events. Three recent examples from our laboratory include measures of immune function, cortisol, and corticotropin-releasing hormone (CRH). The latter two measures represent key molecules in the activation of a coordinated response to stressful events. Our strategy in each case was to examine relations between individual differences in measures of prefrontal activation asymmetry and these other biological indices. In two separate studies (Kang *et al.* 1991; Davidson *et al.* 1999) we examined relations between the prefrontal activation indices and natural killer (NK) cell activity since declines in NK activity have been reported in response to stressful, negative events (Kiecolt-Glaser and Glaser 1981). We predicted that subjects with greater left-sided prefrontal activation would exhibit higher NK activity than their right-activated counterparts because the former type of subject has been found to report more dispositional positive affect, to show higher relative BAS activity, and to respond more intensely to positive emotional stimuli. In each of the two studies conducted with independent samples, we found that left-frontally activated subjects indeed had higher levels of NK activity than their right-frontally activated counterparts (Kang *et al.* 1991; Davidson *et al.* 1999). We also examined the magnitude of change in NK activity in response to stress and found that subjects with greater baseline levels of left prefrontal activation showed the smallest magnitude decline in NK activity in response to stress compared with other subjects (Davidson *et al.* 1999).

One of the concerns with the studies that examine NK function is the fact that this is an *in vitro* assay and its significance for immunocompetence is unclear. To address this concern, we recently completed a study examining relations between prefrontal activation asymmetry and antibody responses to influenza vaccine (Rosenkranz *et al.* 2003) in a sample of 52 middle-aged subjects with an average age of 58 years (evenly divided by sex). In this study, we recorded brain electrical measures in the same way as previously described. We compared individuals in the top and bottom quartile on measures of prefrontal activation asymmetry and found large differences between these extreme groups in antibody titres to influenza vaccine (see Fig. 5.4), with the left-prefrontally activated subjects showing significantly greater antibody titres than their right-activated counterparts.

In collaboration with Kalin, our laboratory has been studying similar individual differences in scalp-recorded measures of prefrontal activation asymmetry in rhesus monkeys (Davidson *et al.* 1992, 1993). Recently, we (Kalin *et al.* 1998) acquired measures of brain electrical activity from a large sample of rhesus monkeys

($n = 50$). EEG measures were obtained during periods of manual restraint. A subsample of 15 of these monkeys was tested on two occasions 4 months apart. We found that the test–retest correlation for measures of prefrontal asymmetry was 0.62, suggesting similar stability of this metric in monkey and human. In the group of 50 animals, we also obtained measures of plasma cortisol during the early morning. We proposed that, if individual differences in prefrontal asymmetry were associated with dispositional affective style, such differences should be correlated with cortisol, since individual differences in baseline cortisol have been related to various aspects of trait-related stressful behaviour and psychopathology (see Gold *et al.* 1988). We found that animals with left-sided prefrontal activation had lower levels of baseline cortisol than their right-frontally activated counterparts (see Fig. 5.5). As can be seen from Fig. 5.5, it is the left-activated animals that are particularly low compared with both middle- and right-activated subjects. Moreover, when blood samples were collected 2 years after our initial testing, animals classified as showing

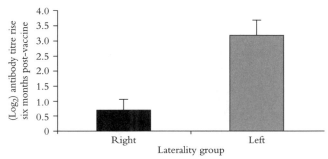

Fig. 5.4 Bar graph of the mean antibody titre rise ($\log_2$) to influenza vaccine 6 months post-vaccine for extreme groups comprising individuals (average age 58 years) in the top and bottom 25th centiles of activation asymmetry at the lateral frontal (F7/8) site. Error bars denote standard error of the mean (SEM). The difference between groups was highly significant ($t(22) = 3.81$, $p < 0.001$). (From Rosenkranz *et al.* (2003).)

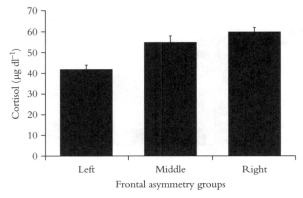

Fig. 5.5 Basal morning plasma cortisol from 1-year-old rhesus monkeys classified as left- ($n = 12$), middle- ($n = 16$), or right-frontally ($n = 11$) activated based upon electrophysiological measurements. Error bars denote SEM. (From Kalin *et al.* (1998).)

extreme left-sided prefrontal activation at age 1 year had significantly lower baseline cortisol levels when they were 3 years of age compared with animals who were classified at age 1 year as displaying extreme right-sided prefrontal activation. Similar findings were obtained with cerebrospinal fluid (CSF) levels of CRH. Those animals with greater left-sided prefrontal activation showed lower levels of CRH (Kalin *et al.* 2000; Fig. 5.6). These findings indicate that individual differences in prefrontal asymmetry are present in non-human primates and that such differences predict biological measures that are related to affective style.

With the advent of neuroimaging, it has become possible to investigate the relation between individual differences in aspects of amygdala function and measures of affective style. We have used PET with fluorodeoxyglucose (FDG) as a tracer to investigate relations between individual differences in glucose metabolism in the amygdala and dispositional negative affect. FDG-PET is well suited to capture trait-like effects since the period of active uptake of tracer in the brain is approximately 30 minutes. Thus, it is inherently more reliable than ^{15}O blood flow measures since the FDG data reflect activity aggregated over a 30 min period. We have used resting FDG-PET to examine individual differences in glucose metabolic rate in the amygdala and its relation to dispositional negative affect in depressed subjects (Abercrombie *et al.* 1998). We acquired a resting FDG-PET scan as well as a structural MR scan for each subject. The structural

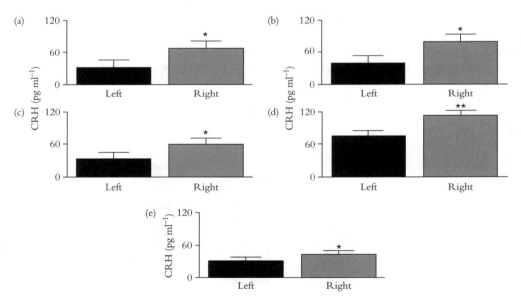

Fig. 5.6 Differences between right- ($n = 9$) and left-prefrontally ($n = 10$) activated animals in CSF measures of CRH at five different ages: (a) 4 months; (b) 8 months; (c) 14 months; (d) 40 months; (e) 52 months. Error bars denote SEM. The original classification of the animals as extreme right- or left-activated was performed on the basis of brain electrical activity data collected when the animals were 13 months of age. (From Kalin *et al.* (2000).)

MR scans are used for anatomical localization by co-registering the two image sets. Thus, for each subject, we used an automated algorithm to fit the MR scan to the PET image. Regions of interest (ROIs) were then drawn on each subject's MR scan to outline the amygdala in each hemisphere. These ROIs were drawn on coronal sections of subjects' MR images and the ROIs were then automatically transferred to the co-registered PET images. Glucose metabolism in the left and right amygdala ROIs was then extracted. The interrater reliability for the extracted glucose metabolic rate is highly significant with intraclass correlations between two independent raters equal to or greater than 0.97. We found that subjects with lower levels of glucose metabolism in the right amygdala reported less dispositional negative affect on the PANAS scale (see Fig. 5.7). These findings indicate that individual differences in resting glucose metabolism in the amygdala

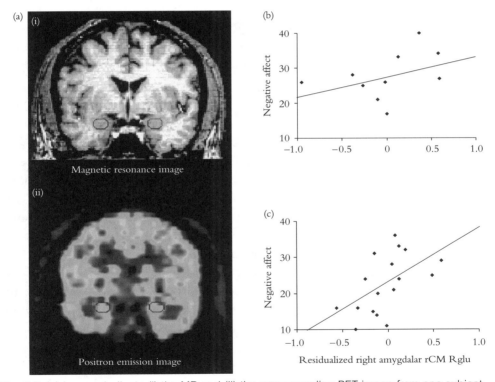

Fig. 5.7 (a) Images indicate (i) the MR and (ii) the corresponding PET image from one subject to illustrate our method of MRI co-registered ROIs around the amygdala. ROIs were individually drawn for each subject around the amygdala, and glucose metabolism was then extracted from the PET image in (ii). See Plate 2 for full colour version. (b), (c) Scatter plots display the relation between glucose metabolic rate (residualized for variations in whole brain metabolism) in the right amygdala and dispositional negative affect assessed with the PANAS for two independent samples of depressed patients tested on two different PET scanners ((b) sample 1; $r = 0.41$; (c) sample 2; $r = 0.56$). Metabolic rate in the left amygdala did not predict PANAS negative affect. The scanner used for the data in (c) had better spatial resolution (GE Advance). (From Abercrombie *et al.* (1998).)

are present and that they predict dispositional negative affect among depressed subjects.

In a small sample of 12 normal subjects, we (Irwin *et al.* 1998) have been able to examine the relation between the magnitude of MR signal change in the amygdala in response to aversive, compared with neutral, pictures and dispositional negative affect on the PANAS scale. We correlated the average value of the voxels with the maximum Student's *t*-value from the left and right amygdala with dispositional negative affect. There was a robust correlation, such that subjects showing the least increase in signal intensity in the right amygdala reported the lowest levels of dispositional negative affect. The findings from the fMRI and PET studies of amygdala function indicate that individual differences in both tonic activation and phasic activation in response to aversive stimuli predict the intensity of dispositional negative affect.

Emotion regulation: a key component of affective style

One of the key components of affective style is the capacity to regulate negative emotion and, specifically, to decrease the duration of negative affect once it arises. We have suggested in several articles that the connections between the PFC and amygdala play an important role in this regulatory process (Davidson 1998*a*; Davidson and Irwin 1999; Davidson *et al.* 2000*d*). In two studies, we (Larson *et al.* 1998; Jackson *et al.* 2003) examined relations between individual differences in prefrontal activation asymmetry and the emotion–modulated startle. In both studies, we presented pictures from the International Affective Picture System (Lang *et al.* 1995) while acoustic startle probes were presented and the electromyography (EMG)-measured blink response from the orbicularis oculi muscle region was recorded (see Sutton *et al.* 1997*a* for basic methods). Startle probes were presented both during the slide exposure as well as at various latencies following the offset of the pictures, on separate trials. We interpreted startle magnitude during picture exposure as providing an index related to the peak of emotional response, while startle magnitude following the offset of the pictures was taken to reflect the automatic recovery from emotional challenge. Used in this way, startle probe methods can potentially provide new information on the time-course of emotional responding. We expected that individual differences during actual picture presentation would be less pronounced than individual differences following picture presentation since an acute emotional stimulus is likely to pull for a normative response across subjects, while individuals are more likely to differ once the stimulus has terminated. Similarly, we predicted that individual differences in prefrontal asymmetry would account for more variance in predicting magnitude of recovery (i.e. startle magnitude post-stimulus) than in predicting startle magnitude during the stimulus. Our findings in each study were consistent with our predictions and indicated that subjects with greater right-sided prefrontal activation show a larger blink magnitude following the offset of the negative stimuli, after the variance in blink magnitude during the negative stimulus was partialled out. Measures of prefrontal asymmetry did not reliably predict startle magnitude during picture presentation. The findings from these studies are consistent with our hypothesis and indicate that individual differences in prefrontal asymmetry are associated with the time-course of affective responding, particularly the recovery following emotional challenge. In a related study, we have found that subjects with greater baseline levels of left prefrontal activation are better able to suppress negative affect voluntarily (see Jackson *et al.* 2000). Moreover, using functional MRI we have demonstrated that, when subjects are instructed to regulate their negative emotion voluntarily, reliable bilateral changes in amygdala MR signal intensity are found (Schaefer *et al.* 2002) and that the magnitude of MR signal decreases in the amygdala during instructions to downregulate negative affect are predicted by increased MR signal in the vmPFC (Urry *et al.* 2003; Fig. 5.8).

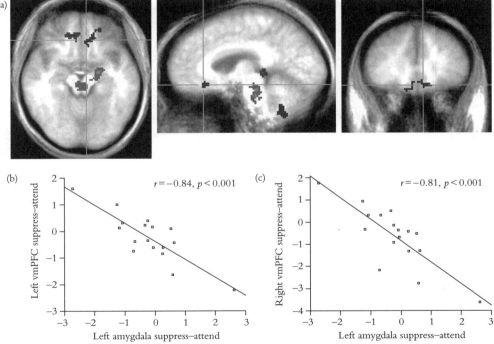

Fig. 5.8 (a) Image display of the associations between activation in the amygdala bilaterally and other regions of the brain. Blue areas denote inverse correlations whereas red areas denote positive correlations. The data reveal an inverse association between activation in vmPFC and signal change in the amygdala following aversive pictures in response to instructions to suppress (downregulate) negative affect, compared with a condition during which subjects were instructed to attend to the stimuli (*n* = 17). (From Urry *et al.* (2003).) See Plate 3 for full colour version. (b) Left vmPFC cluster (including subgenual ACC) with left amygdala (*r* = −0.84; *p* < 0.001); (c) right vmPFC cluster with left amygdala (*r* = −0.81; *p* < 0.001). (From Urry *et al.* (2003).)

The findings from these studies indicate that individual differences in prefrontal activation may play an important role in emotion regulation. Individuals who report less dispositional negative affect and more dispositional positive affect may be those individuals who have increased facility at regulating negative affect and specifically in modulating the intensity of negative affect once it has been activated.

Plasticity in the central circuitry of emotion

The circuits that underlie emotion regulation, in particular the amygdala and PFC, have been the targets of intensive study of plasticity (see Davidson *et al.* 2000*a* for an extensive discussion). In a series of elegant studies in rats, Meaney and colleagues (Francis and Meaney 1999) have demonstrated that an early environmental manipulation in rats—frequency of maternal licking/grooming and arched-back nursing—produces a cascade of biological changes in the offspring that shape the central circuitry of emotion and, consequently, alter the animals' behavioural and biological responsivity to stress. For example, the offspring of mothers high in licking and grooming show increased central benzodiazepine receptor densities in various subnuclei of the amygdala as well as in the locus ceruleus (LC), increased α_2-adrenoreceptor density in the LC, and decreased CRH receptor density in the LC (Caldji *et al.* 1998). In other research, Meaney and co-workers have reported that rats exposed to high licking/grooming mothers exhibited a permanent increase in concentrations of receptors for glucocorticoids in both the hippocampus and the PFC (Meaney *et al.* 1988, 1996; Liu *et al.* 1997). All of these changes induced by early maternal licking/grooming and related behaviour involve alterations in the central circuitry of emotion that result in decreased responsivity to stress later in life.

These findings in animals raise the possibility that similar effects may transpire in humans. There are clearly short-term changes in brain activation that are observed during voluntary emotion regulation, as noted in the previous section. Whether repeated practice in techniques of emotion regulation leads to more enduring changes in patterns of brain activation is a question that has not yet been answered in extant research. There are limited data available that indicate that cognitive behavioural therapy for certain disorders (e.g. obsessive compulsive disorder; simple phobia) produces changes in regional brain activity that are comparable to those produced by medication (Baxter *et al.* 1992; Paquette *et al.* 2003). What are largely absent are data on plastic changes in the brain that might be produced by the practice of methods specifically designed to increase positive affect, such as meditation. In a recent study, we examined changes in brain electrical activity and immune function following an 8 week training programme in mindfulness meditation (Davidson *et al.* 2003). In this study subjects were randomly assigned to a meditation group or a wait-list control group and each of these groups was tested before and after the 8 week training programme, as well as 4 months following the end of the programme. We found that subjects in the meditation group showed significantly larger increases in left-sided anterior activation compared with their counterparts in the control group. Subjects received an influenza vaccine just after the 8 week programme was completed and we found that influenza antibody titres were significantly higher in the meditators compared with the controls. Most remarkably, we observed that those subjects who showed the largest magnitude change in brain activity also showed the largest increase in antibody titres (see Fig. 5.9). These findings suggest that training

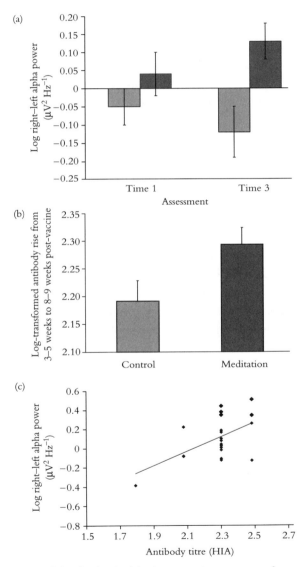

Fig. 5.9 (a) Pre- and post-training brain electrical asymmetry measures from subjects in the meditation and control groups. Note that positive numbers denote greater left-sided activation and negative numbers denote greater right-sided activation. (b) Differences between the meditation and control groups in antibody titres in response to influenza vaccine. Errors bars in (a), (b) denote SEM. (c) The relation between pre-to-post test increases in left-sided activation and rise in antibody titres to influenza vaccine among subjects in the meditation group only (*n* = 25 for meditation group; *n* = 16 for control group). (From Davidson *et al.* (2003).)

procedures designed explicitly to facilitate well-being result in demonstrable and predictable changes in brain and immune function.

The Dalai Lama himself has raised this question in his recent book *The art of happiness* (Dalai Lama and Cutler 1998, pp. 44–5) where he explains that 'The systematic training of the mind—the cultivation of happiness, the genuine inner transformation by deliberately selecting and focusing on positive mental states and challenging negative mental states—is possible because of the very structure and function of the brain. . . . But the wiring in our brains is not static, not irrevocably fixed. Our brains are also adaptable.' We are now at the point in the development of affective neuroscience where we can rigorously address this question by using neuroimaging methods to probe changes in patterns of activation and transmitter function that might be produced by the systematic practice of techniques such as meditation that are designed to promote the cultivation of positive affect. By also examining how changes in the central circuitry of emotion might be related to peripheral biology (endocrine, autonomic and immune function), we can begin to examine mechanistically how well-being may be consequential for our mental and physical health.

Summary and conclusions

The circuitry underlying emotion and affective style was reviewed with an emphasis on the contributions of different sectors of the PFC and amygdala to positive affect and well-being. Individual differences in patterns of prefrontal and amygdala activation are related to behavioural and biological constituents of affective style and emotion regulation. Recent data highlight the role of particular sectors of the PFC in emotion regulation, particularly the regulation of the duration of negative affect and the downregulation of negative emotion once it is elicited. These individual differences are conceptualized as diatheses that alter a person's vulnerability or resilience. Recent evidence in animals underscores the extraordinary plasticity of this circuitry and demonstrates that early social experience in particular has profound consequences for the developing nervous system. The possibilities of transforming this circuitry in adulthood with specific methods designed to cultivate positive affect were considered. New longitudinal research is needed to address these questions.

Acknowledgements

This research was supported by NIMH grants MH43454, MH40747, P50-MH52354, and P50-MH61083, NIA grant PO1-AG021079, and an NIMH training grant T32-MH1893.

References

Abercrombie, H.C. (2000). The effects of pharmacologically manipulated cortisol levels on memory for emotional and neutral information. Unpublished doctoral dissertation, University of Wisconsin.

Abercrombie, H.C., Schaefer, S.M., Larson, C.L., Oakes, T.R., Holden, J.E., Perlman, S.B., Krahn, D.D., Benca, R.M., and Davidson, R.J. (1998). Metabolic rate in the right amygdala predicts negative affect in depressed patients. *Neuroreport* **9**, 3301–7.

Adolphs, R., Damasio, H., Tranel, D., and Damasio, A.R. (1995). Fear and the human amygdala. *J. Neurosci.* **15**, 5879–91.

Adolphs, R., Damasio, H., Tranel, D., and Damasio, A.R. (1996). Cortical systems for the recognition of emotion in facial expressions. *J. Neurosci.* **16**, 7678–87.

Adolphs, R., Tranel, D., and Damasio, A.R. (1998). The human amygdala in social judgment. *Nature* **393**, 470–4.

Aggleton, J.P. (1993). The contribution of the amygdala to normal and abnormal emotional states. *Trends Neurosci.* **16**, 328–33.

Ahern, G.L. and Schwartz, G.E. (1985). Differential lateralization for positive and negative emotion in the human brain: EEG spectral analysis. *Neuropsychologia* **23**, 745–55.

Allen, J.J., Iacono, W.G., Depue, R.A., and Arbisi, P. (1993). Regional electroencephalographic asymmetries in bipolar seasonal affective disorder before and after exposure to bright light. *Biol. Psychiatry* **33**, 642–6.

American Psychiatric Association (1994). *Diagnostic and statistical manual of mental disorders* (-IV), 4th edn. APA Press, Washington, DC.

Angrilli, A., Mauri, A., Palomba, D., Flor, H., Birbaumer, N., Sartori, G., and di Paola, F. (1996). Startle reflex and emotion modulation impairment after a right amygdala lesion. *Brain* **119**, 1991–2000.

Baxter, L.R. (and 11 others) (1992). Caudate glucose metabolic rate changes with both drug and behavior therapy for obsessive–compulsive disorder. *Arch. Gen. Psychiatry* **49**, 681–99.

Bechara, A., Damasio, A.R., Damasio, H., and Anderson, S.W. (1994). Insensitivity to future consequences following damage to human prefrontal cortex. *Cognition* **50**, 7–15.

Bechara, A., Tranel, D., Damasio, H., Adolphs, R., Rockland, C., and Damasio, A.R. (1995). Double dissociation of conditioning and declarative knowledge relative to the amygdala and hippocampus in humans. *Science* **269**, 1115–18.

Bechara, A., Tranel, D., Damasio, H., and Damasio, A.R. (1996). Failure to respond autonomically to anticipated future outcomes following damage to prefrontal cortex. *Cereb. Cortex* **6**, 215–25.

Bechara, A., Damasio, H., Tranel, D., and Damasio, A.R. (1997). Deciding advantageously before knowing the advantageous strategy. *Science* **275**, 1293–5.

Bechara, A., Damasio, H., Damasio, A.R., and Lee, G.P. (1999). Different contributions of the human amygdala and ventromedial prefrontal cortex to decision-making. *J. Neurosci.* **19**, 5473–81.

Birbaumer, N., Grodd, W., Diedrich, O., Klose, U., Erb, E., Lotze, M., Schneider, F., Weiss, U., and Flor, H. (1998). fMRI reveals amygdala activation to human faces in social phobics. *Neuroreport* **9**, 1223–6.

Breiter, H.C., Etcoff, N.L., Whalen, P. J., Kennedy, W.A., Rauch, S.L., Buckner, R.L., Strauss, M.M., Hyman, S.E., and Rosen, B.R. (1996a). Response and habituation of the human amygdala during visual processing of facial expression. *Neuron* **17**, 875–87.

Breiter, H.C. (and 16 others) (1996b). Functional magnetic resonance imaging of symptom provocation in obsessive-compulsive disorder. *Arch. Gen. Psychiatry* **53**, 595–606.

Bremner, J.D. (1999). Does stress damage the brain? *Biol. Psychiatry* **45**, 797–805.

Bremner, J.D., Narayan, M., Anderson, E.R., Staib, L.H., Miller, H.L., and Charney, D.S. (2000). Hippocampal volume reduction in major depression. *Am. J. Psychiatry* **1571**, 115–18.

Broks, P. (and 11 others) (1998). Face processing impairments after encephalitis: amygdala damage and recognition of fear. *Neuropsychologia* **361**, 59–70.

Buchel, C., Morris, J., Dolan, R.J., and Friston, K.J. (1998). Brain systems mediating aversive conditioning: an event-related fMRI study. *Neuron* **20**, 947–57.

Bush, G., Luu, P., and Posner, M.I. (2000). Cognitive and emotional influences in anterior cingulate cortex. *Trends Cogn. Sci.* **4**, 215–22.

Cacioppo, J.T. and Gardner, W.L. (1999). Emotion. *Annu. Rev. Psychol.* **50**, 191–214.

Cahill, L. and McGaugh, J.L. (1998). Mechanisms of emotional arousal and lasting declarative memory. *Trends Neurosci.* **21**, 273–313.

Cahill, L., Weinberger, N.M., Roozendaal, B., and McGaugh, J.L. (1999). Is the amygdala a locus of 'conditioned fear'? Some questions and caveats. *Neuron* **23**, 227–8.

Calder, A.J., Young, A.W., Rowland, D., Perrett, D.I., Hodges, J.R., and Etcoff, N.L. (1996). Facial emotion recognition after bilateral amygdala damage: differentially severe impairment of fear. *Cogn. Neuropsychol.* **135**, 699–745.

Caldji, C., Tannenbaum, B., Sharma, S., Francis, D., Plotsky, P.M., and Meaney, M.J. (1998). Maternal care during infancy regulates the development of neural systems mediating the expression of fearfulness in the rat. *Proc. Natl Acad. Sci. USA* **95**, 5335–40.

Carson, A.J., MacHale, S., Allen, K., Lawrie, S.M., Dennis, M., House, A., and Sharpe, M. (2000). Depression after stroke and lesion location: a systematic review. *Lancet* **356**, 122–6.

Carter, C.S., Botvinick, M.M., and Cohen, J.D. (1999). The contribution of the anterior cingulate cortex to executive processes in cognition. *Rev. Neurosci.* **10**, 49–57.

Carver, C.S. and White, T.L. (1994). Behavioral inhibition, behavioral activation and affective responses to impending reward and punishment: the BIS/BAS scales. *J. Pers. Soc. Psychol.* **67**, 319–33.

Dalai Lama and Cutler, H.C. (1998). *The art of happiness.* Riverhead Books, New York.

Davidson, R.J. (1992). Emotion and affective style: hemispheric substrates. *Psychol. Sci.* **3**, 39–43.

Davidson, R.J. (1993). Cerebral asymmetry and emotion: conceptual and methodological conundrums. *Cogn. Emotion* **7**, 115–38.

Davidson, R.J. (1994). Complexities in the search for emotion-specific physiology. In *The nature of emotion: fundamental questions* (ed. P. Ekman and R.J. Davidson), pp. 237–42. Oxford University Press, New York.

Davidson, R.J. (1995). Cerebral asymmetry, emotion and affective style. In *Brain asymmetry* (ed. R.J. Davidson and K. Hugdahl), pp. 361–87. MIT Press, Cambridge, Massachusetts.

Davidson, R.J. (1998a). Affective style and affective disorders: perspectives from affective neuroscience. *Cogn. Emotion* **12**, 307–20.

Davidson, R.J. (1998b). Anterior electrophysiological asymmetries, emotion and depression: conceptual and methodological conundrums. *Psychophysiology* **355**, 607–14.

Davidson, R.J. and Fox, N.A. (1989). Frontal brain asymmetry predicts infants' response to maternal separation. *J. Abnorm. Psychol.* **98**, 127–31.

Davidson, R.J. and Irwin, W. (1999). The functional neuroanatomy of emotion and affective style. *Trends Cogn. Sci.* **3**, 11–21.

Davidson, R.J. and Tomarken, A.J. (1989). Laterality and emotion: an electrophysiological approach. In *Handbook of neuropsychology*, Vol. 3 (ed. F. Boller and J. Grafman), pp. 419–41. Elsevier, Amsterdam.

Davidson, R.J., Chapman, J.P., Chapman, L.P., and Henriques, J.B. (1990). Asymmetrical brain electrical activity discriminates between psychometrically-matched verbal and spatial cognitive tasks. *Psychophysiology* **27**, 528–43.

Davidson, R.J., Kalin, N.H., and Shelton, S.E. (1992). Lateralized effects of diazepam on frontal brain electrical asymmetries in rhesus monkeys. *Biol. Psychiatry* **32**, 438–51.

Davidson, R.J., Kalin, N.H., and Shelton, S.E. (1993). Lateralized response to diazepam predicts temperamental style in rhesus monkeys. *Behav. Neurosci.* **107**, 1106–10.

Davidson, R.J., Coe, C.C., Dolski, I., and Donzella, B. (1999). Individual differences in prefrontal activation asymmetry predicts natural killer cell activity at rest and in response to challenge. *Brain Behav. Immun.* **13**, 93–108.

Davidson, R.J., Jackson, D.C., and Kalin, N.H. (2000*a*). Emotion, plasticity, context, and regulation: perspectives from affective neuroscience. *Psychol. Bull.* **126**, 890–909.

Davidson, R.J., Jackson, D.C., and Larson, C.L. (2000*b*). Human electroencephalography. In *Principles of psychophysiology* (ed. J.T. Cacioppo, G.G. Bernston, and L.G. Tassinary), pp. 27–52. Cambridge University Press, New York.

Davidson, R.J., Marshall, J.R., Tomarken, A.J., and Henriques, J.B. (2000*c*). While a phobic waits: regional brain electrical and autonomic activity in social phobics during anticipation of public speaking. *Biol. Psychiatry* **47**, 85–95.

Davidson, R.J., Putnam, K.M., and Larson, C.L. (2000*d*). Dysfunction in the neural circuitry of emotion regulation—a possible prelude to violence. *Science* **289**, 591–4.

Davidson, R.J., Pizzagalli, D., Nitschke, J.B., and Putnam, K.M. (2002). Depression: perspectives from affective neuroscience. *Annu. Rev. Psychol.* **53**, 545–74.

Davidson, R.J., Kabat-Zinn, J., Schumacher, J., Rosenkranz, M., Muller, D., Santorelli, S.F., Urbanowski, M.A., Harrington, A., Bonus, K., and Sheridan, J.F. (2003). Alterations in brain and immune function produced by mindfulness meditation. *Psychosom. Med.* **65**, 564–70.

Davis, M. (1992). The role of the amygdala in fear-potentiated startle: implications for animal models of anxiety. *Trends Pharmacol. Sci.* **13**, 35–41.

Davis, M. and Lee, Y.L. (1998). Fear and anxiety: possible roles of the amygdala and bed nucleus of the stria terminalis. *Cogn. Emotion* **12**, 277–306.

Davis, M. and Whalen, P.J. (2001). The amygdala: vigilance and emotion. *Mol. Psychiatry* **6**, 13–34.

Drevets, W.C., Price, J.L., Simpson, J.R.J., Todd, R.D., Reich, T., Vannier, M., and Raichle, M.E. (1997). Subgenual prefrontal cortex abnormalities in mood disorders. *Nature* **386**, 824–7.

Fanselow, M.S. (2000). Contextual fear, gestalt memories, and the hippocampus. *Behav. Brain Res.* **1101–2**, 73–81.

Fanselow, M.S. and LeDoux, J.E. (1999). Why we think plasticity underlying Pavlovian fear conditioning occurs in the basolateral amygdala. *Neuron* **23**, 229–32.

Field, T., Fox, N.A., Pickens, J., and Nawrocki, T. (1995). Relative right frontal EEG activation in 3- to 6-month-old infants of 'depressed' mothers. *Dev. Psychol.* **3**, 358–63.

Francis, D. and Meaney, M.J. (1999). Maternal care and development of stress responses. *Curr. Opin. Neurobiol.* **9**, 128–34.

Frijda, N.H. (1994). Emotions are functional, most of the time. In *The nature of emotion: fundamental questions* (ed. P. Ekman and R.J. Davidson), pp. 112–22. Oxford University Press, New York.

Gainotti, G. (1972). Emotional behavior and hemispheric side of lesion. *Cortex* **8**, 41–55.

Giese-Davis, J. and Spiegel, D. (2003). Emotional expression and cancer progression. In *Handbook of affective neuroscience* (ed. R.J. Davidson, K. Scherer, and H.H. Goldsmith), pp. 1053–82. Oxford University Press, New York.

Gold, P.W., Goodwin, F.K., and Chrousos, G.P. (1988). Clinical and biochemical manifestations of depression: relation to the neurobiology of stress. *New Engl. J. Med.* **314**, 348–53.

Goldman-Rakic, P.S. (1996). The prefrontal landscape: implications of functional architecture for understanding human mentation and the central executive. *Phil. Trans. R. Soc. Lond. B* **351**, 1445–53.

Goldman-Rakic, P.S. (2000). Localization of function all over again. *Neuroimage* **11**, 451–7.

Hamann, S.B., Ely, T.D., Hoffman, J.M., and Kilts, C.D. (2002). Ecstasy and agony: activation of the human amygdala in positive and negative emotion. *Psychol. Sci.* **13**, 135–41.

Henriques, J.B. and Davidson, R.J. (1990). Regional brain electrical asymmetries discriminate between previously depressed subjects and healthy controls. *J. Abnorm. Psychol.* **99**, 22–31.

Henriques, J.B. and Davidson, R.J. (1991). Left frontal hypoactivation in depression. *J. Abnorm. Psychol.* **100**, 535–45.

Henriques, J.B. and Davidson, R.J. (2000). Decreased responsiveness to reward in depression. *Cogn. Emotion* **15**, 711–24.

Henriques, J.B., Glowacki, J.M., and Davidson, R.J. (1994). Reward fails to alter response bias in depression. *J. Abnorm. Psychol.* **103**, 460–6.

Hugdahl, K. (1998). Cortical control of human classical conditioning: autonomic and positron emission tomography data. *Psychophysiology* **35**, 170–8.

Hugdahl, K., Beradi, A., Thompson, W.L., Kosslyn, S.M., Macy, R., Baker, D.P., Alpert, N.M., and LeDoux, J.E. (1995). Brain mechanisms in human classical conditioning: a PET blood flow study. *Neuroreport* **6**, 1723–8.

Irwin, W., Davidson, R.J., Lowe, M.J., Mock, B.J., Sorenson, J.A., and Turski, P.A. (1996). Human amygdala activation detected with echo-planar functional magnetic resonance imaging. *Neuroreport* **7**, 1765–9.

Irwin, W., Davidson, R.J., Kalin, N.H., Sorenson, J.A., and Turski, P.A. (1998). Relations between human amygdala activation and self-reported dispositional affect. *J. Cogn. Neurosci.* (Suppl. S), 109.

Jackson, D.C., Malmstadt, J., Larson, C.L., and Davidson, R.J. (2000). Suppression and enhancement of emotional responses to unpleasant pictures. *Psychophysiology* **37**, 515–22.

Jackson, D.C., Mueller, C.J., Dolski, I., Dalton, K.M., Nitschke, J.B., Urry, H.L., Rosenkranz, M.A., Ryff, C.D., Singer, B.H., and Davidson, R.J. (2003). Now you feel it, now you don't: frontal EEG asymmetry and individual differences in emotion regulation. *Psychol. Sci.* **14**, 612–17.

Jones, N.A. and Fox, N.A. (1992). Electroencephalogram asymmetry during emotionally evocative films and its relation to positive and negative affectivity. *Brain Cogn.* **20**, 280–99.

Kalin, N.H., Larson, C.L., Shelton, S.E., and Davidson, R.J. (1998). Asymmetric frontal brain activity, cortisol, and behavior associated with fearful temperament in rhesus monkeys. *Behav. Neurosci.* **112**, 286–92.

Kalin, N.H., Shelton, S.E., and Davidson, R.J. (2000). Cerebrospinal fluid corticotropin-releasing hormone levels are elevated in monkeys with patterns of brain activity associated with fearful temperament. *Biol. Psychiatry* **47**, 579–85.

Kalin, N.H., Shelton, S.E., Davidson, R.J., and Kelley, A.E. (2001). The primate amygdala mediates acute fear but not the behavioral and physiological components of anxious temperament. *J. Neurosci.* **21**, 2067–74.

Kang, D.H., Davidson, R.J., Coe, C.L., Wheeler, R.W., Tomarken, A.J., and Ershler, W.B. (1991). Frontal brain asymmetry and immune function. *Behav. Neurosci.* **105**, 860–9.

Kiecolt-Glaser, J.K. and Glaser, R. (1981). Stress andimmune function in humans. In *Psychoneuroimmunology* (ed. R. Ader, D.L. Felten, and N. Cohen), pp. 849–67. Academic Press, San Diego, California.

Kirschbaum, C., Wolf, O.T., May, M., Wippich, W., and Hellhammer, D.H. (1996). Stress- and treatment-induced elevations of cortisol levels associated with impaired declarative memory in healthy adults. *Life Sci.* **58**, 1475–83.

Kosslyn, S.M. and Koenig, O. (1992). *Wet mind: the new cognitive neuroscience.* Free Press, New York.

LaBar, K.S., Gatenby, J.C., LeDoux, J.E., and Phelps, E.A. (1998). Human amygdala activation during conditioned fear acquisition and extinction—a mixed-trial fMRI study. *Neuron* **205**, 937–45.

Lang, P.J. (1995). The emotion probe: studies of motivation and attention. *Am. Psychol.* **50**, 372–85.

Lang, P.J., Bradley, M.M., and Cuthbert, B.N. (1995). *International affective picture system IAPS: technical manual and affective ratings.* The Center for Research in Psychophysiology, University of Florida, Gainsville, Florida.

Larson, C.L., Sutton, S.K., and Davidson, R.J. (1998). Affective style, frontal EEG asymmetry and the time course of the emotion-modulated startle. *Psychophysiology* **35**, S52.

LeDoux, J.E. (1996). *The emotional brain: the mysterious underpinnings of emotional lift.* Simon and Schuster, New York.

Levenson, R.W. (1994). Human emotion: a functional view. In *The nature of emotion: fundamental questions* (ed. P. Ekman and R.J. Davidson), pp. 123–6. Oxford University Press, New York.

Liu, D., Diorio, J., Tannenbaum, B., Caldji, C., Francis, D., and Freedman, A. (1997). Maternal care, hippocampal glucocoricoid receptors, and hypothalamic–pituitary–adrenal responses to stress. *Science* **277**, 1659–62.

McEwen, B.S. (1998). Protective and damaging effects of stress mediators. *New Engl. J. Med.* **338**, 171–9.

Meaney, M.J., Aitken, D.H., van Berkel, C., Bhatnagar, S., and Sapolsky, R.M. (1988). Effect of neonatal handling on age-related impairments associated with the hippocampus. *Science* **239**, 766–8.

Meaney, M.J., Bhatnagar, S., Larocque, S., McCormick, C.M., Shanks, N., Sharman, S., Smythe, J., Viau, V., and Plotsky, P.M. (1996). Early environment and the development of individual differences in the hypothalamic–pituitary–adrenal stress response. In *Severe stress and mental disturbance in children* (ed. C.R. Pfeffer), pp. 85–127. American Psychiatric Press, Washington, DC.

Miller, A. and Tomarken, A.J. (2001). Task-dependent changes in frontal brain asymmetry: effects of incentive cues, outcomes expectancies, and motor responses. *Psychophysiology* **38**, 500–11.

Morris, J.S., Frith, C.D., Perrett, D.I., Rowland, D., Young, A.W., Calder, A.J., and Dolan, R.J. (1996). A differential neural response in the human amygdala to fearful and happy facial expressions. *Nature* **383**, 812–15.

Paquette, V., Levesque, J., Mensour, B., Leroux, J.M., Beaudoin, G., Bourgouin, P., and Beauregard, M. (2003). Change the mind and you change the brain: effects of cognitive-behavioral therapy on the neural correlates of spider phobia. *Neuroimage* **18**, 401–9.

Phillips, M.L. (and 11 others) (1997). A specific neural substrate for perceiving facial expressions of disgust. *Nature* **389**, 495–8.

Rauch, S.L., van der Kolk, B.A., Fisler, R.E., Alpert, N.M., Orr, S.P., Savage, C.R., Fischman, A.J., Jenike, M.A., and Pitman, R.K. (1996). A symptom provocation study of posttraumatic stress disorder using positron emission tomography and script-driven imagery. *Arch. Gen. Psychiatry* **535**, 380–7.

Rauch, S.L., Savage, C.R., Alpert, N.M., Fischman, A.J., and Jenike, M.A. (1997). A study of three disorders using positron emission tomography and symptom provocation. *Biol. Psychiatry* **42**, 446–52.

Reid, S.A., Duke, L.M., and Allen, J.J. (1998). Resting frontal electroencephalographic asymmetry in depression: inconsistencies suggest the need to identify mediating factors. *Psychophysiology* **354**, 389–404.

Robinson, R.G., Starr, L.B., and Price, T.R. (1984). A two year longitudinal study of mood disorders following stroke: prevalence and duration at six months follow-up. *Br. J. Psychiatry* **144**, 256–262.

Rolls, E.T. (1999). *The brain and emotion*. Oxford University Press, New York.

Rosenkrantz, M.A., Jackson, D.C., Dalton, K.M., Dolski, I., Ryff, C.D., Singer, B.H., Muller, D., Kalin, N.H., and Davidson, R.J. (2003). Affective style and *in vivo* immune response: neurobehavioral mechanisms. *Proc. Natl Acad. Sci. USA* **100**, 11148–52.

Rusch, B.D., Abercrombie, H.C., Oakes, T.R., Schaefer, S.M., and Davidson, R.J. (2001). Hippocampal morphometry in depressed patients and control subjects: relations to anxiety symptoms. *Biol. Psychiatry* **50**, 960–4.

Sackeim, H.A., Greenberg, M.S., Weiman, A.L., Gur, R.C., Hungerbuhler, J.P., and Geschwind, N. (1982). Hemispheric asymmetry in the expression of positive and negative emotions: neurologic evidence. *Arch. Neurol.* **39**, 210–18.

Schaefer, S.M., Jackson, D.C., Davidson, R.J., Aguirre, G.K., Kimberg, D.Y., and Thompson-Schill, S.L. (2002). Modulation of amygdalar activity by the conscious regulation of negative emotion. *J. Cogn. Neurosci.* **14**, 913–21.

Schaffer, C.E., Davidson, R.J., and Saron, C. (1983). Frontal and parietal EEG asymmetries in depressed and non-depressed subjects. *Biol. Psychiatry* **18**, 753–62.

Schneider, F., Gur, R.E., Alavi, A., Seligman, M.E.P., Mozley, L.H., Smith, R.J., Mozley, P.D., and Gur, R.C. (1996). Cerebral blood flow changes in limbic regions induced by unsolvable anagram tasks. *Am. J. Psychiatry* **153**, 206–12.

Scott, S.K., Young, A.W., Calder, A.J., Hellawell, D.J., Aggleton, J.P., and Johnson, M. (1997). Impaired auditory recognition of fear and anger following bilateral amygdala lesions. *Nature* **385**, 254–7.

Shackman, A.J., Maxwell, J.S., Skolnick, A.J., Schaefer, H.S., and Davidson, R.J. (2003). Exploiting individual differences in the prefrontal asymmetry of approach-related affect: hemodynamic, electroencephalographic, and psychophysiological evidence. Program no. 444.6. (2003). Abstract Viewer/Itinerary Planner. Washington, DC: Society for Neuroscience, online.

Sheline, Y.I., Wang, P.W., Gado, M.H., Csernansky, J.G., and Vannier, M.W. (1996). Hippocampal atrophy in recurrent major depression. *Proc. Natl Acad. Sci. USA* **93**, 3908–13.

Sobotka, S.S., Davidson, R.J., and Senulis, J.A. (1992). Anterior brain electrical asymmetries in response to reward and punishment. *Electroencephalogr. Clin. Neurophysiol.* **83**, 236–47.

Sutton, S.K. and Davidson, R.J. (1997). Prefrontal brain asymmetry: a biological substrate of the behavioral approach and inhibition systems. *Psychol. Sci.* **8**, 204–10.

Sutton, S.K., Davidson, R.J., Donzella, B., Irwin, W., and Dottl, D.A. (1997*a*). Manipulating affective state using extended picture presentation. *Psychophysiology* **34**, 217–26.

Sutton, S.K., Ward, R.T., Larson, C.L., Holden, J.E., Perlman, S.B., and Davidson, R.J. (1997*b*). Asymmetry in prefrontal glucose metabolism during appetitive and aversive emotional states: an FDG-PET study. *Psychophysiology* **34**, S89.

Taylor, S.E. (1991). Asymmetrical effects of positive and negative events: the mobilization–minimization hypothesis. *Psychol. Bull.* **110**, 67–85.

Tomarken, A.J., Davidson, R.J., and Henriques, J.B. (1990). Resting frontal activation asymmetry predicts emotional reactivity to film clips. *J. Pers. Soc. Psychol.* **59**, 791–801.

Tomarken, A.J., Davidson, R.J., Wheeler, R.E., and Doss, R.C. (1992). Individual differences in anterior brain asymmetry and fundamental dimensions of emotion. *J. Pers. Soc. Psychol.* **62**, 676–87.

Urry, H.L., van Reekum, C.M., Johnstone, T., Thurow, M.E., Burghy, C.A., Mueller, C.J., and Davidson.R.J. (2003). Neural correlates of voluntarily regulating negative affect. Program no. 725.18. (2003). Abstract Viewer/Itinerary Planner. Washington, DC: Society for Neuroscience online.

Urry, H.L., Nitschke, J.B., Dolski, I., Jackson, D.C., Dalton, K.M., Mueller, C.J., Rosenkranz, M.A., Ryff, C.D., Singer, B.H., and Davidson, R.J. (2004). Making a life worth living: neural correlates of well-being. *Psychol. Sci.* **15**, 367–72.

Vakili, K., Pillay, S.S., Lafer, B., Fava, M., Renshaw, P.F., and Bonello-Cintron, C.M. (2000). Hippocampal volume in primary unipolar major depression: a magnetic resonance imaging study. *Biol. Psychiatry* **47**, 1087–90.

Vrana, S.R., Spence, E.L., and Lang, P.J. (1988). The startle probe response: a new measure of emotion? *J. Abnorm. Psychol.* **97**, 487–91.

Watanabe, M. (1996). Reward expectancy in primate prefrontal neurons. *Nature* **382**, 629–32.

Watson, D., Clark, L.A., and Tellegen, A. (1988). Developmental and validation of brief measures of positive and negative affect: the PANAS scales. *J. Pers. Soc. Psychol.* **54**, 1063–70.

Watson, D., Clark, L.A., Weber, K., Assenheimer, J.S., Strauss, M.E., and McCormick, C.M. (1995). Testing a tripartite model: I. Evaluating the convergent and discriminant validity of anxiety and depression symptom scales. *J. Abnorm. Psychol.* **104**, 3–14.

Whalen, P., Bush, G., McNally, R.J., Wilhelm, S., McInerney, S.C., Jenike, M.A., and Rauch, S.L. (1998). The emotional counting Stroop paradigm: a functional magnetic resonance imaging probe of the anterior cingulate affective division. *Biol. Psychiatry* **44**, 1219–28.

Wheeler, R.E., Davidson, R.J., and Tomarken, A.J. (1993). Frontal brain asymmetry and emotional reactivity: a biological substrate of affective style. *Psychophysiology* **30**, 82–9.

Wilson, D.S. (1994). Adaptive genetic variation and human evolutionary psychology. *Ethnol. Sociobiol.* **154**, 219–35.

Zald, D.H. and Pardo, J.V. (1997). Emotion, olfaction and the human amygdala: amygdala activation during aversive olfactory stimulation. *Proc. Natl Acad. Sci. USA* **94**, 4119–24.

Zald, D.H., Lee, J.T., Fluegel, K.W., and Pardo, J.V. (1998). Aversive gustatory stimulation activates limbic circuits in humans. *Brain* **121**, 1143–54.

Zalla, T., Koechlin, E., Pietrini, P., Basso, G., Aquino, P., Sirigu, A., and Grafman, J. (2000). Differential amygdala responses to winning and losing: a functional magnetic resonance imaging study in humans. *Eur. J. Neurosci.* **12**, 1764–70.

Zola, S.M. and Squire, L.R. (2000). The medial temporal lobe and the hippocampus. In *The Oxford handbook of memory* (ed. E. Tulving and F.I.M. Craig), pp. 485–500. Oxford University Press, New York.

Stuart J.H. Biddle is professor of exercise and sport psychology and head of the School of Sport and Exercise Sciences at Loughborough University, UK. Stuart is the past president of the European Federation of Sport Psychology, founding editor of *Psychology of Sport and Exercise*, and author of a number of books in the field.

Panteleimon Ekkekakis is an assistant professor of exercise psychology in the Department of Health and Human Performance at Iowa State University, USA. His research examines affective responses to physical activity, including the trait, social–cognitive, and physiological factors that influence them and the implications of these responses for enjoyment and adherence.

Chapter 6

Physically active lifestyles and well-being

Stuart J.H. Biddle and Panteleimon Ekkekakis

Introduction

Few days go past without reference in the mass media to (un)healthy lifestyles in modern society. Diseases of the age that are most often mentioned and have an association with physical inactivity include obesity, diabetes, heart disease, and some cancers. Indeed, obesity seems to be a current health issue most worrying politicians and health professionals, and lack of physical activity is a key element of the energy imbalance that is causing current obesity trends (Bouchard 2000; Bouchard and Blair 1999). However, the beneficial effects of physical activity go far beyond healthy weight management. The 'magic pill' of physical activity is powerful, with effects demonstrated on numerous health outcomes, including positive mental health (Dishman *et al.* 2004). Indeed, physical activity research pioneer, Professor Jeremy Morris, once referred to physical activity as 'today's best buy for public health' (Morris 1994). Nevertheless, it was only in 1988 that one of the present authors (SJHB) was referred to a General Practitioner (GP) at his local primary care health centre who not only smoked (in his consulting room!), but recommended that all one really needed in respect of physical activity was a game of rugby at the weekend! Fortunately, we have come some way since then and now have strong advocacy documents promoting the importance of regular, moderate intensity physical activity on most days of the week (Department of Health 2004; Department of Health and Human Services and Centers for Disease Control and Prevention 1996; Pate *et al.* 1995). In addition, recommendations can be tailored to individual needs, such as those of young people, adults, older adults, or those with certain medical conditions (Department of Health 2004).

In this chapter, we review the evidence that physical activity can contribute to positive well-being as well as prevent or ameliorate disease. We examine the so-called 'feel-good factor'—the psychological benefits that people derive from physical activity. Our current understanding of the mechanisms underlying the beneficial effects of physical activity is presented, challenging some widely held views. A brief examination of the factors underlying why some people are physically active while others are not is followed by a section on successful interventions at both an individual and a societal level. We conclude that, despite the long history of research in this field, there remain some exciting challenges in establishing how physical activity exerts its enhancing effects on well-being and in using this knowledge to develop interventions that would raise levels of physical activity and well-being in the population at large.

Defining key terms

Typically, physical activity includes movement of the body produced by the skeletal muscles that results in energy expenditure (Caspersen *et al.* 1985). This over-arching category can include any form of movement but, for the purposes of health enhancement, we tend to be more interested in more gross motor movements such as walking, cycling, lifting, or large or prolonged do-it-yourself (DIY) activities.

Physical activity can include others types of movement, such as structured exercise and sport. Exercise involves 'planned, structured and repetitive bodily movement' (Caspersen *et al.* 1985, p. 127), often with the objective of fitness maintenance or improvement. An example would be exercising on a treadmill at a fitness club. Sport, on the other hand, is physical activity that is rule-governed, structured, and competitive and involves gross motor movement characterized by physical strategy, prowess, and chance (Rejeski and Brawley 1988), such as golf or tennis. Not all forms of physical activity are necessarily healthy and this may depend on the nature of the activity, how it is performed, and the characteristics of the individual. However, within the normal bounds of safety and appropriate advice, physical activity can have major health benefits.

Behavioural epidemiology framework

The five-phase behavioural epidemiology framework advocated by Sallis and Owen (1999) is a useful way of viewing various processes in the understanding of physical activity and health. Behavioural epidemiology considers the link between behaviours and health and disease, such as why some people are physically active and others are not. In relation to physical activity, this framework has five main phases.

1 *To establish the link between physical activity and health.* This is now well documented for many diverse conditions as well as well-being (Bouchard *et al.* 1994; Dishman *et al.* 2004).

2 *To develop methods for the accurate assessment of physical activity.* This remains a problematic area. Large-scale surveillance of population trends inevitably relies on self-report, a method that is fraught with validity and reliability problems. Recent 'objective' methods, such as movement sensors, heart rate monitors, or

pedometers, are useful but do not necessarily give all of the information required, such as intensity or type of activity or the setting in which activity took place. Until we have better measures of the behaviour itself—i.e. physical activity—the field will struggle to progress in many respects.

3 *To identify factors that are associated with different levels of physical activity.* Given the evidence supporting the beneficial effects of physical activity on health, it is important to identify factors that might be associated with the adoption and maintenance of the behaviour. This area is referred to as the study of 'correlates' or 'determinants' of physical activity. The term correlates is now preferred in order to avoid the assumption of causal links. Much of the evidence at this stage is not able to support causality.

4 *To evaluate interventions designed to promote physical activity.* Once a variable is identified as a correlate of physical activity (e.g. self-efficacy), then interventions can manipulate this variable to test if it is, in fact, a determinant. The number of intervention studies in physical activity is increasing (Kahn *et al.* 2002).

5 *To translate findings from research into practice.* If interventions work, it is appropriate to translate such findings into ecologically valid 'real-world' settings.

It is important to realize that the above sequence is not linear. For example, measures of physical activity are developed and refined alongside tests of outcomes, and often community projects are established prior to convincing evidence, but may include a monitoring and evaluation element to test the efficacy of such an intervention before refining future interventions. The whole process then becomes iterative. Drawing on the behavioural epidemiology framework, this chapter will consider the psychological benefits of physical activity (phase 2 of the framework), the correlates of physical activity (phase 3), and interventions designed to enhance physical activity levels (phase 4).

The feel-good factor: what psychological benefits do people derive from physical activity?

The psychological effects associated with physical activity have been the topic of numerous scientific studies, conducted mainly since the early 1970s. The general conclusion from this research is that physical activity can enhance the participants' sense of well-being. As described in a recent literature review, 'both survey and experimental research . . . provide support for the well publicized statement that "*exercise makes you feel good* " ' (Fox 1999; p. 413, italics in the original). However, although the conceptual and methodological sophistication of the studies has increased over the years, not all reviewers have reached a similarly definitive conclusion. Skeptics have argued that, in many cases, statements about the psychological benefits of physical activity seem to 'anticipate rather than reflect the accumulation of strong evidence' (Salmon 2001, p. 36).

The controversy has been persisting for many years. Over 2 decades ago, based on a few preliminary studies, Morgan (1981, p. 306) asserted that 'the "feeling better" sensation that accompanies regular physical activity is so obvious that it is one of the few universally accepted benefits of exercise.' Yet, at about the same time, Hughes's (1984, p. 76) assessment was that 'the enthusiastic support of exercise to improve mental health has a limited empirical basis and lacks a well-tested rationale.' More recently, a similar contrast is evident in the conclusions reached by different reviewers evaluating the research on the effects of habitual physical activity on depression. Biddle *et al.* (2000, p. 155) stated that 'overall, the evidence is strong enough for us to conclude that there is support for a *causal* link between physical activity and reduced clinically defined depression. This is the first time such a statement has been made.' In contrast, Lawlor and Hopker (2001, p.1) found that 'the effectiveness of exercise in reducing symptoms of depression cannot be determined because of a lack of good quality research on clinical populations with adequate follow up.' Others, characterizing Lawlor and Hopker's conclusion as 'a bit harsh' (Brosse *et al.* 2002, p. 754), acknowledge that, although, if taken together, the extant studies seem to suggest that regular physical activity can reduce depression, the literature still contains a very small number of high-quality clinical trials and, consequently, the quantity and quality of the evidence could not yet satisfy the most stringent of criteria, such as those established for evaluating the effectiveness of prescription drugs.

Despite the conflicting views, the appeal of physical activity as an intervention modality for enhancing psychological well-being is such that the research will surely continue. In an era of rising mental health care costs, physical activity is a potentially viable alternative or adjunct to traditional forms of therapy (i.e. pharmaco- and psychotherapy) and is inexpensive and free of serious side-effects. Moreover, physical activity has some added advantages that other interventions cannot claim. First, besides being potentially effective as a therapeutic modality, physical activity also seems to have great potential value as a preventive modality among healthy individuals. Second, besides its psychological effects, physical activity also has substantial beneficial effects on physical health—and these are supported by a much more extensive and robust evidence base! In the following paragraphs, we provide a brief summary of the research on the psychological effects of physical activity, separating them into the effects of physical fitness or habitual participation in physical activity (e.g. for months or years) and those associated with single bouts of activity.

Psychological effects of habitual physical activity

The effects of habitual physical activity on various aspects of well-being have been examined in numerous studies. These include both cross-sectional studies (i.e. examining differences in aspects of well-being between groups differing in

Table 6.1 Benefits of regular, long-term physical activity for various aspects of well-being, with references to representative literature reviews

Benefit for well-being	Selected references
Reduction in anxiety	Landers and Petruzzello 1994; O'Connor *et al.* 2000; Petruzzello *et al.* 1991; Taylor 2000
Reduction in depression	Brosse *et al.* 2002; Craft and Landers 1998; Mutrie 2000; O'Neal *et al.* 2000
Improved mood states	Arent *et al.* 2000; Biddle 2000
Enhanced health-related quality of life in the elderly and various patient populations	Berger 2004; Berger and Motl 2001; Rejeski *et al.* 1996; Rejeski and Mihalko 2001
Improved physical and general self-worth	Fox 2000*a, b*; Sonstroem 1997
Improved sleep	Kubitz *et al.* 1996; Youngstedt 2000; Youngstedt *et al.* 1997
Reduced reactivity to psychosocial stressors	Dishman and Jackson 2000; Sothmann *et al.* 1996
Improved cognitive function in all populations, including older adults	Colcombe and Kramer 2003; Etnier *et al.* 1997

habitual levels of activity or levels of physical fitness) and experimental studies (i.e. examining the effects of weeks or months of participation). With so many studies whose methodological approaches are quite diverse and none of which can be characterized as definitive, this topic has also been the subject of many reviews, including meta-analytic reviews. These reviews have generally concluded that physical activity is associated with reduced anxiety, reduced depression, improved mood states, enhanced health-related quality of life in the elderly and various patient populations, improved physical and general self-worth, improved sleep, reduced reactivity to psychosocial stressors, and improved cognitive function in all populations including older adults (for references, see Table 6.1).

Collectively, these findings suggest that individuals who are physically active, besides lowering their risk of premature death or chronic disease and physical disability, should experience a sense of psychological well-being. The following additional conclusions can also be drawn. First, consistent with common conceptualizations of well-being as a multifaceted construct, the beneficial effects of physical activity are clearly not restricted to one outcome variable but, instead, appear to be broad (Landers and Arent 2001; McAuley and Katula 1998). Second, the beneficial effects of physical activity on the various aspects of well-being appear to extend to both genders and all age groups. Third, these benefits tend to be larger with longer interventions and for those individuals whose mental health is more compromised at baseline. Having said this, it is important to clarify that individuals without compromised mental health also receive significant benefits, albeit of smaller magnitude, given their narrower available margin for improvement. Fourth, the beneficial effects on well-being are not fully accounted for by the beneficial effects of physical activity on objective measures of physical fitness or physical function.

For example, research on health-related quality of life, particularly among older adults, has demonstrated that 'objective' (e.g. the ability to walk certain distances, climb a flight of stairs, or maintain balance) and 'subjective' measures (e.g. self-reports of pain, physical self-efficacy, or satisfaction with physical function) are not necessarily correlated and do not necessarily show the same time-course in response to physical activity interventions. Other studies have consistently shown that there is no association between activity-induced increases in physical fitness and improvements in anxiety and depression. Collectively, these findings suggest that cognitive self-appraisals of activity-related effects, which may or may not be entirely congruent with actual physical gains, constitute an important mechanism by which physical activity influences quality of life and well-being. Fifth, studies that, based on the assumption that the 'active ingredient' in physical activity interventions is the *aerobic* training, used control groups involving non-aerobic activities, such as stretching and toning (strength training), have shown that reductions in anxiety and depression were similar, regardless of the aerobic or non-aerobic nature of the activity. These findings clearly point to alternative explanations, including the possibility that these psychological benefits are influenced by social interactions or the participants' perception that they are actively taking control of their mental and physical health.

Although, as noted earlier, most literature reviews have concluded that habitual physical activity entails substantial benefits for various aspects of well-being, the research picture is complicated, so it is necessary to examine the findings from a critical standpoint. As a case in point, consider the arguably most complete study of the effects of physical activity on depression (Blumenthal *et al.* 1999). In this study, 156 men and women, 50 years of age or older, who had been diagnosed with major depressive disorder, were randomly assigned to one of three 16-week treatment conditions: (1) exercise (3 sessions per week, lasting for 45 minutes each, at 70–85% of heart rate reserve); (2) antidepressant medication (using the popular serotonin reuptake inhibitor sertraline hydrochloride or Zoloft™); or (3) a combination of the exercise and antidepressant treatments. The drop-out rates at the end of the 16-week period were not significantly different between the three groups (26%, 15%, and 20% for groups 1, 2, and 3, respectively). At the end of the treatment period, both clinician-rated and self-reported levels of depression were reduced compared to baseline, with no significant differences between the groups. A similar result was also found for anxiety, self-esteem, life satisfaction, and dysfunctional attitudes. At the 10-month follow-up (6 months after the conclusion of treatment), self-reported depression scores were also not different across the three groups. However, based on DSM-IV criteria and clinician ratings (a Hamilton Rating Scale score higher than 7), the participants in the exercise group had a lower rate of depression (30%) than those in the medication (52%) and combined-treatment groups (55%). Furthermore, of the participants who were in remission after the initial 16-week treatment period, those who had been assigned to the exercise group were more likely to have partly or fully recovered after 6 months than those in the medication and combined-treatment groups (Babyak *et al.* 2000). The authors discussed these findings stating that exercise helps participants develop 'a sense of personal mastery and positive self-regard', whereas the exclusive reliance on or the inclusion of medication 'may undermine this benefit by prioritising an alternative, less self-confirming attribution for one's improved condition' (Babyak *et al.* 2000, p. 636).

On the one hand, this study overcame several of the methodological shortcomings of earlier studies, having an adequate sample size, including both men and women, and examining individuals who were depressed at baseline rather than a convenience sample. Furthermore, the study involved reasonably long treatment and follow-up periods, two comparison conditions, and more than one standard measure of the main outcome variable (i.e. both self-reports and clinician ratings of depression). Finally, as Lawlor and Hopker (2001) noted in their review, the Blumenthal *et al.* (1999) study did involve an 'intention to treat' analysis, thus accounting for the possible biasing effects of the less-than-perfect adherence and often substantial drop-out rates commonly associated with exercise and medication interventions.

On the other hand, this study could not address other persistent problems, some of which seem to be unavoidable (Morgan 1997). First, there was a selection bias, since the participants were all volunteers who responded to advertisements for a research study of 'exercise therapy for depression'. It has been shown that the expectation of psychological benefits from physical activity can significantly influence the outcome (Desharnais *et al.* 1993). Babyak *et al.* (2000) also commented on the possible presence of an 'anti-medication' bias among some participants. Perhaps associated with volunteerism, the participants were also highly educated and physically healthy. Furthermore, the possibility of 'spontaneous recovery' cannot be excluded since there was no no-treatment control condition. The reason for this is that there can be no true 'placebo' exercise intervention, leaving only control conditions that are of questionable meaningfulness (e.g. wait list), since they fail to control for expectancy. Moreover, since the exercise was conducted in a group environment, it is possible that the beneficial effects of exercise were partly or fully mediated by the factor of social interaction. Finally, there is no way to fully account for treatment cross-overs that can take place during the follow-up period (e.g. participants opting to switch or discontinue treatments). These limitations, which are clearly not trivial, underscore the fact that even large, costly, and well-designed studies that produce seemingly robust results supporting the beneficial role of physical activity should be viewed cautiously.

In addition to the continued efforts to design methodologically stronger randomized clinical trials, research is also being conducted on several other fronts. One important area of research deals with the delineation of the shape of the relationship between the 'dose' of physical activity (i.e. frequency, session duration, intensity) and the psychological response. In the case of depression and anxiety, which are the most intensely studied outcomes of habitual physical activity, the current conclusion is that 'there is little evidence for dose–response effects, though this is largely because of a lack of studies rather than a lack of evidence' (Dunn *et al.* 2001, p. S587).

Psychological effects of single bouts of physical activity

Findings that people report 'feeling better' after they participate in a session of physical activity, regardless of how this 'feel-better' effect is operationally defined and largely regardless of the characteristics of the physical activity stimulus and the participants, are remarkably robust. The most common methodological approach in this line of research has been the assessment of psychological states, using multi-item inventories, before and after a bout of physical activity and some control condition, typically a sedentary one (e.g. reading an article or participating in an arts and crafts class). Compared to these control conditions, physical activity has consistently been shown to be associated with reductions in state anxiety and improvements in various mood states, such as decreases in tension and depression and increases in vigour (Landers and Arent 2001; Tuson and Sinyor 1993; Yeung 1996). The publication of studies that follow this basic paradigm (and produce similar, positive results) has continued unabated for over 3 decades, having generated a literature that now contains literally hundreds of reports.

As was the case with research examining the effects of habitual physical activity, studies on the effects of single sessions of activity also have limitations, both conceptual and methodological. One example is the fact that the aforementioned operational definitions of the 'feel-better' effect (i.e. mainly self-reports of state anxiety and mood states) were chosen not because these had been demonstrated to be the only or the strongest changes associated with physical activity but rather due to the fact that these could be assessed by available self-report measures when this research got under way (i.e. in the early 1970s). Consequently, even though the positive changes in these variables have been replicated in numerous studies, it remains unclear if these are the only or the most experientially salient changes that occur in response to physical activity under various conditions.

Another consequence is that very little attention was paid over the years to distinctions between the various constructs that fall under the affective umbrella, such as prototypical emotions, moods, and core affect (Ekkekakis and Petruzzello 2000; Russell 2003), yet these distinctions are clearly important. Emotions, such as state anxiety or pride, and moods, such as irritability or cheerfulness, rely on cognitive appraisals, whereas core affect, such as tension or calmness, could emanate from the body in a direct, cognitively unmediated fashion. These differences imply that not all facets of the response to a session of physical activity are necessarily subject to the same mechanisms or likely to follow a unified pattern.

Furthermore, as a consequence of using standard self-report measures of state anxiety and mood states that contain a relatively large number of items (typically between 20 and 65 each), change in the outcome variables was assessed from before to various time points after the termination of the activity. Logically, this practice would make sense only if the trajectory of change during the intervening period were linear. However, it has become clear that, in many cases, the changes are nonlinear, particularly as the intensity of the activity increases or the duration progresses (e.g. consisting of a curvilinear decline during the activity, followed by an instantaneous rebound after the end). Thus, a pre-to-post assessment protocol would clearly misrepresent the changes that take place in the interim.

An important consequence of failing to recognize that such dynamic changes might occur has been the inability to identify a consistent pattern of dose–response effects. Despite the fact that, as anyone with any physical activity experience can attest, different intensities and/or durations of physical activity will likely lead a participant to experience a gamut of experientially different responses, possibly including both pleasant and unpleasant ones, 30 years of research have failed to provide reliable evidence for such an effect (Ekkekakis and Petruzzello 1999).

Finally, contrary to the assumption that all or most individuals would respond to physical activity in the same direction (presumably, with changes toward a more pleasant state), it has become apparent that, under certain conditions, it is possible for some participants to respond with changes toward pleasure and others with changes toward displeasure. For example, a study involving cycling for 30 minutes at 60% of the participants' maximal capacity, showed that approximately half reported feeling progressively better and half reported feeling progressively worse (Van Landuyt *et al.* 2000). Nevertheless, given the enormous multitude of interacting influences on affective responses (e.g. physical and social conditions, exercise intensity and duration, and the physiological and psychological traits of the participants), the proportions of individuals who respond positively and negatively under various conditions remain extremely difficult to predict.

As these and other limitations became apparent, more recent studies have started to investigate the affective changes associated with single sessions of physical activity using a new conceptual and methodological platform. The primary characteristics of this new approach have been: (1) the conceptualization and assessment of affective changes not in terms of distinct states but rather in terms of broad dimensions, such as those comprising the circumplex model of affect, namely pleasure–displeasure and activation (Ekkekakis and Petruzzello 1999, 2002); (2) the repeated assessments of these dimensions both during and after the sessions of activity; and (3) the examination of individual patterns of change. These studies have produced the first reliable evidence of a specific relationship between the intensity of physical activity and affective responses, showing that the level of intensity corresponding to the transition from aerobic to anaerobic metabolism appears to be the 'turning point' toward displeasure during physical activity (Acevedo *et al.* 2003; Bixby *et al.* 2001; Ekkekakis *et al.* 2004; Hall *et al.* 2002). This is an important physiological landmark, since the metabolic resources that are available to anaerobic metabolism are limited and exceeding the point of transition entails the inability to maintain a physiological steady state (i.e. a continuous rise in heart rate, oxygen uptake, and lactic acid concentration) and a multitude of physiological adjustments as the body approaches fatigue and exhaustion. Below this transition, affective changes tend to be mostly pleasant. However, above this transition, there is a gradual decrease in pleasure and, ultimately, an increase in displeasure. However, once physical activity is stopped, there is a rapid increase in self-rated pleasure, regardless of whether this response represents a *continuation* of a positive trend or a *rebound* from a negative trend during the activity (Bixby *et al.* 2001). Examples of affective responses to two physical activity stimuli of markedly different intensities, one treadmill walk at a self-chosen pace (lasting 15 minutes) and a treadmill test involving gradual increases of the speed and grade until the point of volitional exhaustion (lasting 11.3 minutes), are shown in Fig. 6.1 (see caption for more details).

Mechanisms underlying the psychological benefits of physical activity

An important area of research in exercise psychology focuses on the mechanism(s) by which physical activity benefits well-being. Establishing one or more plausible mechanisms could help to show that the relationship between physical activity and well-being goes beyond statistical association, providing evidence that physical activity can, in fact, *cause* positive changes in well-being. Unfortunately, this research has traditionally been fragmented, reflecting dualistic thinking, with some researchers seeking explanations in *psychological* mechanisms, others seeking explanations in *physiological* mechanisms, and virtually no efforts aimed at integrating the two.

As noted earlier, cognitive appraisals of agency, mastery, control, or self-efficacy have long been theorized to underlie the beneficial effects of physical activity on various aspects of well-being. For example, studies designed to manipulate the known sources of self-efficacy (i.e. prior accomplishments, vicarious experiences, verbal persuasion, and physiological arousal) have shown congruent changes in affective parameters, both in single sessions of activity and in long-term participation (McAuley 1991).

According to another psychological mechanistic hypothesis, physical activity can enhance well-being by providing a distraction or a 'time-out' from daily hassles and worries. This explanation was based on a finding that exercise, meditation, and quiet rest for equal periods of time were all accompanied by similar reductions in

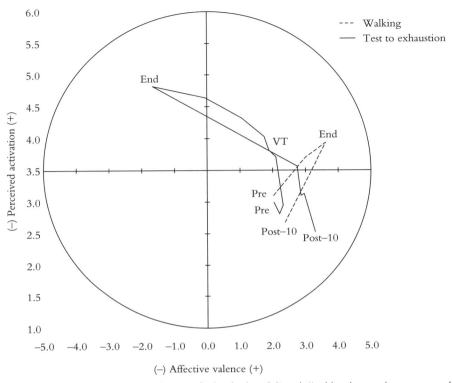

Fig. 6.1 Affective responses to two bouts of physical activity, plotted in circumplex space, where the horizontal dimension represents affective valence, ranging from displeasure (left) to pleasure (right) and the vertical dimension represents perceived activation, ranging from low (bottom) to high (top). Both dimensions are assessed by self-ratings. 'Pre' indicates the beginning of each activity, 'End' indicates its end, and 'Post-10' indicates a time point 10 minutes after the end. A 15 minute treadmill walk at a self-chosen pace on the treadmill results in an activated pleasant state during and immediately after its completion, whereas a 10 minute seated recovery period leads to a low-activation pleasant state. A treadmill test, lasting on average 11.3 minutes, during which the speed and grade are gradually increased until the point of volitional exhaustion, leads to an activated unpleasant state, whereas a subsequent cool-down and 10 minute seated recovery period bring about a return to a low-activation pleasant state. Notice that the ventilatory threshold (indicated by VT), a marker of the transition from aerobic to anaerobic metabolism, appears to be the turning point toward displeasure during the treadmill test (see text for additional information). (Plotted with data from Ekkekakis *et al.* (2000) and Hall *et al.* (2002).)

state anxiety (Bahrke and Morgan 1978). Given that these three tasks had no similarities other than giving participants the opportunity for a break, researchers assumed that this common ingredient was the key to reducing state anxiety. However, if the assessment of the affective outcomes is extended beyond the single variable of state anxiety, it becomes clear that physical activity leads to affective responses that differ substantially from those associated with sedentary tasks, with physical activity leading, at least initially, to a high-activation pleasant state (e.g. energy, excitement) and sedentary or relaxing activities leading to a low-activation pleasant state (e.g. calmness, serenity). Given this qualitative difference, the explanation that physical activity leads to an enhanced affective state simply because it provides a distraction from worries becomes unlikely.

A third psychological explanation is that, given the fact that in most studies physical activity takes place in groups, physical activity can enhance the sense of well-being by providing an opportunity for social interactions. Although this explanation remains possible, it cannot be considered the only explanation, primarily because of studies that have shown positive affective responses to physical activity even if the activity takes place in social isolation and in the relatively dull environment of a laboratory. Furthermore, studies have shown that the presence of others does not necessarily have a positive influence on affect, as individuals concerned about their appearance may respond with increased anxiety when exercising in a social setting (Focht and Hausenblas 2003).

Of the physiological explanations, perhaps the most intuitively appealing proposes that the positive effects of physical activity on well-being are consequences of its well-established effects on physical fitness, including cardiorespiratory endurance, strength, and flexibility. Specifically, it is assumed that a physically active individual will 'feel better' because he or she is able to do more things more efficiently, avoid health problems (such as obesity or chronic debilitating diseases), and, importantly, maintain these abilities into old age. This hypothesis seems particularly relevant to how physical activity benefits health-related quality of life. However, as noted earlier, several studies have failed to provide evidence of an association between objectively quantified physical gains, such as improvements in endurance or strength, and changes in well-being. Therefore, what remains as a viable explanation at this point is that the beneficial effects of physical activity on well-being might be mediated by the perceived, rather than the objective, physical and physiological changes associated with physical activity participation.

The studies that have focused on brain mechanisms have usually taken one of two approaches. In one, physical activity is viewed as an appetitive stimulus (akin to tasty food or addictive drugs) and, consequently, what is being sought is the mechanism by which it is experienced as pleasant or rewarding. Studies that have followed this approach have considered primarily brain areas known to be involved in pleasure and reward, focusing particularly on the mesolimbic dopaminergic pathway projecting from the ventral tegmental area of the midbrain to the nucleus accumbens. In the second approach, physical activity is viewed as an intervention modality capable of correcting imbalances in brain neurotransmission commonly associated with anxiety and depression, in a manner analogous to that of centrally acting drugs. For example, modern antidepressant medications (serotonin-specific reuptake inhibitors or SSRIs) work mainly by blocking the reuptake pumps that collect serotonin from the synaptic cleft back into the releasing neuron, thus leaving more serotonin available to attach to receptors on the receiving neuron. Using techniques such as microdialysis, which allows the collection of extracellular fluid from specific locations in the brain of free-living animals, studies have shown that exercise appears to have an effect similar to that of the SSRIs, raising the levels of serotonin (Meeusen and De Meirleir 1995; Meeusen *et al.* 2001).

But perhaps the most widely known mechanism for the 'feel-better' effects of physical activity is the 'endorphin hypothesis', popularized through numerous press reports over the years. The popularity of this hypothesis can be attributed to the coincident timing of several events. First, the isolation of endogenous opioids was quickly followed by the discovery that the level of these opioid substances is raised following vigorous exercise. Then, this happened to coincide with the developing exercise craze during the second half of the 1970s and anecdotal reports of what became known as the 'runner's high' phenomenon. Eager journalists (and some overzealous scientists) hastily made the connection that the natural opioids must be responsible for the 'high'. From a research standpoint, this connection has proven infinitely more difficult to make and the literature has by no means yielded an unequivocal answer (Hoffmann 1997). What seems clear is that peripherally circulating opioids (beta-endorphin from the pituitary and beta-enkephalin from the chromaffin cells of the adrenal medulla) have limited, if any, central effects and do not reflect the dynamics of brain opioid neurotransmission. On the other hand, brain opioids, which can be experimentally manipulated by blocker agents such as naloxone and naltrexone, may have a role in exercise-associated affective responses. Even if their role is not to directly induce pleasure, central opioids have been shown to have an attenuating effect on cardiovascular and respiratory responses to exercise. Through that, they can potentially suppress symptoms associated with perceived exertion (e.g. heart rate, blood pressure, ventilation), which is, in turn, closely linked to affect at strenuous levels of exercise intensity. Furthermore, descending opioidergic neurons, originating primarily in the periaqueductal grey and activated by high levels of stress, can regulate the flow of bodily sensory cues that ascend the spinal cord and enter the brain.

Finally, a recently proposed hypothesis is aimed at explaining the changing affective responses to increasing levels of physical activity intensity (Ekkekakis 2003). According to this hypothesis, affective responses to physical activity are shaped by evolution and are the product of the continuous interaction of social–cognitive factors (such as physical self-efficacy) and interoceptive factors (such as respiratory and muscular cues). The relative influence of these two factors on affect is hypothesized to change systematically as a function of the intensity of the activity, with the former being dominant at low and mid-range intensities and the latter gaining dominance at high and near-maximal intensities. This hypothesis also has some implications for interventions aimed at controlling some unpleasant responses to physical activity, particularly among beginner exercisers. According to the hypothesis, cognitive techniques, such as attentional dissociation, cognitive reframing, or boosting self-efficacy, can only be expected to be effective when the intensity of the activity presents an appreciable but not yet overwhelming challenge but not when it reaches high or near-maximal levels.

Expanding the focus: the psychosomatic benefits

Although the vast majority of research on the health benefits of physical activity has been conducted along either *physiological* or *psychological* lines, some studies have taken a more interdisciplinary approach by examining health and well-being from a psychosomatic perspective. The influence of psychosocial factors in the pathogenesis of the two leading causes of death in Western societies, namely, coronary heart disease (Krantz and McCeney 2002; Rozanski *et al.* 1999) and cancer (Kiecolt-Glaser and Glaser 1999), is fairly well established. It is also well established that the two psychoneuroendocrine systems, namely, the sympathetic–adrenomedullary (SAM) axis and the hypothalamic–pituitary–adrenocortical (HPA) axis, are the main mediators in the relationship between psychosocial stress and pathogenesis (Chrousos 1998; Chrousos and Gold 1998; Tsigos and Chrousos 2002).

Unlike pharmacological and psychotherapeutic interventions, physical activity appears to offer the unique advantage of being able to produce beneficial changes in both the psychosocial variables (by reducing depression and anxiety and increasing perceived control) and the neuroendocrine variables (by producing a more adaptive pattern of hormonal responses to stressors). Therefore, two lines of research have emerged, one focusing on physical activity-induced changes in neuroendocrine and cardiovascular responses to psychosocial stressors (Sothmann *et al.* 1996) and one focusing on physical activity-induced changes in the relationship between psychosocial stress and immune function (Hong 2000; LaPerriere *et al.* 1994; Perna *et al.* 1997). The starting point for both of these lines of research is the numerous findings in the last 15 years that physical activity offers significant protection from mortality associated with cardiovascular disease and various types of cancer. The question is whether part of this beneficial effect is due to physical activity-induced changes in how stress impacts the pathogenesis and progression of these diseases.

Conceptually, based on studies focusing on neuroendocrine responses to exercise stimuli among trained and untrained individuals, physical activity and fitness should be associated with an adaptive overall pattern of responses to stressors. Such a pattern would consist of an attenuated elevation of catecholamine levels (the end-products of the SAM axis) and a rapid return to baseline, reduced basal levels and stress-induced responses of cortisol (the end-product of the HPA axis), and an enhanced production of endogenous opioid peptides. These changes, if they manifested themselves in response to daily psychosocial stressors, could help physically active and fit individuals buffer the harmful effects of stress (Jonsdottir 2000;

LaPerriere *et al.* 1994; Perna *et al.* 1997). Yet, research has once again shown that the complexity of these relationships is far greater than one might have expected. For example, cortisol does not always have an immunosuppressive effect (McEwen *et al.* 1997). Opioids do not always have an immunoenhancing effect (Risdahl *et al.* 1998). And physical training does not necessarily lead to an attenuated cortisol (Chennaoui *et al.* 2002; Droste *et al.* 2003) or catecholamine (Peronnet and Szabo 1993) response to psychosocial stressors. Nevertheless, the hypothesis that part of the effectiveness of physical activity for reducing mortality from cardiovascular disease and cancer is due to its effects on stress mechanisms remains viable and certainly warrants additional research. A prominent example is a series of studies showing that physical activity reduces anxiety and depression, improves quality of life, and enhances immunocompetence in individuals suffering from HIV (LaPerriere *et al.* 1990; Rojas *et al.* 2003; Stringer *et al.* 1998).

Physical activity: why we do or don't

A fundamental question concerning many health behaviours—and physical activity is no different—is why people do or do not participate. If physical activity is so 'good for you', yet a clear majority of the adult population fail to meet current guidelines of healthy physical activity,[1] it is important to understand the key correlates of an active lifestyle.

Key determinants/correlates

Researchers interested in factors associated with physical activity have typically categorized such 'correlates' or 'determinants' into personal or demographic, psychological, social, and environmental factors (Sallis *et al.* 2000; Trost *et al.* 2002).

Personal/demographic correlates

There are consistent positive trends for leisure-time physical activity in adults to be associated with male gender and higher levels of education and socio-economic status, but negatively associated with non-White ethnicity and age (Trost *et al.* 2002), with similar trends in youth. Such gender differences are highly reproducible and one of the most consistent findings in the literature. Promoting physical activity in girls seems a particular challenge, although trials with adults suggest that more women than men show interest in taking part (Mutrie *et al.* 2002).

Psychological correlates

Psychological correlates of physical activity have been studied quite extensively. There are two main types of studies: those using descriptive approaches whereby

---

[1] Typically, guidelines for adults are to participate in 30 minutes of moderate intensity physical activity on most (5) days of the week, that is 150 min/week (Department of Health 2004; Pate *et al.* 1995).

psychological variables are assessed alongside physical activity and those that use a theoretical model. The latter enable us to build knowledge and understanding of how and why people might be motivated or not ('amotivated') to adopt and/or maintain a physically active lifestyle. Descriptive studies can be helpful in developing more explanatory research designs.

One intuitively obvious motive is enjoyment. Those who are active tend to report higher levels of enjoyment than those who participate less. However, this may mask a number of issues. First, enjoyment can cover many things. Some may report enjoyment because of the social aspects of participating, others for reasons of positive well-being. In addition, research suggests that people exercise less for intrinsic fun, but more for the satisfaction in meeting valued goals, such as weight control, feeling better, or social connectedness (Chatzisarantis *et al.* 2003).

The development of exercise psychology as a thriving research field has led to the proliferation of theories borrowed from other areas of psychology (Biddle and Mutrie 2001). In particular, theories tested in social and health psychology have been utilized. To help make sense of the different approaches, it is useful to view theories as falling into four categories. There are theories focused on: (1) beliefs and attitudes; (2) perceptions of competence; (3), perceptions of control; and (4) decision-making processes. Although these divisions are not always clear-cut, they may help readers better organize the field (Biddle and Nigg 2000).

Belief/attitude theories test the links between beliefs, attitudes, intentions, and physical activity, such as the theory of planned behaviour. Evidence shows that intentions are predicted best by attitudes and perceived behavioural control, and rather less so by subjective (social) norms (Hagger *et al.* 2002). However, research tends to show that intentions are far from perfect predictors of behaviour, and one could argue that greater emphasis is needed on how to translate intentions into behaviour. Competence-based theories focus on perceptions of competence and confidence as a prime driver of behaviour, such as in self-efficacy approaches (Bandura 1997; McAuley and Blissmer 2000). Early attempts in exercise psychology favoured theories of perceived control, such as locus of control (Rotter 1966). These yielded small effects or were inadequately tested so researchers searched for other control-related constructs (Biddle 1999; Biddle and Mutrie 2001). One that has been popular is the self-determination theory advocated by Deci, Ryan, and colleagues (Deci and Flaste 1995; Deci and Ryan 1985; Ryan and Deci 2000*a, b*; Williams *et al.* 2000) and applied to physical activity by others (Chatzisarantis *et al.* 2003). Research shows that motivation for physical activity is likely to be more robust if it involves greater choice and self-determination rather than external control. In addition, such an approach is likely to lead to feelings of higher well-being.

Finally, decision-making theories have recently been favoured, and the transtheoretical model (TTM) of behaviour change has grown in popularity (Marshall and Biddle 2001). The TTM was developed as a comprehensive model of behaviour change and was initially applied to smoking cessation (Prochaska and DiClemente 1982). It incorporates cognitive, behavioural, and temporal aspects of changing behaviour. The TTM applied to physical activity consists of the stages of change, the processes of change, decisional balance (weighing up the pros and cons of change), and self-efficacy. The stage of change is the time dimension along which behaviour change occurs. The stages are:

- precontemplation: no intention to start physical activity on a regular basis;
- contemplation: intending to start physical activity on a regular basis, usually within the next 6 months;
- preparation: immediate intention (within the next 30 days) and commitment to change (sometimes along with small behavioural changes, such as obtaining membership at a fitness club);
- action: engaging in regular physical activity but for less than 6 months;
- maintenance: engaging in regular physical activity for some time (more than 6 months).

The processes of change are the strategies used to progress along the stages of change. The processes are divided into cognitive (thinking) and behavioural strategies. For example, people might seek information on physical activity and mood enhancement (cognitive strategy) or post a note on the refrigerator door to remind them to walk that day (behavioural strategy). We found that both types of strategies tend to be used throughout the change cycle (Marshall and Biddle 2001), probably due to individual preferences.

Social and environmental correlates

Social support appears to be associated with physical activity in adults and youth. Trost *et al.*'s (2002) review suggested that social support from friends/peers and family/spouse was particularly important. In addition, the influence of one's GP (family physician) plays a role, particularly for adults, as may the teacher or trainer in group exercise sessions. The 'motivational climate' created by such a leader may be vital in determining whether people return for future sessions (Ntoumanis and Biddle 1999). Evidence suggests that the most positive climate will be when the exercise leader encourages cooperation and rewards effort over comparative performance.

The influence of school settings and local media may also help or hinder activity. For example, physical activity may be more likely in a school that encourages staff and pupils to be active and has a 'healthy school' policy and positive adult role models, although many of these likely influences lack robust supporting data. In addition, it is not clear to what extent the social setting influences positive mood during or after exercise.

Environmental correlates of physical activity have only been studied quite recently. Trost *et al.*'s (2002) review showed that 11 new environmental variables had been studied since a review of correlates published in 1999. There was weak or mixed support for an association of most variables with physical activity. A recent review of environmental correlates of walking in adults (Owen *et al.* 2004) found 18 studies. From these, it was concluded that walking was associated with aesthetic attributes; convenience of facilities, such as trails; accessibility of destinations, such as shops; and perceptions of traffic and busy roads. However, the authors concluded that the current evidence is 'promising, although at this stage limited' (p. 75). It is likely that facilities, including open spaces and parks, are only part of a solution to increase physical activity levels. Other factors include previous experiences of physical activity and current level of fitness.

Interventions: what works?

The behavioural epidemiological framework suggests that interventions should follow from the identification of correlates or determinants. One reason is that effective interventions are the result of manipulating and changing the antecedent of behaviour as a precursor to actual behaviour change. This is illustrated in Fig. 6.2. Interventions, however, can occur at several levels, including the individual, group, and community. Interventions at societal or national level can also be identified, such as governmental policy initiatives (Bull *et al.* 2004).

Individual-level interventions typically involve advice-giving, such as by your family doctor (GP), or through counselling for behaviour change. These strategies might be based on the TTM with the aim to 'tailor' strategies and advice to fit the individual's stage of decision-making. For example, it would seem unnecessary to give much educational information about the benefits of physical activity to those looking to move from the action to the maintenance stage, whereas such a strategy

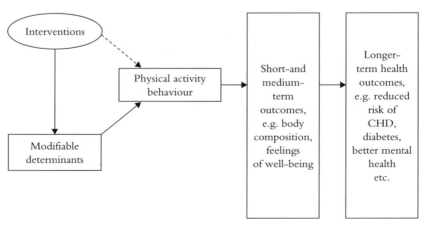

Fig. 6.2 A conceptual framework showing links between an intervention, determinants/correlates, and physical activity behaviours and outcomes. CHD = Coronary heart disease. (Adapted from Kahn *et al.* 2002.)

is likely to be key to moving a pre-contemplator to contemplation. Sometimes it is possible to deliver such types of intervention in small groups.

A popular intervention setting is that of primary health care (PHC). Given the importance attached to advice from health care professionals, and GPs in particular, as well as the regularity with which many visit their PHC facility, it seems an appropriate setting in which to test the effectiveness of individually oriented counselling approaches.

There are lessons to be learned from studies in primary care that have focused on improving health behaviour in general and not just physical activity. For example, Steptoe et al. (1999) reported on a large-scale trial of the use of behaviourally oriented counselling for primary care patients at increased risk of coronary heart disease. The intervention group received counselling from the practice nurse who had been trained in an approach based on the TTM. At both 4 and 12 months, the intervention group, in comparison to the control group, had a favourable reduction in cigarette smoking, reduction in fat intake, and increased physical activity. This study demonstrated that physical activity, and other health behaviours, can be influenced by a counselling approach in the primary care setting. Beneficial changes in physical activity have also been noted in a similar trial in New Zealand (Elley et al. 2003). Physical activity of older adults in Finland has also been shown to be strongly associated with opportunistic advice given by health care professionals (Hirvensalo et al. 2003).

Project PACE (physician-based assessment and counselling for exercise), in the USA, has also used the TTM to design short interventions delivered by GPs (Patrick et al. 1994). The intervention consisted of brief (3–5 minutes) counselling with each patient where pros and cons, self-efficacy, and likely processes (strategies) might be assessed and discussed. The counselling focused on the benefits and barriers to increasing activity, self-efficacy, and gaining social support for increasing activity. The strategies differed depending on the stage of exercise behaviour of each patient; thus the intervention was stage-matched. Physicians themselves find the PACE tools acceptable (Long et al. 1996) and a randomized controlled trial showed that the PACE interventions did increase physical activity, particularly walking (Calfas et al. 1996).

Two major reviews of interventions aimed at PHC or related settings are now available. Riddoch et al. (1998) located 25 papers from the UK. They concluded that 'the majority of studies report some form of improvement in either physical activity or related measures. However, the size of the effect is generally small, and there is no real consistency across studies' (p. 25).

Although not restricted to primary health care, Simons–Morton et al. (1998), in their review of interventions in health care settings, concluded that 'interventions in health care settings can increase physical activity for both primary and secondary prevention. Long-term effects are more likely with continuing intervention and multiple intervention components such as supervised exercise, provision of equipment, and behavioral approaches' (p. 413).

Kahn *et al.* (2002) also reviewed what they called 'individually adapted' programmes. The interventions were often delivered by people through outlets such as phone or mail. Reviewing 18 such studies, they concluded that there was strong evidence for effectiveness and good applicability across diverse settings and populations. They located some evidence for economic effectiveness but warned that successful interventions of this type require careful planning, well-trained staff, and adequate resources.

Perhaps somewhat surprisingly given their profile in the media, personal trainers have been poorly studied as a motivational tool. This may be due to their probably small influence at the level of public health. Few can afford such intense interventions and they remain a marginal strategy for population health gains.

Social–environmental interventions

Ecological models of health suggest that behaviour is determined by multiple influences, including individual, social, and environmental factors (Sallis and Owen 2002). To date, most interventions target the individual or social (e.g. community walking groups) level. However, more recently, interest has grown in the potential of effective interventions at the level of the environment (Owen *et al.* 2000). Sallis *et al.* (1998) suggest that environmental interventions might include those aimed at both the natural and constructed environment. The former might include providing additional indoor facilities for physical activity in places where the weather may preclude or inhibit participation outside, or the lighting of walking or ski paths in the winter. Constructed environment factors might include interventions aimed at the transport infrastructure, suburban environments for walking, the workplace, or specialized facilities.

One of the most successful environmental interventions is the placing of signs in public places to encourage stair climbing (Kahn *et al.* 2002). One example is the study by Blamey *et al.* (1995) in which they aimed to discover if Scottish commuters would respond to motivational signs encouraging them to 'Stay healthy, save time, use the stairs'. The signs were placed in a Glasgow city centre underground station where stairs (30 steps) and escalators were adjacent. Eight observation weeks were split into four stages: a 1-week baseline; a 3-week period when the sign was present; a 2-week period immediately after the sign was removed; and two 1-week follow-ups, during the fourth and twelfth weeks after the intervention. Observers recorded the number of adults using the escalators and stairs and categorized them by gender. Those carrying luggage or with pushchairs were excluded. A comparison was made between the baseline week stair use and each of the seven subsequent observation weeks.

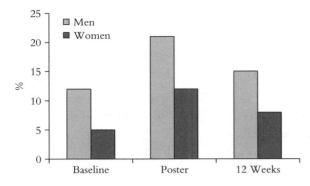

Fig. 6.3 Stair climbing before (baseline), during (poster), and after (12 weeks) a poster intervention encouraging stair use in favour of escalator use (data from Blamey *et al.* 1995).

Stair use during the 1-week baseline period was around 8%. This increased to the order of 15–17% during the 3 weeks that the sign was present (Fig. 6.3). Stair use significantly increased after the signs were in place and continued to increase during the three intervention weeks. A sudden decrease in stair use occurred once the sign was removed. At the 12-week follow-up, stair use remained significantly higher than at baseline. There is, however, an obvious downward trend suggesting a possible eventual return to baseline levels. The results do show, however, that a motivational sign positively influenced stair use, and similar studies have shown supportive findings (Kerr *et al.* 2001).

Future directions for research

Great progress has been made in the past few years in the field of exercise psychology. However, inevitably there remain a number of significant research issues that require attention. These include the following.

1 Larger and better designed trials are required to investigate the effects of physical activity on various indices of psychological well-being.

2 There must be continued efforts to improve on the methodologies adopted for such trials, including controlling for expectancy effects and other likely influencing factors.

3 Further efforts are required to delineate likely mechanisms of the 'feel good effect' of physical activity and integrating psychological with somatic-based mechanisms.

4 More evidence is required on the effects of physical activity on the psychological well-being of children and adolescents.

5 More needs to be known about whether there is a dose–response relationship between physical activity and psychological well-being. For example, the relationship between intensity of exercise and psychological effects needs further

exploration, and other characteristics of exercise, such as type, frequency, and duration, are also understudied.

6 Research needs to advance understanding of how physical activity-induced changes can enhance our ability to combat harmful effects of psychosocial stress.

7 Research must continue to study the diverse individual differences reported in psychological responses to exercise.

8 Aspects of well-being, such as health-related quality of life, happiness, and optimism, have been understudied by exercise psychologists.

9 Exercise psychology has been dominated by research concerning correlates of participation and needs to devote more energies to the study of interventions concerning participation.

10 Less is known about environmental correlates of physical activity and how they might interact with other potentially influencing factors, such as socio-economic status or physical fitness.

11 Interventions need wider evaluation, such as in respect to cost-effectiveness.

Summary and conclusions

Physical activity is a significant health behaviour requiring serious research and public policy attention. Lack of physical activity has major public health consequences with huge personal, societal, and economic costs (McGinnis 1992). However, another focus is to show that physical activity has many health benefits— in other words, it is health-*enhancing* as well as disease-preventing. In a book on well-being it is important to recognize this. As a result, in this chapter, we have outlined the important psychological consequences of physical activity, what factors might be associated with physical activity participation, and how physical activity levels can be changed. Emphasis needs to continue to be placed on the importance of physically active lifestyles for the health and well-being of society.

References

Acevedo, E.O., Kraemer, R.R., Haltom, R.W., and Trynjecki, J.L. (2003). Perceptual responses proximal to the onset of blood lactate accumulation. *J. Sports Med. Phys. Fitness* **43**, 267–73.

Arent, S.M., Landers, D.M., and Etnier, J.L. (2000). The effects of exercise on mood in older adults: a meta-analytic review. *J. Aging Phys. Activity* **8**, 407–30.

Babyak, M., Blumenthal, J.A., Herman, S., Khatri, P., Doraiswamy, M., Moore, K., *et al.* (2000). Exercise treatment for major depression: maintenance of therapeutic benefit at 10 months. *Psychosom. Med.* **62**, 633–8.

Bahrke, M.S. and Morgan, W.P. (1978). Anxiety reduction following exercise and meditation. *Cogn. Ther. Res.* **2**, 323–33.

Bandura, A. (1997). *Self-efficacy: the exercise of control.* W.H. Freeman, New York.

Berger, B.G. (2004). Subjective well-being in obese individuals: the multiple roles of exercise. *Quest* **56**, 50–76.

Berger, B.G. and Motl, R. (2001). Physical activity and quality of life. In *Handbook of sport psychology*, 2nd edn (ed. R.N. Singer, H.A. Hausenblas, and C.M. Janelle), pp. 636–71. John Wiley, New York.

Biddle, S.J.H. (1999). Motivation and perceptions of control: tracing its development and plotting its future in exercise and sport psychology. *J. Sport Exerc. Psychol.* **21**, 1–23.

Biddle, S.J.H. (2000). Emotion, mood and physical activity. In *Physical activity and psychological well-being* (ed. S.J.H. Biddle, K.R. Fox, and S.H. Boutcher), pp. 63–87. Routledge, London.

Biddle, S.J.H. and Mutrie, N. (2001). *Psychology of physical activity: determinants, well-being and interventions*. Routledge, London.

Biddle, S.J.H. and Nigg, C.R. (2000). Theories of exercise behavior. *Int. J. Sport Psychol.* **31**, 290–304.

Biddle, S.J. H., Fox, K.R., Boutcher, S.H., and Faulkner, G.E. (2000). The way forward for physical activity and the promotion of psychological well-being. In *Physical activity and psychological well-being* (ed. S.J.H. Biddle, K.R. Fox, and S.H. Boutcher), pp. 154–68. Routledge, London.

Bixby, W.R., Spalding, T.W., and Hatfield, B.D. (2001). Temporal dynamics and dimensional specificity of the affective response to exercise of varying intensity: differing pathways to a common outcome. *J. Sport Exerc. Psychol.* **23**, 171–90.

Blamey, A., Mutrie, N., and Aitchison, T. (1995). Health promotion by encouraged use of stairs. *Br. Med. J.* **311**, 289–90.

Blumenthal, J.A., Babyak, M.A., Moore, K.A., Craighead, W.E., Herman, S., Khatri, P., *et al.* (1999). Effects of exercise training on older patients with major depression. *Arch. Intern. Med.* **159**, 2349–56.

Bouchard, C. (2000). Introduction. In *Physical activity and obesity* (ed. C. Bouchard), pp. 3–19. Human Kinetics, Champaign, Illinois.

Bouchard, C. and Blair, S.N. (1999). Introductory comments for the consensus on physical activity and obesity. *Med. Sci. Sports Exerc.* **31**(11, Suppl.), S498–S501.

Bouchard, C., Shephard, R.J., and Stephens, T. (Eds.) (1994). *Physical activity fitness and health: International proceedings and consensus statement*. Human Kinetics, Champaign, Illinois.

Brosse, A.L., Sheets, E.S., Lett, H.S., and Blumenthal, J.A. (2002). Exercise and the treatment of clinical depression in adults. Recent findings and future directions. *Sports Med.* **32**, 741–60.

Bull, F.C., Bellew, B., Schoppe, S., and Bauman, A.E. (2004). Developments in national physical activity policy: an international review and recommendations towards better practice. *J. Sci. Med. Sport* **7** (1, Suppl.), 93–104.

Calfas, K.J., Long, B., Sallis, J., Wooten, W., Pratt, M., and Patrick, K. (1996). A controlled trial of physician counseling to promote the adoption of physical activity. *Prev. Med.* **25**, 225–33.

Caspersen, C.J., Powell, K.E., and Christenson, G.M. (1985). Physical activity, exercise and physical fitness: definitions and distinctions for health-related research. *Public Health Rep.* **100**, 126–31.

Chatzisarantis, N.L.D., Hagger, M.S., Biddle, S.J.H., Smith, B., and Wang, C.K. J. (2003). A meta-analysis of perceived locus of causality in exercise, sport, and physical education contexts. *J. Sport Exerc. Psychol.* **25**, 284–306.

Chennaoui, M., Gomez Merino, D., Lesage, J., Drogou, C., and Guezennec, C.Y. (2002). Effects of moderate and intensive training on the hypothalamo–pituitary–adrenal axis in rats. *Acta Physiol. Scand.* **175**, 113–21.

Chrousos, G.P. (1998). Stressors, stress, and neuroendocrine integration of the adaptive response: The 1997 Hans Selye memorial lecture. *Ann. NY Acad. Sci.* **851**, 311–35.

Chrousos, G.P. and Gold, P.W. (1998). A healthy body in a healthy mind—and vice versa: the damaging power of "uncontrollable" stress. *J. Clin. Endocrinol. Metab.* **83**, 1842–5.

Colcombe, S. and Kramer, A.F. (2003). Fitness effects on the cognitive function of older adults: a meta-analytic study. *Psychol. Sci.* **14** (2), 125–30.

Craft, L.L. and Landers, D.M. (1998). The effect of exercise on clinical depression and depression resulting from mental illness: a meta-analysis. *J. Sport Exerc. Psychol.* **20**, 339–57.

Deci, E.L. and Flaste, R. (1995). *Why we do what we do: understanding self-motivation.* Penguin, New York.

Deci, E.L. and Ryan, R.M. (1985). *Intrinsic motivation and self-determination in human behavior.* Plenum Press, New York.

Department of Health. (2004). *At least five a week: evidence on the impact of physical activity and its relationship to health: a report from the Chief Medical Officer.* London.

Department of Health and Human Services and Centers for Disease Control and Prevention. (1996). *Physical activity and health: a report of the Surgeon General.* Atlanta, Georgia.

Desharnais, R., Jobin, J., Cote, C., Levesque, L., and Godin, G. (1993). Aerobic exercise and the placebo effect: a controlled study. *Psychosom. Med.* **55**, 149–54.

Dishman, R.K. and Jackson, E.M. (2000). Exercise, fitness, and stress. *Int. J. Sport Psychol.* **31**, 175–203.

Dishman, R.K., Washburn, R.A., and Heath, G.W. (2004). *Physical activity epidemiology.* Human Kinetics, Champaign, Illinois.

Droste, S.K., Gesing, A., Ulbricht, S., Müller, M.B., Linthorst, A.C. E., and Reul, M.H. M. (2003). Effects of long-term voluntary exercise on the mouse hypothalamic–pituitary–adrenocortical axis. *Endocrinology* **144**, 3012–23.

Dunn, A.L., Trivedi, M.H., and O'Neal, H.A. (2001). Physical activity dose–response effects on outcomes of depression and anxiety. *Med. Sci. Sports Exerc.* **33**(6, Suppl.), S587–S597.

Ekkekakis, P. (2003). Pleasure and displeasure from the body: perspectives from exercise. *Cogn. Emotion* **17**, 213–39.

Ekkekakis, P. and Petruzzello, S.J. (1999). Acute aerobic exercise and affect: current status, problems and prospects regarding dose–response. *Sports Med.* **28**, 337–74.

Ekkekakis, P. and Petruzzello, S.J. (2000). Analysis of the affect measurement conundrum in exercise psychology: I. Fundamental issues. *Psychol. Sport Exerc.* **1**, 71–88.

Ekkekakis, P. and Petruzzello, S.J. (2002). Analysis of the affect measurement conundrum in exercise psychology: IV. A conceptual case for the affect circumplex. *Psychol. Sport Exerc.* **3**, 35–63.

Ekkekakis, P., Hall, H.E., Van Landuyt, L.M., and Petruzzello, S.J. (2000). Walking in (affective) circles: can short walks enhance affect? *J. Behav. Med.* **23**, 245–75.

Ekkekakis, P., Hall, E.E., and Petruzzello, S.J. (2004). Practical markers of the transition from aerobic to anaerobic metabolism during exercise: rationale and a case for affect-based exercise prescription. *Prev. Med.* **38**, 149–59.

Elley, C.R., Kerse, N., Arroll, B., and Robinson, E. (2003). Effectiveness of counselling patients on physical activity in general practice: cluster randomised controlled trial. *Br. Med. J.* **326**, 793–8.

Etnier, J.L., Salazar, W., Landers, D.M., Petruzzello, S.J., Han, M., and Nowell, P. (1997). The influence of physical fitness and exercise upon cognitive functioning: a meta-analysis. *J. Sport Exerc. Psychol.* **19**, 249–77.

Focht, B.C. and Hausenblas, H.A. (2003). State anxiety responses to acute exercise in women with high social physique anxiety. *J. Sport Exerc. Psychol.* **25**, 123–44.

Fox, K.R. (1999). The influence of physical activity on mental well-being. *Public Health Nutr.* **2**, 411–18.

Fox, K.R. (2000a). The effects of exercise on self-perceptions and self-esteem. In *Physical activity and psychological well-being* (ed. S.J.H. Biddle, K.R. Fox, and S.H. Boutcher), pp. 88–117. Routledge, London.

Fox, K.R. (2000b). Self-esteem, self-perceptions and exercise. *Int. J. Sport Psychol.* **31**, 228–40.

Hagger, M.S., Chatzisarantis, N.L.D., and Biddle, S.J.H. (2002). A meta-analytic review of the theories of reasoned action and planned behaviour in physical activity: predictive validity and the contribution of additional variables. *J. Sport Exerc. Psychol.* **24**, 3–32.

Hall, E.E., Ekkekakis, P., and Petruzzello, S.J. (2002). The affective beneficence of vigorous exercise revisited. *Br. J. Health Psychol.* **7**, 47–66.

Hirvensalo, M., Heikkinen, E., Lintunen, T., and Rantanen, T. (2003). The effect of advice by health care professionals on increasing physical activity of older people. *Scand. J. Med. Sci. Sports* **13**, 231–6.

Hoffmann, P. (1997). The endorphin hypothesis. In *Physical activity and mental health* (ed. W.P. Morgan), pp. 163–77. Taylor and Francis, Washington, DC.

Hong, S. (2000). Exercise and psychoneuroimmunology. *Int. J. Sports Med.* **31**, 204–27.

Hughes, J.R. (1984). Psychological effects of habitual aerobic exercise: a critical review. *Prev. Med.* **13**, 66–78.

Jonsdottir, I.H. (2000). Neuropeptides and their interaction with exercise and immune function. *Immunol. Cell Biol.* **78**, 562–70.

Kahn, E.B., Ramsey, L.T., Brownson, R.C., Heath, G.W., Howze, E.H., Powell, K.E., et al. (2002). The effectiveness of interventions to increase physical activity: a systematic review. *Am. J. Prev. Med.* **22** (4S), 73–107.

Kerr, J., Eves, F., and Carroll, D. (2001). Six-month observational study of prompted stair climbing. *Prev. Med.* **33**, 422–7.

Kiecolt-Glaser, J.K. and Glaser, R. (1999). Psychoneuroimmunology and cancer: fact or fiction? *Eur. J. Cancer* **35**, 1603–7.

Krantz, D.S. and McCeney, M.K. (2002). Effects of psychological and social factors on organic disease: critical assessment of research on coronary heart disease. *Ann. Rev. Psychol.* **53**, 341–69.

Kubitz, K.A., Landers, D.M., Petruzzello, S.J., and Han, M. (1996). The effects of acute and chronic exercise on sleep: a meta-analytic review. *Sports Med.* **21**, 277–91.

Landers, D.M. and Arent, S.M. (2001). Physical activity and mental health. In *Handbook of sport psychology*, 2nd edn (ed. R.N. Singer, H.A. Hausenblas and C.M. Janelle), pp. 740–65. John Wiley, New York.

Landers, D.M. and Petruzzello, S.J. (1994). Physical activity, fitness and anxiety. In *Physical activity fitness and health* (ed. C. Bouchard, R.J. Shephard and T. Stephens), pp. 868–82. Human Kinetics, Champaign, Illinois.

LaPerriere, A.R., Antoni, M.H., Schneiderman, N., Ironson, G., Klimas, N., Caralis, P., et al. (1990). Exercise intervention attenuates emotional distress and natural killer cell decrements following notification of positive serological status for HIV-1. *Biofeedback Self-regul.* **15**, 229–42.

LaPerriere, A.R., Ironson, G., Antoni, M.H., Schneiderman, N., Klimas, N., and Fletcher, M.A. (1994). Exercise and psychoneuroimmunology. *Med. Sci. Sports Exerc.* **26**, 182–90.

Lawlor, D.A. and Hopker, S.W. (2001). The effectiveness of exercise as an intervention in the management of depression: Systematic review and meta-regression analysis of randomised controlled trials. *Br. Med. J.* **322**(7289), <http://www.bmj.com/cgi/content/full/322/7289/7763>.

Long, B., Calfas, K.J., Wooten, W., Sallis, J.F., Patrick, K., Goldstein, M., *et al.* (1996). A multisite field test of the acceptibility of physical activity counseling in primary care: Project PACE. *Am. J. Prev. Med.* **12 (2)**, 73–81.

Marshall, S.J. and Biddle, S.J.H. (2001). The transtheoretical model of behavior change: a meta-analysis of applications to physical activity and exercise. *Ann. Behav. Med.* **23**, 229–46.

McAuley, E. (1991). Efficacy, attributional, and affective responses to exercise participation. *J. Sport Exerc. Psychol.* **13**, 382–93.

McAuley, E. and Blissmer, B. (2000). Self-efficacy determinants and consequences of physical activity. *Exerc. Sport Sci. Rev.* **28**, 85–8.

McAuley, E. and Katula, J. (1998). Physical activity interventions in the elderly: influence on physical health and psychological function. *Ann. Rev. Gerontol. Geriatr.* **18**, 111–54.

McEwen, B.S., Biron, C.A., Brunson, K.W., Bulloch, K., Chambers, W.H., Dhabhar, F.S., *et al.* (1997). The role of adrenocorticoids as modulators of immune function in health and disease: Neural, endocrine and immune interactions. *Brain Res. Rev.* **23**, 79–133.

McGinnis, J.M. (1992). The public health burden of a sedentary lifestyle. *Med. Sci. Sports Exerc.* **24** (6, Suppl.), S196–S200.

Meeusen, R. and De Meirleir, K. (1995). Exercise and brain neurotransmission. *Sports Med.* **20**, 160–88.

Meeusen, R., Piacentini, M.F., and De Meirleir, K. (2001). Brain microdialysis in exercise research. *Sports Med.* **31**, 965–83.

Morgan, W.P. (1981). Psychological benefits of physical activity. In *Exercise in health and disease* (ed. F.J. Nagle and H.J. Montoye), pp. 299–314. Charles C. Thomas, Springfield, Illinois.

Morgan, W.P. (1997). Methodological considerations. In *Physical activity and mental health* (ed. W.P. Morgan), pp. 3–32. Taylor and Francis, Washington, DC.

Morris, J.N. (1994). Exercise in the prevention of coronary heart disease: today's best buy in public health. *Med. Sci. Sports Exerc.* **26**, 807–14.

Mutrie, N. (2000). The relationship between physical activity and clinically defined depression. In *Physical activity and psychological well-being* (ed. S.J. H. Biddle, K.R. Fox, and S.H. Boutcher), pp. 46–62. Routledge, London.

Mutrie, N., Carney, C., Blamey, A., Crawford, F., Aitchison, T., and Whitelaw, A. (2002). "Walk in to work out": a randomised controlled trial of self help intervention to promote active commuting. *J. Epidemiol. Community Health* **56**, 407–12.

Ntoumanis, N. and Biddle, S. (1999). A review of motivational climate in physical activity. *J. Sports Sci.* **17**, 643–65.

O'Connor, P.J., Raglin, J.S., and Martinsen, E.W. (2000). Physical activity, anxiety, and anxiety disorders. *Int. J. Sport Psychol.* **31**, 136–55.

O'Neal, H.A., Dunn, A.L., and Martinsen, E.W. (2000). Depression and exercise. *Int. J. Sport Psychol.* **31**, 110–35.

Owen, N., Leslie, E., Salmon, J., and Fotheringham, M.J. (2000). Environmental determinants of physical activity and sedentary behavior. *Exerc. Sport Sci. Rev.* **28**, 153–158.

Owen, N., Humpel, N., Leslie, E., Bauman, A., and Sallis, J.F. (2004). Understanding environmental influences on walking: review and research agenda. *Am. J. Prev. Med.* **27** (1), 67–76.

Pate, R.R., Pratt, M., Blair, S.N., Haskel, W.L., Macera, C.A., Bouchard, C., *et al.* (1995). Physical activity and public health: a recommendation from the Centers for Disease Control and Prevention and the American College of Sports Med. *J. Am. Med. Assoc.* **273**, 402–7.

Patrick, K., Sallis, J.F., Long, B.J., Calfas, K.J., Wooten, W.J., and Heath, G. (1994). PACE: Physician–based assessment and counseling for exercise, background and development. *Physician Sports Med.* **22**, 245–55.

Perna, F.M., Schneiderman, N., and LaPerriere, A. (1997). Psychological stress, exercise, and immunity. *Int. J. Sports Med.* **18**, S78–S83.

Peronnet, F. and Szabo, A. (1993). Sympathetic response to psychosocial stressors in humans: linkage to physical exercise and training. In *Exercise psychology: the influence of physical exercise on psychological processes* (ed. P. Seraganian), pp. 172–217. John Wiley, New York.

Petruzzello, S.J., Landers, D.M., Hatfield, B.D., Kubitz, K.A., and Salazar, W. (1991). A meta-analysis on the anxiety-reducing effects of acute and chronic exercise: outcomes and mechanisms. *Sports Med.* **11**, 143–82.

Prochaska, J.O. and DiClemente, C.C. (1982). Transtheoretical therapy: toward a more integrative model of change. *Psychother: Theory Res. Pract.* **19**, 276–88.

Rejeski, W.J. and Brawley, L.R. (1988). Defining the boundaries of sport psychology. *Sport Psychologist* **2**, 231–42.

Rejeski, W.J. and Mihalko, S.L. (2001). Physical activity and quality of life in older adults. *J. Gerontol.* **56A** (2 Special), 23–35.

Rejeski, W.J., Brawley, L.R., and Shumaker, S.A. (1996). Physical activity and health-related quality of life. *Exerc. Sport Sci. Rev.* **24**, 71–108.

Riddoch, C., Puig-Ribera, A., and Cooper, A. (1998). *Effectiveness of physical activity promotion schemes in primary care: a review.* Health Education Authority, London.

Risdahl, J.M., Khanna, K.V., Peterson, P.K., and Molitor, T.W. (1998). Opiates and infection. *J. Neuroimmunol.* **83**, 4–18.

Rojas, R., Schlicht, W., and Hautzinger, M. (2003). Effects of exercise training on quality of life, psychological well-being, immune status, and cardiopulmonary fitness in an HIV-1 positive population. *J. Sport Exerc. Psychol.* **25**, 440–55.

Rotter, J.B. (1966). Generalised expectancies for internal versus external control of reinforcement. *Psychol. Monogr.* **80** (whole no. 609), 1–28.

Rozanski, A., Blumenthal, J.A., and Kaplan, J. (1999). Impact of psychological factors on the pathogenesis of cardiovascular disease and implications for therapy. *Circulation* **99**, 2192–217.

Russell, J.A. (2003). Core affect and the psychological construction of emotion. *Psychol. Rev.* **110**, 145–72.

Ryan, R.M. and Deci, E.L. (2000a). Intrinsic and extrinsic motivations: classic definitions and new directions. *Contemp. Educ. Psychol.* **25**, 54–67.

Ryan, R.M. and Deci, E.L. (2000b). Self-determination theory and the facilitation of intrinsic motivation, social development, and well-being. *Am. Psychol.* **55**, 68–78.

Sallis, J.F. and Owen, N. (1999). *Physical activity and behavioral medicine.* Sage, Thousand Oaks, California.

Sallis, J.F. and Owen, N. (2002). Ecological models of health behavior. In *Health behavior and health education: theory research and practice*, 3rd edn (ed. K. Glanz, B. Rimer, and F. Lewis), pp. 403–24. Jossey-Bass, San Francisco.

Sallis, J.F., Bauman, A., and Pratt, M. (1998). Environmental and policy interventions to promote physical activity. *Am. J. Prev. Med.* **15**, 379–97.

Sallis, J.F., Prochaska, J., and Taylor, W. (2000). A review of correlates of physical activity of children and adolescents. *Med. Sci. Sports Exerc.* **32** (5), 963–75.

Salmon, P. (2001). Effects of physical exercise on anxiety, depression, and sensitivity to stress: a unifying theory. *Clin. Psychol. Rev.* **21**, 33–61.

Simons-Morton, D.G., Calfas, K.J., Oldenburg, B., and Burton, N.W. (1998). Effects of interventions in health care settings on physical activity or cardiorespiratory fitness. *Am. J. Prev. Med.* **15**, 413–30.

Sonstroem, R.J. (1997). Physical activity and self-esteem. In *Physical activity and mental health* (ed. W.P. Morgan), pp. 127–43. Taylor and Francis, Washington, DC.

Sothmann, M.S., Buckworth, J., Claytor, R.P., Cox, R.H., White-Welkley, J.E., and Dishman, R.K. (1996). Exercise training and the cross-stressor adaptation hypothesis. *Exerc. Sport Sci. Rev.* **24**, 267–87.

Steptoe, A., Docherty, S., Rink, E., Kerry, S., Kendrick, T., and Hilton, S. (1999). Behavioural counselling in general practice for the promotion of healthy behaviour amoung adults at increased risk of coronary heart disease: randomised trial. *Br. Med. J.* **319**, 943–8.

Stringer, W.W., Berezovskaya, M., O'Brien, W., Beck, C.K., and Casaburi, R. (1998). The effect of exercise training on aerobic fitness, immune indices, and quality of life in HIV+ patients. *Med. Sci. Sports Exerc.* **30**, 11–16.

Taylor, A.H. (2000). Physical activity, anxiety, and stress. In *Physical activity and psychological well-being* (ed. S.J.H. Biddle, K.R. Fox, and S.H. Boutcher), pp. 10–45. Routledge, London.

Trost, S.G., Owen, N., Bauman, A.E., Sallis, J.F., and Brown, W. (2002). Correlates of adults' participation in physical activity: review and update. *Med. Sci. Sports Exerc.* **34** 1996–2001.

Tsigos, C. and Chrousos, G.P. (2002). Hypothalamic–pituitary–adrenal axis, neuroendocrine factors and stress. *J. Psychosom. Res.* **53**, 865–71.

Tuson, K.M. and Sinyor, D. (1993). On the affective benefits of acute aerobic exercise: taking stock after twenty years of research. In *Exercise psychology: the influence of physical exercise on psychological processes* (ed. P. Seraganian), pp. 80–121. Wiley, New York.

Van Landuyt, L.M., Ekkekakis, P., Hall, E.E., and Petruzzello, S.J. (2000). Throwing the mountains into the lakes: on the perils of nomothetic conceptions of the exercise–affect relationship. *J. Sport Exerc. Psychol.* **22**, 208–34.

Williams, G.C., Frankel, R.M., Campbell, T.L., and Deci, E.L. (2000). Research on relationship-centered care and healthcare outcomes from the Rochester Biopsychosocial Program: a self-determination theory integration. *Fam. Syst. Health* **18**, 79–90.

Yeung, R.R. (1996). The acute effects of exercise on mood state. *J. Psychosom. Res.* **40**, 123–41.

Youngstedt, S.D. (2000). The exercise–sleep mystery. *Int. J. Sport Psychol.* **31**, 241–55.

Youngstedt, S.D., O'Connor, P.J., and Dishman, R.K. (1997). The effects of acute exercise on sleep: a quantitative synthesis. *Sleep* **20**, 203–14.

Bernard Gesch is a senior research scientist in the Department of Physiology, University of Oxford. He is also the director of the research charity, Natural Justice, which investigates causes of antisocial and criminal behaviour.

This chapter is dedicated to the memory of Bishop Hugh Montefiore, inspirational Chairman of Natural Justice.

Chapter 7*

The potential of nutrition to promote physical and behavioural well-being

Bernard Gesch

Introduction

Good nutrition is considered to be seminal to our physical well-being throughout our lives. It is widely accepted as a major factor in the prevention of chronic disease. Since the brain also has to be nourished, there are increasingly good reasons to believe that the health benefits of good nutrition extend into positive effects on our behaviour. Yet it is curious that most of the authoritative reference works on diet and international dietary standards focus on health issues but hardly mention behavioural well-being. It is argued that such standards should be reassessed to take into account brain function and behaviour.

Food is a meeting point of the social and physical worlds. The availability of food, food types, and food choices will interact with a wide range of socio–economic factors. Yet, no species can flourish without food and water. This global necessity highlights vast inequalities. Many of the World's population starve while at the same time we are witnessing an epidemic of obesity. This may be the visible tip of an iceberg and, to extend that analogy further, it is the 90% that goes unseen below the water that is the most dangerous. The impacts of various changes in modern diets such as increased intakes of salt, saturated fats, and refined sugars are already apparent in global health. This chapter will extend the discussion to examine the relationship with *behavioural well-being*.

The nutritional ascent of man

The human brain is exceptionally large for a terrestrial animal and has facilitated an enhanced capacity for adaptation and problem-solving (Crawford *et al.* 1999). Anthropological science has taught us how important nutrition was during the evolution of our brain (e.g. Martin 1983; Blumenschine 1991; Foley and Lee 1991). The brain is a fatty organ and it is believed that the availability of two highly unsaturated fatty acids, arachidonic acid (AA) and docosahexaenoic acid (DHA), from which our central nervous system is primarily constructed is likely to have been a limiting factor in brain size (e.g. Crawford *et al.* 1972; Broadhurst *et al.* 2002) and complexity (Fernstrom 1999). The omega-3 highly unsaturated fatty acids that are most important for brain function, eicosapentaenoic acid (EPA) and DHA, are only found in appreciable quantities in oily fish and seafood, primarily because these creatures consume the algae that synthesize EPA and DHA. The dietary omega-3 precursor of EPA and DHA is alpha-linolenic acid (ALA). It is found in dark green leafy vegetables and some nuts and seeds, but its conversion to DHA in particular may be limited (Burdge and Wootton 2002). AA on the other hand is an omega-6 fatty acid and this group is plentiful in modern diets. Omega-6 fatty acids are typically found in vegetable oils, but AA is found in meat and dairy produce. The fact that the savannah is not a rich environment for DHA has led scientists to reconsider whether *Homo sapiens* really did evolve on the plains (Broadhurst *et al.* 2002; Crawford *et al.* 1999) rather than near water in coastal or lakeside habitats (Stringer 2000; Tobias 2002; Morgan 1990). However, DHA could have come from eating the fresh brains of other terrestrial animals on the savannah. So, either humans used their stone tools to break open the strong skulls of freshly slaughtered beasts, or they could have simply reached down on the shore and popped fresh shellfish into their mouths; for an intelligent species the latter has greater elegance. Certainly, some of the earliest fossil remains of our species have been found on the shoreline of the Cape of Good Hope (Stringer 2000; Rightmire and Deacon 1991) and it is gaining credence in anthropology that it was around coastal and inland shorelines that our brain mass evolved to about twice the size of that of our predecessor, *Homo habilis* (Crawford *et al.* 1999; Broadhurst *et al.* 2002).

The evolution of such a metabolically expensive organ is remarkable because in the adult it consumes around 20% of available energy (Leonard and Robertson 1997) and receives 12% of basal cardiac output to supply it with nutrients (Shepherd *et al.* 1983). A recent article concluded that middle Palaeolithic people (Neanderthals) were less effective at exploiting coastal and lakeside food resources than *Homo sapiens* in the upper Palaeolithic era (Klein *et al.* 2004). This change in exploitation occurred as recently 50 000 years ago and it is argued this may have been the decisive change in our behaviour that facilitated our success over Neanderthals (Klein *et al.* 2004). Furthermore, it has been pointed out that, in coastal hunter–gatherer cultures, women tend to collect shellfish and this may have been true of our ancestors as well (Parkington 1998). A pregnant mother would thus have ready access to large quantities of the nutrients essential for the development of her fetus's brain and central nervous system, which receives 70% of the energy supplied (Cunnane *et al.* 2000). Shellfish are also a rich source of protein and animal studies have demonstrated that lower maternal protein intakes reduce delivery of DHA from the mother to the fetus, which may impair the development and function of the fetal brain (Burdge and Wootton 2002). Klein and colleagues (2004) suggest that the Palaeolithic evidence points to the collection of shellfish as the key source of nutrients rather than deep water fishing as there is a lack of evidence for fishing technology. Certainly, it appears to be the case that shellfish were plentiful in the areas where early fossils of our ancestors have been found (Stynder *et al.* 2001).

Moving into the fast lane

It should be recognized, however, that anthropology is an uncertain science involving the meticulous piecing together of evidence to form a picture, but many of the conclusions about the importance of nutrients on brain development are beginning to be supported by experimental evidence, indicating that nutrients rich in sea food are as relevant to brain development now as they appear to have been to our forebears (Uauy *et al.* 2001). Indeed, shellfish not only supply DHA and EPA but are also particularly rich in trace elements such as zinc, iron, and iodine when compared to deep-water fish. These nutrients are also essential for cognitive development, offering further support to anthropological evidence that these were the food sources that facilitated our brain development. Zinc, for instance, may also turn out to be limiting factor in brain development as intake positively correlates with fetal head circumference (Ward *et al.* 1990) and has been shown experimentally to alter essential fatty acid composition (Wauben *et al.* 1999; Ayala and Brenner 1983, 1987; Dieck *et al.* 2005). Animal studies have shown that that low zinc (Oteiza *et al.* 1990) and also low protein intakes (Bennis-Taleb *et al.* 1999) impair brain development. Fortunately for us, our highly evolved brain appeared to pay off as evidence suggests our ancestors moved north and overwhelmed existing populations of Neanderthals (Klein *et al.* 2004). Indeed, contemporary findings from 10 of 13 randomized trials suggest that significant improvements occur in nonverbal intelligence when participants are given micronutrient supplementation (Benton 2001) but these effects depend on the quality of dietary baselines; broadly the worse the diet, the greater the effect. In a population of rural Thai children where zinc, iron, and iodine deficiencies are prevalent, supplementation of these nutrients treated the deficiency with concomitant improvements in cognitive functioning (Manger *et al.* 2004). Since nutrients appear to positively impact cognitive functioning nowadays, it is feasible that better nutrition raised our ancestors' functioning to a new level.

Food for thought

An underlying question remains as to which is the causal factor. Thus, did our diet actually shape our behaviour, was it the other way round, or, indeed, a little of each? The environment in which a living organism evolved is likely to have been a determinant of which elements became essential for life. The phospholipid membrane structure of the brain, as well as the use of trace elements as enzyme catalysts, is considered to have originated in a marine environment (Broadhurst *et al.* 2002; Nielsen 2000). This is consistent with evidence that is suggestive that a coastal diet provided a suitable nutritional environment to support the physical enlargement of our brains. While nutrition is widely accepted as influencing long-term health, we somehow manage to decouple that relationship from behaviour with the assumption that our behaviour is purely a matter of free will. This is despite the fact that we cannot by any means decouple nutrition from actual brain function.

Nevertheless, Derek Bryce-Smith in his Royal Society of Chemistry Lecture (Bryce-Smith 1986) reviewed evidence that poor nutrition or exposure to lead (a neurotoxin and nutrient antagonist) might influence behaviour, personality, and mentation. He reasoned that, since nutrients support the operation of our senses, an individual couldn't normally sense a lack of nutrients or exposure to neurotoxins ambient in the environment. Hence, Bryce-Smith proposed that there could be potent effects on our behaviour that (unlike alcohol) act without our knowledge, e.g. an association between exposure to lead and violence (Needleman *et al.* 1996; Masters 2000; Stretesky and Lynch 2001). Essential nutrients are involved in protecting the body from the deleterious effects of neurotoxins. Furthermore, if an individual is unwittingly undermined by poor nutrition, those around them are unlikely to know about it either, and would tend to attribute any inappropriate behaviour to deficits in the perpetrator's personality.

Changing dietary baselines

If this scenario is correct, our forebears' diet may have been a recipe for success. Curiously then, there is surprisingly little overlap between estimates of what our ancestor ate and what we eat nowadays: the intakes of the essential omega-3 and omega-6 fatty acids were 1:1 rather than the 1:15 ratio found nowadays, perhaps the most extreme example of dietary change (Eaton and Konner 1995; Eaton and Eaton 2000; Eaton *et al.* 1997): Palaeolithic diets are considered to provide more fibre and less salt; carbohydrates would come chiefly from uncultivated vegetables, fruit, and perhaps honey rather than the grains and refined sugar consumed nowadays. The focus of further discussions will be on these dietary factors.

There are, however, many other factors in addition. We can be sure that our ancestors did not eat a highly processed diet. The Palaeolithic diet would probably have provided more phytochemicals, which we are beginning to appreciate are also important for health (Eaton and Eaton 2000). Palaeolithic diets would, by current standards, be considered organic, additive- and pesticide-free (Curl *et al.* 2003; Egger *et al.* 1985; Schab and Trinh 2004). Our forebears are likely to have expended more energy in physical activity (Eaton and Eaton 2003), and we are well aware that sedentary lifestyles are associated with obesity and undesirable consequences for health (US Department of Health and Human Services 1996; WHO 1998). Diet also interacts with our genes (see Simopoulos and Pavlow 2001).

While we cannot be certain of the Palaeolithic diets and estimates have been revised over the years, it would nevertheless be prudent to test experimentally the relative merits of modern diets with such differing dietary baselines—particularly as it is argued that our genes are attuned to the Palaeolithic diet (Cordain *et al.* 2000). How quickly can we expect our brain to adapt to its new nutritional environment? Perhaps our forebears were so closely engaged in the collection of food that they never took food security for granted. Indeed, it is useful to reflect if it is the hallmark of an intelligent primate that we are only now recognizing the seminal role of the aquatic environment for our well-being, just as concern is growing about the exhaustion of fish stocks through wasteful overexploitation (Clover 2004) and the contamination of fish stocks (Hightower and Moore 2003) through ill-considered treatment of the world's oceans by industrialized nations.

Organized food production

We may be tempted to look back to our hunter–gatherer origins through rose-tinted glasses, but the increased security of food availability from agriculture and industrialization is likely to have facilitated the expansion of our population. If food shaped our behavioural strategies about 50 000 years ago, then it seemed to happen again around 10 000 years ago when we began organizing our society around agricultural food production, which played an important role in shaping our present social structures and higher population densities (Crawford *et al.* 1999). Curiously, despite the likely advantages of better food security, it is argued that this change in food harvesting may have resulted in a decline in life span from that of the hunter–gatherer (Angel 1984). This could well be related to increased population density and the spread of infectious diseases. It is also possible that another element was the change in diet. Comparisons of active wild animals with farmed animals illustrate important changes in body tissue composition. Wild animals contain a higher percentage of protein than farmed animals (15% versus 10%) and a lower percentage of fat (5% versus 30%). Fat provides 9 calories for every gram consumed, so the wild animal would provide only 45 calories as fat to every 270 calories as fat from the intensively reared animal (Crawford 1968). The types of fats also differ (e.g. Ledger, 1968; FAO 1994). In wild animals the ratio of non-essential to essential fatty acids is typically 3 to 1, whereas in modern beef the same ratio is reported to be 50 to 1 (Crawford 1968). Animal studies have also shown that both nutrient restriction and overnourishment during pregnancy suppress placental cell proliferation and vascularity, which has an important impact on fetal development and thereby on neonatal mortality and morbidity (Redmer *et al.* 2004). Indeed, estimates of life span based on that of modern hunter–gatherers are around 40 years of age, whereas life expectancy from the onset of agriculture was estimated to be relatively stable at around 20–25 years until the onset of industrial-ization in about 1800 (Eaton *et al.* 2002*a, b*).

The dawn of dietary complacency

A major benefit of industrialization appears to have been the extension of life expectancy. This is generally attributed to better food security and sanitation. There is, however, a possible downside, as incidence of degenerative disease increased. Longer life expectations and lack of exercise doubtless contributed to this, but the link with dietary change is also strong. As our social structures became more sophisticated, Westernized countries became increasingly distant from our food supplies. Most recently, with the globalization of markets, we may have surrendered control to the food industries. In the process, we seem to have forgotten that food and water are the physical basis for our survival.

With the onset of industrialization from around 200 years ago, it seems, from the perspective of what we eat, that a miracle must have occurred, because, irrespective of how much salt, saturated fat, hydrogenated fats, and refined sugar we started eating, we seemed to assume it had no implications for our well-being. Was this the ascent of man from primordial soup to the couch potato? Can it really be assumed that our nutrition was crucial in our evolution but has no relevance now? Or could something as simple as paying more attention to what we eat potentially be a major resource in promoting human well-being?

Diet and physical well-being

Food inequality in developing countries

No species can flourish without adequate food and water. Sadly, for many of the world's population, this is the cruel reality as, according to the UN Food and Agriculture Organization (FAO 2003) report, there were 840 million people in the world who did not have enough to eat. The laudable target set at the millennium World Food Summit to halve global hunger by 2015 is already off target and, according to the FAO, the present rate of decline of hunger will need to be increased 12-fold to reach that target (FAO 2003). Indeed, the overall decrease in global hunger is largely due to improvements in China. The scale of the Chinese achievement obscures the fact that hunger has got worse in 47 countries (Short 2002).

Diet-related low birthweights affect approximately 30 million children born each year, which has implications for their mental and physical development as well their survival (WHO 2002; de Onis *et al.* 1998) and chronic diseases later in life (e.g. Barker 1995; Barker *et al.* 1989, 2001). For the two-thirds of the world's absolute poor who still do not have food security, there is an irrefutable case that better food would promote their physical and social well-being. The hunger of these people puts our global economic culture into perspective. How precious even the most humble offer of food and water must appear when faced with such tragic circumstances. The World Health Organization (WHO) has published perhaps the definitive studies on changes in global dietary practices and, according to the WHO (2003, section 2.2),

> Hunger and malnutrition remain among the most devastating problems facing the majority of the world's poor and needy people, and continue to dominate the health of the world's poorest nations. Nearly 30% of humanity is currently suffering from one or more of the multiple forms of malnutrition (WHO 2000). The tragic consequences of malnutrition include death, disability, stunted mental and physical growth, and as a result, retarded national socioeconomic development. Some 60% of the 10.9 million deaths each year among children aged under five years in the developing world are associated with malnutrition.

If the anthropological perspective on our dietary origins is correct, it could be predicted that patterns of nutritional deficiencies seen nowadays would be related to our movement away from our historic coastal and lakeside food sources. It is noteworthy that the most important nutrient deficiencies globally are nutrients that are typically richest in shellfish and seafood. Iodine deficiency is estimated to affect more than 700 million people and is considered the greatest single preventable cause of brain damage and poor cognitive function (WHO/UNICEF 1999; Driutskaia and Riabkova 2004). It is estimated that 250 million young children are affected by subclinical vitamin A deficiency, which is considered the greatest preventable cause of childhood blindness and increased risk of infections and death (WHO/UNICEF 1995). Over 2000 million people are estimated to suffer from anaemia (WHO/UNICEF 2001). Unfortunately, dietary standards have only recently considered deficiencies in fatty acid intake and its sequelae.

In addition to food insecurity, many developing countries are experiencing rapid changes from traditional diets. These changes typically reflect trends towards Western dietary practices of high saturated fat and energy-dense foods. Increasingly, malnourishment persists with rapidly emerging incidence of chronic disease attributed to modern diets—so much so that by 2020 it is projected that, in the developing countries, diet will account for 71% of mortality from ischaemic heart disease and 75% of deaths due to stroke. Diabetes is projected to increase from 84 million in 1995 to 228 million in 2025 (Aboderin et al. 2001). On a global basis, 60% of the burden of chronic diseases will occur in developing countries.

If there are ways of ensuring that the typical circumstances of an entire population can be induced to favour distinctly positive health, surely finding a more equitable method of food distribution must be among them.

Food excess in Western countries

Western diets in particular have increasingly focused on energy-dense foods like saturated fats and sugar. According to the WHO (2003), 'the increasing Westernization, urbanization, and mechanization in most countries around the World is associated with changes in diets towards one of high-fat, high-energy foods and sedentary lifestyles.'

Obesity

Obesity is a function of both energy intake and exercise but the evidence linking it to energy-dense foods is compelling. For instance, a meta-analysis of 16 studies testing high-fat versus low-fat diets concluded that a 10% reduction in fat intake over typically 2 months resulted in a loss of around 3 kg in body weight (Astrup *et al.* 2000). Similarly, consumption of energy-dense sugary drinks is associated with increased weight when examined by cross-sectional, longitudinal, and cross-over studies (Tordoff and Alleva 1990; Harnack *et al.* 1999; Ludwig *et al.* 2001). China has just reported that the numbers of obese people in its population has doubled to 60 million between 1992 and 2002, almost equalling the numbers in America in 10 short years. Obesity and depression have been reported as a major factor in maternal death in Britain (Brettingham 2004). Even economically poor countries are witnessing a disturbing increase in incidents of childhood obesity (de Onis and Blössner 2000). The best known comorbidities of obesity are diabetes and heart disease.

Diabetes

Diets high in saturated fats are an important risk factor for diabetes (Tuomilehto *et al.* 2002; Knowler *et al.* 2002). Australian Aboriginals have an inherent tendency to insulin dependence but this only seems to manifest itself when they consume Western diets (O'Dea 1991).

Cancers

WHO (2003) estimates that only a negligible proportion of the world's population consumes the recommended intakes of fruit and vegetables, which is an important part of dietary variation. Diet is estimated to account for 30% of cancers in industrialized countries (Doll and Peto 1996) and 20% in developing countries (Willet 1995). Epidemiological studies provide circumstantial evidence that incidence of cancers follow the nutritional transitions of Latin America, Africa, North America, and Asia (Popkin *et al.* 1993).

Heart disease

Evidence of the role of saturated animal fats in cardiovascular disease is strong and has been demonstrated in animal experiments, epidemiological studies, and clinical trials in various populations (Kris-Etherton P *et al.* 2001). Heart rate variability is considered to be a good indicator of heart condition where beat-to-beat variations of heart-beat are measured using spectral estimation techniques (DePetrillo *et al.* 1999). Lowered heart rate variability has been associated with major depressive disorders (Carney *et al.* 1995; Tulen *et al.* 1996), minor depression (Carney *et al.* 1995), and 'depression self-rating' in non-depressed subjects (e.g. Krittayaphong *et al.* 1997).

Processed foods

Other aspects of modern processed diets have also been found to be highly questionable from a health perspective such as trans fatty acids from the industrial hardening of oils and increasingly from high-temperature fried foods. This process introduces hydrogen into polyunsaturated oils, resulting in a more extended fatty acid carbon chain similar to that of saturated fatty acids. This de-odorizes the oil and improves shelf life: qualities that are attractive to the food industry. Unfortunately, there is no evidence to suggest any health benefits from consuming trans fatty acids—quite the contrary. Since the early 1990s evidence has suggested that trans fats adversely affect fetal and infant growth by interfering with the biosynthesis of AA and DHA, the main structural fats in the central nervous system (Koletzko 1992; van Houwelingen and Hornstra 1994) and more recently in preschool children (Innis *et al.* 2004). Trans fats have also been shown to increase the rate of heart disease in large cohort studies (Oomen *et al.* 2001; Willett *et al.* 1993). The US Food and Nutrition Board (2002) famously suggested a 'tolerable upper intake level' of *zero*! High sodium intake is also a matter of concern in Western diets as a major risk factor associated with high blood pressure. This relationship has been demonstrated experimentally, by epidemiological studies, controlled clinical trials, and in population studies on restricted sodium intake (Gibbs *et al.* 2000).

The positive benefits for health of better diets

The positive flip side of this evidence implies that eating a better diet would be protective from these chronic diseases and certainly that would be the position argued by WHO. A meta-analysis of 32 trials (Cutler *et al.* 1997) concluded that a daily reduction of sodium intake reduced systolic and diastolic blood pressure in those with high blood pressure. For instance, studies that reintroduced dietary factors that would be plentiful in the Palaeolithic diet, such as fish oils and vitamin E, found a reduction in mortality by cardiovascular disease of 30% after a 3.5 year follow-up (GISSI-Prevenzione investigators 1999). Fish consumption correlates as a protective factor against chronic diseases in China (Wang *et al.* 2003). Double-blind placebo-controlled clinical trials of omega-3 fatty acids have demonstrated improved heart rate variability of adult subjects given fatty acids during 24-hour monitoring (Christensen *et al.* 1996, 1997, 1999). There is evidence that fish oils provide a protective effect in certain groups of cancers such as breast cancers (Palakurthi *et al.* 2000), an illness that is more common in industrialized countries. Numerous studies have shown that the consumption of fruit and vegetables, and in some cases pulses, is protective against cancers (World Cancer Research Fund 1997). Zinc supplementation has been shown in a double blind trial to reduce childhood mortality in small for gestational age infants (Sazawal *et al.* 2001).

A need for balance

If we are seeking to raise many of the world population from the shadow of chronic disease to distinctly positive health, then it is arguable that finding a dietary middle-way between the extremes of starvation and excess should be a priority.

Healthy food choices

At the same time that there is concern that we are eating too much sugar and saturated and trans fats, we also do not seem to be eating enough of the healthy foods such as fruits and vegetables. According to the US Centers for Disease Control (CDC 2002), for instance, 79% of US high school students had eaten less than the recommended five servings of fruits and vegetable a day.

Even the nutritional qualities of the healthy staple foods may also have altered. A comparison (Mayer 1997) of the nutritional values of fruit and vegetables first published in 1936 in the seminal work, *The nutritive value of fruits, vegetables and nuts* (McCance *et al*. 1936), with revised figures published in the fifth edition (Holland *et al*. 1991) showed there was significantly less calcium, magnesium, copper, and sodium in vegetables and magnesium, iron, copper, and potassium in fruit. The reason given for the update was because 'the nutritional value of many of the more traditional foods has changed'. A comparative study highlighted changes to national diets in the UK in the last 50 years. A comparison of diets consumed by 4599 4-year-old children in 1950 was made with 493 children from the 1992/93 National Diet and Nutrition Survey. The authors (Prynne *et al*. 1999) concluded that the post-war diet, with its reliance on staple foodstuffs such as bread and vegetables, might well have been beneficial to the health of young people. Thus, the increase in food choice we hold dear nowadays may not necessarily be beneficial from a nutritional perspective. With the availability of such a variety of healthy foods to choose from today it is intriguing to consider why we do not manage to make better food choices. A possible answer is that, typically, it is the energy-dense foods that are often heavily promoted so perhaps this is driven by conscious choices. After all, the food industry has invested a great deal of effort in packaging foods to be attractive (Nestle 2003). Nestle (2002) characterises the food industry claims as 'there is no such thing as good or bad food [except when their products are considered good]'. Alternatively, it has been proposed that energy density influences energy intake due to weak satiety signals that fail to compensate for very energy-dense foods (Prentice and Jebb 2004). It could be a matter of taste. Our sense of taste is influenced by zinc status for instance (Bryce-Smith and Simpson 1984). The addition of sugar, salt, and fats results in very stimulating foods that may be more palatable if the sense of taste is less sensitive. A survey found that only 56% of the US population had adequate zinc intakes (Breifel *et al*. 2000). According to the International Bibliographic Information on Dietary

Supplements (IBIDS) database of the Office of Dietary Supplements (ODS) at the National Institutes of Health, 98% of zinc is removed when sugar is refined, a common feature of modern dietary practices.

Another possibility is that high fat and sugar diets influence our brain chemistry directly and may have addictive qualities. While researching factors in obesity, animal studies suggest that mu opioid stimulation enhances the appetite for high fat and sugar consumption (Zhang and Kelley 2000) and vice versa. High glucose intakes of 25% added to feeding chow (Colantuoni *et al.* 2001) have been shown to stimulates mu–1 and dopamine D1 receptors much like drugs of abuse. It is argued that there may be a genetic adaptation towards energy-dense foods that developed at a time before we had unlimited supplies of these foods (Kelley *et al.* 2002). When animals are intermittently withdrawn from sugar intakes of 25% they exhibit signs of withdrawal (Colantuoni *et al.* 2002). So far, such findings do not seem to have been replicated on human populations. Nevertheless, breakfast cereals can contain 40% sugars and some drinks contain 100% of calories as free sugars.

This brings us back to the disparity between the estimates of the Palaeolithic diet where free sugar intakes are estimated to be far lower than in modern times. If our physiology and genetic make-up is more fully adapted to the diet of the last 50 000 years rather than that of the last 200 years as has been suggested (Cordain *et al.* 2000), we need to take notice, because our dietary baselines seem to have shifted considerably. These changes may map out into our behaviour. Following the publication of the US Food and Nutrition Board statement (2002) suggesting that we minimize consumption of trans fats, evidence has emerged from an animal study that dietary trans fatty acids may act on the endogenous neurotransmitter levels during brain development (Acar *et al.* 2002). Since these same neurotransmitter pathways are also implicated in mental illnesses such as schizophrenia (Davis *et al.* 1991), we are presumably left to guess if the introduction of trans fats has similarly affected human brains and if there are any consequent mental health implications. The fact is we don't know and it should be a concern that evidence emerges after the widespread introduction of such dietary changes. Are the improved keeping qualities of hydrogenated oils really a good trade-off for the functionality of our brains?

While the adoption of such unhealthy food products is doubtless profitable, perhaps it would be a good idea to systematically test new food types for their implications on our brains before introducing them.

Diet and mental health

It is curious that many of the authoritative reference works on diet and inter-national dietary standards focus on the health issue but hardly mention mental health. Unless we still believe that mind and body are separate, we might expect that what we eat would influence brain function if only because of its high metabolic activity. The blood–brain barrier protects the brain's nutrient supply, but it is not invulnerable.

Animal studies suggest that the brain's composition does not change as quickly in response to diet as the liver for instance, but smaller changes were noted nevertheless (Crawford *et al.* 1976). Many of the nutrients supplied to the brain are classed as essential. Indeed, diseases of nutritional deficiencies such as pellagra can present as mental confusion and delusions that can give the appearance of mental illness. Increasingly, there are good reasons to believe that the health benefits of good nutrition extend into effects on mental health. This brief discussion will touch on some of the evidence that nutrients can also positively impact on mental illness, perhaps holding out the possibility of a degree of prevention in the future.

Patterns of mental illness appear to be similar to the cross-national patterns of chronic disease in that, broadly, they are increasing in prevalence and appear to follow modern dietary practices, but it is recognized that we are only gradually beginning to appreciate what is a very complex picture.

Depression

A multinational epidemiological study that examined incidents of depression and schizophrenia found strong correlations with national dietary practices (Peet 2004). The most striking finding was that sugar consumption was strongly associated with worse outcomes in schizophrenia and prevalence of depression, while fish consumption was inversely associated with depression. Like any epidemiological data, these findings have to be interpreted with care. Such data cannot demonstrate a causal relationship but the strength of the correlations does, nevertheless, strongly suggest a need for follow-up clinical trials. Evidence is converging from treatment studies to indicate that the same nutrients that would have been plentiful in estimated Palaeolithic diets are protective against depression. The fish oil EPA was found highly effective in a case of severe treatment-resistant depression (Puri *et al.* 2001, 2002) and randomized controlled trials have confirmed that it can significantly reduce symptoms of unipolar depression (Nemets *et al.* 2002; Peet *et al.* 2002) and similar benefits have been found for omega-3 treatment in bipolar mood disorder (Stoll *et al.* 1999). Peet demonstrated that a one-gram dose of EPA was more effective than a larger dose (Peet and Horrobin 2002) in treating depression in patients who remained depressed despite adequate medication. This suggests that there are optimal dosages, which is reassuring from an evolutionary perspective as moderate intakes are more easily achievable on a daily basis. In addition to consuming less omega-3 fatty acids nowadays, it has also been noted that we may not eat sufficient green vegetables, a good dietary source of folate (Gerrior and Zizza 1994). There is strong evidence of the protective relationship between folate intake and depression, as set out in a Cochrane review (Taylor *et al.* 2003). A number of clinical studies have found low zinc status in depressed patients compared to controls (e.g. Maes *et al.* 1994, 1997; McLoughlin and Hodge 1990). Furthermore, the severity of unipolar depression was negatively correlated with the serum level of zinc. A pilot randomized controlled trail of 14 patients with unipolar depression on antidepressant medication found that those who received 25 mg zinc daily significantly augmented the reduction in Beck Depression Inventory (BDI) scores by 40% after 12 weeks of treatment when compared with placebo supplementation (Nowak *et al.* 2003). The authors appropriately argued that this should be followed up with larger trials.

Schizophrenia

There is a history of treating schizophrenia with various dietary interventions but, sadly, this has received little attention until comparatively recently. Within schizophrenia there is considerable heterogeneity, so there is a need for caution in generalizing about this condition but abnormalities of fatty acid and phospholipid metabolism have been consistently reported, including excessive activity of PLA2 enzymes, increased lipid peroxidation, and nuclear magnetic resonance (NMR) spectroscopy evidence of increased brain membrane phospholipid turnover (Peet *et al.* 1999). Controlled trial treatments with omega-3 fatty acids have been shown to be effective in reducing hallucinations and delusions as well as anhedonia, inattention, or lack of volition, in both medicated and unmedicated patients (Puri *et al.* 1998; Peet *et al.* 2001). In patients given 2 g/day EPA there were improvements on the positive and negative affect scale (PANAS) and its subscales, but there was also a large placebo effect in patients on typical and new atypical antipsychotics and no difference between active treatment and placebo. In patients on clozapine, however, there was little placebo response, but a clinically important and statistically significant effect of EPA on all rating scales. This effect was greatest at 2 g/day rather than at a higher dose of EPA (Peet *et al.* 2002). This suggests that there are optimal dosages and that nutrients can augment the effectiveness of drugs. Clinical improvements following treatment with EPA have also been shown to correlate with improved blood fatty acid status, reduced membrane phospholipid turnover, and reversal of cerebral atrophy (Puri *et al.* 2000; Richardson *et al.* 2000). Negative results have been reported in only one of five controlled trials to date where a 3 g/day intervention was used (Fenton *et al.* 2001). These data on fish oils indicate that it is EPA rather than DHA that is effective but it has to be recognized that these are relatively small studies and more large-scale clinical trials are required, particularly as this is a low-risk approach.

Overall, these findings are highly encouraging but a great deal more research is needed in this area, particularly focusing on the possibility that a better diet is protective from such illness. The World Health Organization is predicting a 50% rise in child mental disorders by 2020. This strongly suggests that the relationship between diet and mental health should be added to formal considerations of dietary adequacy.

Nutrition and behavioural well-being

It may be possible to extrapolate from the health benefits of diet to our behaviour. The primary concern will be to focus on the possibility that diet can influence behaviour directly and hence that dietary interventions may even be possible for behavioural disorders.

Childhood developmental disorders

Evidence is emerging that diet may be linked with childhood developmental disorders. Many of the features associated with attention deficit hyperactivity disorder (ADHD), dyslexia, dyspraxia, and autistic spectrum disorders are consistent with lack of or imbalances in highly unsaturated fatty acids (Richardson and Ross 2000). Fatty acid abnormalities could also help to account for some of the key cognitive and behavioural features of these conditions, such as anomalous visual, motor, attentional, or language processing, as well as some of the associated difficulties with mood, appetite or digestion, temperature regulation, and sleep (Richardson and Puri 2000). Reviews of clinical and experimental studies support the idea that lack of highly unsaturated fatty acids (HUFA) may play a role in these overlapping developmental conditions (Richardson and Ross 2000). Thus, physical signs of fatty acid deficiency, such as excessive thirst, frequent urination, rough, dull, or dry hair and skin, and soft or brittle nails, have been clearly linked with ADHD, dyslexia, and autistic spectrum disorders, as have reduced blood concentrations of HUFA and iron (Konofal et al. 2004).

There are relatively few controlled trials to test the effects of fatty acids on ADHD. One of the first involved evening primrose oil (providing the omega-6 fatty acid, gamma-linolenic acid (GLA)), but this showed little clear benefit (Aman et al. 1987; Arnold et al. 1989). With the accumulation of evidence that it is the omega-3 fatty acids that may be washed out in modern diets, emphasis shifted to fish oils (providing both EPA and DHA), which have been demonstrated to reduce behavioural and learning difficulties in both ADHD and dyslexia (Burgess 1998; Richardson and Puri 2002). However, supplementation with pure DHA was twice found to be ineffective in ADHD (Voigt et al. 2001; Hirayama et al. 2004). A further pilot randomized study of effects of fatty acids on children with inattention, hyperactivity, and other disruptive behaviours included biochemical analyses that demonstrated that both the omega-3 fatty acids and vitamin E were correlated with reductions assessed by the disruptive behaviour disorders (DBD) rating scale. Vitamin E is commonly added to fish oil as an antioxidant but was not always declared as an active constituent in earlier studies using EPA and DHA. Thus, the results of this pilot study suggest the need to declare it as an active ingredient and for further research with both n-3 fatty acids and vitamin E in children with behavioural disorders (Stevens et al. 2003). Zinc has been shown to be an effective adjunct to methylphenidate in treating ADHD at 15 mg/day in a randomized study of 44 children (Akhondzadeh *et al.* 2004). Zinc at 40 mg/day has been shown to be an effective monotherapy for ADHD in a randomized controlled trial of 400 Turkish children (Bilici *et al.* 2004). Although this dosage is above physiological requirements, Turkish children of lower socio-economic status have been shown to have significantly lowered zinc status (Tanzer *et al.* 2004).

Linkage of heart and brain function

These findings are consistent with evidence from the treatment of depression and schizophrenia that EPA, not DHA, may be the key omega-3 fatty acid in functional disturbances of attention, cognition, or mood. This in itself is surprising as DHA, not EPA, is found in the brain. One alternative is that EPA may improve brain blood flow through inhibition of cyclooxygenase and vasoconstrictive eicosanoids (Ellis *et al.* 1992). More recently it has been found that the phospholipase A2–arachidonic acid pathway and 20-hydroxyeicosatetraenoic acid production may have a regulatory role in cerebral blood flow (Mulligan and MacVicar 2004). This blood flow carries the brain's energy supply, which may constitute a limiting factor to the brain's information-processing capacity (Peppiatt and Attwell 2004). Certainly there is a body of literature that suggests that heart function affects brain function and behaviour; this will be discussed below. It is equally possible that the therapeutic effects of fatty acids are influenced by the dietary intakes of other nutrients such as vitamin E (Stevens *et al.* 2003) or zinc as these nutrients appear to interact with fatty acid metabolism (Bekaroglu *et al.* 1996; Wauben *et al.* 1999). These data nevertheless suggest that there is a plausible relationship between diet and a range of developmental conditions with clearly described associated behaviours. More research is required to delineate that relationship. If diet is a protective factor from these disorders then it can only be hoped that such requirements are eventually taken into account when assessing standards of dietary adequacy.

Sugar and behaviour

If the amount of energy supplied to the brain puts a limit on its information-processing capacity, it might be deduced that high sugar consumption would be helpful to brain performance. There have been a number of randomized studies conducted to assess the behavioural and cognitive effects of sugar. An experiment using a counterbalanced Latin square design involving 48 children, compared the behavioural effects of sugar, aspartame, and saccharine as a placebo in blind dietary sequence changes every 3-weeks. The study concluded, 'Even when intake exceeds typical dietary levels, neither dietary sucrose nor aspartame affects children's behaviour or cognitive function' (Wolraich *et al.* 1994). A study of 48 children may not, however, be a strong basis on which to form such robust conclusions. Small studies do not have the numbers for appropriate randomized control for broader extrapolation and may lack the power to detect smaller effects that may be cumulative in the population at large. A literature review acknowledged that 'Sugar clearly does not induce psychopathology where there was none before, but it may on occasion aggravate an existing behaviour disorder' but again the largest study quoted was 76 selected children (Kinsbourne 1994). To further illustrate the complications of dietary studies, the Wolraich study used vitamin C and riboflavin to assess compliance in the active cells at dosages comparable to those reported in the literature to improve cognition and behaviour (e.g. Benton *et al.* 1995; Heseker *et al.* 1995). A recent dose-ranging study of 67 adult students (Flint and Turek 2003) with normal glucose metabolism suggests that large doses of glucose that increase blood glucose levels do not influence attention, but that a moderate dose (100 mg/kg) of glucose selectively impairs measures of impulsivity or disinhibition. Furthermore, the glycaemic index (broadly how quickly the carbohydrate can be utilized) of carbohydrate sources also appears to influence cognitive effects in human and animal studies (Benton *et al.* 2003). Refining of sugar removes many of the micronutrients required in carbohydrate metabolism; hence the term empty calories. It has been argued that the lack of these micronutrients is the reason that behaviour might be affected by sugar consumption (Schoenthaler 1994). Given the pervasive use of refined sugar in modern diets and its interactions with health concerns, there is a strong case for large randomized studies to investigate sugar and behaviour in more detail.

Mood

While many of us might be ostensibly healthy, we may wonder if taking better care of our diet would enhance our mood. Researchers have examined the effects of diet on mood in apparently healthy subjects but, as with other areas reported, the effects may depend on the dietary baseline, i.e. the quality of the exiting diet. Such factors may be regional. Intakes of many trace elements are influenced by the soil types from which the food originates; for example, UK soil is low in selenium. In a study of 11 healthy men who where fed either a low or a high selenium diet for 15 weeks, raising selenium intakes beyond the typical US recommended dietary allowance (RDA) of 70–356 mg had no measured effect on mood but low baseline erythrocyte selenium correlated with changes in hostile–agreeable and depressed–elated scales using the profile of mood state (POMS)-B10 assessment subscales (Hawkes and Hornbostel 1996). The authors concluded that, because US dietary selenium baselines are higher than in the UK, these findings were con-sistent with a UK double-blind experimental study (Benton and Cook 1991). Here 50 subjects significantly decreased ratings of anxiety, depression, and tiredness over 5 weeks with 100 mg of selenium per day. This was particularly true of those subjects who had estimated intakes of 28–62 mg per day, while those in the higher intake ranges of 63–280 mg per day showed less change in mood. It is noteworthy that the lower range is more typical of UK dietary selenium intakes (MAFF 1997).

Similarly, low iron status is relatively common and may contribute to lethargy. Studies of athletes suggest that the lack of energy that results from iron deficiency anaemia can be a serious disincentive to exercise (e.g. Nielson and Nachtigall 1998). It should be noted that nutritional regimes are routinely applied with training to enhance athletic excellence. Other factors of the diet that have been considered in relation to mood are the intakes of carbohydrates. Benton and Donohoe (1998) assert that carbohydrate intake is associated with improved mood, while poor mood stimulates the eating of 'comfort foods' such as chocolate. Carbohydrate metabolism predicted results in tests of memory and cognition in healthy subjects (Donohoe and Benton 2000). The specific mood response to carbohydrate intake is complex as this has been shown to be influenced by many aspects of the dietary baseline: when food is consumed; the time delay between meals; the amount consumed; the fibre content; the glycaemic index of the food; the level of accompanying protein (Benton 2002; Benton *et al.* 2003).

The relationship between diet and mood has so far not benefited from the considerable international investment to investigate the relationship between diet and health. Mood reflects how well we feel in ourselves and hence the quality of life experienced. This would also be a worthy area to add to the considerations of dietary adequacy.

Diet and antisocial behaviour

A question remains, however, as to whether diet might affect behaviour to which we assign free will. This is a common underlying assumption in the criminal justice system, so this discussion will focus on criminal and antisocial behaviour.

Firstly, let us define what we mean by crime and the broader category of antisocial behaviour. Central to any form of criminal justice is the notion that culpability can be attributed. Culpability is distinct from simply establishing guilt as the degree of individual liability is judged in relation to an action. The classical form of justice assumes that man is an agent of free will and can choose to commit an offence; hence culpability can be fully attributed. Crime is judged in relation to a body of criminal law that sets out offences, that is, acts for which a legal penalty will apply. Straightforwardly, a crime is deemed to have been committed when one of these laws is judged to have been broken. When other forms of social rules are judged to have been broken, such as within an institution, it will be described as antisocial behaviour. While this serves as a useful distinction, it is recognized that in both cases it is assumed that the perpetrator would know that such behaviours attract a sanction, albeit of differing severity.

A dramatic illustration of a possible relationship between diet and behaviour is the following diet consumed by a persistent criminal offender. He had been sentenced by UK courts on 13 occasions for stealing trucks in the early hours of the morning and on the last three occasions he was imprisoned. This was all he ever ate.

- Breakfast: Nothing (asleep).
- Mid-morning: Nothing (asleep).
- Lunch time: 4–5 cups of coffee with milk and 2.5 heaped sugars.
- Mid-afternoon: 3–4 cups of coffee with milk and 2.5 heaped sugars.
- Tea: Fries, egg, ketchup, and 2 slices of white bread; 5 cups of tea or coffee, with milk and 2.5 heaped sugars.
- Evening: 5 cups of tea or coffee, with milk and 2.5 heaped sugars; 20 cigarettes; £2 worth of sweets, cake, and (if money available) 3–4 pints of beer.

The court was exasperated by his behaviour because on release from prison he had chosen to steal a truck to return home! Notice that there are no fruits or green vegetables whatsoever. There are no obvious sources of omega-3 fatty acids. There are at best modest sources of protein and fibre. Much of the calories come from sugary drinks and cake. Consider how you might feel if you lived on his diet for 3 months. With the approval of the court he was given dietary education to improve his eating habits. He eventually trained as a chef. His probation officer reported that he had not re-offended in 15 years.

Nutritional status of offenders

National dietary surveys have not to date differentiated between offenders and non-offenders, so the possibility remains that the diets of offenders are somehow atypical. The most accurate method is to determine this is by assessments of nutritional status from blood. A study that compared the nutritional status of offenders with that of non-offending controls found no significant differences in nutritional status. However, there were differences in sugar metabolism (Gans *et al.* 1990). This needs wider population assessments, as such conclusions cannot be taken as typical given regional variations in diet. It would be a reasonable point to make that there must be criminals who consume a healthy diet. Perhaps so, but since national dietary standards were never established with brain function or behaviour in mind, we really don't know yet if a diet is adequate from the perspective of the brain.

Reinterpreting crime trends

If nutrition plays a role in our behaviour, then effects from nutrition would not only have to be in force within individuals but presumably should be capable of helping shape patterns of social behaviour. In the case of crime, these changes are considerable over the last century. Using the UK as an example, in 1900 there were 77 934 offences reported which rose to 5 170 843 in 2000, albeit reaching a peak in 1992, followed by a decline during the 1990s and then a levelling off.

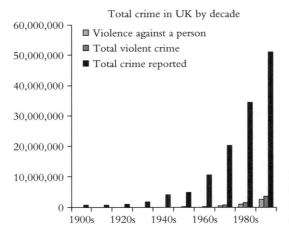

Fig. 7.1 Total crime in UK by decade (Data supplied by National Statistics Office 2002).

Even when adjusting for population, notifiable offences (broadly more serious) rose almost 10-fold per person in the UK from the 1950s to the mid-1990s (Home Office 1997). Rapid changes in reported crime, as illustrated in Fig. 7.1, would be difficult to explain in terms of genetics for instance. There is no doubting the relevance of other changes such as socio-economic factors, legislation, and police practice in these trends. Nevertheless, the scale of change over the decades is considerable. Taking the foregoing into account, a scholarly review of such trends (Rutter *et al.* 1998) commented, 'The overall rate of crime has risen severalfold over the last fifty years, a rise so rapid that it can only be due to some impact in the environment. Clearly if society has been so spectacularly successful in causing the levels of crime to increase there must be the potential for the right sort of interventions to be equally effective in causing it to decrease!'

One element of this might be that it is our brain's environment that is being impacted. We have after all already considered a range of changes that have occurred in modern diets. There is evidence that violence is influenced by genetic susceptibility factors (e.g. Caspi *et al.* 2002). These effects may themselves be influenced by diet as there is a growing awareness of nutrient-gene interactions. Zinc for instance has been shown to be involved in the regulation of gene expression (Dieck *et al.* 2003; Blanchard *et al.* 2000) and maintaining the configuration of mammalian gene transcription proteins (e.g. Hanas *et al.* 1983; Vallee *et al.* 1981; Dieck *et al.* 2005). In addition, exposure to lead (a nutrient antagonist) correlates with increased delinquency (Needleman *et al.* 1996) and this may have increased in the UK during this period until the introduction of unleaded fuel in the early 1990s (e.g. Thomson *et al.* 1989). It has to be recognized, however, that such historical data have limitations and the connection with changing social patterns going back many decades is speculative. There is, however, more direct evidence that raises the possibility that such influences could have increased their grip at a time when social behaviour has deteriorated.

Evidence linking diet and antisocial behaviour

A comparison of conventional and nutrition education approaches to rehabilitating 102 offenders over 12 months in the community found that the re-offending rates (11.9%) of the group given nutrition education were almost a third of that of the controls (33.8%) receiving conventional probation programmes (Schauss 1978). An experimental study of 3399 imprisoned juveniles, where refined and sugary foods, snack foods, and drinks available to the inmates were replaced with unsweetened fruit juices and popcorn, found that, over a 12-month period, there were 21% fewer serious antisocial acts, a 25% reduction in assaults, a 100% reduction in suicides, and a 75% reduction in the use of restraints (Schoenthaler 1983). This study lacked randomized allocation or placebo control. Schoenthaler subsequently conducted a placebo-controlled double-blind randomized experimental trial using nutritional supplements based around the US RDA of minerals and three times the US RDA of vitamins on 62 13–17-year-old male and female incarcerated juveniles. The active group committed 28% (15–41%, 95% confidence interval (CI)) fewer rule violations compared to controls ($p = 0.005$) (Schoenthaler et al. 1997). It was reported that this effect was most marked in 16 subjects who had significantly improved their blood vitamin status during the trial from a low baseline of the following vitamins: C, thiamin, niacin, pantothenic acid, pyridoxine, and folate. A further randomized placebo-controlled study of 468 school children aged 6–8 years was conducted using a food supplement formulation of 50% of the US RDA of vitamins and minerals. Of these children, only 80 had a school discipline record at baseline. It was reported that antisocial incidents fell significantly ($p < 0.02$) by 47% (29–65%, 95% CI) for those 40 children on active supplementation who had a discipline record at baseline (Schoenthaler and Bier 2000). There are more appropriate statistical methods to model these data to deal with the floor effect using a Poisson regression. However, the broad conclusion that significant improvements in behaviour can also be seen in younger children from a simple dietary adjunct should be followed up. Hamazaki et al. (1996) found that Japanese university students undertaking exams while receiving supplemental DHA on a randomized double-blind basis did not show an increase in hostility during a frustration test, while those on placebo increased hostility. This effect was however, not observed under non-stressful conditions (Hamazaki et al. 1998). A randomized placebo-controlled pilot study of 30 women with borderline personality disorder found that EPA significantly reduced ratings of aggression and depression compared to controls (Zanarini and Frankenburg 2003). While these experimental studies offer broadly encouraging results, they need to be replicated with larger randomized designs.

With the co-operation of the UK Home Office an empirical study was conducted to test if poor nutrition is a cause of antisocial behaviour (Gesch et al. 2002). The study was simply designed to provide a powerful test for a general effect on behaviour, where statistical power was reported as 92% to detect a change in the rate of proven disciplinary offences in a prison. Statistical significance was set at a conservative 1%. A prison was chosen as all sources of food are known. It was found that, typically, prisoners did not make appropriate food choices resulting in many of their dietary intakes falling below UK government dietary standards (Eves and Gesch 2003). Using a double-blind placebo-controlled randomized stratified design, 231 young adult prisoners (18–21 years of age) were studied to see whether nutrient supplementation had any influence on proven disciplinary offences committed by the prisoners. The active nutritional supplements provided broadly the daily adult requirements of vitamins, minerals, and omega-3 and -6 essential fatty acids (see Gesch *et al.* 2002). On an intention to treat basis compared to placebo, those who received active nutritional supplements committed an average of 26.3% (8.3–44.33%, 95% CI) fewer offences ($p = 0.03$, two-tailed). This analysis included everyone recruited to the trial including those who had participated for as little as 3 days but the average was 5 months (see Fig. 7.2).

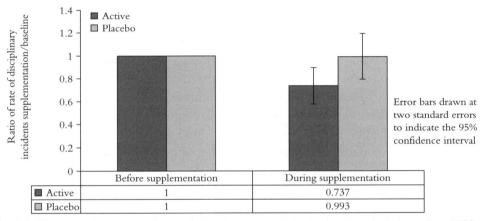

Fig. 7.2 Difference between active and placebo groups: intent to treat analysis based on 1132 offences. Test for null hypothesis: $\gamma_{Placebo} = \gamma_{Active}$.

Having rejected the null hypothesis on an intent to treat basis in a between-groups comparison, a more sensitive test for the actual effect of treatment (in this analysis, each individual becomes their own control as we have established that both groups are matched at baseline by rates of offending) was conducted for the effect on those taking active supplements for a minimum of 2 weeks ($n = 172$) (see Fig. 7.3). The result was an average 35.1% reduction of offences (95% CI 16.3–53.9%, $p < 0.001$, two-tailed), whereas those taking placebos remained within standard error, proving no evidence of change. The greatest reduction occurred for the most serious incidents (including violence) dealt with by governor reports. Based on an analysis of 338 governor reports, the active group committed 37% (11.6–62.4%, 95% CI) fewer governor reports ($p < 0.005$, two-tailed), whereas the placebo group remained within standard error ($p > 0.1$, two-tailed) at 10.1% fewer governor reports ($-16.9–37.1\%$, 95% CI), again providing no evidence of change.

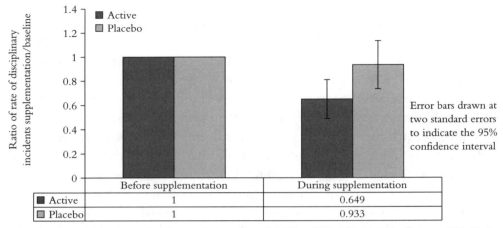

Fig. 7.3 Effect on those treated for more than 2 weeks (N = 172). Number of offences 754. Test for effect: $\gamma_{Placebo} = 1$, $\gamma_{Active} < 1$.

Thus, with placebos there was no evidence for a reduction in offending over time spent in the prison, while, in marked contrast, those receiving active nutritional supplements committed significantly fewer disciplinary offences over the same period.

The differences in the rate of offending in the prison could not be explained by ethnic or social factors, or variations in the administration of governor reports, or opportunity to offend as they should have been controlled for by the randomized design: the compliance of both randomly allocated groups was closely matched as were all baseline measures of behaviour and diet, so it had to be the nutrients in the capsules that caused the change in behaviour. The issue then is the nutritional quality of the existing dietary baseline. The dietary baseline had been modified (the independent variable) using nutritional supplements. The supplements were really an analogue for a better diet but they have the advantage that the nutritional value is precisely known and allowed for the use of placebo. Ideally, such improvements in nutritional intake should come from the diet (for a more detailed discussion, see Eves and Gesch 2003). The dosages provided were physiological rather than high dose and might be readily achievable with even a modest increase in expenditure on diet coupled with dietary education.

Given the wide implications, these findings need to be widely replicated. The participants did not guess accurately what sort of capsules they had been given either, so the importance of Bryce-Smith's idea becomes clear; here we have a potent effect on behaviour that can be measured but not sensed. The findings are all the more surprising because this age group of prisoners has proved to be notoriously resistant to behavioural improvements (Kershaw 1999). A limitation of the study was that biochemical measures were not available; they will be required in any replication to explore the utilization of nutrients and also mediating mechanisms. Thus, having empirically demonstrated an effect on antisocial behaviour, we are only at the start of understanding the potential of this intervention. It is quite possible that, with a better understanding of the range and balance of nutrients required, the protective effect could be further improved.

Health implications of an adequate diet

It should come as no surprise that ensuring the prisoners' diets reached UK government dietary standards did not result in adverse reactions. Indeed, it could be argued that the real experiment was the ongoing default position where prisoners were not reaching these dietary standards. The only referral for a possible adverse event was a young man who was surprised that his evacuations had increased in frequency from once a fortnight to once every 2 days: the prison doctor rightly considered this an indicator rather than an adverse event! Here we have criminals in a maximum-security institution not appreciating what their bodies are telling them. Some of the participants had not heard of vitamins, let alone knew which foods contained them; they were not even remotely equipped to make healthy food choices. Being pragmatic, a combination of supplementation to reinstate nutrients first coupled with dietary education may be prudent in such populations. Notwithstanding the seriousness of their crimes, many of them took the trouble to express gratitude for being involved in the trial.

Implications of a protective effect on behaviour

These findings suggest that diet can affect behaviour to which we assign free will. Indeed, it is difficult to see how free will could be exercised without involving brain function. Furthermore, these effects may be far more potent than we realized. This raises the question as to what life would have held for these 231 young men if they had grown up with better nourishment. The only studies so far published that provide any indication are longitudinal studies conducted in Mauritius. An experimental enrichment programme, comprising nutrition, education, and physical exercise for 83 children aged 3 to 5 years, yielded significantly lower scores for schizotypal personality and antisocial behaviour at age 17 years and for criminal behaviour at age 23 years compared with 355 matched controls (Raine *et al.* 2003). The beneficial effects of the intervention were greatest for children who showed signs of malnutrition at 3 years of age. A second birth cohort of 1795 were assessed for signs of malnutrition at age 3 years. Cognitive measures were assessed at ages 3 and 11 years, while antisocial, aggressive, and hyperactive behaviour was assessed at ages 8, 11, and 17 years. The authors concluded that there was a dose–response relationship between the degree of malnutrition and degree of externalizing behaviour at ages 8 and 17 (Liu *et al.* 2004). They argue that these findings suggest that reducing early malnutrition may help reduce later antisocial and aggressive behaviour. Given the implications, this needs to be replicated. It may be simplistic, but if there are fewer crimes committed there will also be fewer victims.

Observational evidence

The experimental studies above are supported by observational studies that indicate relationships between antisocial behaviour and a number of nutrients but it is recognized that these relationships will require more research to delineate. Violent offenders were found to have lowered levels of omega-3 and omega-6 but raised omega-9 essential fatty acids compared to age-matched non-offending controls (Corrigan *et al.* 1994). A cross-national epidemiological analysis found a significant inverse correlation between seafood consumption and incidence of homicides (Hibbeln 2001). Hibbeln *et al.* (1998) found an inverse relationship between omega-3 plasma essential fatty acids and levels of the metabolites of serotonin and dopamine taken from cerebrospinal fluid in violent offenders, but this was not found with non-offending controls. Stevens *et al.* (1996) found an increased number of behavioural problems in 6–12-year-old boys with lower plasma concentrations of omega-3 fatty acids. Virkkunen *et al.* (1987) reported abnormalities in plasma phospholipid concentrations of fatty acids in offenders with a history of alcohol abuse. Violent offenders were found to have significantly lower cholesterol levels than non-violent subjects who were matched by age, sex, alcohol indices, and education (Golomb *et al.* 2000). Walsh *et al.* (1997) reported significantly elevated copper/zinc ratios in assaultive young males compared to non-assaultive controls. Plasma homocysteine has been positively associated with ratings of hostility and anger (Stoney and Engebretson 2000). Regulation of homocysteine is dependent on folate, cobalamin (vitamin B_{12}), and pyridoxine (vitamin B_6). Depressed patients, for instance, were found to have raised homocysteine levels and lowered cobalamin and folate (Bottiglieri *et al.* 2001). There may also be a case for addressing individual dietary requirements. A review of 207 subjects selected from 258 consecutive cases suggested that self-reports of violence were significantly reduced compared to baseline by individualized nutritional therapy. Such data are difficult to interpret without controls but the authors appropriately argued for the need to follow up with randomized trials to test efficacy (Walsh *et al.* 2004).

There may even be behavioural implications for trace elements not yet accepted as essential for human health. A study found significantly fewer homicides, suicides, and rapes in counties of Texas with naturally occurring levels of lithium in drinking water in the range 70–170 μg/L compared to rates in those counties with little or no lithium; these dosages are far lower than those used in medicine. Similarly, violent offenders were found to have less lithium in their scalp hair than controls. These studies simply demonstrate correlations but a double-blind randomized trial of 24 former drug addicts supplemented with 400 μg lithium for a month found significant improvements in mood, reflecting changes in happiness, friendliness, and energy. The author argued a case for the essentiality of lithium in humans at dosages of 1000 μg a day (Schrauzer 2002).

Physiological markers

Physiological markers linked to nutrition also seem to be predictive of antisocial behaviour (Christensen *et al.* 1996). It has already been reported that the output of the heart supplies the brain with energy, the amount of which may be the limiting factor on information-processing, and one might presume has an impact on behaviour. This appears to be so. Low heart rate variability is thought to be a sensitive marker of low autonomic nervous system function (Akselrod and Gordon 1985) and is considered to measure autonomic adaptation; as such it is considered representative of more basic brain functioning. Wadsworth (1976) reported a stronger correlation between delinquency and lowered resting pulse rate than between delinquency and social deprivation. Children aged 6–7 years with mental health symptoms could be distinguished from controls by low reactivity in parasympathetic and sympathetic systems (Boyce *et al.* 2001). In a 15-year longitudinal study of the siblings of juvenile delinquents, abnormalities in autonomic function identified at age 15 years predicted with 75% accuracy which subjects would be incarcerated by the age of 29 (Raine *et al.* 1995). Low heart rate variability also predicted psychopathology and hostility in a 2-year longitudinal analysis of urban boys at risk of delinquency (Pine *et al.* 1998). Similarly, heightened autonomic nervous system responsiveness was associated with a reduced likelihood of criminal convictions in subjects at social high risk for criminal behaviour (Brennan *et al.* 1997).

Physiological mechanisms

The research suggests that nutrition influences behaviour, but the mechanisms by which this occurs are not fully understood. It is, however, widely accepted that antisocial behaviours are influenced by the regulation of serotonin (e.g. Linnoila *et al.* 1983; Stanley *et al.* 2000; Virkkunen *et al.* 1995), so this will be used as an illustration. Serotonin is derived from the dietary amino acid tryptophan. To demonstrate that dietary tryptophan depletion results in increased aggressive responses, researchers had to control for other possible explanations such as food depletion causing an increase in aggression (Bjork *et al.* 1999). This illustrates that nutrients tend to interact with many mechanisms operating in parallel (see also Stacey *et al.* 1994). Providing piglets with supplemental AA and DHA (the two fatty acids considered so influential in our brain development) significantly increased concentrations of serotonin, tryptophan, and dopamine in the frontal cortex of their brains (de la Presa Owens and Innis 1999). Vitamin E deficient diets have been shown in animal studies to lower serotonin in rat striatum (Castano *et al.* 1993). Zinc is known to modulate the function of a number of neurotransmitter gated ion channels, including recombinant murine 5-hydroxytryptamine (3A) (5-HT(3A)) receptors, and it potentiates serotonin-induced responses (Hubbard and Lummis 2000). Furthermore, animal studies have shown that low iron and zinc intakes predispose towards aggression (Halas and Reynolds 1977; Munro 1987). Pyridoxine has been demonstrated in monkey brains to have a regulatory effect on the conversion of 5-hydroxy-L-trytophan (5-HTTP), a precursor of serotonin (Hartvig *et al.* 1995). Oral doses of pyridoxine were found to influence blood serotonin and pyridoxal phosphate levels in hyperactive patients (Bhagavan *et al.* 1975). It is not suggested that these studies offer proof of mechanisms that are at work in antisocial behaviour, as this will require a great deal more research, but they do illustrate the potential for plausible physical pathways to be identified by which levels of nutrients could affect levels of antisocial behaviour.

Mediation of dietary effects on antisocial behaviour

How such effects are mediated is another important question. The picture is far from clear at this stage but inhibitory mechanisms are considered fundamental components of the attentional and higher cognitive processes that form the key dysfunctions of ADHD, for example, which is a common precursor of antisocial personality disorder (ASP; e.g. Taylor 1998; Rubia *et al.* 1998). Impulse control is a current focus of interest for how nutrients might impact behaviour. Adults with ASP, conduct disorder, and aggressive impulsiveness have been shown to have neuropsychological deficits in frontal-lobe-mediated paradigms (Pennington and Bennetto 1993; Brower and Price 2001). Moreover, low executive functioning has been shown to correlate with measures of physically aggressive behaviour (Giancola *et al.* 1998; Foster *et al.* 1993; LaPierre *et al.* 1995). Violent offenders have also been found to be deficient in shifting attention from one category to another when compared to normal control subjects (Bergvall *et al.* 2001). Poor development of self-directedness and cooperative traits in forensic populations may be associated with deficits in neuropsychological functioning (Bergvall *et al.* 2003).

Replication

Large ($N = 700$) clinical trials are planned within the UK and Dutch prison systems to investigate the relationship between a participant's nutritional status (measured from blood samples) and changes in specific nutrient groups with the following outcomes, antisocial behaviours, violence, drug abuse, self-harm, impulse control, attention, planning, interpersonal relating, and food choice, with concomitant implications for reassessing dietary standards and our understanding of such behaviours in the community.

Integrating nutrition with existing risk factors for antisocial behaviour

There do appear to be interactions between our social and physical environment as Bryce-Smith (1986) predicted. While this may seem daunting initially, there are many positive aspects to this, as diet could provide an additional means to promote behavioural well-being at a time when criminal justice resources are under stress. These findings also may improve our understanding of established socio-economic risk factors such as low income, family interactions, and peer-groups that correlate with offending (Blackburn 2002). There will also always be a biophysical component to any social situation, as we all need to eat. This will interact with both individual and environmental theories of crime (Blackburn 2002). There is a great deal of research into 'risk factors' that affect the behaviour of juveniles (Smith 1995; Rutter *et al.* 1998) including, for example, the breakdown of families (Heiss 1995). However, one of the social functions of families is to provide food, so it would be illuminating to investigate the extent to which diets interact with such breakdowns. It is well established that diet is related to income (e.g. WHO 2003). If we no longer produce our own food, the costs per calorie of healthy foods like fruits and green leafy vegetables are typically much higher because they contain so few calories. The healthy energy-dense foods like lean meats and fish are much more expensive than more highly refined energy-dense foods derived from saturated fats, white flour, and refined sugars. These cheap, energy-dense foods are likely to seem more attractive and affordable if you have a limited income. Those on low incomes are likely to face the dual pressures of socio-economic and physiological stress as a consequence of not easily affording adequate diets. The cycle is therefore self-reinforcing. If we grow up with friends in this environment, this may conceivably increase the wayward influence of peer pressure as the behaviour of our friends is more likely to be undermined by poor nutrition. We will presumably also be subject to the vagaries of a poor social environment. However, this would not explain why, in general, more boys get into trouble than girls and this reaches a peak in late adolescence. One might wonder how diet could be involved in these trends. The link is speculative at this stage but, for instance, differences have been found in DHA metabolism between males and females (Burdge and Wootton 2002; Burdge *et al.* 2003). A factor in the peak age of offending may turn out be the more pronounced growth spurt in young males, which brings with it a range of physiological changes such as rapid growth that may place the brain under increased nutritional stress (Gesch 2002). If this scenario is correct, it highlights the importance of providing healthy meals to schoolchildren to help break this self-reinforcing cycle. The British Government recently concluded that just 6% of schoolchildren made healthy food choices at school meals and, according to the *Times* Educational Supplement, 14 April 2004, 'The Government has been forced to pump £342 million into school behaviour improvement programmes.' In view of the foregoing, these two facts may not be coincidence. Would it not be prudent to apply the precautionary principle and to try our best to ensure that all our children at least reach existing dietary standards? Large studies of the effects of diet in schools are being planned.

Implications in criminal justice

It is *not* suggested that nutrition is the only explanation of antisocial behaviour, only that it might form a significant part. Unlike in medicine, however, most of the 'risk factors' targeted in criminal justice interventions are typically correlations, which leaves questions of volition, culpability, prevention, and rehabilitation open if cause and effect are not formally demonstrated with appropriate randomized experimental designs. This leaves crucial questions unanswered. What causes someone to commit a crime or behave in an antisocial manner when others do not? How do you predict or prevent offending unless you know what causes it? How do you accurately focus resources unless you know about causes? For instance, the public costs of testing cognitive skills approaches in English prisons was reported in the *London Times* (18 November 2003) to cost £150 000 000 and this approach was found to be ineffective (Cann *et al.* 2003). In contrast, *The Economist* of 29 June 2002 reported the cost of the nutritional approach to be as little as 0.2% of that expended on custody. Faced with escalating costs, there is a climate of increasing honesty in the US that federal criminal justice policies are not underpinned by rigorous evidence, with statements such as 'The effectiveness of most crime prevention strategies will remain unknown until the nation invests more in evaluating them' (Sherman *et al.* 1997), 'It's easy to fool yourselves about efficacy if you haven't done a proper clinical trial' (Marshall 2000), and 'Progress is often thwarted by Government programmes and strategies that are not based on rigorous evidence' (Baron *et al.* 2003). The report by the US Department for Justice and Collation for Evidence-Based Policy (Baron *et al.* 2003) recommended that in future interventions should be evaluated by randomized designs as the sole basis for strong evidence of efficacy with which to underpin criminal justice policy. Evidence is, after all, only as good as the means used to obtain it! Ironically, the finding that a better diet can positively impact on behaviour exceeds the standards suggested as a strong basis of evidence to underpin criminal justice strategy but this approach is not yet assimilated into criminal justice thinking. These studies will need to be widely replicated and, if so, we may have finally demonstrated a causal factor in antisocial behaviour. Further studies are also planned in the UK to test effects of nutrition on crime in the community. If the finding that diet can positively influence behaviour directly is widely replicated, it questions the underlying assumptions of the classical form of justice where it is assumed that behaviour is entirely a matter of free will. The picture will then be more complex but the solution perhaps more straightforward. Some of these factors will act in ways that we can see and some of them *will not*, so we will need a broader interpretation of the causes of antisocial behaviour in criminal justice, where physical and social functioning are both considered relevant to culpability.

Awareness offers hope because, if this scenario is correct, this process can be reversed if we choose to nourish our children rather than condemn them for influences on their behaviour that we have not taken into account. In de choose if we prefer to lock up ever more of our nation's children or them properly. Indeed, one of the difficulties in criminal justice is knowing when to intervene. Pre-emptive involvement in the criminal justice system runs the risk of being prejudicial to those who might not offend and has been shown to accelerate criminal careers through association, while intervening too late can also result in escalation. In marked contrast, how is providing a healthy diet going to be prejudicial at any stage? The Dutch Ministry of Justice is actively investigating this approach and made a telling comment: 'There are only positive side-effects to providing offenders with a better diet!'

According to a UK Home Office report (Brand and Price 2000), 'The total cost of crime to England and Wales in 1999/2000 is estimated at around £60 billion, although this figure is still far from comprehensive, as it does not include important costs such as fear of crime or quality of life impacts.' If diet could achieve even a modest impact on these costs the benefits could be vast. There is strong case to conduct further large randomized trials to explore if diet can reduce crime in the community and antisocial behaviours in prisons and other institutions such as schools.

Discussion

The global implications for well-being

A range of evidence has been reviewed that suggests that the dietary requirements for good health are also supportive of positive social behaviour. These findings need to be widely replicated, as potentially there are global implications. The interventions that have been reviewed simply provide nutrients at physiological dosages that are already known to be essential for health. Providing a healthy diet does absolutely no harm, yet we have arrived at a situation where many of the human population do not obtain even this. Millions starve while millions suffer from diseases of excess and yet we spend untold amounts on medical treatment to treat illness and prisons to curb behaviour. What if much of this suffering could be prevented?

Good nutrition is essential irrespective of race, age, legislative boundaries, if you are in prison, or in a rich or developing country. This is a basic need that unites us all: hence it is not where you eat that is important but what you eat. Humans are, after all, both social and physical beings—the hardware and software of life, so to speak. Nutrition can impact on both aspects and thus spans both nature and nurture. How we interact with food is likely to be a complex function of choice, family, peer-group, physical activity, economics, food distribution, and physiology. Food plays an important role in our social structures. Much of traditional family routine is based around meals. A mother's love, courtship, and friendship are often expressed through giving food. We incorporate food into medicine and our

_.ous practices such as offerings to deities or sacrament. Yet for every social situation there will always be a biophysical analogy. Irrespective of what we eat, getting food into one's mouth will be socially mediated but one's mouth is the gateway to the physical world. Crucially, our physiological responses to our diet are likely to be the most finite component of this equation, as they seem to be primarily derived from our evolution. This may explain why longitudinal studies can link a child's diet (or, for instance, childhood exposure to lead) with crime as an adult. This is not a reinterpretation of crude determinism because this is a feedback cycle where choice will inform what we put into our mouths, while the nutrients in the food will support the operation of the senses that inform our choice (for instance the retina is 30% AA and DHA). It is worth reiterating that some of these factors will act in ways that we can see and some of them will not, so we need a broader interpretation of what promotes well-being, where physical and social functioning are both considered relevant. The brain needs to be nourished in two ways: love, nurturing, education as well as the nutrients for the brain to function properly. Not one, not the other, but both. That applies to both the rich and poor nations.

If we really are what we eat then changing our diet will change us.

We seem to have made major changes to modern diets in a relatively short space of time with little or no systematic examination for potential impacts on brain function or behaviour. Hence, evidence of such impact is only emerging on a retrospective basis. More should be done to prospectively assess such impact. This not only relates to the food industries. Such a basic issue also seems to have passed largely unnoticed between the boundaries of academic disciplines and the authorities that construct dietary standards. There is a need for contiguous scientific disciplines to cooperate to investigate the broader picture. Because nutrition is fundamental for life and also the composition of our brains, its effects are likely to be pervasive. The health consequences of modern diets are already visible but we are only beginning to appreciate them in brain chemistry, mental health, cognition, and behaviour. We are likely to interpret behavioural problems in terms of what we can see, so it is feasible that, on a societal scale, the effects of poor diet are liable to be greatly amplified, as greater numbers of social interactions are subject to these hitherto unnoticed influences on our behaviour. These effects may even shift the socially acceptable norms of behaviour without our knowledge. It may sound far-fetched but few would have predicted the potency of effect shown in the prison study. There is even research being planned to explore the role of nutrition in artistic expression. Physiological factors like nutrition may also turn out to be important in understanding the tragic and irrational acts that litter news broadcasts and seem to defy a rational explanation. They may add to, rather than supplant existing explanations. Compared to the myriad of socio-economic problems found in deprived areas, nutrition may actually be one of the more straightforward factors to change. An intriguing aspect of our British prison study is that the prisoners received three meals a day and, despite making poor food choices, their diets were possibly better than those consumed by many young men of the same age in the community. Yet the improvement in behaviour from boosting their diets was highly significant. We do not know if the improvement in behaviour would have come from raising the prisoners to existing dietary standards or because some would have exceeded them slightly. We have limited knowledge of what the optimum ranges of nutrients are from a behavioural perspective because dietary standards have not really considered this. The bottom line here is that *dietary standards need to be reassessed* to take into account mental health, behavioural, developmental, and cognitive parameters. It is recognized that this will be a major undertaking involving extensive research. However, an implication is that, for a great number of people, their ability to behave sociably could be improved by what they eat, without them even being aware of it. It is hard to imagine anything that might impact on the future well-being of our species so directly.

Research is emerging that suggests that good nutrition is potentially inexpensive, humane, and effective at promoting health as well as social behaviour. Most importantly, people's positive potential might be realized if such an approach is taken at a time when global resources are under stress. To make an analogy, no amount of energy spent on software will resolve a hardware problem. Nature offers us a clue about the way forward by providing natural foodstuffs that grow in harmony with their environment and invariably contain a range of nutrients in dosages to which our metabolism is attuned. It should come as no surprise, therefore, that many vitamins, minerals, and essential fatty acids seem to promote aspects of well-being whether it be health or behaviour—particularly those nutrients that modern dietary practices have rendered less available. If nutrition was truly seminal in raising our ancestors to a new level of functioning, then let us rethink our present-day attitude to food. We need to develop our understanding of the dosage and range of nutrients required to positively impact on the human condition. It may be a recipe that goes beyond individual well-being; it may be a recipe for peace.

Acknowledgements

Bernard is honoured to work as part of a broad multidisciplinary collaboration comprising of many eminent academic colleagues as well as the distinguished trustees of Natural Justice, all of whom have contributed with wisdom and generosity. The work is therefore a synthesis of these important contributions. Bernard would particularly like to thank Professor Michael Crawford, Dr Sean Hammond, Professor Malcolm Peet, Dr Alex Richardson, and Professor John Stein for their kind contributions to the text.

References

Aboderin, I., *et al.* (2001). *Life course perspectives on coronary heart disease, stroke and diabetes: key issues and implications for policy and research.* World Health Organization, Geneva.

Acar, N., Chardigny, J.M., Berdeaux, O., Almanza, S., and Sebedio, J.L. (2002). Modification of the monoaminergic neurotransmitters in frontal cortex and hippocampus by dietary trans alpha-linolenic acid in piglets. *Neurosci. Lett.* **331** (3), 198–202.

Akhondzadeh, S., Mohammadi, M.R., and Khademi, M. (2004). Zinc sulfate as an adjunct to methylphenidate for the treatment of attention deficit hyperactivity disorder in children: a double blind and randomized trial. *BMC Psychiatry* **4** (1), 9.

Akselrod, S. and Gordon, D. (1985) Hemodynamic regulation: investigation by spectral analysis. *Am. J. Physiol.* **249**, 867–75.

Aman, M.G., Mitchell, E.A., and Turbott, S.H. (1987). The effects of essential fatty acid supplementation by Efamol in hyperactive children. *J. Abnorm. Child Psychol.* **15** (1), 75–90.

Angel, J.L. (1984). Health as a crucial factor in the changes from hunting to developed farming in the eastern Mediterranean. In *Paleopathology at the origins of agriculture* (ed. M.N. Cohen and G.J. Armelagos), pp. 51–74. Academic Press, New York.

Arnold, L.E., Kleykamp, D., Votolato, N.A., Taylor, W.A., Kontras, S.B., and Tobin, K. (1989). Gamma-linolenic acid for attention-deficit hyperactivity disorder: placebo-controlled comparison to D-amphetamine. *Biol. Psychiatry* **25** (2), 222–8.

Astrup, A., *et al.* (2000). The role of low-fat diets in body weight control: a meta-analysis of ad libitum dietary intervention studies. *Int. J. Obesity* **24**, 1545–52.

Ayala, S. and Brenner, R.R. (1983). Essential fatty acid status in zinc deficiency. Effect on lipid and fatty acid composition, desaturation activity and structure of microsomal membranes of rat liver and testes. *Acta Physiol. Lat. Am.* **33** (3), 193–204.

Ayala, S. and Brenner, R.R. (1987). Effect of zinc deficiency on the *in vivo* biosynthesis of fatty acids of the linoleic series in the rat. *Acta Physiol. Pharmacol. Latinoam.* **37** (3), 321–30.

Barker, D.J.P. (1995). Fetal origins of coronary heart disease. *Br. Med. J.* **311**, 171–4.

Barker, D.J.P., Osmond, C., Winter, P.D., Margetts, B., and Simmonds, S.J. (1989). Weight in infancy and death from ischaemic heart disease. *Lancet* **2**, 577–80.

Barker, D.J.P., Forsén, T., Uutela, A., Osmond, C., and Eriksson, J.G. (2001). Size at birth and resilience to the effects of poor living conditions in adult life: longitudinal study. *Br. Med. J.* **323**, 1273–6.

Baron, J., *et al.* (2003). *Bringing evidence based progress to crime and substance-abuse policy: a recommended federal strategy.* Coalition for Evidence-based Policy, sponsored by the Council for Excellence in Government, Washington, DC.

Bekaroglu, M., *et al.* (1996). Relationships between serum free fatty acids and zinc, and attention deficit hyperactivity disorder *J. Child Psychol. Psychiatry* **37**, 225–7.

Bennis-Taleb, N., Remacle, C., Hoet, J.J., and Reusens, B. (1999). A low-protein isocaloric diet during gestation affects brain development and alters permanently cerebral cortex blood vessels in rat offspring. *J. Nutr.* **129**, 1613–19.

Benton, D. (2001). Micro-nutrient supplementation and the intelligence of children. *Neurosci. Biobehav. Rev.* **25**, 297–309.

Benton, D. (2002). Carbohydrate ingestion, blood glucose and mood. *Neurosci. Biobehav. Rev.* **26** (3), 293–308.

Benton, D. and Cook, R. (1991). Selenium supplementation improves mood in a double-blind crossover trial. *Biol. Psychiatry* **29**, 1092–8.

Benton, D. and Donohoe, R.T. (1999). The effects of nutrients on mood. *Public Health Nutr.* **2** (3a), 403–9.

Benton, D., Haller, J., and Fordy, J. (1995). Vitamin supplementation for 1 year improves mood. *Neuropsychobiology* **32** (2), 98–105.

Benton, D., Ruffin, M.P., Lassel, T., Nabb, S., Messaoudi, M., Vinoy, S., Desor, D., and Lang, V. (2003). The delivery rate of dietary carbohydrates affects cognitive performance in both rats and humans. *Psychopharmacology (Berl.)* **166** (1), 86–90.

Bergvall, A., Wessely, H., Forsman, A., and Hansen, S. (2001). A deficit in attentional set-shifting of violent offenders. *Psychol. Med.* **31** (6), 1095–105.

Bergvall, A.H., Nilsson, T., and Hansen, S. (2003). Exploring the link between character, personality disorder, and neuropsychological function. *Eur. Psychiatry* **18** (7), 334–44.

Bhagavan, H.N., Coleman, M., and Coursin, D.B. (1975). The effect of pyridoxine hydrochloride on blood serotonin and pyridoxal phosphate contents in hyperactive children. *Pediatrics* **55** (3), 437–41.

Bilici, M., Yildirim, F., Kandil, S., Bekaroglu, M., Yildirmis, S., Deger, O., Ulgen, M., Yildiran, A., and Aksu, H. (2004). Double-blind, placebo-controlled study of zinc sulfate in the treatment of attention deficit hyperactivity disorder. *Prog. Neuropsychopharmacol. Biol. Psychiatry* **28** (1), 181–90.

Bjork, J.M., Dougherty, D.M., Moeller, F.G., *et al.* (1999). The effects of tryptophan depletion and loading on laboratory aggression in men: time course and a food-restricted control. *Psychopharmacology* **142**, 24–30.

Blackburn, R. (2002). *The psychology of criminal conduct: theory, research and practice.* Wiley and Sons, Chichester.

Blanchard, R.K. and Cousins, R.J. (2000). Regulation of intestinal gene expression by dietary zinc: induction of uroguanylin mRNA by zinc deficiency. *J. Nutr.* **130**, 1393S–1398S.

Blumenschine, R.J. (1991). Hominid carnivory and foraging strategies, and the socio–economic function of early archaeological sites. *Phil. Trans. R. Soc. Lond. B* **334**, 211–21.

Bottiglieri, T., Laundy, M., Crellin, R., Toone, B.K., Carney, M.W., and Reynolds, E.H. (2001). Homocysteine, folate, methylation and monoamine metabolism in depression. *J. Neurol. Neurosurg. Psychiatry* **70** (3), 419.

Boyce, W.T., Quas, J., *et al.*: MacArthur Assessment Battery Working Group of the MacArthur Foundation Research Network on Psychopathology and Development (2001). Autonomic reactivity and psychopathology in middle childhood. *Br. J. Psychiatry* **179**, 144–50.

Brand, S. and Price, R. (2000). *The economic and social costs of crime*, Home Office Research study 217. Home Office, London.

Brennan, P., Raine, A., Schulsinger, F., Kirkegaard-Sorensen, L., Knop, J., Hutchings, B., Rosenberg, R., and Mednick, S. (1997). Psychophysiological protective factors for male subjects at high risk for criminal behavior. *Am. J. Psychiatry* **154**, 853–5.

Brettingham, M. (2004). Depression and obesity are major causes of maternal death in Britain. *Br. Med. J.* **329**, 1205.

Briefel, R.R., Biolostosky, K., Kennedy-Stevenson, J., *et al.* (2000). Zinc intake of the US population: findings from the Third National Health and Nutrition Examination Surveys 1988–1994. *J. Nutr.* **130**, 1367–73.

Broadhurst, C.L., Wang, Y., Crawford, M.A., Cunnane, S.C., Parkington, J.E., and Schmidt, W.F. (2002). Brain-specific lipids from marine, lacustrine, or terrestrial food resources: potential impact on early African *Homo sapiens. Comp. Biochem. Physiol. B Biochem. Mol. Biol.* **131** (4), 653–73.

Brower, M.C. and Price, B.H. (2001). Neuropsychiatry of frontal lobe dysfunction in violent and criminal behaviour: a critical review. *J. Neurol. Neurosurg. Psychiatry* **71**, 720–6.

Bryce-Smith, D. (1986). Environmental chemical influences on behaviour, personality, and mentation, Royal Society of Chemistry John Jeyes Lecture. *Int. J. Biosoc. Res.* **8** (2), 115–50.

Bryce-Smith, D. and Simpson, R.I. (1984). Case of anorexia nervosa responding to zinc sulphate. *Lancet* **2** (8398), 350.

Burdge, G.C. and Wootton, S.A. (2002). Conversion of alpha-linolenic acid to eicosapentaenoic, docosapentaenoic and docosahexaenoic acids in young women. *Br. J. Nutr.* **88** (4), 411–20.

Burdge, G.C., Delange, E., Dubois, L., Dunn, R.L., Hanson, M.A., Jackson, A.A., and Calder, P.C. (2003). Effect of reduced maternal protein intake in pregnancy in the rat on the fatty acid composition of brain, liver, plasma, heart and lung phospholipids of the offspring after weaning. *Br. J. Nutr.* **90** (2), 345–52.

Burgess, J.R. (1998). Attention deficit hyperactivity disorder: observational and interventional studies. In *NIH workshop on omega-3 essential fatty acids and psychiatric disorders*, September 2–3, 1998. National Institutes of Health, Bethesda, Maryland.

Cann, J., Falshaw, L., Nugent, F., and Friendship, C. (2003). *Understanding what works: accredited cognitive skills programmes for adult men and young offenders*, Home Office findings 226. Home Office, London.

Carney, R.M., Saunders, R.D., et al. (1995). Association of depression with reduced heart rate variability in coronary artery disease. *Am. J. Cardiol.* **76**, 562–4.

Caspi, A., McClay, J., Moffitt, T.E., Mill, J., Martin, J., Craig, I.W., Taylor, A., and Poulton, R. (2002). Role of genotype in the cycle of violence in maltreated children. *Science* **297** (5582), 851–4.

Castano, A., Herrera, A.J., Cano, J., and Machado, A. (1993). Effects of a short period of vitamin E-deficient diet in the turnover of different neurotransmitters in substantia nigra and striatum of the rat. *Neuroscience* **53** (1), 179–85.

CDC (Centers for Disease Control) (2002). Youth risk behaviour surveillance. *Morb. Mortal. Wkly Rep.* **51** (SS04), 1–64.

Christensen, J.H., Gustenhoff, P., et al. (1996). Effect of fish oil on heart rate variability in survivors of myocardial infarction: a double blind randomized controlled trial. *Br. Med. J.* **312**, 677–8.

Christensen, J.H., Korup, E., et al. (1997). Fish consumption, n–3 fatty acids in cell membranes, and heart rate variability in survivors of myocardial infarction with left ventricular dysfunction. *Am. J. Cardiol.* **79**, 1670–3.

Christensen, J.H., Christensen, M.S., et al. (1999). Heart rate variability and fatty acid content of blood cell membranes: a dose–response study with n–3 fatty acids. *Am. J. Clin. Nutr.* **70**, 331–7.

Clover, C. (2004). *The end of the line: how over-fishing is changing the world and what we eat.* Ebury Press, London.

Colantuoni, C., Schwenker, J., McCarthy, J., Rada, P., Ladenheim, B., Cadet, J.L., Schwartz, G.J., Moran, T.H., and Hoebel, B.G. (2001). Excessive sugar intake alters binding to dopamine and mu-opioid receptors in the brain. *Neuroreport* **12** (16), 3549–52.

Colantuoni, C., Rada, P., McCarthy, J., Patten, C., Avena, N.M., Chadeayne, A., and Hoebel, B.G. (2002). Evidence that intermittent, excessive sugar intake causes endogenous opioid dependence. *Obesity Res.* **10** (6), 478–88.

Cordain, L., Miller, J.B., Eaton, S.B., and Mann, N. (2000). Macronutrient estimations in hunter–gatherer diets. *Am. J. Clin. Nutr.* **72** (6), 1589–92.

Corrigan, F.M., Gray, R., Strathdee, A., et al. (1994). Fatty acid analysis of blood from violent offenders. *J. Forens. Psychiatry* **5** (1), 83–92.

Crawford, M.A. (1968). Fatty-acid ratios in free-living and domestic animals. Possible implications for atheroma. *Lancet.* **22**:1(7556), 1329–33.

Crawford, M.A. and Sinclair, A.J. (1972). Nutritional influences in the evolution of the mammalian brain. In *Lipids, malnutrition and the developing brain*, A Ciba Foundation Symposium 19–21 October 1971 (ed. K. Elliot and J. Knight), pp. 267–92. Elsevier, Amsterdam.

Crawford, M.A., Casperd, N.M., and Sinclair, A.J. (1976). The long chain metabolites of linoleic acid and linolenic acids in liver and brain in herbivores and carnivores. *Comp. Biochem. Physiol. B* **54** (3), 395–401.

Crawford, M.A., Bloom, M., Broadhurst, C.L., Schmidt, W.F., Cunnane, S.C., Galli, C., Gehbremeskel, K., Linseisen, F., Lloyd-Smith, J., and Parkington, J. (1999). Evidence for the unique function of docosahexaenoic acid during the evolution of the modern hominid brain. *Lipids* **34** (Suppl.), 39–47.

Cunnane, S.C., Francescutti, V., Brenna, J.T., and Crawford, M.A. (2000). Breast-fed infants achieve a higher rate of brain and whole body docosahexaenoate accumulation than formula fed infants not consuming dietary docosahexaenoate. *Lipids* **35**, 105–11.

Curl, C.L., Fenske, R.A., Elgethun, K. (2003). Organophosphorus pesticide exposure of urban and suburban preschool children with organic and conventional diets. *Environ. Health Perspect.* **111** (3), 377–82.

Cutler, J.A., Follmann, D., and Allender, P.S. (1997). Randomized trials of sodium reduction: an overview. *Am. J. Clin. Nutr.* **65**, 643–51.

Davis, K.L., Kahn, R.S., Ko, G., and Davidson, M. (1991). Dopamine in schizophrenia: a review and reconceptualization. *Am. J. Psychiatry* **148** (11), 1474–86.

de la Presa Owens, S. and Innis, S.M. (1999). Docosahexaenoic and arachidonic acid prevent a decrease in dopaminergic and serotoninergic neurotransmitters in frontal cortex caused by a linoleic and alpha-linolenic acid deficient diet in formula-fed piglets. *J. Nutr.* **129**, 2088–93.

de Onis, M. and Blössner, M. (2000). Prevalence and trends of overweight among preschool children in developing countries. *Am. J. Clin. Nutr.* **72**, 1032–9.

de Onis, M., Blössner, M., and Villar, J. (1998). Levels and patterns of intrauterine growth retardation in developing countries. *Eur. J. Clin. Nutr.* **52**, S5–S15.

DePetrillo, P., Speers, d'A.E., and Ruttimann, U. (1999). Determining the Hurst exponent of fractal time series and its application to electrocardiographic analysis. *Comput. Biol. Med.* **29**, 393–406.

Dieck, H., Doring, F., Roth, H.P., and Daniel, H. (2003). Changes in rat hepatic gene expression in response to zinc deficiency as assessed by DNA arrays. *J. Nutr.* **133**, 1004–10.

Dieck, H., Doring, F., Fuchs, D., Roth, H.P., and Daniel, H. (2005). Transcriptome and proteome analysis identifies the pathways that increase hepatic lipid accumulation in zinc-deficient rats. *J. Nutr.* **135**(2), 199–205.

Doll, R. and Peto, R. (1996). Epidemiology of cancer. In *Oxford textbook of medicine* (ed. D.J. Weatherall, J.G.G. Ledingham, and D.A. Warrell), pp. 197–221. Oxford University Press, Oxford.

Donohoe, R.T. and Benton, D. (2000). Glucose tolerance predicts performance on tests of memory and cognition. *Physiol. Behav.* **71** (3–4), 395–401.

Driutskaia, S.M. and Riabkova, V.A. (2004). Hygienic assessment of iodine deficiency in the Khabarovsk Territory. [In Russian]. *Gig. Sanit.* Jul–Aug (4), 15–18.

Eaton, S.B. and Eaton, S.B. (2000). Paleolithic vs. modern diets—selected pathophysiological implications. *Eur. J. Nutr.* **39** (2), 67–70.

Eaton, S.B. and Eaton, S.B. III (2003). An evolutionary perspective on human physical activity: implications for health. *Comp. Biochem. Physiol. A* **136**, 153–9.

Eaton, S.B. and Konner, M. (1985). Paleolithic nutrition. A consideration of its nature and current implications. *New Engl. J. Med.* **312**, 283–9.

Eaton, S.B., Eaton, S.B. III, and Konner, M.J. (1997). Paleolithic nutrition revisited: a twelve-year retrospective on its nature and implications. *Eur. J. Clin. Nutr.* **51**, 207–16.

Eaton, S.B., Cordain, L., and Lindeberg, S. (2002a). Evolutionary health promotion: a consideration of common counterarguments. *Prev. Med.* **34** (2), 119–23.

Eaton, S.B., Strassman, B.I., *et al.* (2002b). Evolutionary health promotion. *Prev. Med.* **34**, 109–18.

Egger, J., Carter, C.M., Graham, P.J., Gumley, D., and Soothill, J.F. (1985). Controlled trial of oligoantigenic treatment in the hyperkinetic syndrome. *Lancet* **1**, 540–5.

Ellis, E.F., *et al.* (1992). Effect of dietary n-3 fatty acids on cerebral microcirculation, *Am. J. Physiol.* **258**, H1780–H1785.

Eves, A. and Gesch, C.B. (2003). Food provision and the nutritional implications of food choices made by young adult males in a young offenders' institution. *J. Hum. Nutr. Dietet.* **16**, 167–79.

FAO (2003). *The state of food insecurity in the world.* UN Food and Agriculture Organisation Geneva.

FAO/WHO (1994). *Expert consultation on the role of dietary fats and oils in human nutrition.* FAO, Rome.

Fenton, W.S., Dickerson, F., Boronow, J., Hibbeln, J.R., and Knable, M. (2001). A placebo-controlled trial of omega-3 fatty acid (ethyl eicosapentaenoic acid) supplementation for residual symptoms and cognitive impairment in schizophrenia. *Am. J. Psychiatry* **158** (12), 2071–4.

Fernstrom, J.D. (1999). Effects of dietary polyunsaturated fatty acids on neuronal function. *Lipids* **34**, 161–9.

Flint, R.W. and Turek, C. (2003). Glucose effects on a continuous performance test of attention in adults. *Behav. Brain Res.* **142**, 217–28.

Foley, R.A. and Lee, P.C. (1991). Ecology and energetics of encephalization in hominid evolution. *Phil. Trans. R. Soc. Lond. B* **334**, 223–32.

Food and Nutrition Board (2002). Letter report on dietary reference intakes for *trans* fatty acids. Drawn from *Dietary reference intakes for energy, carbohydrate, fiber, fat, fatty acids, cholesterol, protein, and amino acids (macronutrients).* National Academies Press, Washington, DC.

Foster, H.G., Hillbrand, M., and Silverstein, M. (1993). Neuropsychological deficit and aggressive behavior: a prospective study. *Prog. Neuropsychopharmacol. Biol. Psychiatry.* **17**, 939–46.

Gans, D.A., Harper, A.E., Bachorowski, J-A., Newman, J.P., Shrago, E.S., and Taylor, S. (1990). Sucrose and delinquency: oral sucrose tolerance test and nutritional assesment. *Pediatrics* **86** (2), 254–62.

Gerrior, S.A. and Zizza, C. (1994). *Nutrient content of the U.S. food supply 1909–1990*, Home Economics Research Report No. 52. US Department of Agriculture, Washington, D C.

Gesch, C.B. (2002). A recipe for peace: the role of nutrition in social behaviour. In *Foodstuff. Living in an age of feast and famine* (ed. J. Holden, L. Howland, and D. Stedman Jones), pp. 111–16. DEMOS, London.

Gesch, C.B., Hammond, S.M., Hampson, S.E., Eves, A., and Crowder, M.J. (2002). Influence of supplementary vitamins, minerals and essential fatty acids on the antisocial behaviour of young adult prisoners. *Br. J. Psychiatry* **181**, 22–8.

Giancola, P.R., Mezzich, A.C., and Tarter, R.E. (1998). Executive cognitive functioning, temperament, and antisocial behavior in conduct-disordered adolescent females. *J. Abnorm. Psychol.* **107**, 629–41.

Gibbs, C.R., Lip, G.Y., and Beevers, D.G. (2000). Salt and cardiovascular disease: clinical and epidemiological evidence. *J. Cardiovasc. Risk* **7**, 9–13.

GISSI-Prevenzione Investigators (1999). Dietary supplementation with n-3 polyunsaturated fatty acids and vitamin E after myocardial infarction: results of the GISSI-Prevenzione trial. Gruppo Italiano per lo Studio della Sopravvivenza nell'Infarto miocardico. *Lancet* **354**, 447–55.

Golomb, B.A., Stattin, H., and Mednick, S. (2000). Low cholesterol and violent crime. *J. Psychiatr. Res.* **34** (4–5), 301–9.

Halas, E.S., Reynolds, G.M., and Sandstead, H.H. (1977). Intra-uterine nutrition and its effects on aggression. *Physiol. Behav.* **19**, 653–61.

Hamazaki, T., Sawazaki, S., Itomura, M., *et al.* (1996). The effect of docosahexaenoic acid on aggression in young adults. *J. Clin. Invest.* **97**, 1129–33.

Hamazaki, T., Sawazaki, S., Nagao, Y., Kuwamori, T., Yazawa, K., Mizushima, Y., and Kobayashi, M. (1998). Docosahexaenoic acid does not affect aggression of normal volunteers under nonstressful conditions. A randomized, placebo-controlled, double-blind study. *Lipids* **33**, 663–7.

Hanas, J.S., Hazuda, D.J., Bogenhagen, D.F., Wu, F.Y., and Wu, C.W. (1983). Xenopus transcription factor A requires zinc for binding to the 5 S RNA gene. *J. Biol. Chem.* **258**, 14120–5.

Harnack, L., Stang, J., and Story, M. (1999). Soft drink consumption among US children and adolescents: nutritional consequences. *J. Am. Dietet. Assoc.* **99**, 436–41.

Hartvig, P., Lindner, K.J., Bjurling, P., Laengstrom, B., and Tedroff, J. (1995). Pyridoxine effect on synthesis rate of serotonin in the monkey brain measured with positron emission tomography. *J. Neural. Transm. Gen. Sect.* **102** (2), 91–7.

Harvey, P.H. and Clutton-Brock, T.H. (1985). Life history variation in primates. *Evolution* **39**, 557–81.

Hawkes, W.C. and Hornbostel, L. (1996). Effects of dietary selenium on mood in healthy men living in a metabolic research unit. *Biol. Psychiatry* **39** (2), 121–8.

Heiss, L.E. (1995). Changing family patterns in Western Europe: opportunity and risk factors for adolescent development. In *Psychosocial disorders in young people. time, trends and their causes* (ed. M. Rutter and D.J. Smith), pp. 104–93. John Wiley and Sons, New York.

Heseker, H., Kubler, W., Pudel, V., and Westenhofer, J. (1995). Interaction of vitamins with mental performance. *Bibl. Nutr. Dieta.* (52), 43–55.

Hibbeln, J.R. (2001). Seafood consumption and homicide mortality. A cross-national ecological analysis. *World Rev. Nutr. Diet* **88**, 41–6.

Hibbeln, J.R., Umhau, J.C., and Linnoila, M. (1998). A replication study of violent and nonviolent subjects: cerebrospinal fluid metabolites of serotonin and dopamine are predicted by plasma essential fatty acids. *Biol. Psychiatry* **44**, 243–9.

Hightower, J.M. and Moore, D. (2003). Mercury levels in high-end consumers of fish. *Environ. Health Perspect.* **111** (4), 604–8.

Hirayama, S., *et al.* (2004). Effect of docosahexaenoic acid-containing food administration on symptoms of attention–deficit/hyperactivity disorder—a placebo controlled double-blind study. *Eur. J. Clin. Nutr.* **58** (3), 467–73.

Holland, B., Welch A.A., Unwin, I.D., *et al.* (eds.) (1991). *McCance and Widdowson's the composition of foods*, 5th edn. Royal Society of Chemistry and Ministry of Agriculture, Fisheries, and Food, Cambridge.

Home Office (1997). *Criminal statistics for England and Wales (1996)*. HMSO, London.

Hubbard, P.C. and Lummis, S.C. (2000). Zn(2+) enhancement of the recombinant 5-HT(3) receptor is modulated by divalent cations. *Eur. J. Pharmacol.* **394** (2–3), 189–97.

Innis, S.M., Vaghri, Z., and King, D.J. (2004). N-6 Docosapentaenoic acid is not a predictor of low docosahexaenoic acid in Canadian pre-school children. *Am. J. Clin. Nutr.* **80**, 768–73.

Kelley, A.E., Bakshi, V.P., Haber, S.N., Steininger, T.L., Will, M.J., and Zhang, M. (2002). Opioid modulation of taste hedonics within the ventral striatum. *Physiol. Behav.* **76** (3), 365–77.

Kershaw, C. (1999). *Reconvictions of offenders sentenced or discharged from prison in 1994, England and Wales*, The Home Office Statistics Bulletin 5/99. Research Development and Statistics Directorate, London.

Kinsbourne, M. (1994). Sugar and the hyperactive child. *New Engl. J. Med.* **330**, 355–6.

Klein, R.G., Avery, G., Cruz-Uribe, K., Halkett, D., Parkington, J.E., Steele, T., Volman, T.P., and Yates, R. (2004). The Ysterfontein 1 Middle Stone Age site, South Africa, and early human exploitation of coastal resources. *Proc. Natl Acad. Sci. USA* **101** (16), 5708–15.

Knowler, W.C., *et al.* (2002). Reduction in the incidence of type 2 diabetes with lifestyle intervention of metformin. *New Engl. J. Med.* **346**, 393–403.

Koletzko, B. (1992). Trans fatty acids may impair biosynthesis of long-chain polyunsaturates and growth in man. *Acta Paediatr.* **81**, 302–6.

Konofal, E., Lecendreux, M., Arnulf, I., and Mouren, M.C. (2004). Iron deficiency in children with attention-deficit/hyperactivity disorder. *Arch. Pediatr. Adolesc. Med.* **158** (12), 1113–15.

Kris-Etherton, P. *et al.* (2001). Summary of the scientific conference on dietary fatty acids and cardiovascular health: conference summary from the nutrition committee of the American Heart Association. *Circulation* **103**, 1034–9.

Krittayaphong, R., Cascio, W.E., *et al.* (1997). Heart rate variability in patients with coronary artery disease: differences in patients with higher and lower depression scores. *Psychosom. Med.* **59**, 231–5.

LaPierre, D., Braun, C.M.J., and Hodgins, S. (1995). Ventral frontal deficits in psychopathy: neuropsychological test findings. *Neuropsychologia* **131**, 139–51.

Ledger, H.P. (1968). Body composition as a basis for a comparative study of some East African mammals. In *Comparative nutrition of wild animals* (ed. M.A. Crawford), pp. 289–310, Symposiums of the Zoological Society, London, no. 21. Zoological Society, London.

Leonard, W.R. and Robertson, M.L. (1997). Comparative primate energetics and hominid evolution. *Am. J. Phys. Anthropol.* **102**, 265–81.

Linnoila, M.V., Virkkunen, M., *et al.* (1983). Low cerebrospinal fluid 5-hydroxyindoleacetic acid concentration differentiates impulsive from nonimpulsive violent behavior. *Life Sci.* **33**, 2609–14.

Liu, J., Raine, A., Venables, P.H., and Mednick, S.A. (2004). Malnutrition at age 3 years and externalizing behavior problems at ages 8, 11, and 17 years. *Am. J. Psychiatry* **161** (11), 2005–13.

Ludwig, D.S., Peterson, K.E., and Gortmaker, S.L. (2001). Relation between consumption of sugar-sweetened drinks and childhood obesity: a prospective, observational analysis. *Lancet* **357**, 505–8.

Maes, M., D'Haese, P.C., Scharpe, S., D'Hondt, P.D., Cosyns, P., and De Broe, M.E. (1994). Hypozincemia in depression. *J. Affect. Dis.* **31**, 135–40.

Maes, M., Vandoolaeghe, E., Neels, H., Demedts, P., Wauters, A., Meltzer, H.Y., Altamura, C., and Desnyder, R. (1997). Lower serum zinc in major depression is a sensitive marker of treatment resistance and of the immune/inflammatory response in that illness. *Biol. Psychiatry* **42**, 349–58.

MAFF (UK Ministry of Agriculture, Fisheries, and Food) (1997). *Food surveillance information sheet*, No. 126. Joint Food Safety and Standards Group, MAFF, London.

Manger, M.S., Winichagoon, P., Pongcharoen, T., Gorwachirapan, S., Boonpraderm, A., McKenzie, J., and Gibson, R.S. (2004). Multiple micronutrients may lead to improved cognitive function in NE Thai schoolchildren. *Asia Pac. J. Clin. Nutr.* **13**, 46.

Marshall, E. (2000). The shots heard round the world. *Science* **289**, 570–4.

Martin, R.D. (1983). *Human brain evolution in an ecological context*, fifty-second James Arthur Lecture on the Evolution of the Human Brain. American Museum of Natural History, New York.

Masters, R.D., Coplan, M.J., Hone, B.T., and Dykes J.E. (2000). Association of silicoflouride treated water with elevated lead. *Neurotoxicology* **21** (6), 1091–100.

Mayer, A.M. (1997). Historical changes in the mineral content of fruits and vegetables. *Br. Food J.* **99** (6), 207–211.

McCance, R.A., Widdowson, E.M., and Shackleton, L.R.B. (1936). *The nutritive value of fruits, vegetables and nuts*, Medical Research Council Special Report Series No. 213. HMSO, London.

McLoughlin, I.J. and Hodge, J.S. (1990). Zinc in depressive disorder. *Acta Psychiatr. Scand.* **82**, 451–3.

Morgan, E. (1990). *The scars of evolution*. Oxford University Press, Oxford.

Mulligan, S.J. and MacVicar, B.A. (2004). Calcium transients in astrocyte endfeet cause cerebrovascular constrictions. *Nature* **431** (7005), 195–9.

Munro, N. (1987). A three year study of iron deficiency and behavior in rhesus monkeys. *Int. J. Biosoc. Res.* **9**, 35–62.

Needleman, H.L., Riess, J.A., Tobin, M.J., Biesecker, G.E., and Greenhouse, J.B. (1996). Bone lead levels and delinquent behaviour. *J. Am. Med. Assoc.* **275** (5), 363–9.

Nemets, B., Stahl, Z., and Belmaker, R.H. (2002). Addition of omega-3 fatty acid to maintenance medication treatment for recurrent unipolar depressive disorder. *Am. J. Psychiatry* **159** (3), 477–9.

Nestle, M. (2002). Big food: politics and nutrition in the United States. In *Foodstuff. Living in an age of feast and famine* (ed. J. Holden, L. Howland, and D. Stedman Jones), pp. 83–90. DEMOS, London.

Nestle, M. (2003). The ironic politics of obesity. *Science* **299**, 781.

Nielsen, F.H. (2000). Evolutionary events culminating in specific minerals becoming essential for life. *Eur. J. Nutr.* **39** (2), 62–6.

Nielson, P. and Nachtigall, D. (1998). Iron supplementation in athletes. Current recommendations. *Sports Med.* **26**, 207–16.

Nowak, G., Siwek, M., Dudek, D., Zieba, A., and Pilc, A. (2003). Effect of zinc supplementation on antidepressant therapy in unipolar depression: a preliminary placebo-controlled study. *Pol. J. Pharmacol.* **55** (6), 1143–7.

O'Dea, K. (1991) Cardiovascular risk factors in Australian aborigines. *Clin. Exp. Pharmacol. Physiol.* **18**, 85–8.

Oomen, C.M., *et al.* (2001). Association between trans fatty acid intake and 10-year risk of coronary heart disease in the Zutphen Elderly Study: a prospective population-based study. *Lancet* **357**, 746–51.

Oteiza, P.I., Hurley, L., Lonnerdal, B., and Keen, C. (1990). Effects of marginal zinc deficiency on microtubule polymerization in the developing rat brain. *Biol. Trace Elm. Res.* **23**, 13–23.

Palakurthi, S.S., Flückiger, R., Aktas, H., Changolkar, A.K., Shahsafaei, A., Harneit, S., Kilic, E., and Halperin, J.A. (2000). Inhibition of translation initiation mediates the anticancer effect of the n-3 polyunsaturated fatty acid eicosapentaenoic acid. *Cancer Res.* **60**, 2919–25.

Parkington, J.E. (1998). *The impact of the systematic exploitation of marine foods on human evolution*. Colloquia of the Dual Congress of the International Association of the Study of Human Paleontology and the International Study of Human Biologists, Sun City, South Africa, 28 June to 4 July, 1998.

Peet, M. (2004). International variations in the outcome of schizophrenia and the prevalence of depression in relation to national dietary practices: an ecological analysis. *Br. J. Psychiatry* **184**, 404–8.

Peet, M. and Horrobin, D.F. (2002). A dose-ranging study of the effects of ethyl-eicosapentaenoate in patients with ongoing depression despite apparently adequate treatment with standard drugs. *Arch. Gen. Psychiatry* **59** (10), 913–19.

Peet, M., Glen, I., and Horrobin, D.F. (eds.) (1999). *Phospholipid spectrum disorder in psychiatry.* Marius Press, Carnforth.

Peet, M., Brind, J., Ramchand, C.N., Shah, S., and Vankar, G.K. (2001). Two double-blind placebo-controlled pilot studies of eicosapentaenoic acid in the treatment of schizophrenia. *Schizophr. Res.* **49** (3), 243–51.

Peet, M., Horrobin, D.F.; E-E Multicentre Study Group. (2002). A dose-ranging exploratory study of the effects of ethyl-eicosapentaenoate in patients with persistent schizophrenic symptoms. *J. Psychiatr. Res.* **36** (1), 7–18.

Pennington, B.F. and Bennetto, L. (1993). Main effects of transactions in the neuropsychology of conduct disorder. *Dev. Psychopathol.* **5** (1–2), 153–64.

Peppiatt, C. and Attwell, D. (2004). Neurobiology: feeding the brain. *Nature* **431** (7005), 137–8.

Pine, D.S., Wasserman, G.A., Miller, L., Coplan, J.D., Bagiella, E., Kovelenku, P., Myers, M.M., and Sloan, R.P. (1998). Heart period variability and psychopathology in urban boys at risk for delinquency. *Psychophysiology* **35** (5), 521–9.

Popkin, B.M., Ge, K., Zhai, F., Xuguang, G., and Ma, H. (1993). The nutritional transition in China: a cross sectional analysis. *Eur. J. Clin. Nutr.* **47**, 333–46.

Prentice, A. and Jebb, S. (2004). Energy intake/physical activity interactions in the homeostasis of body weight regulation. *Nutr. Rev.* **62** (7 Pt, 2), S98–104.

Prynne, C.J., Paul, A.A., Price, G.M., Day, K.C., Hilder, W.S., and Wadsworth, M.E. (1999). Food and nutrient intake of a national sample of 4-year-old children in 1950, comparison with the 1990s. *Public Health Nutr.* **2** (4), 537–47.

Puri, B.K. and Richardson, A.J. (1998). Sustained remission of positive and negative symptoms of schizophrenia following treatment with eicosapentaenoic acid. *Arch. Gen. Psychiatry* **55** (2), 188–9.

Puri, B.K., Richardson, A.J., Horrobin, D.F., Easton, T., Saeed, N., Oatridge, A., *et al.* (2000). Eicosapentaenoic acid treatment in schizophrenia associated with symptom remission, normalisation of blood fatty acids, reduced neuronal membrane phospholipid turnover and structural brain changes. *Int. J. Clin. Pract.* **54** (1), 57–63.

Puri, B.K., Counsell, S.J., Hamilton, G., Richardson, A.J., and Horrobin, D.F. (2001). Eicosapentaenoic acid in treatment-resistant depression associated with symptom remission, structural brain changes and reduced neuronal phospholipid turnover. *Int. J. Clin. Pract.* **55** (8), 560–3.

Puri, B.K., Counsell, S.J., Richardson, A.J., and Horrobin, D.F. (2002). Eicosapentaenoic acid in treatment-resistant depression. *Arch. Gen. Psychiatry* **59** (1), 91–2.

Raine, A., Venables, P.H., *et al.* (1995). High autonomic arousal and electrodermal orienting at age 15 years as protective factors against criminal behavior at age 29 years. *Am. J. Psychiatry* **152** (11), 1595–600.

Raine, A., Mellingen, K., Liu, J., Venables, P., and Mednick, S.A. (2003). Effects of environmental enrichment at ages 3–5 years on schizotypal personality and antisocial behavior at ages 17 and 23 years. *Am. J. Psychiatry* **160**, 1627–35.

Redmer, D.A., Wallace, J.M., and Reynolds, L.P. (2004). Effect of nutrient intake during pregnancy on fetal and placental growth and vascular development. *Domest. Anim. Endocrinol.* **27** (3), 199–217.

Richardson, A.J. and Puri, B.K. (2000). The potential role of fatty acids in attention-deficit/hyperactivity disorder. *Prostaglandins Leukot. Essent. Fatty Acids* **63** (1–2), 79–87.

Richardson, A.J. and Puri, B.K. (2002). A randomized double-blind, placebo-controlled study of the effects of supplementation with highly unsaturated fatty acids on ADHD-related symptoms in children with specific learning difficulties. *Prog. Neuropsychopharmacol. Biol. Psychiatry* **26** (2), 233–9.

Richardson, A.J. and Ross, M.A. (2000). Fatty acid metabolism in neurodevelopmental disorder: a new perspective on associations between attention-deficit/hyperactivity disorder, dyslexia, dyspraxia and the autistic spectrum. *Prostaglandins Leukot. Essent. Fatty Acids* **63** (1–2), 1–9.

Richardson, A.J., Easton, T., and Puri, B.K. (2000). Red cell and plasma fatty acid changes accompanying symptom remission in a patient with schizophrenia treated with eicosapentaenoic acid. *Eur. Neuropsychopharmacol.* **10** (3), 189–93.

Rightmire, G.P. and Deacon, H.J. (1991). Comparative studies of Late Pleistocene human remains from Klasies River Mouth, South Africa. *J. Hum. Evol.* **20**, 131–56.

Rubia, K., Oosterlaan, J., Sergeant, J., Brandeis, D., and van Leeuwen, T. (1998). Inhibitory dysfunction in hyperactive boys. *Behav. Brain Res.* **94** (1), 25–32.

Rutter, M., Giller, H., and Hagell, H. (1998). *Antisocial behaviour by young people.* Cambridge University Press, Cambridge.

Sazawal, S., Black, R.E., Menon, V.P., Dinghra, P., Caulfield, L.E., Dhingra, U., and Bagati, A. (2001). Zinc supplementation in infants born small for gestational age reduces mortality: a prospective, randomized, controlled trial. *Pediatrics* **108** (6), 1280–6.

Schab, D.W. and Trinh, N.H. (2004). Do artificial food colors promote hyperactivity in children with hyperactive syndromes? A meta-analysis of double-blind placebo-controlled trials. *J. Dev. Behav. Pediatr.* **25** (6), 423–34.

Schauss, A.G. (1978). Differential outcomes among probationers comparing orthomolecular approaches to conventional casework/counselling. *J. Orthomol. Psychiatry* **8** (3), 158–68.

Schoenthaler, S.J. (1983). The Northern California diet and Behaviour Program: an empirical examination of 3000 incarcerated juveniles in Stanilaus County Juvenile Hall. *Int. J. Biosoc. Res.* **5** (2), 99–106.

Schoenthaler, S.J. (1994). Sugar and children's behavior. *New Engl. J. Med.* **330** (26), 1901–4.

Schoenthaler, S.J. and Bier, I.D. (2000). The effect of vitamin and mineral supplementation juvenile delinquency among American schoolchildren: a randomised double blind placebo controlled study. *J. Altern. Complement. Med.* **6** (1), 7–17.

Schoenthaler, S.J., Amos, S., Doraz, W., *et al.* (1997). The effect of randomised vitamin-mineral supplementation on violent and non-violent antisocial behavior among incarcerated juveniles. *J. Nutr. Environ. Med.* **7**, 343–52.

Schrauzer, G.N. (2002). Lithium: occurrence, dietary intakes, nutritional essentiality. *J. Am. Coll. Nutr.* **21** (1), 14–21.

Shepherd, J.T., Abboud, F.M., and Geiger, S.R. (eds.) (1983). *Handbook of physiology.* Section 2, Vol. 3. *Cardiovascular system.* American Physiological Society, Bethesda, Maryland.

Sherman, L., *et al.* (1997). Preventing crime: what works, what doesn't, what's promising, a report to the United States Congress, prepared for the National Institute of Justice.

Shipman, P. and Walker, A. (1989). The costs of becoming a predator. *J. Hum. Evol.* **18**, 373–92.

Short, C. (2002). The rich diet, and the poor go hungry. In: *Foodstuff. Living in an age of feast and famine* (ed. J. Holden, L. Howland, and D. Stedman Jones), pp. 131–7. DEMOS, London.

Shreeve, J. (1995). *The Neandertal enigma: solving the mystery of modern human origins.* William Morrow, New York.

Simopoulos, A.P. and Pavlow, K.N. (eds.) (2001). Nutrition and fitness: diet, genes, physical activity and health, Proceedings of 4th International Conference on Nutrition and Fitness, Athens, May (2000). *World Rev. Nutr. Diet.* **89**.

Smith, D.J. (1995). Youth crime and conduct disorders: trends, patterns and causal explanations. In *Psychosocial disorders in young people. time, trends and their causes* (ed. M. Rutter and D.J. Smith), pp. 389–489. John Wiley and Sons, New York.

Stacey, M., Curnow, R.N., and Ward, N. (1994). Analyzing the relationship between reading ability and the concentrations of trace elements in hair and saliva. *J. Nutr. Med.* **4**, 25–31.

Stanley, B., Molcho, A, Stanley, M., Winchel, R., Gameroff, M.J., Parsons, B., and Mann J.J. (2000). Association of aggressive behavior with altered serotonergic function in patients who are not suicidal. *Am. J. Psychiatry* **157** (4), 609–14.

Stevens, L.J., Zentall, S.S., and Abate, M.L. (1996). Omega-3 fatty acids in boys with behavior, learning, and health problems. *Physiol. Behav.* **59**, 915–20.

Stevens, L., Zhang, W., Peck, L., Kuczek, T., Grevstad, N., Mahon, A., Zentall, S.S., Arnold, L.E., and Burgess, J.R. (2003). EFA supplementation in children with inattention, hyperactivity, and other disruptive behaviors. *Lipids* **38** (10), 1007–21.

Stoll, A.L., Severus, W.E., Freeman, M.P., Rueter, S., Zboyan, H.A., Diamond, E., *et al.* (1999). Omega 3 fatty acids in bipolar disorder: a preliminary double-blind, placebo-controlled trial. *Arch. Gen. Psychiatry* **56** (5), 407–12.

Stoney, C.M. and Engebretson T.O. (2000). Plasma homocysteine concentrations are positively associated with hostility and anger. *Life Sci.* **66** (23), 2267–75.

Stretesky, P.B. and Lynch, M.J. (2001). The relationship between lead exposure and homicide. *Arch. Pediatr. Adolesc. Med.* **155** (5), 579–82.

Stringer, C. (2000). Palaeoanthropology. Coasting out of Africa. *Nature* **405** (6782), 24–5, 27.

Stynder, D.D., Moggi-Cecchi, J., Berger, L.R., and Parkington, J.E. (2001). Human mandibular incisors from the late Middle Pleistocene locality of Hoedjiespunt 1, South Africa. *J. Hum. Evol.* **41** (5), 369–83.

Tanzer, F., Yaylaci, G., Ustdal, M., and Yonem, O. (2004). Serum zinc level and its effect on anthropometric measurements in 7–11 year-old children with different socioeconomic backgrounds. *Int. J. Vitamin Nutr. Res.* **74** (1), 52–6.

Taylor, E. (1998). Clinical foundations of hyperactivity research. *Behav. Brain Res.* **94** (1), 11–24.

Taylor, M.J., Carney, S., Geddes, J., and Goodwin, G. (2003). Folate for depressive disorders. *Cochrane Database Syst. Rev.* (02) CD003390.

Thomson, G.O.B., Raab, G.M., Hepburn, W.S., Hunter, R., Fulton, M., and Laxen, D.P.H. (1989). Blood lead levels and children's behaviour—results from the Edinburgh lead study. *J. Child Psychol. Psychiatry* **30**, 515–28.

Tobias, P.V. (2002). Some aspects of the multifaceted dependence of early humanity on water. *Nutr. Health* **16** (1), 13–17.

Tordoff, M.G. and Alleva, A.M. (1990). Effect of drinking soda sweetened with aspartame or high-fructose corn syrup on food intake and body weight. *Am. J. Clin. Nutr.* **51**, 963–9.

Tulen, J.H., Bruijn, J.A., *et al.* (1996) Anxiety and autonomic regulation in major depressive disorder: an exploratory study. *J Affect Disord.* **40**, 61–71.

Tuomilehto, J., *et al.* (2002). Prevention of type 2 diabetes mellitus by changes in lifestyle among subjects with impaired glucose tolerance. *New Engl. J. Med.* **344**, 1343–50.

Uauy, R., Hoffman, D.R., Peirano, P., Birch, D.G., and Birch, E.E. (2001). Essential fatty acids in visual and brain development. *Lipids* **36** (9), 885–95.

US Department of Health and Human Services (1996). *Physical activity and health: a report of the Surgeon General*, pp. 85–172. Centers for Disease Control and Prevention, Atlanta, Georgia.

Vallee, B.L. and Falchuk, K.H. (1981). Zinc and gene expression. *Philos. Trans. R. Soc. Lond. B Biol. Sci.* **294**, 185–97.

van Houwelingen, A.C. and Hornstra, G. (1994). Trans fatty acids in early human development. *World Rev. Nutr. Diet.* **75**, 175–8.

Virkkunen, M.E., Horrobin, D.F., Jenkins, D.K., and Manuku, M.S. (1987). Plasma phospholipids, essential fatty acids and prostaglandins in alcoholic, habitually violent and impulsive offenders. *Biol. Psychiatry* **22**, 1087–96.

Virkkunen, M., Goldman, D., et al. (1995). Low brain serotonin turnover rate (low CSF 5-HIAA) and impulsive violence. *J. Psychiatry Neurosci.* **20**, 271–5.

Voigt, R.G., Llorente, A.M., Jensen, C.L., Fraley, J.K., Berretta, M.C., and Heird, W.C. (2001). A randomized, double-blind, placebo-controlled trial of docosahexaenoic acid supplementation in children with attention-deficit/hyperactivity disorder. *J. Pediatr.* **139** (2), 189–96.

Wadsworth, M.E.J. (1976). Delinquency, pulse rates and early emotional deprivation, *Br. J. Criminol.* **16** (3), 245–56.

Walsh, W.J., Isaacson, H.R., Rehman, F., and Hall, A. (1997). Elevated blood copper/zinc ratios in assaultive young males. *Physiol. Behav.* **62**, 327–9.

Walsh, W.J., Glab, L.B., and Haakenson, M.L. (2004). Reduced violent behavior following biochemical therapy. *Physiol. Behav.* **82** (5), 835–9.

Wang, Y., Crawford, M.A., Chen, J., Li, J., Ghebremeskel, K., Campbell, C., Fan, W., Parker, R., and Leyton, J. (2003). Fish consumption, blood docosahexaenoic acid and chronic diseases in Chinese rural populations. *Comp. Biochem. Physiol. A Mol. Integr. Physiol.* **136** (1),127–40.

Ward, N.I., Durrant, S., Sankey, R.J., Bound, J.P., and Bryce-Smith, D. (1990). Elemental factors in human fetal development. *J. Nutr. Med.* **1**, 19–26.

Wauben, I.P., Xing, H.C., and Wainwright, P.E. (1999). Neonatal dietary zinc deficiency in artificially reared rat pups retards behavioral development and interacts with essential fatty acid deficiency to alter liver and brain fatty acid composition. *J. Nutr.* **129** (10), 1773–81.

WHO (1998). *Obesity: preventing and managing the global epidemic*, WHO Technical Report. Series 894, pp. 1–276. World Health Organization, Geneva.

WHO (2000). *A global agenda for combating malnutrition: progress report*, WHO/NHD/00.6. World Health Organization, Geneva.

WHO (2002). *Programming of chronic disease by impaired fetal nutrition: evidence and implications for policy and intervention strategies*, documents WHO/NHD/02.3 and WHO/NPH/02.1. World Health Organization, Geneva.

WHO (2003). *Diet, nutrition and the prevention of chronic disease*, Report of a Joint FAO/WHO Expert Consultation, Technical report Series 916. World Health Organization, Geneva.

WHO/UNICEF (1995). *Global prevalence of vitamin A deficiency*, MDIS Working Paper No. 2, document WHO/NUT/95.3. World Health Organization, Geneva.

WHO/UNICEF/International Council for the Control of Iodine Deficiency Disorders (1999). *Progress towards the elimination of iodine deficiency disorders (IDD)*, document WHO/NHD/99.4. World Health Organization, Geneva.

WHO/UNICEF/United Nations University (2001). *Iron deficiency anaemia assessment, prevention and control: a guide for programme managers*, document WHO/NHD/01.3. World Health Organization, Geneva.

Willet, M.C. (1995). Diet, nutrition, and avoidable cancer. *Environ. Health Perspect.* **103** (Suppl. 8), 165–70.

Willett, W.C., *et al.* (1993). Intake of trans fatty acids and risk of coronary heart disease among women. *Lancet* **341**, 581–5.

Wolraich, M.L., Lindgren, S.D., Stumbo, P.J., Stegink, L.D., Appelbaum, M.I., and Kiritsy, M.C. (1994). Effects of diets high in sucrose or aspartame on the behavior and cognitive performance of children. *New Engl. J. Med.* **330** (5), 301–7.

World Cancer Research Fund in association with American Institute for Cancer Research (1997). *Food, nutrition and the prevention of cancer: a global perspective.* American Institute for Cancer Research, Washington, DC.

Zanarini, M.C. and Frankenburg, F.R. (2003). Omega-3 fatty acid treatment of women with borderline personality disorder: a double-blind, placebo-controlled pilot study. *Am. J. Psychiatry* **160** (1), 167–9.

Zhang, M. and Kelley, A.E. (2000). Enhanced intake of high-fat food following striatal mu-opioid stimulation: microinjection mapping and fos expression. *Neuroscience* **99** (2), 267–77.

Part 3

Psychology of well-being

Barbara L. Fredrickson, PhD, is an associate professor of psychology at the University of Michigan, USA. Her research centres on the social, cognitive, physiological, and health consequences of positive emotions and is supported by the US National Institute of Mental Health. In 2000, she was awarded the the Templeton Positive Psychology Prize, which is the world's largest prize in psychology.

Chapter 8*

The broaden-and-build theory of positive emotions

Barbara L. Fredrickson

Introduction

At first blush, it might appear that positive emotions are important to the science of well-being simply because positive emotions are markers of optimal well-being. Certainly, moments in people's lives characterized by experiences of positive emotions—such as joy, interest, contentment, love, etc.—are moments in which they are not plagued by negative emotions, such as anxiety, sadness, anger, and the like. Consistent with this intuition, the overall balance of peoples' positive to negative emotions has been shown to contribute to their subjective well-being (Diener *et al.* 1991). In this sense, positive emotions signal optimal functioning, but this is far from the whole story. I argue that positive emotions also produce optimal functioning, not just within the present, pleasant moment, but over the long-term as well. The bottom-line message is that people should cultivate positive emotions in themselves and in those around them, not just as an end states in themselves, but also as a means to achieving psychological growth and improved psychological and physical well-being over time.

History of research on positive emotions

This view of positive emotions represents a significant departure from traditional approaches to the study of positive emotions. In this section I provide a brief, selective review of the history of research on positive emotions.

Positive emotions have been neglected relative to negative emotions

Relative to the negative emotions, positive emotions have received little empirical attention. There are several interrelated reasons for this. One reason, which has plagued psychology more generally (Seligman and Csikszentmihalyi 2000), is the traditional focus on psychological problems alongside remedies for those problems. Negative emotions—when extreme, prolonged, or contextually inappropriate— produce many grave problems for individuals and society, ranging from phobias and anxiety disorders, aggression and violence, depression and suicide, eating disorders and sexual dysfunction, to a host of stress-related physical disorders. Although positive emotions do at times pose problems (e.g. mania, drug addiction), these problems have often assumed lower priority among psychologists and emotion researchers. So, in part as a result of their association with problems and dangers, negative emotions have captured most research attention.

Another reason positive emotions have been sidelined is the habit among emotion theorists of creating models of emotions in general. Such models are typically built to the specifications of those attention-grabbing negative emotions (e.g. fear and anger), with positive emotions squeezed in later, often seemingly as an afterthought. For instance, key to many theorists' models of emotion is the idea that emotions are, by definition, associated with specific action tendencies (Frijda 1986; Frijda *et al.* 1989; Tooby and Cosmides 1990; Lazarus 1991; Levenson 1994; Oatley and Jenkins 1996). Fear, for example, is linked with the urge to escape, anger with the urge to attack, disgust with the urge to expel, and so on. No theorist argues that people invariably act out these urges when feeling particular emotions rather, people's ideas about possible courses of action narrow in on a specific set of behavioural options. A key idea in these models is that having a specific action tendency come to mind is what made an emotion evolutionarily adaptive: these were among the actions that worked best in getting our ancestors out of life-or-death situations. Another key idea is that specific action tendencies and physiological changes go hand-in-hand. So, for example, when you have an urge to escape when feeling fear, your body reacts by mobilizing appropriate autonomic support for the possibility of running by redirecting blood flow to large muscle groups.

Although specific action tendencies have been invoked to describe the form and function of positive emotions as well, the action tendencies identified for positive emotions are notably vague and underspecified (Fredrickson and Levenson 1998). Joy, for instance, is linked with aimless activation, interest with attending, and contentment with inactivity (Frijda 1986). These tendencies are far too general to be called specific (Fredrickson 1998). Although a few theorists had earlier noted that fitting positive emotions into emotion-general models posed problems (Lazarus 1991; Ekman 1992), this acknowledgement was not accompanied by any new or revised models to better accommodate the positive emotions. Instead, the difficulties inherent in 'shoehorning' the positive emotions into emotion-general models merely tended to marginalize them further. Many theorists, for instance, minimize challenges to their models by maintaining their focus on negative emotions, paying little or no attention to positive emotions.

Positive emotions are often confused with related affective states

Perhaps because they have received less direct scrutiny, the distinctions among positive emotions and other closely related affective states, like sensory pleasure and positive mood, have often been blurred instead of sharpened. Although working definitions of emotions vary somewhat across researchers, consensus is emerging that emotions (both positive and negative) are best conceptualized as multicomponent response tendencies that unfold over relatively short time spans. Typically, emotions begin with an individual's assessment of the personal meaning of some antecedent event: what Lazarus (1991) called the person–environment relationship or adaptational encounter. Either conscious or unconscious, this appraisal process triggers a cascade of response tendencies manifest across loosely coupled component systems, such as subjective experience, facial expressions, and physiological changes.

Sometimes various forms of sensory pleasure (e.g. sexual gratification, satiation of hunger or thirst) are taken to be positive emotions because they share with positive emotions a pleasant subjective feel and include physiological changes, and because sensory pleasure and positive emotions often co-occur (e.g. sexual gratification within a loving relationship). However, emotions differ from physical sensations in that emotions require cognitive appraisals or meaning assessments to be initiated. In contrast to positive emotions, pleasure can be caused simply by changing the immediate physical environment (e.g. eating or otherwise stimulating the body). Moreover, whereas pleasure depends heavily on bodily stimulation, positive emotions more often occur in the absence of external physical sensation (e.g. joy at receiving good news or interest in a new idea). Pleasurable sensations, then, are best considered automatic responses to fulfilling bodily needs. In fact, Cabanac (1971) suggested that people experience sensory pleasure with any external stimulus that 'corrects an internal trouble'. A cool bath, for instance, is only pleasant to someone who is overheated (who thus needs to be cooled). Likewise, food is pleasant to the hungry person, but becomes less pleasant—even unpleasant—as that person becomes satiated.

Positive emotions are also often confused with positive moods. However, emotions differ from moods in that emotions are about some personally meaningful circumstance (i.e. they have an object) and are typically short-lived and occupy the foreground of consciousness. By contrast, moods are typically free-floating or objectless, more long-lasting, and occupy the background of consciousness (Oatley and Jenkins 1996; Rosenberg 1998). These distinctions between emotions and moods, however, are guarded more at theoretical than empirical levels. In research practice, virtually identical techniques are used for inducing positive moods and positive emotions (e.g. giving gifts, viewing comedies).

Functions of positive emotions are linked to urges to approach or continue

Most commonly, the function of all positive emotions has been identified as facilitating approach behaviour (Cacioppo *et al.* 1993; Davidson 1993; Frijda 1994) or continued action (Carver and Scheier 1990; Clore 1994). From this perspective, experiences of positive emotions prompt individuals to engage with their environments and partake in activities, many of which were evolutionarily adaptive for the individual, its species, or both. This link between positive emotions and activity engagement provides an explanation for the often-documented positivity offset, or the tendency for individuals to experience mild positive affect frequently, even in neutral contexts (Diener and Diener 1996; Cacioppo *et al.* 1999). Without such an offset, individuals would most often be unmotivated to engage with their environments. However, with such an offset, individuals exhibit the adaptive bias to approach and explore novel objects, people, or situations.

Although positive emotions do often appear to function as internal signals to approach or continue, they share this function with other positive affective states as well. Sensory pleasure, for instance, motivates people to approach and continue consuming whatever stimulus is biologically useful for them at the moment (Cabanac 1971). Likewise, free-floating positive moods motivate people to continue along any line of thinking or action that they have initiated (Clore 1994). As such, functional accounts of positive emotions that emphasize tendencies to approach or continue may capture only the lowest common denominator across all affective states that share a pleasant subjective feel. This traditional approach leaves additional functions that are unique to positive emotions uncharted.

The broaden-and-build theory of positive emotions

Traditional approaches to the study of emotions have tended to ignore positive emotions, squeeze them into purportedly emotion-general models, confuse them with closely related affective states, and describe their function in terms of generic tendencies to approach or continue. Sensing that these approaches do not do justice to positive emotions, I have developed an alternative model for positive emotions that better captures their unique effects. I call this the broaden–and–build theory of positive emotions because positive emotions appear to broaden people's momentary thought–action repertoires and build their enduring personal resources (Fredrickson 1998, 2001).

I contrast this new model with traditional models based on specific action tendencies. Specific action tendencies work well to describe the form and function of negative emotions, and should be retained for models of this subset of emotions. Without loss of theoretical nuance, a specific action tendency can be re-described as the outcome of a psychological process that narrows a person's momentary thought–action repertoire by calling to mind an urge to act in a particular way (e.g. escape, attack, expel). In a life-threatening situation, a narrowed thought–action repertoire promotes quick and decisive action that carries direct and immediate benefit: specific action tendencies called forth by negative emotions represent the sort of actions that worked best to save our ancestors' lives and limbs in similar situations.

However, positive emotions seldom occur in life-threatening situations. As such, a psychological process that narrows a person's momentary thought–action repertoire to promote quick and decisive action may not be needed. Instead, positive emotions have a complementary effect: relative to neutral states and routine action, positive emotions broaden people's momentary thought–action repertoires, widening the array of the thoughts and actions that come to mind. Joy, for instance, creates the urge to play, push the limits, and be creative—urges evident not only in social and physical behaviour, but also in intellectual and artistic behaviour. Interest, a phenomenologically distinct positive emotion, creates the urge to explore, take in new information and experiences, and expand the self in the process. Contentment, a third distinct positive emotion, creates the urge to sit back and savour current life circumstances, and integrate these circumstances into new views of self and of the world. Love—viewed as an amalgam of distinct positive emotions (e.g. joy, interest, and contentment) experienced within contexts of safe, close relationships—creates recurring cycles of urges to play with, explore, and savour our loved ones. These various thought–action tendencies—to play, to explore, or to savour and integrate— each represent ways in which positive emotions broaden habitual modes of thinking or acting. (For descriptions of pride and elevation from the perspective of the broaden-and-build theory see Fredrickson and Branigan 2001; for a description of gratitude see Fredrickson 2004.)

In contrast to negative emotions, which carry direct and immediate adaptive benefits in situations that threaten survival, the broadened thought–action repertoires triggered by positive emotions are beneficial in other ways. Specifically, broadened mindsets carry indirect and long-term adaptive benefits because broadening builds enduring personal resources.

Take play as an example. Specific forms of chasing play evident in juveniles of a species—like running into a flexible sapling or branch and catapulting oneself in an unexpected direction—are re-enacted in adults of that species exclusively during predator avoidance (Dolhinow 1987). Such correspondences between juvenile play manoeuvres and adult survival manoeuvres suggest that juvenile play builds enduring physical resources (Caro 1988; Boulton and Smith 1992). Play also builds enduring social resources: social play, with its shared amusement and smiles, builds lasting social bonds and attachments (Lee 1983; Simons *et al.* 1986; Aron *et al.* 2000) that can become the locus of subsequent social support. Childhood play also builds enduring intellectual resources, by increasing creativity (Sherrod and Singer 1989), creating theory of mind (Leslie 1987), and fuelling brain development (Panksepp 1998). Similarly, the exploration prompted by the positive emotion of interest creates knowledge and intellectual complexity, and the savouring prompted by contentment produces self-insight and alters world views. So each of these phenomenologically distinct positive emotions shares the feature of augmenting an individual's personal resources, ranging from physical and social resources to intellectual and psychological resources (for more detailed reviews see Fredrickson 1998, 2001; Fredrickson and Branigan 2001).

Importantly, the personal resources accrued during states of positive emotions are durable. They outlast the transient emotional states that led to their acquisition. By consequence, then, the often incidental effect of experiencing a positive emotion is an increase in one's personal resources. These resources can be drawn on in subsequent moments and in different emotional states. Through experiences of positive emotions, then, people transform themselves, becoming more creative, knowledgeable, resilient, socially integrated, and healthy individuals.

In short, the broaden–and–build theory describes the form of positive emotions in terms of broadened thought–action repertoires, and describes their function in terms of building enduring personal resources. In doing so, the theory provides a new perspective on the evolved adaptive significance of positive emotions. Those of our ancestors who succumbed to the urges sparked by positive emotions—to play, explore, and so on—would have as a consequence accrued more personal resources. When these same ancestors later faced inevitable threats to life and limb, their greater personal resources would have translated into greater odds of survival and, in turn, greater odds of living long enough to reproduce. To the extent then that the capacity to experience positive emotions is genetically encoded, this capacity, through the process of natural selection, would have become part of our universal human nature.

Summary of current research findings

Empirical support for several key propositions of the broaden–and–build theory can be drawn from multiple subdisciplines within psychology, ranging from work on cognition and intrinsic motivation, to attachment styles and animal behaviour (reviewed in Fredrickson 1998). This evidence suggests that positive emotions broaden the scopes of attention, cognition, and action, and that they build physical, intellectual, and social resources. However, much of this evidence, because it pre-dated the broaden–and–build theory, provides only indirect support for the model. Here, I briefly describe recent direct tests of hypotheses drawn from the broaden–and–build theory.

Positive emotions broaden thought–action repertoires

Foundational evidence for the proposition that positive emotions broaden people's momentary thought–action repertoires comes from 2 decades of experiments conducted by Isen and colleagues (reviewed in Isen 2000). They have documented that people experiencing positive affect show patterns of thought that are notably unusual (Isen *et al.* 1985), flexible (Isen and Daubman 1984), creative (Isen *et al.* 1987), integrative (Isen *et al.* 1991), open to information (Estrada *et al.* 1997), and efficient (Isen and Means 1983; Isen *et al.* 1991). They have also shown that those experiencing positive affect show increased preference for variety and accept a broader array of behavioural options (Kahn and Isen 1993). In general terms, Isen has suggested that positive affect produces a 'broad, flexible cognitive organization and ability to integrate diverse material' (Isen 1990, p. 89), effects linked to increases in brain dopamine levels (Ashby *et al.* 1999). So, although Isen's work does not target specific positive emotions or thought–action tendencies *per se*, it provides the strongest evidence that positive affect broadens cognition. Whereas negative emotions have long been known to narrow people's attention, making them miss the forest for the trees, more recent work suggests that positive affect also expands attention (Derryberry and Tucker 1994). The evidence comes from studies that use global–local visual processing paradigms to assess biases in attentional focus. Negative states—like anxiety, depression, and failure—predict local biases consistent with narrowed attention, whereas positive states—like subjective well-being, optimism, and success—predict global biases consistent with broadened attention (Derryberry and Tucker 1994; Basso *et al.* 1996).

These findings provide initial empirical footing for the hypothesis, drawn from the broaden-and-build theory, that distinct types of positive emotion serve to broaden people's momentary thought–action repertoires, whereas distinct types of negative emotions serve to narrow these same repertoires. Together with Christine Branigan, I tested this broaden hypothesis by showing research participants short emotionally evocative film clips to induce the specific emotions of joy, contentment, fear, and anger. We also used a non-emotional film clip as a neutral control condition. Immediately following each film clip, we measured the breadth of participants' thought–action repertoires. We asked them to step away from the specifics of the film and imagine being in a situation themselves in which similar feelings would arise. Given this feeling, we asked them to list what they would like to do right then. Participants recorded their responses on up to 20 blank lines that began with the phrase 'I would like to . . . '.

Tallying the behaviours each participant listed, we found support for the broaden hypothesis. Participants in the two positive emotion conditions (joy and contentment) identified more things that they would like to do right then relative to those in the two negative emotion conditions (fear and anger) and, more importantly, relative to those in the neutral control condition. Those in the two negative emotion conditions also named fewer behaviours than those in the neutral control condition (Fig. 8.1; Fredrickson and Branigan 2005).

In several other experiments, we assessed broadened thinking by measuring the degree to which people see the 'big picture' or focus on smaller details. We do this by using what are called global–local visual processing tasks. An example item from one such task is shown in Fig. 8.2. A participant's task is to judge which of

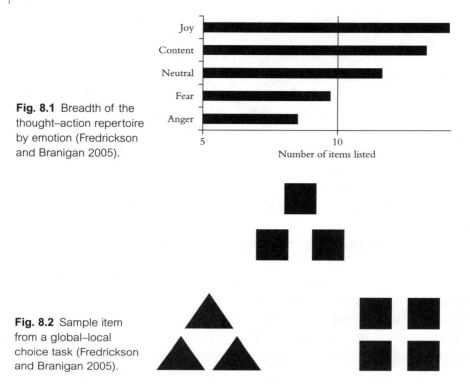

Fig. 8.1 Breadth of the thought–action repertoire by emotion (Fredrickson and Branigan 2005).

Fig. 8.2 Sample item from a global–local choice task (Fredrickson and Branigan 2005).

two comparison figures (bottom) is more similar to a standard figure (top). Neither choice is right or wrong. But one comparison figure resembles the standard in global configuration (lower left), and the other in local detail elements (lower right). Using this and similar measures, we have found that, compared with those in negative or neutral states, people who experience positive emotions—as assessed either by self-report or by electromyographic signals coming from the face—show evidence of broadened thinking (Fredrickson and Branigan 2005; K.J. Johnson, C.E. Waugh, B.L. Fredrickson, and T. Wager, unpublished data).

These data provide preliminary evidence that two distinct types of positive emotion—a high activation state of joy and a low activation state of contentment—each produce a broader attentional scope and thought–action repertoire than does a neutral state. Likewise, two distinct types of negative emotion—fear and anger—each produce a narrower attentional scope and thought–action repertoire than does a neutral state. This pattern of results supports a core proposition of the broaden-and-build theory: that distinct positive emotions widen the array of thoughts and actions that come to mind. By contrast, distinct negative emotions, as models based on specific action tendencies would suggest, would shrink this same array. So far, seven different studies from our laboratory support the broaden hypothesis (Fredrickson and Branigan 2005; K.J. Johnson and B.L. Fredrickson, in press; K.J. Johnson, C.E. Waugh, B.L. Fredrickson, and T. Wager, unpublished data; C.E. Waugh and

B.L. Fredrickson, unpublished data). Supportive evidence from other laboratories is also emerging (Gasper and Clore 2002; Bolte *et al.* 2003).

Positive emotions undo lingering negative emotions

Evidence for the broaden hypothesis has clear implications for the strategies that people use to regulate their experiences of negative emotions. If negative emotions narrow the momentary thought–action repertoire and positive emotions broaden this same repertoire, then positive emotions ought to function as efficient antidotes for the lingering effects of negative emotions. In other words, positive emotions might 'correct' or 'undo' the after-effects of negative emotions; we call this the 'undo hypothesis' (Fredrickson and Levenson 1998; Fredrickson *et al.* 2000). The basic observation that positive emotions (or key components of them) are somehow incompatible with negative emotions is not new, and has been demonstrated in earlier work on anxiety disorders (e.g. systematic desensitization; Wolpe 1958), motivation (e.g. the opponent–process theory; Solomon and Corbit 1974), and aggression (e.g. the principle of incompatible responses; Baron 1976). Even so, the precise mechanism ultimately responsible for this incompatibility has not been adequately identified. The broaden function of positive emotions may play a role. By broadening a person's momentary thought–action repertoire, a positive emotion may loosen the hold that a negative emotion has gained on that person's mind and body by dismantling or undoing the preparation for a specific action.

One marker of the specific action tendencies associated with negative emotions is increased cardiovascular activity, which redistributes blood flow to relevant skeletal muscles. In the context of negative emotions, then, positive emotions should speed recovery from—or undo—this cardiovascular reactivity, returning the body to more mid-range levels of activation. By accelerating cardiovascular recovery, positive emotions create the bodily context suitable for pursuing the broader array of thoughts and actions called forth.

My collaborators and I tested this undo hypothesis by first inducing a high-activation negative emotion in all participants (Fredrickson and Levenson 1998; Fredrickson *et al.* 2000). In the latter study, we used a time-pressured speech preparation task. In just 1 minute, participants prepared a speech on 'why you are a good friend' believing that their speech would be videotaped and evaluated by their peers. This speech task induced the subjective experience of anxiety along with increases in heart rate, peripheral vasoconstriction, and systolic and diastolic blood pressure. Into this context of anxiety-related sympathetic arousal, we randomly assigned participants to view one of four films. Two films elicited mild positive emotions (joy and contentment) and a third served as a neutral control condition. Notably, these three films, when viewed after a resting baseline, elicit virtually no cardiovascular reactivity (Fredrickson *et al.* 2000). So the two positive films used in this study are indistinguishable from neutrality with respect to

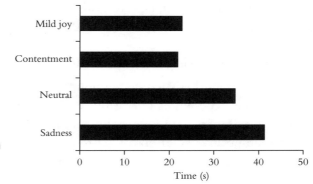

Fig. 8.3 Speed of cardiovascular recovery by emotion (Fredrickson *et al.* 2000).

cardiovascular changes. Our fourth film elicited sadness. We chose sadness as an additional comparison because, among the negative emotions, it has not been definitively linked to a high-energy action tendency, and thus could be a contender for speeding cardiovascular recovery.

The undo hypothesis predicts that those who experience positive emotions on the heels of a high-activation negative emotion will show the fastest cardiovascular recovery. We tested this by measuring the time elapsed from the start of the randomly assigned film, until the cardiovascular reactions induced by the negative emotion returned to baseline levels. In three independent samples, participants in the two positive emotion conditions (joy and contentment) exhibited faster cardiovascular recovery than those in the neutral control condition, and faster than those in the sadness condition, who exhibited the most protracted recovery (Fig. 8.3; Fredrickson and Levenson 1998; Fredrickson *et al.* 2000).

Recalling that the two positive emotion films and the neutral film did not differ in what they do to the cardiovascular system, these data suggest that they do differ in what they can undo within this system. Two distinct types of positive emotion— mild joy and contentment—share the ability to undo the lingering cardiovascular after-effects of negative emotions. Although the precise cognitive and physiological mechanisms of this undo effect remain unknown, the broaden-and-build theory suggests that broadening at the cognitive level mediates undoing at the cardiovascular level. Phenomenologically, positive emotions may help people place the events in their lives in broader context, lessening the resonance of any particular negative event.

Positive emotions fuel psychological resiliency

Evidence for the undo effect of positive emotions suggests that people might improve their psychological well-being, and perhaps also their physical health, by cultivating experiences of positive emotions at opportune moments to cope with negative emotions (Fredrickson 2000). Folkman and colleagues have made similar claims that experiences of positive affect during chronic stress help people cope (Lazarus *et al.* 1980; Folkman 1997; Folkman and Moskowitz 2000). Evidence supporting this claim can be drawn from experiments showing that positive affect facilitates attention to negative, self-relevant information (Trope and Neter 1994; Reed and Aspinwall 1998; Trope and Pomerantz 1998; for a review see Aspinwall 1998). Extrapolating from these findings, Aspinwall (2001) describes how positive affect and positive beliefs serve as resources for people coping with adversity (see also Aspinwall and Taylor 1997; Taylor *et al.* 2000).

It seems plausible that some individuals, more than others, might intuitively understand and use the benefits of positive emotions to their advantage. One candidate individual difference is psychological resilience. Resilient individuals are said to 'bounce back' from stressful experiences quickly and efficiently, just as resilient metals bend, but do not break (Lazarus 1993; Carver 1998).

The association between resilience and positive emotions is supported by the network of correlates of resilience discovered across a range of self-report, observational, and longitudinal studies. This converging evidence suggests that resilient people have optimistic, zestful, and energetic approaches to life, are curious and open to new experiences, and are characterized by high positive emotionality (Block and Kremen 1996; Klohnen 1996). Although positive emotions are no doubt at times an outcome of resilient coping (Block and Kremen 1996), other evidence suggests that resilient people may also use positive emotions to achieve their effective coping, indicating reciprocal causality. For instance, resilient people have been found to use humour (Werner and Smith 1992; Wolin and Wolin 1993; Masten 1994), creative exploration (Cohler 1987), relaxation, and optimistic thinking (Murphy and Moriarty 1976; Anthony 1987) as ways of coping. This diverse set of coping strategies has in common the ability to cultivate one or more positive emotions, such as amusement, interest, contentment, or hope, respectively. Strikingly, resilient people not only cultivate positive emotions in themselves to cope, but they are also skilled at eliciting positive emotions in others (i.e. care-givers early in life and companions later on), which creates a supportive social context that also facilitates coping (Demos 1989; Werner and Smith 1992; Kumpfer 1999).

Conceptualizing resilience as the ability to 'bounce back' after adversity suggests that, relative to their less resilient peers, resilient individuals would exhibit faster cardiovascular recovery following a high-activation negative emotion. Additionally, the broaden-and-build theory suggests that this ability to 'bounce back' to cardio-vascular baseline may be fuelled by experiences of positive emotion.

Michele Tugade and I tested these two hypotheses about resilient individuals, using the same time-pressured speech preparation task (described earlier) to induce a high-activation negative emotion. We measured psychological resilience using Block and Kremen's (1996) self-report scale. Interestingly, resilience did not predict the levels of anxiety that participants reported experiencing during the speech task or the magnitude of their cardiovascular reactions to the stressful task, both of which were considerable. Resilience did, however, predict participants' reports of positive emotions. Before the speech task was even introduced, more resilient individuals reported higher levels of pre-existing positive affect on an initial mood measure. When later asked how they felt during the time-pressured speech preparation phase, more resilient individuals reported that, alongside their high anxiety, they also experienced higher levels of happiness and interest.

As predicted by the theoretical definition of psychological resilience, more resilient participants exhibited significantly faster returns to baseline levels of cardiovascular activation following the speech task. Moreover, as predicted by the broaden-and-build theory, this difference in time to achieve cardiovascular recovery was accounted for by differences in positive emotions (Tugade and Fredrickson 2004).

These data suggest that positive emotions may fuel psychological resilience. In effect, then, resilient individuals may be—wittingly or unwittingly—expert users of the undo effect of positive emotions (Tugade and Fredrickson 2002). A prospective field study of American college students before and after the terrorist attacks of 11 September 2001 provided consistent evidence. Relative to their less resilient peers, resilient individuals were less likely to become depressed and more likely to experience post-crisis growth after the attacks. More importantly, the greater positive emotions that resilient people experienced fully accounted for each of these beneficial effects (Fredrickson *et al.* 2003).

Positive emotions build personal resources

Evidence suggests, then, that positive emotions may fuel individual differences in resilience. Noting that psychological resilience is an enduring personal resource, the broaden-and-build theory makes the bolder prediction that experiences of positive emotions might also, over time, build psychological resilience, not just reflect it. That is, to the extent that positive emotions broaden the scopes of attention and cognition enabling flexible and creative thinking, they should also augment people's enduring coping resources (Isen 1990; Aspinwall 1998, 2001; Fredrickson and Joiner 2002).

Together with my students, I recently completed an experimental test of the build effect of positive emotions. Each evening for 1 month, college students logged on to a secure web site, reported the emotions they had experienced in the past 24 hours, and then wrote about the best, worst, or a seemingly ordinary event of their day (topics randomly assigned within participants). Using a between-groups design, we induced a subset of these students to feel more positive emotions over the month by asking them to find the positive meaning and long-term benefits within their best, worst, and seemingly ordinary experiences each day. At the end of the month, compared with those who did not make this daily effort to find positive meaning, those who did showed increases in resilience. Moreover, these increases in resilience were completely accounted for by the greater positive emotions garnered by the daily habit of finding positive meaning (Fredrickson *et al.* 2004). These data support the causal direction posited by the broaden-and-build theory: positive emotions produce increments in personal resources. Beyond providing the first direct evidence for this causal claim, this new finding is important for two additional reasons. First, our past work has shown that resilience, as measured in this experiment, is a consequential trait that predicts both psychological well-being and growth, and physiological recovery (Fredrickson *et al.* 2003; Tugade and Fredrickson 2004). Second, it suggests how people might begin to harness the beneficial effects of positive emotions to optimize their own well-being: by regularly finding positive meaning within the daily ups and downs of life (Fredrickson 2000).

Positive emotions fuel psychological and physical well-being

By broadening people's mindsets and building their psychological resources, over time positive emotions should also enhance people's emotional and physical well-being. Consistent with this view, studies have shown that people who experience positive emotions during bereavement are more likely to develop long-term plans and goals. Together with positive emotions, plans and goals predict greater psychological well-being 12 months post-bereavement (Stein *et al.* 1997; for related work see Bonanno and Keltner 1997; Keltner and Bonanno 1997). One way in which people experience positive emotions in the face of adversity is by finding positive meaning in ordinary events or within the adversity itself (Affleck and Tennen 1996; Folkman and Moskowitz 2000; Fredrickson 2000). Importantly, the relationship between positive meaning and positive emotions is considered reciprocal: finding positive meaning not only triggers positive emotion, but also positive emotions—because they broaden thinking—should increase the likelihood of finding positive meaning in subsequent events (Fredrickson 2000).

These suspected reciprocal relations among positive emotions, broadened thinking, and positive meaning suggest that, over time, the effects of positive emotions should accumulate and compound: the broadened attention and cognition triggered by earlier experiences of positive emotion should facilitate coping with adversity, and this improved coping should predict future experiences of positive emotion. As this cycle continues, people build their psychological resilience and enhance their emotional well-being.

The cognitive literature on depression had already documented a downward spiral in which depressed mood and the narrowed, pessimistic thinking it engenders influence one another reciprocally, over time leading to ever-worsening moods, and even clinical levels of depression (Beck 1979; Peterson and Seligman 1984). The broaden-and-build theory suggests a complementary upward spiral in which positive emotions and the broadened thinking they engender also influence one another reciprocally, leading to appreciable increases in emotional well-being over time. Positive emotions may trigger these upward spirals, in part, by building resilience and influencing the ways that people cope with adversity. (For a complementary discussion of upward spirals, see Aspinwall 1998, 2001.)

Together with Thomas Joiner, I conducted an initial prospective test of the hypothesis that, through cognitive broadening, positive emotions produce an upward spiral towards enhanced emotional well-being. We assessed positive and negative emotions, as well as a concept that we call broad-minded coping, at two time points, 5 weeks apart. Broad-minded coping was tapped by items such as 'think of different ways to deal with the problem' and 'try to step back from the situation and be more objective'.

Our data revealed evidence for at least a fragment of an upward spiral. Individuals who experienced more positive emotions than others over time became more resilient to adversity, as indexed by increases in broad-minded coping. These enhanced coping skills, in turn, predicted increased positive emotions over time (Fredrickson and Joiner 2002). These findings suggest that positive emotions and broad-minded coping mutually build on one another: positive emotions not only make people feel good in the present, but also—by broadening thinking and building resources—positive emotions increase the likelihood that people will feel good in the future.

What are the long-term consequences of such upward spirals? A recent longitudinal study that spanned 7 decades suggests that the pay-off may be longer lives. The data come from a study of 180 Catholic nuns who pledged their lives not only to God but also to science. As part of a larger study of ageing and Alzheimer's disease, these nuns agreed to give scientists access to their archived work and medical records (and to donate their brains at death). The work archives included autobiographies hand-written when the nuns were in their early twenties. Researchers scored these essays for emotional content, recording instances of positive emotions such as happiness, interest, love, and hope and negative emotions such as sadness, fear, and disinterest. No association was found between negative emotional content and mortality, perhaps because it was rather rare in these essays. But a strong association was found between positive emotional content and mortality: those nuns who expressed the most positive emotions lived on average 10 years longer than those who expressed the least positive emotions (Danner *et al.* 2001). This is not an isolated finding. Several other researchers have found the same solid link between feeling good and living longer, even when accounting for age, gender, health status, social class, and other possible confounds (Ostir *et al.* 2000, 2001; Levy *et al.* 2002; Moskowitz 2003).

Complex dynamics triggered by positive emotions

The broaden-and-build theory challenges existing paradigms because it casts positive emotions in a far more consequential role in the story of human welfare. Whereas traditional perspectives have suggested that positive emotions mark or signal health and well-being (Kahneman 1999; Diener 2000), the broaden-and-build theory suggests that positive emotions also produce health and well-being (Fredrickson 2001). Put differently, to the extent that the broaden-and-build effects of positive emotions accumulate and compound over time, positive emotions carry the capacity to transform individuals for the better, making them healthier and more socially integrated, knowledgeable, effective, and resilient. In short, the theory suggests that positive emotions fuel human flourishing.

Flourishing describes a state of optimal human functioning, one that simultaneously implies growth, goodness, resilience, and generativity (Keyes 2003; B.L. Fredrickson and M. Losada, in press). Flourishing can be contrasted, not just with pathology, but also with languishing, which has been described as a disorder on the mental health continuum experienced by people who describe their lives as 'hollow', 'empty', or 'stuck in a rut'. Although distinct from mental illness, languishing has been linked with comparable levels of emotional distress, limitations in daily activities, psychosocial impairment, and economic cost from lost workdays (Keyes 2003). Building on my past work, I argue that positive emotions—by broadening people's mindsets and building their enduring resources—can alleviate human languishing and seed human flourishing.

How much positivity is needed to flourish? A nonlinear dynamical model developed to describe flourishing business teams suggests an answer. Losada (1999) observed 60 management teams in 1 hour meetings as they crafted their annual strategic plans. Behind one-way mirrors, trained coders rated every speech act on three opposing pairs: positive–negative; inquiry–advocacy; and other–self. Utterances were coded as 'positive' if speakers showed support, encouragement, or appreciation, and as 'negative' if they showed disapproval, sarcasm, or cynicism. They were coded as 'inquiry' if they offered questions aimed at exploring or examining a position, and as 'advocacy' if they offered arguments in favour of the speaker's viewpoint. They were coded as 'self' if they referred to the person speaking, the group present, or the company, and as 'other' if they referenced a person or group not present or not part of the company.

Later, Losada identified which teams were flourishing, defined as showing uniformly high performance across three indicators: profitability; customer satisfaction; and evaluations by superiors, peers, and subordinates. Other teams had mixed or uniformly low performance. Analyses of the time-series of the observed data, as well as their lead–lag relationships, led Losada (1999) to develop a nonlinear dynamics model to capture the interaction patterns observed within the different levels of team performance.

The complex dynamics of flourishing business teams followed the classic 'butterfly' trajectory of the Lorenz system, first discovered in the 1960s to represent the complex dynamics underlying weather patterns (Losada 1999; Losada and

Heaphy 2004). For flourishing teams, the dynamic structure showed the highest ratio of positivity to negativity and the broadest range of inquiry and advocacy.

The dynamics of medium performance teams were different. Although they begin with a complex butterfly structure that mirrored that of the flourishing teams, albeit at a much lower positivity ratio and a narrower range of inquiry and advocacy, they did not show enough behavioural flexibility to be resilient to adversity. In fact, the dynamics of medium performance teams calcified after any encounter with extreme negativity. After peak negativity, these teams lost behavioural flexibility and their ability to question and ended up languishing in a limit cycle centred on self-absorbed advocacy (Losada 1999; Losada and Heaphy 2004).

The dynamics of low performance teams were different still. They never showed the complex and generative dynamics of high performance teams, but instead were stuck in self-absorbed advocacy from the start. But worse than being stuck in the endless loop of a limit cycle, their dynamics showed the properties of a fixed point attractor: they eventually lost behavioural flexibility altogether as they spiralled down to a dead stop.

The nonlinear dynamical system that emerged from Losada's in-depth study of business teams resonates well with the broaden-and-build theory (B.L. Fredrickson and M. Losada, unpublished data). Just as predicted by the broaden-and-build theory, Losada's work shows that higher levels of positivity are linked with: (1) broader behavioural repertoires; (2) greater flexibility and resilience to adversity; and (3) optimal functioning or flourishing. Losada also found that higher levels of positivity are linked with greater social resources, as indexed by the degree of connectivity among team members (Losada and Heaphy 2004).

In fact, the most potent single variable within Losada's mathematical model is the ratio of positivity to negativity. If this ratio is known, the model can predict whether the complex dynamics of flourishing will be evident. Developing Losada's mathematicalmodel further, B.L. Fredrickson and M. Losada (unpublished data) identified the positivity ratio at which the dynamical structure bifurcates between a limit cycle of languishing and the complex dynamics of flourishing. This turns out to be a ratio of positivity to negativity of about 3:1. We hypothesize that only at or above this ratio is positivity in sufficient supply to seed human flourishing (B.L. Fredrickson and M. Losada, unpublished data).

We sought to test this hypothesis with observed data on human flourishing at multiple levels of analysis. We first drew from archival data gathered by Fredrickson et al. (2004) in which college students first took a survey to identify flourishing mental health (Keyes 2002). Participants who scored above the median on six out of 11 symptoms of positive psychological and social functioning were classified as flourishing, and the remaining were classified as languishing. Then, each day for a month, participants indicated the degree to which they experienced each of several positive and negative emotions. We calculated the ratio of positive to negative emotions experienced over the month. This ratio for flourishing individuals was 3.2:1, whereas for languishing individuals it was 2.3:1. As predicted, these ratios fall on either side of the hypothesized ratio of about 3:1 (B.L. Fredrickson and M. Losada, unpublished data).

Data from Gottman's longitudinal studies of marriage are also relevant. He and his colleague observed 79 couples, married an average of 5 years, as they discussed an area of continuing conflict in their relationship. They measured positivity and negativity using two coding schemes: one focused on positive and negative speech acts, and another focused on observable positive and negative emotions. Gottman (1994) reported that among marriages that last and that both partners find to be satisfying—what we call flourishing marriages—the mean positivity ratio was 4.9. By contrast, among marriages identified as being on cascades towards dissolution—languishing marriages at best—the mean positivity ratio was 0.8. These ratios also flank the predicted ratio of about 3:1 (B.L. Fredrickson and M. Losada, unpublished data).

At three levels of analysis—for flourishing individuals, flourishing marriages, and flourishing business teams—we find positivity ratios above 3:1. Likewise, for individuals, marriages, or business teams that do not function so well, for those we identify as languishing, we find positivity ratios below 3:1. Remarkable coherence has thus emerged among theory, mathematics, and observed data for positivity and human flourishing. First, Fredrickson's (1998, 2001) broaden-and-build theory of positive emotions describes the psychological mechanisms through which positivity can fuel human flourishing. Second, Losada's nonlinear dynamic model (B.L. Fredrickson and M. Losada, unpublished data; Losada 1999; Losada and Heaphy 2004) describes the mathematical relationship between certain positivity ratios and the complex dynamics of human flourishing. And, third, fine-grained empirical observations at three levels of analysis—within individuals, within couples, and within business teams—support Fredrickson's theory and Losada's mathematics.

Concluding remarks

The broaden-and-build theory underscores the ways in which positive emotions are essential elements of optimal functioning, and therefore an essential topic within the science of well-being. The theory, together with the research reviewed here, suggests that positive emotions: (1) broaden people's attention and thinking; (2) undo lingering negative emotional arousal; (3) fuel psychological resilience; (4) build consequential personal resources; (5) trigger upward spirals towards greater well-being in the future; and (6) seed human flourishing. The theory also carries an important prescriptive message: people should cultivate positive emotions in their own lives and in the lives of those around them, not just because doing so makes them feel good in the moment, but also because doing so transforms people for the better and sets them on paths toward flourishing and healthy longevity.

When positive emotions are in short supply, people get stuck. They lose their degrees of behavioural freedom and become painfully predictable. But when positive emotions are in ample supply, people take off. They become generative, creative, resilient, ripe with possibility, and beautifully complex. The broaden-and-build theory conveys how positive emotions move people forward and lift them to the higher ground of optimal well-being.

Acknowledgements

The author thanks the University of Michigan, the National Institute of Mental Health (MH53971 and MH59615), and the John Templeton Foundation for supporting the research described in this chapter.

References

Affleck, G. and Tennen, H. (1996). Construing benefits from adversity: adaptational significance and dispositional underpinnings. *J. Pers.* **64**, 899–922.

Anthony, E.J. (1987). Risk, vulnerability, and resilience: an overview. In *The invulnerable child* (ed. E.J. Anthony and B.J. Cohler), pp. 3–48. Guilford Press, New York.

Aron, A., Norman, C.C., Aron, E.N., McKenna, C., and Heyman, R.E. (2000). Couple's shared participation in novel and arousing activities and experienced relationship quality. *J. Pers. Soc. Psychol.* **78**, 273–84.

Ashby, F.G., Isen, A.M., and Turken, A.U. (1999). A neuropsychological theory of positive affect and its influence on cognition. *Psychol. Rev.* **106**, 529–50.

Aspinwall, L.G. (1998). Rethinking the role of positive affect in self-regulation. *Motivation and Emotion* **22**, 1–32.

Aspinwall, L.G. (2001). Dealing with adversity: self-regulation, coping, adaptation, and health. In *The Blackwell handbook of social psychology*, Vol. 1. *Intraindividual processes* (ed. A. Tesser and N. Schwarz), pp. 591–614. Blackwell, Malden, Massachusetts.

Aspinwall, L.G. and Taylor, S.E. (1997). A stitch in time: self-regulation and proactive coping. *Psychol. Bull.* **121**, 417–36.

Baron, R.A. (1976). The reduction of human aggression: a field study of the influence of incompatible reactions. *J. Appl. Soc. Psychol.* **6**, 260–74.

Basso, M.R., Schefft, B.K., Ris, M.D., and Dember, W.N. (1996). Mood and global–local visual processing. *J. Int. Neuropsychol. Soc.* **2**, 249–55.

Beck, A.T. (1979). *Cognitive therapy of depression*. Guilford Press, New York.

Block, J. and Kremen, A.M. (1996). IQ and ego-resilience: conceptual and empirical connections and separateness. *J. Pers. Soc. Psychol.* **70**, 349–61.

Bolte, A., Goschke, T., and Kuhl, J. (2003). Emotion and intuition: effects of positive and negative mood on implicit judgments of semantic coherence. *Psychol. Sci.* **14**, 416–21.

Bonanno, G.A. and Keltner, D. (1997). Facial expressions of emotion and the course of conjugal bereavement. *J. Abnorm. Psychol.* **106**, 126–37.

Boulton, M.J. and Smith, P.K. (1992). The social nature of play fighting and play chasing: mechanisms and strategies underlying cooperation and compromise. In *The adapted mind: evolutionary psychology and the generation of culture* (ed. J. H. Barkow, L. Cosmides, and J. Tooby), pp. 429–44. Oxford University Press, New York.

Cabanac, M. (1971). Physiological role of pleasure. *Science* **173**, 1103–7.

Cacioppo, J.T., Priester, J.R., and Berntson, G.G. (1993). Rudimentary determinants of attitudes. II. Arm flexion and extension have differential effects on attitudes. *J. Pers. Soc. Psychol.* **65**, 5–17.

Cacioppo, J.T., Gardner, W.L., and Berntson, G.G. (1999). The affect system has parallel and integrative processing components: form follows function. *J. Pers. Soc. Psychol.* **76**, 839–55.

Caro, T.M. (1988). Adaptive significance of play: are we getting closer? *Trends Ecol. Evol.* **3**, 50–4.

Carver, C.S. (1998). Resilience and thriving: issues, models, and linkages. *J. Soc. Issues* **54**, 245–66.

Carver, C.S. and Scheier, M.F. (1990). Origins and functions of positive and negative affect: a control–process view. *Psychol. Rev.* **97**, 19–35.

Clore, G.L. (1994). Why emotions are felt. In *The nature of emotion: fundamental questions* (ed. P. Ekman and R. Davidson), pp. 103–11. Oxford University Press, New York.

Cohler, B.J. (1987). Adversity, resilience, and the study of lives. In *The invulnerable child* (ed. A.E. James and B.J. Cohler), pp. 363–424. Guilford Press, New York.

Danner, D.D., Snowdon, D.A., and Friesen, W.V. (2001). Positive emotions in early life and longevity: findings from the nun study. *J. Pers. Soc. Psychol.* **80**, 804–13.

Davidson, R.J. (1993). The neuropsychology of emotion and affective style. In *Handbook of emotion* (ed. M. Lewis and J.M. Haviland), pp. 143–54. Guilford Press, New York.

Demos, E.V. (1989). Resiliency in infancy. In *The child of our times: studies in the development of resiliency* (ed. T.F. Dugan and R. Cole), pp. 3–22. Brunner/Mazel, Philadelphia.

Derryberry, D. and Tucker, D.M. (1994). Motivating the focus of attention. In *The heart's eye: emotional influences in perception and attention* (ed. P. M. Neidenthal and S. Kitayama), pp. 167–96. Academic Press, San Diego, California.

Diener, E. (2000). Subjective well-being: the science of happiness and a proposal for a national index. *Am. Psychol.* **55**, 34–43.

Diener, E. and Diener, C. (1996). Most people are happy. *Psychol. Sci.* **7**, 181–5.

Diener, E., Sandvik, E., and Pavot, W. (1991). Happiness is the frequency, not the intensity, of positive versus negative affect. In *Subjective well-being: an interdisciplinary perspective* (ed. F. Strack), pp. 119–39. Pergamon Press, Oxford.

Dolhinow, P.J. (1987). At play in the fields. In *The natural history reader in animal behavior* (ed. H. Topoff), pp. 229–37. Columbia University Press, New York.

Ekman, P. (1992). An argument for basic emotions. *Cogn. Emotion* **6**, 169–200.

Estrada, C.A., Isen, A.M., and Young, M.J. (1997). Positive affect facilitates integration of information and decreases anchoring in reasoning among physicians. *Org. Behav. Hum. Decision Processes* **72**, 117–35.

Folkman, S. (1997). Positive psychological states and coping with severe stress. *Soc. Sci. Med.* **45**, 1207–21.

Folkman, S. and Moskowitz, J.T. (2000). Positive affect and the other side of coping. *Am. Psychol.* **55**, 647–54.

Fredrickson, B.L. (1998). What good are positive emotions? *Rev. Gen. Psychol.* **2**, 300–19.

Fredrickson, B.L. (2000). Cultivating positive emotions to optimize health and well-being. *Prevention and Treatment* 3. http://journals.apa.org/prevention/volume3/pre0030001a.html. Accessed 17 September 2003.

Fredrickson, B.L. (2001). The role of positive emotions in positive psychology: the broaden–and–build theory of positive emotions *Am. Psychol.* **56**, 218–26.

Fredrickson, B.L. (2004). Gratitude, like other positive emotions, broadens and builds. In *The psychology of gratitude* (ed. R.A. Emmons and M.E. McCullough), pp. 145–66. Oxford University Press, New York.

Fredrickson, B.L. and Branigan, C. (2001). Positive emotions. In *Emotion: current issues and future directions* (ed. T.J. Mayne and G.A. Bonnano), pp. 123–51. Guilford Press, New York.

Fredrickson, B.L. and Branigan, C. (2005). Positive emotions broaden the scope of attention and thought–action repertoires. *Cogn. Emotion* **19**, 313–32.

Fredrickson, B.L. and Joiner, T. (2002). Positive emotions trigger upward spirals toward emotional well-being. *Psychol. Sci.* **13**, 172–5.

Fredrickson, B.L. and Levenson, R.W. (1998). Positive emotions speed recovery from the cardiovascular sequelae of negative emotions. *Cogn. Emotion* **12**, 191–220.

Fredrickson, B.L. and Losada, M. (in press). Positive affect and the complex dynamics of human flourishing. *Am. Psychol.*

Fredrickson, B.L., Mancuso, R.A., Branigan, C., and Tugade, M. (2000). The undoing effect of positive emotions. *Motivation and Emotion* **24**, 237–58.

Fredrickson, B.L., Tugade, M.M., Waugh, C.E., and Larkin, G. (2003). What good are positive emotions in crises?: a prospective study of resilience and emotions following the terrorist attacks on the United States on 11 September 2001. *J. Pers. Soc. Psychol.* **84**, 365–76.

Fredrickson, B.L., Brown, S. Cohn, M.A., Conway, A., Crosby, C., McGivern, M., and Mikels, J. (2004). Finding positive meaning and experiencing positive emotions builds resilience. Symposium presented at *The Functional Significance of Positive Emotions: Symposium presented at the fifth annual meeting of the Society for Personality and Social Psychology, Austin, Texas, 29–31 January 2004* (Chairs S. Brown and K.J. Johnson).

Frijda, N.H. (1986). *The emotions.* Cambridge University Press, Cambridge.

Frijda, N.H. (1994). Emotions are functional, most of the time. In *The nature of emotion: fundamental questions* (ed. P. Ekman and R. Davidson), pp. 112–22. Oxford University Press, New York.

Frijda, N.H., Kuipers, P., and Schure, E. (1989). Relations among emotion, appraisal, and emotional action readiness. *J. Pers. Soc. Psychol.* **57**, 212–28.

Gasper, K. and Clore, G.L. (2002). Attending to the big picture: mood and global versus local processing of visual information. *Psychol. Sci.* **13**, 34–40.

Gottman, J.M. (1994). *What predicts divorce: the relationship between marital processes and marital outcomes.* Lawrence Erlbaum, New York.

Isen, A.M. (1990). The influence of positive and negative affect on cognitive organization: some implications for development. In *Psychological and biological approaches to emotion* (ed. N. Stein, B. Leventhal, and T. Trabasso), pp. 75–94. Erlbaum, Hillsdale, New Jersey.

Isen, A.M. (2000). Positive affect and decision making. In *Handbook of emotions*, 2nd edn (ed. M. Lewis and J.M. Haviland-Jones), pp. 417–35. Guilford Press, New York.

Isen, A.M. and Daubman, K.A. (1984). The influence of affect on categorization. *J. Pers. Soc. Psychol.* **47**, 1206–17.

Isen, A.M. and Means, B. (1983). The influence of positive affect on decision-making strategy. *Soc. Cogn.* **2**, 18–31.

Isen, A.M., Johnson, M.M.S., Mertz, E., and Robinson, G.F. (1985). The influence of positive affect on the unusualness of word associations. *J. Pers. Soc. Psychol.* **48**, 1413–26.

Isen, A.M., Daubman, K.A., and Nowicki, G.P. (1987). Positive affect facilitates creative problem solving. *J. Pers. Soc. Psychol.* **52**, 1122–31.

Isen, A.M., Rosenzweig, A.S., and Young, M.J. (1991). The influence of positive affect on clinical problem solving. *Med. Decision Making* **11**, 221–7.

Johnson, K.J. and Fredrickson, B.L. (in press). "We all look the same to me": Positive emotions eliminate the own-race bias in face recognition. *Psychol. Sci.*

Kahn, B.E. and Isen, A.M. (1993). The influence of positive affect on variety seeking among safe, enjoyable products. *J. Consumer Res.* **20**, 257–70.

Kahneman, D. (1999). Objective happiness. In *Well-being: the foundations of hedonic psychology* (ed. D. Kahneman, E. Diener, and N. Schwarz), pp. 3–25. Russell Sage, New York.

Keltner, D. and Bonanno, G.A. (1997). A study of laughter and dissociation: distinct correlates of laughter and smiling during bereavement. *J. Pers. Soc. Psychol.* **73**, 687–702.

Keyes, C.L.M. (2002). The mental health continuum: from languishing to flourishing in life. *J. Health Soc. Behav.* **43**, 207–22.

Keyes, C.L.M. (2003). Complete mental health: an agenda for the 21st century. In *Flourishing: positive psychology and the life well-lived* (ed. C.L.M. Keyes and J. Haidt), pp. 293–312. American Psychological Association, Washington, DC.

Klohnen, E.C. (1996). Conceptual analysis and measurement of the construct of ego-resiliency. *J. Pers. Soc. Psychol.* **70**, 1067–79.

Kumpfer, K.L. (1999). Factors and processes contributing to resilience: the resilience framework. In *Resilience and development: positive life adaptations* (ed. M.D. Glantz and J.L. Johnson), pp. 179–224. Kluwer/Plenum, New York.

Lazarus, R.S. (1991). *Emotion and adaptation.* Oxford University Press, New York.

Lazarus, R.S. (1993). From psychological stress to the emotions: a history of changing outlooks. *Annu. Rev. Psychol.* **44**, 1–22.

Lazarus, R.S., Kanner, A.D., and Folkman, S. (1980). Emotions: a cognitive–phenomenological analysis. In *Theories of emotion* (ed. R. Plutchik and H. Kellerman), pp. 189–217. Academic Press, New York.

Lee, P.C. (1983). Play as a means for developing relationships. In *Primate social relationships* (ed. R.A. Hinde), pp. 82–9. Blackwell, Oxford.

Leslie, A.M. (1987). Pretense and representation: the origins of 'theory of mind'. *Psychol. Rev.* **94**, 412–26.

Levenson, R.W. (1994). Human emotions: a functional view. In *The nature of emotion: fundamental questions* (ed. P. Ekman and R. Davidson), pp. 123–6. Oxford University Press, New York.

Levy, B.R., Slade, M.D., Kunkel, S.R., and Kasl, S.V. (2002). Longevity increased by positive self-perceptions of aging. *J. Pers. Soc. Psychol.* **83**, 261–70.

Losada, M. (1999). The complex dynamics of high performance teams. *Math. Comput. Model.* **30**, 179–92.

Losada, M. and Heaphy, E. (2004). The role of positivity and connectivity in the performance of business teams: a nonlinear dynamics model. *Am. Behav. Sci.* **47**, 740–65.

Masten, A.S. (1994). Resilience in individual development: successful adaptation despite risk and adversity. In *Educational resilience in inner-city America: challenges and prospects* (ed. M.C. Wang and E.W. Gordon), pp. 3–25. Erlbaum, Hillsdale, New Jersey.

Moskowitz, J.T. (2003). Positive affect predicts lower risk of AIDS mortality. *Psychosom. Med.* **65**, 620–6.

Murphy, L.B. and Moriarty, A. (1976). *Vulnerability, coping and growth: from infancy to adolescence.* Yale University Press, New Haven, Connecticut.

Oatley, K. and Jenkins, J.M. (1996). *Understanding emotions*. Blackwell, Cambridge, Massachusetts.

Ostir, G.V., Markides, K.S., Black, S.A., and Goodwin, J.S. (2000). Emotional well-being predicts subsequent functional independence and survival. *J. Am. Geriatr. Soc.* **48**, 473–8.

Ostir, G.V., Markides, K.S., Peek, K., and Goodwin, J.S. (2001). The associations between emotional well-being and the incidence of stroke in older adults. *Psychosom. Med.* **63**, 210–15.

Panksepp, J. (1998). Attention deficit hyperactivity disorders, psychostimulants, and intolerance of childhood playfulness: a tragedy in the making? *Curr. Dir. Psychol. Sci.* **7**, 91–8.

Peterson, C. and Seligman, M.E. P. (1984). Causal explanations as a risk factor for depression: theory and evidence. *Psychol. Rev.* **91**, 347–74.

Reed, M.B. and Aspinwall, L.G. (1998). Self-affirmation reduces biased processing of health-risk information. *Motivation and Emotion* **22**, 99–132.

Rosenberg, E.L. (1998). Levels of analysis and the organization of affect. *Rev. Gen. Psychol.* **2**, 247–70.

Seligman, M.E.P. and Csikszentmihalyi, M. (2000). Positive psychology: an introduction. *Am. Psychol.* **55**, 5–14.

Sherrod, L.R. and Singer, J.L. (1989). The development of make-believe play. In *Sports, games and play* (ed. J. Goldstein), pp. 1–38. Lawrence Erlbaum, Hillsdale, New Jersey.

Simons, C.J.R., McCluskey-Fawcett, K.A., and Papini, D.R. (1986). Theoretical and functional perspective on the development of humor during infancy, childhood, and adolescence. In *Humor and aging* (ed. L. Nahemow, K.A. McCluskey-Fawcett, and P.E. McGhee), pp. 53–77. Academic Press, San Diego, California.

Solomon, R.L. and Corbit, J.D. (1974). An opponent–process theory of motivation. I. Temporal dynamics of affect. *Psychol. Rev.* **81**, 119–45.

Stein, N.L., Folkman, S., Trabasso, T., and Richards, T.A. (1997). Appraisal and goal processes as predictors of psychological well-being in bereaved caregivers. *J. Pers. Soc. Psychol.* **72**, 872–84.

Taylor, S.E., Kemeny, M.E., Reed, G.M., Bower, J.E., and Gruenewald, T.L. (2000). Psychological resources, positive illusions, and health. *Am. Psychol.* **55**, 99–109.

Tooby, J. and Cosmides, L. (1990). The past explains the present: emotional adaptations and the structure of ancestral environments. *Ethol. Sociobiol.* **11**, 375–424.

Trope, Y. and Neter, E. (1994). Reconciling competing motives in self-evaluation: the role of self-control in feedback seeking. *J. Pers. Soc. Psychol.* **66**, 646–57.

Trope, Y. and Pomerantz, E.M. (1998). Resolving conflicts among self-evaluative motives: positive experiences as a resource for overcoming defensiveness. *Motivation and Emotion* **22**, 53–72.

Tugade, M.M. and Fredrickson, B.L. (2002). Positive emotions and emotional intelligence. In *The wisdom of feelings: psychological processes in emotional intelligence* (ed. L. Feldman-Barrett and P. Salovey), pp. 319–40. Guilford Press, New York.

Tugade, M. and Fredrickson, B.L. (2004). Resilient individuals use positive emotions to bounce back from negative emotional arousal. *J. Pers. Soc. Psychol.* **86**, 320–33.

Werner, E. and Smith, R.S. (1992). *Overcoming the odds: high risk children from birth to adulthood*. Cornell University Press, Ithaca, New York.

Wolin, S.J. and Wolin, S. (1993). *Bound and determined: growing up resilient in a troubled family*. Villard, New York.

Wolpe, J. (1958). *Psychotherapy by reciprocal inhibition*. Stanford University Press.

Nick Baylis has lectured positive psychology at Cambridge University since 2001, and works as a coach and therapist taking an holistic approach to individual life-development. Through his book, *Learning from Wonderful Lives*, and a weekly column in *The Times* newspaper, Nick tries to make the study of well-being clear and helpful to a wide readership.

Photograph courtesy of Magdalena Pietka.

Chapter 9*

Relationship with reality and its role in the well-being of young adults

Nick Baylis

Introduction

This chapter proposes that our characteristic cognitive–behavioural strategies of dealing with everyday real life—whether we are prone to reality-investing or quick-fixes or reality-evading—can play a crucial role in our well-being (physical, psychological, and social). This characteristic style of coping I refer to as our 'relationship with reality' (RwR).

RwR theory endeavours to connect and explain the motivation behind a diverse range of thoughts and behaviours that have not previously been linked by theory or empirical study. That is to say, many everyday and commonplace thoughts and behaviours (such as fantasizing, television watching, and drinking alcohol) that may seem insignificant and unrelated can, on closer consideration, be seen to be united by their common goal in terms of reality-evasion, quick-fixes, or reality-investment. These activities can accumulate to form our characteristic style of dealing with real life, and thereby have substantial effects on our well-being (Baylis 2005).

RwR theory arose out of my personal life experience and subsequent empirical investigations. Much of my research revisits the potential importance of two phenomena almost universally ignored in the mainstream of psychological and educational literature: *escapist fantasizing* about absolute impossibilities and *wishful daydreaming* about wished-for future scenarios and outcomes. I propose that these universal and seemingly benign thoughts may, far more often than is currently recognized, be so overused by some individuals as to cause serious detriment to their real-life well-being.

I argue that consideration of our RwR is a crucial task, particularly pertinent in a Western world characterized by ever-increasing means to digitally distort our reality, to communicate via phones and internets, and to alter our moods and perceptions through ingesting drugs. It seems vital that psychology consider the distinctive strengths and benefits of using, *and not using*, such technological developments. How highly do we prize reality, clarity of consciousness, and having someone's physical presence in comparison to the technological alternatives? In short, what sort of life do we wish to lead?

The first part of this chapter presents the RwR theory. The second part reviews the prior literature that served as a cornerstone of the RwR research. The third part recounts some key findings from my own empirical investigations among a broad spectrum of young adults (ranging from imprisoned young offenders, through to highly accomplished university students, and elite members of the armed forces), and the fourth part goes beyond the individual consequences of the RwR concept, to consider the societal implications. I begin with a case study so as to illustrate some key features of the concept.

Joshua's story: a case study

At age 16 years, there appeared no more affable, humorous, and popular a middle-class schoolboy than Joshua. He was socially extravert, physically robust, handsome, and non-delinquent (i.e. there was no substance abuse, crime, or truancy). Josh, now aged 40 years, said in retrospect: 'I put on the show of being happy, but was profoundly sad most of the time, and if I wasn't sad it was only because I was fantasizing about living another existence.'

Joshua's father was a chronic alcoholic who emotionally bullied his three children on a near daily basis. Despite the partially protective influence of his mother, Josh recalls: 'Throughout my teenage years, many hours a day and whole evenings could go on escapist fantasy, because I had been made to feel so painfully worthless and doomed to terrible failure. Fantasy was my home.'

Josh was well-clothed and well-fed and the family appeared happy enough to all outside observers. But the reality was quite different, and it is indicative of the difficulties in the home environment that Josh's younger sibling became, by the age of 25, a clinically diagnosed alcoholic, while Josh's older sibling had, by the same age, renounced all connection with the family and emigrated to a distant continent.

Faced with this noxious home environment, what my theory calls 'reality-evading' characterized Joshua's attempt at coping. Excessive use of 'escapist fantasy' (i.e. impossible but pleasing scenarios) was the mainstay of this defensive strategy, though he also escaped into many hours of television, or listened to music to provide a sound-track to his escapist imaginary scenarios.

Unfortunately, Joshua's distrust of real life and his resort to evasive strategies had the side-effect of isolating him from close relationships with others. This meant that, throughout his teenage years and early twenties, this heterosexual and apparently confident chap never had a girlfriend who lasted more than a half a dozen dates, and even his best school buddies were not privy to his emotional pain.

Josh recalls, 'As a teenager, I had no concept of a future save for my somehow ending up in the gutter, which meant that longer-term investment in anyone or anything, seemed pointless.'

By age 18, Josh had scored very poorly in his end of school exams and, while 98% of his fellow school-leavers went off to higher education, Josh embarked on 8 years of poorly paid shop-assistant work. 'I didn't feel capable of making ambitious things happen in real life. I knew I wasn't worth it. I knew I would fail.'

Throughout those growing-up years, it was never suggested that Josh needed to see a qualified therapist, nor was he regarded as abnormal or maladjusted. His parents, teachers, and friends all dismissed his approach to the world as lazy, unambitious, or distractable. The true motivations for his syndrome of behaviours were quite invisible and unsuspected.

Only very gradually, after several years of rented rooms and semi-skilled labour, did Joshua's self-confidence develop just sufficiently for Josh to take himself to university at age 26. 'It was the first deliberate and significant investment in myself that I had ever made.' And, once he discovered the fertile and encouraging real life environment of the small and friendly classes in his provincial university, the transition in his coping strategy from 'reality-evading' to 'reality-investing' took a matter of only a few weeks. Because he felt 'in meaningful control of my life and, as a consequence, realistically hopeful of a good future', Josh enthusiastically invested his energies in his coursework and in collaborative relations with his peer group and faculty members. He quickly excelled at undergraduate level and went straight on to gain an Ivy League master's degree. He also became fluent in the foreign language of the girl with whom he had fallen in love in his first year at university. Josh then proceeded with her to a happy marriage, home ownership, highly profitable and skilled self-employment, and a profound joy in parenthood. Aged 40, he now regards himself as a fervent 'reality investor' and feels his quality of life to be 'very satisfying'. 'Yes, I still occasionally find myself fantasizing when things get difficult, for perhaps 20 minutes per week in all, but now I know them for what they are. Reality is so much more interesting to me than fantasy, I wouldn't want to trade it.'

Key features of RwR raised by this case study

◆ It was not the emotional pains that inhibited Joshua's life-development; it was his impoverished relationship with reality. It was this that provided the medium by which the real damage was done. Reality is an environment that can put up a lot of resistance to our making progress and, consequently, it can build the mental skills for problem-solving. By contrast, escapist fantasy is an environment in which anything is possible, so our problem-solving skills begin to atrophy if we spend too long there. Hence, this unwitting strategy of escaping into fantasy and resorting to quick-fixes when reality felt difficult, was to prove extremely inhibiting of Josh's healthy social development. This phenomenon of an initial problem leading to serious consequences because of an unhelpful coping strategy, is a common pattern in the developmental path of psychological distress and pathology. It reminds us that it is only our *response* to problems that determines their net effect upon us, not the problems themselves. If we are able to apply *helpful* coping strategies that develop our skills and confidence, then those very same problems and challenges can lead us to thrive and flourish in the face of adversity, not just survive it.

♦ Richard Lazarus, a seminal researcher in the field of coping strategies, writes, 'Most important of all, we should examine which coping patterns succeed or fail in the short and long run, and in what ways . . . and how these strategies come together and are synthesized into an overarching coping style' (Lazarus 1993, p. 272). The RwR model suggests how an individual's seemingly unconnected cognitive–behavioural characteristics are, in fact, strongly linked by their intention regarding real life: whether to invest in it, quick-fix it, or evade it.

♦ An individual's RwR can be difficult to identify or can be misinterpreted because many of its component features are invisible and either private or covert. Wilson and Barber (1983, p. 366) report how their study among 26 so-called 'fantasy addicts' disclosed a similar extreme of secrecy about their fantasy use.

♦ Joshua's life history is typical of the 'late bloomers' whom I interviewed during my empirical work. These were individuals whose teenage years and much of their twenties were blighted by a strongly negative RwR, which rapidly became distinctly positive once the individual encountered sufficiently nurturing circumstances. These circumstances were most often characterized by deeply rewarding personal relationships, readily achievable meaningful goals, and some first-hand experience of their ability to improve real life. The work of Putner (2000), Snyder (2002), and Bandura (1997), respectively, supports the strongly beneficial potential of those three factors.

Defining our relationship with reality

This chapter proposes that our relationship with reality comprises three types of cognitive–behavioural processes. These are: reality-investing, quick-fixes, and reality-evading. Our relative use of these coping strategies defines our overall RwR, which can be positive or negative. A *negative* RwR is one that is biased towards the unhelpful deployment of quick-fixes and reality-evasion and can impair healthy development in one or more arenas of an individual's life. A *positive* RwR would, in ordinary circumstances, be biased towards reality-investing, and can promote healthy development. The theory also suggests that the characteristic pattern of RwR can, in at least some individuals, aggregate to form the most distinctive and overriding feature of their personality, for better or for worse. To help understand the essence of the concept, it is useful to consider one's own life through the RwR lens.

Please consider what thoughts and behaviours you deploy with the intention of bringing about lasting improvements in your real life circumstances? RwR theory calls this style of coping with the world, 'reality-investing' and it is likely to include the sort of step-by-step planning or rehearsal that aims for sustainable improvement in real-life performances. Now consider what thoughts and behaviours you use to feel better immediately, with only secondary regard for the medium or longer-term consequences. These thoughts and behaviours can either be 'quick-fixes' (exemplified by such activities as consuming alcohol or fast food to reduce stress) or they can be 'reality-evading' (exemplified by excessive television watching or generating escapist fantasies of seemingly impossible scenarios). The defining characteristic of these quick-fixes and evasions is that they bring short-term improvements in our subjective experience of reality, or sheer relief from reality, but only at the expense of jeopardizing or undermining our longer-term well-being. However, this distinction between short- and long-term benefits is not meant to imply that reality-investing is only interested in the future. A skilful investment in real life not only fulfils its purpose of benefiting longer-term outcomes, but is also enjoyable in the present. Playing team sport is usually a good example of this: it is pleasurable in the present, while also building social ties and physical health for the future.

RwR theory is not aiming to demonize certain thinking types or behaviours as pro- or anti-reality. For instance, daydreaming and fantasy can have highly beneficial uses (which will be detailed below). Moreover, too great an emphasis on future-oriented thinking can lead to workaholism, anxiety, and depression (see the time perspective theory of Zimbardo and Boyd 1999). RwR theory only advocates that we be more aware of our characteristic patterns and ratios of investing, quick-fixing, and evading, and be versatile in deploying our mental and behavioural tools to meet different life-demands more adaptively. Figure 9.1 is a flow diagram indicating how our relationship with reality is formed.

Thinking about thinking (metacognition) has proven itself one of the most profitable routes to understanding and improving our quality of life (for instance, the cognitive therapy of Beck 1976, the 'learned optimism' of Seligman 1990, the 'self-efficacy' of Bandura 1997, the triarchic model of intelligence of Sternberg 1997, and the time perspective of Zimbardo and Boyd 1999). Understanding an individual's relationship with reality is one more lens through which to interpret our behaviour, just as 'optimism' describes a positive relationship with the future, and 'rumination' describes a negative relationship with the past. Using this new lens, RwR theory aims to improve our understanding of the psychological dynamics and external behaviour of a wide range of individuals, from those who

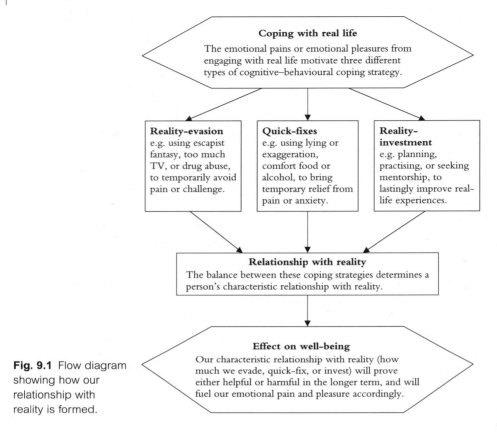

Coping with real life
The emotional pains or emotional pleasures from engaging with real life motivate three different types of cognitive–behavioural coping strategy.

Reality-evasion
e.g. using escapist fantasy, too much TV, or drug abuse, to temporarily avoid pain or challenge.

Quick-fixes
e.g. using lying or exaggeration, comfort food or alcohol, to bring temporary relief from pain or anxiety.

Reality-investment
e.g. planning, practising, or seeking mentorship, to lastingly improve real-life experiences.

Relationship with reality
The balance between these coping strategies determines a person's characteristic relationship with reality.

Effect on well-being
Our characteristic relationship with reality (how much we evade, quick-fix, or invest) will prove either helpful or harmful in the longer term, and will fuel our emotional pain and pleasure accordingly.

Fig. 9.1 Flow diagram showing how our relationship with reality is formed.

lead lives overshadowed by chronic underachievement, depression, addiction, or crime, through to those who lead exceptionally satisfying, healthy, and accomplished lives.

At the time of writing, there appears little akin to this RwR dimension in any of the mainstream literature in the educational and psychological professions. As a consequence perhaps, it is notable that the words 'daydream' and 'fantasy' that are so strongly associated with RwR theory simply do not appear in the indexes of the leading practitioner texts in psychology and psychiatry, no matter whether their focus is adolescence, delinquency, happiness, or high achievement. It is in respect of this gap that my work has looked particularly at individuals' use of daydreaming and fantasy and the role these tools might play in their RwR.

Definition of other key terms

◆ 'Well-being' is a positive and sustainable state in which we can thrive and flourish. At its best, the science of well-being is about exploring how good life can become and how good we can become at living it.

- 'Reality' and 'real life' are used interchangeably to refer to objective aspects of the physical and social environment.

- 'Wishful daydreams' are the pleasurable images and narratives about anticipated future scenarios that the daydreamer thinks are quite possible and even likely. However, the extent to which daydreamers actively pursue such possibilities in real life depends on their RwR.

- 'Escapist fantasies' are pleasurable images and narratives about desirable scenarios that the fantasist never expects to happen in real life. These escapist fantasies are in no way delusional or hallucinatory in any clinical sense because the individual is fully aware of their total unreality.

- 'Imagination' (which might also be called 'creativity') is the all-purpose mental process we use to conjure up possibilities and solve problems. It might be used to generate fantasies or daydreams, but equally it might be used to solve real-life problems. Imagination's distinctly positive potential is well observed by the quantum physicist and Nobel Laureate, Max Planck, who wrote in his autobiography that 'the scientist must have a vivid intuitive imagination, for new ideas are not generated by deduction, but by an artistically creative imagination. (Planck 1949, p. 109). Likewise, imaginal rehearsal, i.e. the detailed practising of tasks in our imagination as opposed to actual movement, has been shown to improve skilled performance almost as much as real-life practice (Robertson 2000). Imagination in the form of 'therapist-guided positive imagery' also serves an invaluable role in clinical practice (e.g. Erickson in Rosen 1991; Yapko 2003; Griffin and Tyrrell 2003).

RwR theory

The three categories of relationship with reality

The following offers an illustrative but by no means exhaustive list of the characteristic thoughts and behaviours that can be collated into three main categories according to their relationship with reality: reality-investing; quick-fixes; reality-evading. RwR theory is not proposing that any particular type of thought or action is intrinsically negative or positive, because it is only the specific context in which they are used that can determine this. This will be seen particularly in reference to fantasy and daydreaming, which can have both adaptive and maladaptive uses. The categorization of characteristics below indicates only that they are most commonly associated with investing, quick-fixing, or evading. Furthermore, an individual's characteristic patterns may be deliberate, or be quite unconscious and automatic, or may even be the result of simply misjudging the net outcome of an activity. In any event, the categorization can be a useful insight into the link between coping strategy and life outcomes.

Reality-investing

This is characterized by thinking and behaviours that serve to bring about sustainable improvement in the experience of real life. This may or may not bring about immediate improvement, but the investor expects a net gain over the longer term. Typical reality-investing thoughts include:

◆ step-by-step planning, the goals of which are likely to be achieved (e.g. making a realistic 'to do' list at the beginning of the day);

◆ day-to-day problem-solving: negotiating the physical, psychological, and social demands of carrying out the to do list, in a way that does not compromise or jeopardize longer-term well-being;

◆ mental rehearsal with the intention of improving the real-life performance of a future task;

◆ wishful daydreaming for self-motivation, i.e. generating pleasurable thoughts about anticipated possible future scenarios so that those daydreams are pursued with more determination in real life;

◆ escapist fantasies of impossible scenarios being used simply to rest or refresh the user's mental energies, so as to better engage with real-life action;

◆ creative fantasies of seemingly impossible scenarios that eventually form the catalyst or inspiration for some real-life goals or solutions that had not previously been considered when the fantasy initially took place.

Typical reality-investing behaviours include:

◆ day-to-day practical activities that aim for longer-term benefits (e.g. physically practising skills);

◆ self-improving health regimes (e.g. relating to nutrition, exercise, and adequate sleep);

◆ self-improving educational regimes to acquire useful knowledge and qualifications;

◆ consuming fiction and other forms of art so as to illuminate, inform, and motivate our reality-investment (e.g. reading fiction to inspire one's real life action or using music to self-motivate).

In summary, reality-investing is characteristic of those personalities who considerably advantage themselves through engaging with real life in ways that eventually pay a net profit of rewarding real-life outcomes.

Quick-fixes

These are characterized by thoughts and behaviours that temporarily blur or skew or distort our perception of reality, though they stop short of disengaging us from it. Immediate gratification is the primary concern of the user and, though this may bring temporary symptom relief from our emotional pains or frustrations, the downside is that they may jeopardize or damage our medium and longer term real–life well-being. Typical quick-fix thoughts include:

- wishful daydreaming that doesn't lead to action. The user deliberately generates pleasurable thoughts about anticipated possible future scenarios or outcomes, but with little or no attempt to make the daydream a reality;

- lying or exaggerating, thereby creating the temporary illusion of real life being better than it is;

- ignoring a challenging situation so as to avoid dealing with the emotional discomfort.

Typical quick-fix behaviours include:

- excessive consumption of high-calorie but low-nutrition 'junk foods', or eating beyond one's nutritional requirements to an extent that risks obesity;

- ill-considered sexual relations or excessive masturbation;

- a bulimic or anorexic relationship with food;

- abuse of high-risk body-building steroids to rapidly gain muscle bulk and strength, despite the physical and psychological health risks;

- abuse of high-risk, mood-enhancing substances such as alcohol, speed, ecstasy, or cocaine, so as to enhance the user's immediate experience of real life;

- impulsive forms of gambling, crime, and aggression.

In summary, quick-fix characteristics are an unconstructive engagement with real life that reflects a short-termism that, in the longer run, generates only increased pain and set-backs.

Reality-evading

This is characterized by a predominance of thoughts and behaviours inadvertently or deliberately deployed to block out or escape the first-hand experience of real life. Typical reality-evading thoughts include:

- escapist fantasies that never lead back to real life. These fantasies are tailor-made to meet the individual's emotional needs, albeit temporarily;

- consistent denial, to oneself or others, of one's real life circumstances. This denial is being used to evade the emotional pain generated by owning up to those circumstances.

Typical reality-evading behaviours include:

- consuming fiction materials with the purpose of escaping real life. These might include television, videos, cinema, novels, glamour magazines, fantasy computer games, and pornography. These materials appear to provide vicarious pleasures, and can help fuel the individual's own fantasies;

- listening to music with the purpose of blocking out reality, and/or to support a self-generated internal fantasy story or fantasy world (i.e. music becomes a soundtrack for the fantasy movie playing in the mind's eye of the user);

- habitual procrastination so as not to have to deal with problems;
- keeping real people at arms length by not developing any close or intimate relationships, because close relationships bind one to reality;
- abusing substances with a strong potential for reality-disengaging effects, such as heroin and LSD, or drinking oneself to oblivion.

In summary, the core strategy of someone with reality-evading traits is to disengage as completely as possible from an emotionally painful, threatening, or simply unsatisfying real life. Suicide, perhaps, is the ultimate act of reality-evasion, as it offers permanent escape.

Some key clarifications of RwR theory

The defining characteristic of a positive relationship with reality is *a sustainable improvement in well-being*. Within this objective, a person might deploy reality-evading (e.g. escapist fantasy) for short periods so as to protect his or her mental well-being from an overly demanding real life, and this strategy may well allow the person to return mentally refreshed to far more effective engagement with reality. Similarly, a modest consumption of alcohol might be used to enhance social engagement or for relaxation. But note how the ultimate intention and/or likely effect is to enable an enhanced performance in the real world, as opposed to not caring about the consequences. It is this caring about sustainable progress in real life that is the hallmark of a person that has a positive RwR.

By contrast, a negative RwR would be characterized by someone simply trying to improve their immediate feelings with no care for the longer-term impact upon their real-life well-being. RwR theory suggests that sometimes the trade-off is conscious (e.g. by abusing alcohol), but sometimes it is unwitting (e.g. by abusing fantasy).

It was not evident in my research (to be described in more detail later) that particular emotions were typical of any particular category of RwR. For instance, a reality-investing personality could be optimistic and hopeful, but so might a personality prone to the quick-fixes of making wildly unrealistic plans. Likewise, someone fearful could become a reality evader prone to fantasy, but could just as easily use reality-investing strategies to acclimatize themselves to their fears. In a similar way, anger could motivate either vengeful fantasies or be channelled into constructive action. So, in short, there are no good or bad emotions; they should be evaluated in terms of their usefulness or otherwise in terms of motivating and fuelling our RwR.

Finally, a positive RwR should not be confused with implying that an individual is necessarily 'pro-social'. For instance, a 'master criminal' might exhibit a strongly reality-investing style (e.g. rigorously planning and practising a crime), but might nonetheless act in highly antisocial ways (e.g. committing an armed bank-robbery).

The origins of an individual's RwR

Emotionally hurtful or deeply unsatisfying lives are probably the major reason for withdrawing into ourselves by one means or another. Very often it seems that reality-evasion in young adults is an inadvertent habit or unhelpful strategy left over from a daunting childhood. But, as Vaillant (1995a) argues to be the case for much alcohol dependency, the abuse of a drug can accidentally take a hold without there being any prior emotional problems of a magnitude to bring about pathology. Likewise, RwR theory proposes that inadvertently acquired habits of evasion or quick-fixing may be quite capable of initiating their own downward spiral of real-life problems.

Development of a negative RwR

Stage 1 Inherent in real life are occasional setbacks, and these setbacks tend to result in some form of emotional pain, for instance, a sense of loneliness, frustration, fear, shame, impotency, and/or hopelessness.

Stage 2 Emotional pain makes the individual more apprehensive of real life scenarios.

Stage 3 To ease or escape emotional pain and/or the fear of it, the individual increasingly deploys quick-fix and reality-evading strategies.

Stage 4 These strategies further lessen the individual's confidence and capability in engaging effectively with reality.

Stage 5 The increasing diet of quick-fix and evading strategies only serves to widen the gap between the individual's perception of reality, compared with what genuine reality can offer.

Hence, the cycle repeats itself, as stage 5 feeds back into stage 3 in a downward spiral.

Development of a positive RwR

The initial three stages are the same as above, but it is stage 3 that most crucially determines whether an RwR develops positively or negatively. The positive versions of stages 3–5 are as follows.

Stage 3 Despite the emotional pains inherent in real life setbacks, the individual has the resilience to pursue a strategy of persisting in the positive development of their reality-investing skills.

Stage 4 Reality becomes less and less painful as the individual grows accustomed to coping with it. Moreover, the sheer practice of negotiating the demands of reality means that the individual becomes more proficient at managing it and perhaps eventually enjoying the changes.

Stage 5 The increasing proficiency leads to increasing real-life rewards and this makes the individual even more prepared to invest in real life.

Hence, stage 5 feeds cyclically back into positive stage 3 in an upward spiral.

Note how a positive RwR is one of the mechanisms through which an individual can achieve not only 'resilience in the face of adversity', but can positively thrive, all depending on how the individual responds at stage 3 to the emotional pains of life. This resilience might be fostered by mentors, or educative role-models, or caring friends. (See Werner and Smith 1992 for evidence on the role of these factors.)

Addiction, localized deficits, and disguised evasions

Any deficits in a young person's relationship with reality will most often present an educational challenge rather than a clinical or psychiatric one. However, excessive and debilitating use of quick-fixes and evasion might meet the following criteria for being an addiction (see American Psychiatric Association 1994).

- ◆ Psychosocial stress is the likely precipitant.
- ◆ Many traits are socially acceptable though maladaptive.
- ◆ Dependency and tolerance can develop.
- ◆ There is considerable deterioration in social, physical, and economic functioning.
- ◆ The behavioural and/or cognitive syndrome can grow to dominate the individual's life.

A rather more common phenomenon than addiction occurs when an individual displays a reality-investing style in some arenas of their life (e.g. academic and working life), but displays a quick-fix or evading style in other arenas (e.g. in sexual and romantic life). I refer to this as 'local deficits' in the individual's RwR.

There are also less apparent ways to evade reality than those I have listed above. For instance, the sort of obsessive working that compromises one's health or closest relationships; or the sort of perfectionism that means important projects either don't get started, don't get finished, or in their doing almost destroy one's well-being; or simply being too secretive about a serious problem (such as an illness). These three traits—workaholism masquerading as passion; perfectionism masquerading as determination; and self-isolating secrecy masquerading as independence—are all, in essence, reality-evasions, and our real life will suffer very severely if we persist in them.

Conclusion

This chapter proposes that relationship with reality is an ever-present and quite stable cognitive–behavioural dimension, i.e. a characteristic coping style. The goal of RwR theory is to promote a healthy and versatile ratio between our investing, quick-fixes, and evading, and I argue that such a goal is worth far more consideration than it has been hitherto afforded.

Reviewing the literature

Although there is nothing akin to RwR theory in the literature of the psychological and educational professions, many of its individual component phenomena have attracted attention. For instance, most of the common cognitions and behaviours described above as investments, quick-fixes, or evasions are already being extensively investigated by psychologists (e.g. planning ahead, substance abuse, and television usage). For this reason, the following review concentrates on 'wishful daydreaming' and 'escapist fantasy'. As will become apparent below, these two particular psychological phenomena are still very much neglected by social scientists, yet I argue that they serve as good examples of how commonplace, private, and seemingly trivial aspects of an individual's thoughts can form part of their RwR and thereby significantly impact upon important life outcomes.

A profile of fantasy and daydreaming

By far the most substantial empirical reference point for the potential role of fantasy and daydreaming in the lives of the general population, is Jerome Singer (1966, 1975, 1981), whose pioneering clinical and research work led to the short imaginal processes inventory or SIPI (Huba *et al.* 1982). However, when it came to my own empirical research, I had three reasons for devising my own inventory of thought-types. First, Huba *et al.* used a definition of daydreaming that essentially embraced any thought that wasn't task-oriented, i.e. loose associations, well-developed fantasies, plans for the future, memories of the past, worries, wishes, and fears were all examples of daydreaming (see Singer 1966, p. 37). By comparison, RwR theory regards that approach as insufficiently discriminating, because RwR theory regards daydreams and fantasies as quite particular types of future oriented thinking, exactly as they were defined in the previous section of this chapter. Secondly, the SIPI was not concerned with the eventual real-life outcomes of daydreaming or fantasy, whereas I argue this is a key criterion by which to judge our RwR. Thirdly, the SIPI makes no attempt to incorporate the phenomena into a larger cognitive–behavioural pattern. By contrast, RwR theory conceives of daydreams and fantasies as only two elements in our much broader relationship with reality.

Another substantial work in the field is the popular psychology book called *The force of fantasy*, by Columbia University clinical psychiatrist, Ethel Person (1997). However, Person offers a classification of fantasy that never attempts to consider fantasy's role in terms of its impact on an individual's real life. Instead, Person describes only the purpose and emotional effects of fantasy. This dearth of interest in the possible correlation between fantasy use and real-life outcomes characterizes the literature that currently exists in the field.

The likely functions of fantasy and daydreaming

There seems to be a consensus among the handful of researchers who have explored this specific field that the major purpose of fantasy and daydreaming is to control the user's level of psychological arousal, as if fantasy and daydreaming serve as psychological shock-absorbers for emotions such as loneliness, frustration, anger, sexual impulse, fear, and boredom (see Freud 1957; Adler 1927, pp. 49, 57–59; Singer 1975). The long-time director of the Harvard Study of Adult Development, George Vaillant, is representative of this view when he writes, 'the plain brown wrapper of our imagination . . . serves to obliterate the overt expression of aggressive, dependent, and sexual impulses towards others' (Vaillant 1997, pp. 47–48).

Vaillant psychoanalyses the lives of men tracked prospectively for more than 60 years by The Grant Study of Adult Development (which is a subgroup of the larger Harvard Study), and from this analysis he concludes that fantasy is one of 18 ego-defence mechanisms (Vaillant 1995b, 2002). He describes it as a primarily unconscious, involuntary, and immature defence mechanism for dealing with stress, and this prompts my making three comparisons with RwR theory. First, the RwR model is consistent with Vaillant's view of fantasy and other defensive traits as being long-held and pervasive and largely context-inflexible and his view that there is a correlation between defensive style and adverse life outcomes. However, RwR theory conceives of fantasy and daydreaming as often being deployed deliberately as a conscious coping mechanism in the face of real-life challenges (rather than necessarily being the unconscious defence that Vaillant proposes). Furthermore, Vaillant's categorization regards fantasy in individuals aged 16 and over as an entirely *negative* coping strategy, whereas RwR theory argues that fantasy can be either helpful or unhelpful depending on the context of its use.

What determines fantasy-proneness?

Singer (1981, p. 133) writes,

> Even in three and four year olds we can discern the beginnings of strong individual differences in modes of using make-believe as a way of dealing with the world. By adolescence this predisposition to resort to fantasy as a resource is well established and may play an important later role in the lifestyle of the individual.

Vaillant offers an additional insight. Drawing once again on the Grant Study, he observes that 'More than any other adaptive style, fantasy correlated with bleak childhoods' (Vaillant 1995b, p. 171). That assessment echoes the psychiatrists Goertzel *et al.* in their study of the biographies and autobiographies of 300 eminent people, in which they wrote the following about the fiction writers and actors among their cohort, 'All those who traffic in fantasy remember their homes as having tragic elements' (Goertzel *et al.* 1978, p. 11).

Despite these case studies, we must be cautious not to presume that using fantasy necessarily indicates a bleak or tragic childhood, because we simply don't have sufficient evidence to confirm that.

How fantasy and daydreaming affect real life

The positive associations

Daydreaming and fantasy are predominantly regarded as normal, necessary, useful, healthy, and adaptive in all age ranges. (see Singer ibid.; also Klinger 1990; Wilson and Barber 1983; Storr 1988, p. 72; Person 1997, p. 52). In brief, the positive roles are perceived as follows.

- Fantasy can provide a safe and freely available medium for highly experimental ideas and roles; a means of tailor-made mental relaxation; a catharsis for taboo or unattainable emotions and desires; and psychological self-protection from a potentially damaging reality (e.g. hostages might usefully escape into fantasy worlds when held under extreme conditions).

- Daydreaming can be self-motivating if the user imaginatively conjures up some desired goal and how it might feel to achieve it.

- Fiction and art (which can be regarded as the products of fantasy) can serve to illustrate alternative opportunities in life, to model or inspire new perspectives, thoughts, and behaviours, to offer substitute experiences for uncommon or unattainable emotions, or to simply provide a refreshing retreat from reality.

The negative associations

By comparison to its positive potential, and notwithstanding the seminal work of Vaillant, very little sustained attention has been paid to the possibility of the negative effects of fantasy and daydreaming in a normal population. However, the history of the clinical field offers glimpses into the potential problems of misuse. Sigmund Freud's (1957) view was that happy people do not fantasize, only unsatisfied ones do, and that the motive of fantasies is unsatisfied wishes, and that every single fantasy is the correction of an unsatisfying reality. Freud's contemporary, the Austrian psychoanalyst Alfred Adler also introduced a note of caution, 'At a certain stage in their [children's] development, their powers of fantasy may become a way of avoiding the realities of life. Fantasy may be misused as a rejection of reality . . .' (Adler 1927, p. 49). However, neither Freud nor Adler paid the phenomenon much attention beyond an occasional paragraph.

Even Singer, whose works are otherwise overwhelmingly supportive of the positive attributes of fantasy, notes that fantasies may 'lead us into expectations that are not likely to be fulfilled and little by little we feel inner despair or rage' (Singer 1981, p. 203) and Singer and Pope (1978, p. 90) speak of 'fantasy run rampant' and of its 'addictive qualities'. The British psychiatrist Robin Skynner who was a pioneer of family therapy in the UK, is even more critical of fantasy: 'Respect for reality and the truth is everything, but is easily lost . . . denial causes all the problems because it leads us to avoid facing reality and to live in fantasy instead' (Skynner and Cleese 1993, p. 346).

It was two clinical psychologists, Sheryl Wilson and Theodore Barber, who, while pioneering the scientific study of hypnotism, coined the term 'fantasy prone personality' and wrote about 'fantasy addicts' who set aside several hours a day for fantasy and regard it as integral to their life (Wilson and Barber 1983). They estimated such addicts to be 4–6% of the population, but the great majority of such addicts were described by the researchers as 'well-adjusted' and 'high-functioning'. Five years later, Lynn and Rhue (1988) found similarly unremarkable results when they screened for high-fantasizers among 6000 colleges students. From that cohort, they identified a subgroup of 156 fantasy-prone participants, and estimated that up to one-third were not only high-fantasizers but also exhibiting signs of psychopathology as judged by the self-report scores of the Minnesota Multiphasic Personality Inventory. The remaining two-thirds of the 156 high-fantasizers reported having as many close friends and equally high grade-point averages as low-fantasizers, and appeared to be coping well with college life with no signs of psychopathology. (It is notable that the backgrounds of the maladjusted individuals had certain aspects in common: excessive physical punishment during childhood, or sexual abuse, or loneliness.) So, according to the Lynn and Rhue (1988) study, barely 1% of the original 6000 cohort of students were high-fantasizers who were also showing maladjustment.

However, the findings of the above studies, all of which were cross-sectional, differ markedly from Vaillant's longitudinal prospective work, which is recounted in *Adaptation to life* (Vaillant 1977, 1995b). From age 20 and for nearly 60 years, the 268 Harvard College men of the Grant Study had received questionnaires every 2 years, physical examinations every 5 years, and were interviewed for 2 hours about every 15 years. Information on the participants was also gathered from their wives and children. Drawing on this database, Vaillant convincingly raises the possibility of a much greater prevalence of high fantasy being correlated with decidedly negative life-outcomes. He writes gravely of a subsample of 95 men, drawn from The Grant Study cohort, whom he personally interviewed and evaluated when they were in their late forties: 'Nine of the ninety-five men that I studied used fantasy often. None of them engaged in games with others, none had close friends, and only four stayed in touch with their parents or siblings' (Vaillant 1995b, p. 168). Vaillant characterizes those nine high users of fantasy as lonely, self-isolating, and unfulfilled individuals (Vaillant 1995b, pp. 168–70). This observation that fantasy-use by adults in their late forties correlated with dysfunctional behaviour is all the more pertinent when we consider that all of those 95 middle-aged men had originally been selected for study when circa 20 years of age (from among 10 times their number of Harvard peers) on account of their being deemed by medical, psychiatric, and educational evaluation to be the individuals most likely to lead healthy and successful lives. It seems plausible that, if psychosocial stress is one major source of fantasy (as all the available literature has suggested), then a less privileged and promising social group than those Harvard undergraduates might contain a far higher proportion of high fantasy-users who are psychologically disturbed.

Despite the above submissions by Vaillant, research psychologist Erick Klinger is strongly representative of the current received wisdom of those few researchers who have studied in the clinical field when he claims that '. . . there is no evidence that any amount of daydreaming can . . . bring on . . . psychological disorder . . . In summary: frequent and fanciful daydreamers are clearly no worse off psychologically—and possibly better off—than those who daydream less' (Klinger 1987, p. 67). Judging by the complete omissions from potentially relevant texts, this view has prevailed well into the twenty-first century. Yet, the substantial discrepancies between the evaluations of different researchers (e.g. Vaillant compared to Lynn and Rhue) clearly indicate the need for prospective longitudinal data from representative samples if we are to procure a clearer picture of the field.

Fantasy in the forensic setting

In the forensic field, the potentially malignant role of fantasy in negative life-outcomes has yet to gain widespread acknowledgement. Arguably, it was the seminal work by MacCulloch *et al.* (1983) that first alerted the forensic world to the likely role of fantasy, very specifically in sexual offending. MacCulloch and two other forensic psychiatrists interviewed 16 teenagers convicted of serious sexual offences. They concluded that: 'in only 3 cases were the crimes explicable in terms of external circumstances and personality traits. The offences of the remaining 13 cases became comprehensible only when the offender's internal circumstances were explored.' (MacCulloch *et al.* 1983, p. 25).

The authors hypothesized that, faced with real-world failure, 'it becomes easier and more pleasurable for the individual to live predominantly in his fantasy world.' Though MacCulloch *et al.* (1983, p. 2000) make no reference to the possibility of the significance of fantasies in other, non-sexual offending behaviour, the forensic adolescent psychiatrist, Sue Bailey (1995) notes that deviant daydreaming is very frequently correlated with extremely serious criminal behaviour. It would seem in these instances that fantasy is a crucible in which the fantasies grow until they are eventually explored in real-life try-outs that can lead to assault and murder. It is pertinent to RwR theory that Bailey concludes that there had been grave consequences of carers and clinicians neglecting to explore the early fantasies of the youngsters with whom she eventually worked.

Conclusion

Of those few pioneering clinicians who have explicitly attributed negative real-life outcomes to an abuse of fantasy, Vaillant (1977, 1995b, 1997, 2002) was the first to support his assertions with rigorously acquired and analysed prospective longitud-inal data from a large number of lifetimes, albeit it among a far from representative sample of the general population. Alas, his work in this arena is little cited. The prevailing view is that fantasies and daydreams are quite harmless and insignificant in any but some rare clinical cases or very serious forensic ones.

Empirical investigation underlying RwR theory

The roots of this research

While serving as a creative writing tutor in the high-security environment of Feltham Young Offender's Prison in London during 1994, I immediately recognized among the young inmates (aged 18 to 21) the same pervasive use of quick-fixing and evading thoughts and behaviours that had characterized my own unhelpful relationship with reality in my teens and early twenties. I was equally struck by the absence of anything akin to this phenomenon documented in the mainstream of educational or psychological literature.

I resolved to embark upon a systematic investigation of this discrepancy, and conducted retrospective qualitative semi-structured interviews of between 1 and 2 hours in length with a sample of nine convicted young prisoners serving long sentences for serious crimes of theft, robbery, and physical violence (Baylis 1995). For comparison with this imprisoned group, I selected a group of young men of the same age who had grown up in grimly impoverished socio-economic circumstances. These were 10 enlisted men (i.e. the lowest rank of non-officers) from the Grenadier Guards, who are an elite British army regiment famous for their combat excellence, red uniforms, bearskin hats, and guarding the royal residences in times of peace.

The interview data illuminated a startling contrast between the early teenage years of these two groups. Those young men who became the elite Grenadier Guardsmen claimed they hadn't resorted to high levels of quick-fixing and evading, whether via fantasy, fiction, or substance abuse. By contrast, the prisoners self-reported that their resort to quick-fixing and reality-evading had been a dominant trait in their lives for many years *before* their conviction, though they were adamant that the loneliness and stress of incarceration had certainly exacerbated such pursuits. The prisoners' autobiographical accounts also suggested that, although their strategy of quick-fixing and evading as children and young teenagers was initially in response to emotional pain and/or a sense of inadequacy, these same strategies were themselves contributing to those negative feelings.

By contrast to the prisoners, it seemed that, for the Grenadier Guardsmen, the real-life prospect of belonging to a world-renowned British army regiment was sufficiently attractive that they strategically deployed fiction and daydreaming in a reality-investing way to help motivate themselves to adhere to healthier and non-delinquent lifestyles. When they did use fantasy, it was to relieve boredom rather than to escape emotional pain, and it was notable too that their alcohol use, though heavy, was for social enhancement. This was in stark contrast to the way the prisoners had always used alcohol or drugs to evade their reality rather than enhance it. As for an explanation of these life-changing differences, I noted that in all 10 cases the Guardsman spoke of an adult, non-parental mentor who helped initiate the ambition and then took a benevolent interest in their achieving it. It seemed that this informal mentoring relationship might be a key factor in those boys pursuing the adaptive life-goal of becoming Guardsmen, and then helping them adhere to their goal-oriented regimes over several years.

The strongly patterned results of that preliminary qualitative study encouraged me to make a more thorough investigation in the form of a PhD thesis, and the following sections recount some key features of that research (Baylis 1999).

The main study

My hypothesis was that differences between individuals' self-reported relationship with reality would correlate with substantial differences in a rudimentary index of their observable life-outcomes. Put more specifically: I predicted that individuals prone to quick-fixes and reality-evading would have fared considerably worse in some key arenas of life, when compared to individuals who were characterized by a higher ratio of reality-investment.

The empirical investigation comprised two parts: one-to-one interviews lasting on average 2 hours; followed by self-report questionnaires. Participants were young adults, aged 18 to 24, selected to represent high, medium, or low attainment. They were members of one of three distinct types of institution:

- an elite university or an equivalent elite institution in the arts or in the armed forces;
- a standard university with modest entrance requirements;
- a high-security prison for young offenders.

The long interviews

The interviews were one-to-one, semi-structured, largely qualitative, and lasted for an average of 2 hours. Only hand-written notes were taken to reassure confidentiality because of the private and arguably taboo nature of the themes being discussed (e.g. fantasy use, substance abuse). The aim was to identify and explore the sorts of themes and questions that would subsequently make up a largely quantitative questionnaire.

The following two quotes are highly illustrative of the high-achievers' ($n = 50$) use of fantasy and daydreaming.

◆ At 18, Jim declined the opportunity to become a concert-level musician and went on to take an Oxbridge first class degree, before joining one of the world's top investment banks. At age 22 he said, 'Fantasizing? no I don't do that.'

◆ Rick is an Oxbridge PhD in science who then went on to a highly successful decade in an extremely competitive corporate career. 'I might do half an hour's escapist fantasy in a week. It's been that way since I was 13 or so.'

Note: Despite these quotes, I noted there was a considerable discrepancy between the interview and questionnaire self-report data of the high-achievers, the latter showing considerably higher amounts of time attributed to escapist fantasies. This suggests that people may be too embarrassed to admit to very much fantasizing, and this finding may help explain why it has been underrepresented in clinical and educational literature and research.

The following quotes are strongly illustrative of the group of 18 imprisoned young men when thinking back over their growing-up years as teenage schoolboys long before prison.

◆ Young prisoner A: 'It got to the point where I'd come home from school and just fantasize for hours. I was fantasizing too much and I wasn't actually doing anything. When I got bored, I'd automatically switch and start thinking about this. I couldn't keep it out. Real life couldn't compete.'

◆ Young prisoner B: 'I had quite a few different fantasy worlds. No one realized what was going on.'

Sixteen of the 18 imprisoned young offenders I interviewed also self-reported that their fantasy use increased in prison in both frequency and intensity and its themes became more extreme and ever-more divorced from likelihood of enactment in reality. In addition to which, fiction materials such as TV or pornographic magazines assumed an inflated value to the user. I asked the question, 'How does your imagination change in prison?'

◆ Young prisoner C: 'My imagination has got more intense and bizarre. I do it pretty much most of the day. Because I'm doing a lot of it, it has to get a little bit stronger each time to keep the buzz up.'

It may not be surprising to us that imprisoned youngsters testify to using extreme amounts of fantasy and daydreams. After all, on a bad day, a prisoner can spend all but 1 hour confined in his cell, and many days of solitary confinement are still used as a punishment in young offender prisons. However, it is certainly surprising how these phenomena have never received more than a paragraph or so of comment in the mainstream criminological literature. It is, too, a tragic irony that, while these youngsters need to become more involved in positive social relationships (see Sampson and Laub 1993), what prison produces in them instead is the pain of fear, anger, shame, and sensory deprivation that so fuels the social detachment of escapist fantasy.

The questionnaire

My participants were circa 20 years of age and comprised four groups.

- High academic achievers (HAs): Oxford or Cambridge undergraduates ($n = 82$). A place at an Oxbridge university commonly requires at least three A grades (which is close to the maximum achievable mark). The vast majority of these individuals had also publicly excelled in some other performance arena such as sport or music.

- High physical achievers: Royal Marine Commandos (RMCs; $n = 14$). A commando is a small force of soldiers operating covertly, and 'The Corps' as it is known is widely regarded as an elite, special forces regiment. These were enlisted men, i.e. non-officer ranks, who were physically oriented, non-academic high-achievers. I chose these young men because they were very likely to come from unprivileged socio-economic backgrounds ostensibly comparable to those of the UAs.

- Moderate-achievers (MAs; $n = 75$). All were undergraduates drawn from a university with modest admissions requirements (two C-grades); and these individuals did not self-report having yet obtained a high level of achievement in any other performance arena such as sport or music.

- Under-achievers (UAs, i.e. imprisoned young offenders; $n = 68$). All were serving sentences ranging from 1 to 4 years for crimes that ranged from drug-dealing to armed bank robbery. They were individually assisted with self-completing all the questionnaires, as they were less accustomed to such a task than the other groups, and it meant that any reading difficulties could be mitigated.

Participants completed a questionnaire developed in the light of my in-depth interview data, comprising 54 questions using a 0–6 Likert scale response format. It took on average 20 minutes to complete. Many of the questions focused on escapist fantasy and wishful daydreaming as I regarded the deployment of these as useful indicators of an individual's RwR, though my discussion of results will also look at other elements of RwR.

Amount of time spent in fantasy and daydreaming

Participants were given the following questionnaire instructions: 'How much of the different types of thinking do you do in a day, from waking-up first thing to falling asleep last thing at night.'

- How much 'escapist fantasy'? (about things that you know will never happen, but you imagine it purely for pleasure. An example might be: *fantasizing about dating a favourite film star.*)

- How much 'wishful daydreaming' about things you think will happen? An example might be: thinking how good it's going to feel to drive the particular type of car you intend to have one day.

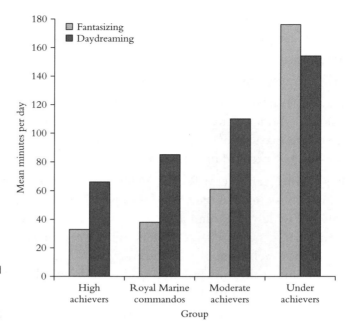

Fig. 9.2 Self-reported mean time (mins) spent fantasizing and daydreaming per day.

The results are shown in Fig. 9.2.

There are striking differences in self-reported timings between the groups, and this finding would support the proposition that excessive fantasy and daydreaming will be inversely related to positive life-outcomes. High-achievers (no matter whether they be academics or commandos) reported doing half as much fantasy as MAs and one-fifth as much as UAs. When it came to daydreaming, HAs reported half as much as MAs and two-fifths as much as UAs. There were, of course, wide standard deviations that indicate that there were considerable differences even between individuals within the same group. An analysis of variance by gender revealed negligible differences. Indeed, men and women in the HA and MA groups apparently used fantasy and daydreaming in much the same way.

An analysis of academic high-achievers as compared to non-academic high-achievers (i.e. Oxford and Cambridge undergraduates compared to Royal Marine Commandos) revealed that, although these two sets of high-achievers came from very different socio-economic backgrounds and had excelled in very different ways, they nonetheless had very similar RwR profiles. What is more, these high-achiever profiles were very substantially different from those of moderate- and under-achievers.

It is possible that verbal or analytic intelligence is the underlying variable determining the amounts of self-reported fantasy and daydreaming, so it is pertinent that Singer notes that 'Whenever intelligence tests have been employed, the degree of intelligence does not especially appear to influence the reported frequency of daydreaming or any special pattern of fantasy' (Singer 1981, p. 66).

I conclude that, although self-reported time spent in fantasy and daydreaming is a crude measure of their total duration, it may nonetheless give us a valuable insight into how an individual is spending his or her thinking time.

Perceived role of fantasy and daydreaming

Guided by the findings of the long interviews, the questionnaire also explored what role participants felt that their fantasy and daydreaming played in their present-day life. The responses revealed the following for the majority of the young adults across all of the groups.

- Fantasies and daydreams were a vital 'safety-valve' for their unfulfilled needs and desires.
- The individual had almost irresistible urges to fantasize or daydream about certain things.
- They sometimes felt they did too much daydreaming and fantasy for their own good.
- They worried that sometimes their fantasies were too extreme.
- Too much daydreaming and fantasy made it hard for reality to live up to them.
- They had dreamt so high, imagined so much, that real life (real lovers, real jobs) seemed quite a let down.

The questionnaire also asked whether there were any features that made real life better than fantasy and fiction. Their responses could be distilled into five characteristics:

- The thing that makes real life special is a good relationship with someone with whom we can share our experience of life.
- We can directly affect other people by our real-life activities, but not by our fantasies and fiction.
- Real life is interestingly unpredictable.
- Real-life activities generated a much superior strength and durability of feelings.
- Real life permits very pleasurable 'reminiscing', whereas we tend not to reminisce about fantasy scenarios.

The above results appear to support the hypothesis that wishful daydreaming and escapist fantasy may play an important role in many young people's lives.

Other behaviours comprising our RwR

In the first section of this chapter, it was proposed that an individual's RwR was the common factor driving a whole range of seemingly unconnected cognitive and behavioural characteristics. It's pertinent that Adler warned that 'the understanding of human nature can never be learned by examining isolated phenomena that have been taken out of their whole psychological context. We must always see them in relation to social life' (Adler 1927, p. 155). With this in mind, my research

Table 9.1 A list of activities and the percentage of high achievers ($n = 48$), moderate achievers ($n = 42$), and underachievers ($n = 50$) who strongly agreed that they 'sometimes used that activity to escape real life for a while'

Activity	High achievers (%)	Moderate achievers (%)	Underachievers (%)
Drinking alcohol	30	36	8
Watching TV	27	38	NA
Listening to music	29	47	54
Reading fiction	22	33	36
Computers/internet	15	12	20
Taking drugs	8	24	44

NA, Not applicable

also explored whether the phenomena of fantasy and daydreaming were part of a wider relationship with reality that embraced other thoughts and behaviours.

Participants were asked to rate their agreement with the statement: 'We sometimes use one or more of the following to escape real life for a while (perhaps because our real life feels too uncomfortable or too tiring)'. The list in Table 9.1 indicates the percentages of each group who considered that this proposition was fairly, very, or extremely true of a particular activity.

Note in Table 9.1 the low alcohol and high drug prevalence among the UAs, which is an anomaly reflecting the peculiarities of their custodial environment in which alcohol is scarce yet other drugs are rife. Note too, how the UA group scores highest on almost every other escapist-use of an activity. Also consistent with the RwR model is the pattern of MAs scoring higher frequencies than HAs on every factor other than 'computers'; indeed very considerably more MAs than HAs use TV, drugs, fiction, and music to escape reality.

Another question asked, 'How many hours of TV and video did you watch on an average weekday between getting up in the morning and going to bed?' There was a very significant statistical difference between the HAs and MAs ($p < 0.001$) with the MAs most likely to say over 4 hours, and the HAs more likely to say over 2 hours. (The UAs were not asked this question because TV watching is managed for them in their custodial environment.)

Factor analysis of the RwR data

A factor analysis was conducted to look at the relationship between questionnaire items. It's important to note that, for the purposes of the factor analysis, *all* of the questions came from a section of the questionnaire that asked participants to give answers according to the 2 year period *before* the participants went either to a university (if HAs or MAs) or to prison (if a UA), rather than asking about the present day. The rationale was that these 2 years pre-prison/pre-university were the 'watershed' period in their lives—the period that most determined whether they ended up in an elite institution, a standard one, or in prison.

Four components emerged from the factor analysis, and taken together these explained 52% of the variance between the three groups of HA, MA, and UA.

1 'Reality-evading thinking' explained 23% of variance. This component is comprised of fantasizing, daydreaming, and the sorts of thinking that were self-reported as 'getting in the way of real life'. On this component, HAs did highly significantly less than MAs ($p < 0.005$) and MAs did highly significantly less than UAs ($p < 0.005$).

2 'Reality-investing thinking' explained 15% of the variance. This component is comprised of self-control of escapist thinking, combined with investment in reality through planning. There were no significant differences between any of the groups. One plausible explanation for this unexpected finding is that the UAs and even the MAs may not be good at distinguishing between what is a 'realistic plan and reality-oriented' compared to what is simply 'wishful daydreaming' with little hope of success.

3 'Reality-evading behaviours' explained 8% of the variance. This component is comprised largely of activities such as listening to music, drinking alcohol, watching TV, and using computers. HAs did less than UAs ($p < 0.05$).

4 'Feel you are addicted to escapist thoughts', e.g. 'fantasy-addicted rather than drug-addicted', explained 6% of the variance. HAs said yes very significantly less often than MAs ($p < 0.01$). HAs said yes highly significantly less often than UAs ($p < 0.001$).

This factor analysis begins to suggest that RwR may be a valid and helpful construct that can usefully link a whole range of seemingly unconnected thoughts and behaviours.

Caveats and conclusions to the research

Overall, these findings suggest that escapist fantasy and wishful daydreaming are self-reported as common and important tools by a substantial number of the wide variety of young adults among whom I researched, and that a characteristic RwR is quite recognizable among those same young adults and bears a strong relation to their life outcomes.

Cross-sectional studies can at best demonstrate correlations between levels of one factor and levels of another. For instance, it is possible that 'present life-circumstances' is the independent variable explaining all of the results presented above and is severely influencing self-reported RwR, albeit in a distinctively patterned and predictable way. Only prospective longitudinal studies and interventions with representative sample groups can hope to demonstrate whether or not RwR can play a *causal* role in life-outcomes. Vaillant's prospective longitudinal assessment of the Grant Study Harvard undergraduates as their lives unfolded over 60 years, strongly suggested that the use of excessive fantasy was as much a cause of serious psychosocial ills as reflective of them. More specifically, 9 out of 95 of those Harvard graduates used excessive fantasy as a coping strategy and led seriously diminished lives because of it (see Vaillant 1995b).

The implications of RwR theory

Why has our relationship with reality gone unrecognized?

I can suggest at least three reasons why something akin to the proposed RwR theory has not previously attracted the attention of educators or clinical practitioners.

- Unless the method of inquiry is extremely sensitive to the relatively private and often taboo nature of reality-evasion and quick-fix thoughts and activities, as well as offering specific prompts and an anonymous means of self-report, I think it unlikely that participants will accurately volunteer their experiences.

- The overt characteristics of quick-fix and reality-evasion, such as alcohol consumption, TV watching, and playing computer games, all have a wide social acceptance. For instance, the majority of adults watch about 3 hours television per day, and teenagers often watch considerably more (Kubey 2004). It is only if we view these various activities through a lens that relates them to each other on account of their underlying intention (i.e. to quick-fix or escape reality) that we might consider their accumulated effect as being profoundly unhelpful.

- There may be an era effect, whereby it may have only been in recent years that quick-fix and evasion thoughts and behaviours have greatly increased in the general population, very likely as a means by which to compensate for the demands and stresses of an increasingly complex modern-day reality. More specifically, we live in a twenty-first century Western world characterized by rapid technological developments, longer and more intense working hours, chronic deficits in sleep, the dislocating nature of social and geographical mobility, and the pervasiveness and power of media messages. All of these modern-day demands are exacerbated by the steep decline in an individual's social supports such as the extended family, the church, marriage, and full, stable employment.

Self-awareness, self-monitoring, and self-management

This chapter has advocated that RwR may be one means through which a person's healthy development in any or all aspects of his or her life can be enhanced or inhibited. Consequently, I advocate further consideration of how individuals can be helped towards a well-informed choice about how to invest their finite amounts of cognitive and behavioural energy. For instance, it seems very likely that sustainable well-being may best be achieved through developing a flexible and complementary ratio between investment, quick-fix, and evasion strategies, so as to create a highly adaptive RwR. The Nobel Laureate psychologist, Daniel

Kahneman, said 'happiness is a skill' (in his keynote speech at the summit in Positive Psychology, Washington DC, October 2003), and likewise, as a means to greater happiness, I regard our RwR as a skill, and thus amenable to practice and improvement. But to what extent can or should an individual proactively exert self-control over the focus of his/her conscious mind? Psychologists have long since researched how different types or styles of thinking might considerably influence important outcomes. Here are some examples.

- Ross and Fabiano (1985) argue that social cognitive skills should be taught specifically and systematically as early as possible in a child's life and should not be regarded as natural by-products of academic education.

- Sternberg and Ruzgis (1994, p. 156) write, 'If we wish to fully understand the success of children in school (and, we would speculate, adults on the job), we need not only to look at abilities and achievements, but at the interface between intelligence and personality—at thinking styles.'

- Csikszentmihalyi, Rathunde, and Whalen (1993, p. 83) advocate that, in order to maximize our experience of what the authors call 'flow' (a highly productive state of enhanced involvement with an activity), 'the challenge consists in direct control over consciousness itself'. Moreover, the authors conclude that such 'attentional habits' cultivated in early years form something akin to a personality trait (Csikszentmihalyi *et al.* 1993; p. 96).

Thus, there exist good precedents for the proposition that an individual's particular RwR could be a thinking style that can be systematically fostered so as to affect the individual's success in life. A problematic RwR may be amenable to change using a variety of well-established techniques. Here are some examples.

- Beck's cognitive therapy is founded on the proposition that 'distorted or dysfunctional thinking (which influences the person's mood and behaviour) is common to all psychological disturbances. Realistic evaluation and modification of thinking produce an improvement in mood and behaviour' (Beck 1995). An individual's RwR should respond well to such a meta-cognitive approach.

- Putnam's (2000) studies of 'social capital' and Diener's (2000) studies of 'subjective well-being' both indicate that the single most important source of lasting happiness and well-being is the depth and breadth of our social relationships. It is very significant that a key side-effect of quick-fix and evasion strategies is the way in which they inhibit the user from developing fulfilling relationships with real life individuals. It is likely, however, that good mentorship, or role-models, or rewarding close friendships can all help promote a positive RwR.

- Bandura's (1997) notion of 'personal efficacy' (which is the belief that our thoughts and actions have the power to bring about improvements) is likely to be a key construct in determining how much we reality invest and how much we quick-fix or evade reality. This is because our level of motivation to take action towards real life goals, is largely governed by our expectancy that such action will procure success.

◆ Related to the above point, Seligman (1990) has advocated the learning of a 'flexible optimism' whereby the deliberate overestimation that a positive outcome will result from one's actions will itself encourage an individual's increased investment in their endeavours. By these means, optimism (in conjunction with a sense of self-efficacy) can act as a determinant of the type of reality-investing that would characterize a positive RwR.

◆ More recently, 'hope therapy' (Lopez *et al.* 2000) has offered clients a systematic set of strategies for increasing their feelings of optimistic self-efficacy. These include electing clear and attainable goals, envisaging multiple pathways to those goals, maintaining self-motivation en route, and reframing problems as challenges to be circumvented. Snyder (2002) has observed that hope flourishes when people develop a strong bond to one or more people in whom they can confide their dreams and goals.

◆ In terms of the sort of character strengths proposed by Peterson and Seligman (2004), a positive RwR could, like optimism and hope, be regarded not only as a universal character strength in its own right, but also as one of the underlying mechanisms by which other human strengths are achieved, such as building friendships, resilience in the face of adversity, and acquiring expertise, all of which require reality-investment.

The broader implications of society's RwR

The RwR concept can also be deployed to consider the well-being of whole societies and the prevailing Zeitgeist of the twenty-first century.

◆ Diener (2000) has investigated cross-national data on the self-reported happiness of citizens. How, then, do countries compare as regards their RwR? For instance, does the ethos of 'the American dream' and the USA's vast TV and film industries create a different national RwR than that found in less media-oriented countries?

◆ Layard (2003) has advocated that increasing happiness rather than income may be a better means of increasing a population's level of well-being. Likewise, there could be macroeconomic implications of encouraging a reality-investing population—a population that is more apt to deploy their thoughts and behaviours with the aim of sustainable well-being rather than in pursuit of the sorts of short-term pleasures or total evasions that can prove compromising.

◆ In what direction is Western society going as regards its RwR? The general population's increasing appetite for vicarious pleasures via fictional materials has been illustrated by the vast uptake in TV channels and internet pornography. We might well ask whether twenty-first century folk increasingly prefer voyeurism, virtual reality, and immediate gratification, and as a consequence are less interested in actively participating in lasting relationships with the real world around them. Putnam's (2000) work on social capital's rise and fall over the past 100 years reminds us that macrosocial trends may be cyclical, so it is possible there will be a backlash against quick-fixes and reality-evasions, and that this backlash will be accompanied by a cultural renaissance in which a more investing relationship with reality becomes the preferred and prioritized mode of living.

◆ In view of the above, we need so much more than a psychology of well-being. What we need is a new generation of 'well-being scientists' working holistically to integrate all the possible sources of well-being from not only our psychology, but also our physiology, sociology, and the built and the natural environment. We also need the creative arts and humanities and investigative news media to explore and portray how positively human life can be led. Such a comprehensive and profoundly integrated study of well-being requires our research institutions and policy-making organizations to be deliberately designed to foster cross-disciplinary collaboration. Only by these means can we hope for an holistic understanding of human well-being and for a greater symbiosis with the life around us.

Conclusion

RwR theory and research have built upon an empirical investigation that was begun almost half a century ago by some pioneering researcher–clinicians, most notably Jerome L. Singer and George E. Vaillant. I hope that RwR can begin to serve as a new means to understand twenty-first century life on an individual and societal level. For, as digital and pharmaceutical technologies become ever more powerful and pervasive, our best hope of them genuinely enhancing our quality of life is for us to be ever more ingenious in how we deploy the innovations. As highly social animals, how can we use technology to enhance our real life relationships, rather than allowing it to diminish or dilute the strength, breadth, and intimacy of our interpersonal connections? I echo Putnam's (2000) work on social capital and his advocacy that we harness technology to physically bring people together, rather than keep us apart.

We need to be equally ingenious in how we use our fantasies and fictions to inspire new ways of engaging with the real world, rather than allowing our imaginations simply to replace or undermine our activities. As Rudyard Kipling wrote, 'If you can dream—and not make dreams your master'. Our imagination is driven to weave fabulous fireside stories and to invent new ways of doing things, but let's be sure to use this faculty in the service of the full-blooded, in-the-flesh relationships that can so enhance individual and societal well-being.

Acknowledgements

I dedicate this chapter in loving memory of Beryl Baylis. The theory and empirical work reported here reflect the principal tenets of my Cambridge University PhD thesis titled 'Learning from young people's lives', which was supervised by Richard Green and examined by Donald J. West (Cambridge) and George E. Vaillant (Harvard). My heartfelt thanks also go to Sarah Fitzharding, Anja Minnich, Cerstin Henning, Shane Lopez, and Felicia Huppert for invaluable contributions at various stages in the development of these ideas.

References

Adler, A. (1927, 1992 translation). *Understanding human nature*. One World Books, Oxford.

American Psychiatric Association (APA) (1994). *Diagnostic and statistical manual of mental disorders*, 4th edn. APA, Washington, DC.

Bailey, S. (1995). *Sadistic and violent acts in the young*. Mental Health Services–Adolescent Forensic Service, Manchester.

Bandura, A. (1997). *Self-efficacy: the exercise of control*. Freeman, New York.

Baylis, N.V.K. (1995). Reality, fantasy, fiction and offending. Unpublished MA thesis, Cambridge University.

Baylis, N.V.K. (1999). Learning from young people's lives. Unpublished PhD thesis, Cambridge University.

Baylis, N.V.K. (2005). *Learning from wonderful lives: lessons from the study of well-being*. Cambridge Well-Being Books. Cambridge.

Beck, A. (1976). *Cognitive therapy and emotional disorders*. New American Library, New York.

Beck, J. (1995). *Cognitive therapy: basics and beyond*. Guilford Press, New York.

Csikszentmihalyi, M., Rathunde, K., and Whalen, K. (1993). *Talented teenagers*. Cambridge University Press, Cambridge.

Diener, E. (2000). Subjective well-being: the science of happiness and a proposal for a national index. *Am Psychol*. **55** (1), 34–41.

Freud, A. (1957). *The ego and the mechanisms of defence* (revised edn). Hogarth Press, London.

Griffin, G. and Tyrrell, I. (2003). *Human givens: the new approach to emotional health and clear thinking*. Human Givens Publishing Ltd, Chalvington.

Goertzel, M., Goertzel, V., and Goertzel, T. (1978). *Three hundred eminent personalities*. Jossey-Bass, San Francisco.

Huba, G., Singer, J., Aneshensenl, C., and Antrobus, J. (1982). *Short imaginal process inventory*. Sigma Assessment Systems, Stanford, California.

Klinger, E. (1987). The power of daydreams. *Psychol. Today* **21** (10), 37–44.

Klinger, E. (1990). *Structure and function of fantasy*. John Wiley, New York.

Kubey, R. (2004). *Creating television: conversations with the people behind 50 years of American TV*. Lawrence Erlbaum Associates Inc, Englewood Cliffs, New Jersey.

Layard, R. (2003). *Happiness: has social science a clue?*, Lionel Robbins Memorial Lectures at the London School of Economics. London School of Economics, London.

Lazarus, R.S. (1993). Coping theory and research: past, present and future. *Psychometr. Med.* **55**, 234–47.

Lopez, S.J., Floyd, K.R., Ulven, J.C., and Snyder, C.R. (2000). Hope therapy: helping clients build a house of hope. In *Handbook of hope: theory, measures, and applications* (ed. C.R. Snyder), pp. 123–66. Academic Press, San Diego.

Lynn, S. and Rhue, J. (1988). Fantasy proneness. *Am Psychol.* **43**, 35–44.

MacCulloch, M., Snowden, P., Wood, P., and Mills, H. (1983). Sadistic fantasy, sadistic behaviour and offending. *Br. J. Psychiatry* **143**, 20–9.

Person, E. (1997). *The force of fantasy.* Harper Collins, London.

Peterson, C. and Seligman, M. (2004). *Character, strengths, and virtues: a handbook and classification.* Amercian Psychological Association and Oxford University Press, New York.

Planck, M. (1949). *Scientific autobiography and other papers* [translation by F. Gaynor]. Philosophical Library, New York.

Putnam, R. (2000). *Bowling alone: the collapse and revival of American community.* Simon and Shuster, New York.

Robertson, I. (2000). *Mind sculpture.* Bantam Book, London.

Rosen, S. (ed.) (1991). *My voice will go with you: the teaching tales of Milton H. Erickson.* W.W. Norton & Company, New York.

Ross, R.R. and Fabiano, E.A. (1985). *Time to think: a cognitive model of delinquency prevention and offender rehabilitation.* Institute of Social Sciences and Arts, Johnson City, Tennessee.

Sampson, R. and Laub, J. (1993) *Crime in the making: pathways and turning points through life.* Harvard University Press, Cambridge, Massachusetts.

Seligman, M.E.P. (1990). *Learned optimism.* A.A. Knopf, New York.

Singer, J.L. (1966). *Daydreaming: an introduction to the experimental study of inner experience.* Random House, New York.

Singer, J.L. (1975). *The inner world of daydreaming.* Harper and Row, New York.

Singer, J.L. (1981). *Daydreaming and fantasy.* Oxford University Press, Oxford.

Singer, J.L. and Pope, K.S. (Eds.) (1978). *The power of human imagination: new methods in psychotherapy.* Plenum Press, New York.

Skynner, R. and Cleese, J. (1993). *Life and how to survive it.* Methuen, London.

Snyder, C. (2002). Hope theory: rainbows in the mind. *Psychol. Inquiry* **13**, 249–75

Sternberg, R.J. (1997). *Successful intelligence.* Plume, New York.

Sternberg, R. and Ruzgis, P. (1994). *Personality and intelligence.* Cambridge University Press, New York.

Storr, A. (1988). *The school of genius.* André Deutsch, London.

Vaillant, G. (1977). *Adaptation to life.* Little Brown, Boston.

Vaillant, G. (1995*a*). *The natural history of alcoholism revisited.* Harvard University Press, Cambridge, Massachusetts.

Vaillant, G.E. (1995*b*). *Adaptation to life* (new edn). Harvard University Press, Cambridge, Massachusetts.

Vaillant, G. (1997). *Wisdom of the ego.* Harvard University Press, Cambridge, Massachusetts.

Vaillant, G.E. (2002). *Aging well: surprising guideposts to a happier life from the landmark Harvard study of adult development.* Little, Brown and Company, London.

Werner, E. and Smith, R. (1992). *Overcoming the odds: high-risk children from birth to adulthood.* Cornell University Press, Ithaca, New York.

Wilson, S. and Barber, T. (1983). The fantasy-prone personality: implications for understanding imagery, hypnosis, and parapsychological phenomena. In *Imagery, current theory, research and application* (ed. A.A. Sheikh), pp. 340–90. Wiley, New York.

Yapko, M.D. (2003). *Trancework: an introduction to the practice of clinical hypnosis* (3rd edn). Brunner-Routledge, New York.

Zimbardo, P. and Boyd, J. (1999). Putting time in perspective. *J. Pers. Soc. Psychol.* **77** (6), 1271–88.

Martin E.P. Seligman PhD is director of the Positive Psychology Center and the Fox Leadership Professor of Psychology at the University of Pennsylvania, USA. In 1996, he was elected president of the American Psychological Association. Among his books are *Learned optimism*; *what you can change and what you can't, The optimistic child, Helplessness*, and, most recently, *Authentic happiness*.

Acacia C. Parks is studying for her PhD in psychology at the University of Pennsylvania. Her research focuses on disentangling the roles that different kinds of happiness (pleasure, engagement, and meaning) might play in buffering individuals against depression, and in comparing the efficacy of cognitive interventions to happiness-focused interventions for preventing depression.
Photograph courtesy of Amanda Parks.

Tracy Steen is a therapist and personal coach in Philadelphia, Pennsylvania. She received her PhD in psychology (clinical area) from the University of Michigan in 2003, and recently completed a postdoctoral fellowship in positive psychology at the University of Pennsylvania. She is now Clinical Director at the Charles O'Brien Center for Addiction Treatment, University of Pennsylvania. Philadelphia.

Chapter 10*

A balanced psychology and a full life

Martin E.P. Seligman, Acacia C. Parks, and
Tracy Steen

Introduction

American psychology prior to the Second World War had three missions: The first
was to cure mental illness, the second was to make relatively untroubled people
happier, and the third was to study genius and high talent. All but the first fell by
the wayside after the war. Researchers turned to the study of mental disorders
because that was where the funding was. The biggest grants were coming from
the newly founded National Institute of Mental Health (NIMH), whose purpose
was to support research on mental illness, not mental health. At the same time,
practitioners suddenly became able to earn a good living treating mental illness
as a result of the Veterans Administration Act of 1946. Psychopathology became
a primary focus of psychology in America because it made sense at that time. Many
very distressed people were left in the wake of the Second World War and the high
incidence of mental disorders had become a pressing and immediate problem.

A wealth of excellent research resulted from this chain of events. In 1946, there
were no effective treatments for any of the psychological disorders, whereas now
we can cure two and treat another 14 via psychotherapy or pharmacology
(Seligman 1993). Furthermore, the intensive study of psychopathology has given
rise to methods of classifying the mental disorders (*International classification of
diseases*, 9th revision (ICD-9; World Health Association 1977) and *Diagnostic and
statistical manual of mental disorders*, 4th edn (DSM-IV; American Psychiatric
Association 1994)), and these methods have allowed clinical psychologists to
produce diagnoses with acceptable accuracy and to reliably measure symptoms
that were once quite difficult to pinpoint. After 50 years and 30 billion dollars of
research, psychologists and psychiatrists can boast that we are now able to
make troubled people less miserable, and that is surely a significant scientific
accomplishment.

The downside of this accomplishment is that a 50-year focus on disease and pathology has taken its toll on society and on science. In our efforts to fix the worst problems that people face, we have forgotten about the rest of our mission as psychologists. Approximately 30% of people in the US suffer from a severe mental disorder at one time or another (Kessler *et al.* 1994) and we have done an excellent job of helping that 30%. It is time now to turn to the other 70%. While these people may not be experiencing severe pathology, there is good evidence to indicate that the absence of maladies does not constitute happiness (Diener and Lucas 2000). Even if we were asymptotically successful at removing depression and anxiety and anger, that would not result in happiness. For we believe 'happiness' is a condition over and above the absence of unhappiness.

That said, we know very little about how to improve the lives of the people whose days are largely free of overt mental dysfunction but are bereft of the things that make life good. We do not know much about what makes a person optimistic, kind, giving, content, engaged, purposive, or brilliant. To address this, the first author proposed during his term as president of the American Psychological Association in 1998 that psychology be just as concerned with what is right with people as it is with what is wrong. As a supplement to the vast research on the disorders and their treatment, we suggest that there be an equally thorough study of strengths and virtues, and that we work towards developing interventions that can help people become lastingly happier.

Towards this goal, our first order of business is to determine what it was we are trying to increase, and why we are increasing it. First, what is happiness? More words have been written about this great philosophical question than perhaps any other. Science can no more presume to answer this question than other classic philosophical questions, such as 'What is the meaning of life?' But science can illuminate components of happiness and investigate empirically what builds those components. We will review one such empirically derived framework of happiness in the following section.

In addition to defining happiness, we must determine whether happiness—however we define it—is worth increasing. Just as there are negative consequences of depression, such as increased vulnerability to further depression, stigma, and social withdrawal, there are positive consequences of being happy, which we will outline below.

Happiness: what is it and why is it a good thing?

A review of the literature led us to identify three constituents of happiness, which were initially proposed by Seligman (2002): (1) pleasure (or positive emotion); (2) engagement; and (3) meaning. We define these three routes to happiness in the paragraphs that follow.

The first route to greater happiness is hedonic, increasing positive emotion. When people refer in casual conversation to being happy, they are often referring to this route. Within limits, we can increase our positive emotion about the past (e.g. by cultivating gratitude and forgiveness), our positive emotion about the present (e.g. by savoring and mindfulness), and our positive emotion about the future (e.g. by building hope and optimism).

There has been a good deal of research on the benefits of experiencing positive emotions such as joy and contentment. Research done by Fredrickson and colleagues indicates that experiencing positive emotions often leads to a variety of positive benefits (Fredrickson 1998). For example, experiencing positive emotions in the present makes future positive emotion more likely (Fredrickson and Joiner 2002). In addition, positive emotions have physical benefits (Fredrickson 2000). Fredrickson and Levenson (1998) found that inducing contentment or amusement in people who had just experienced a negative mood induction (and a resulting decrease in cardiovascular activation) led to a faster recovery to baseline cardio-vascular activity. Positive emotions also lead to greater ability to think creatively and efficiently, and to integrate information (Fredrickson 2001).

Lastly, positive emotions lead to greater resilience, and a heightened ability to 'bounce back' from negative experiences. In one study, Tugade and Fredrickson (2004) found that positive emotions contribute to one's ability to find meaning in negative experiences rather than despairing when things go wrong. In another study, Fredrickson *et al.* (2003) found that resilience against depression was completely mediated by the experience of positive emotions such as gratitude, love, and interest.

In short, pleasure and the experience of immediate positive emotion seem to be worthwhile. However, unlike the other two routes to happiness, the route relying on positive emotions has clear limits. Positive affectivity is heritable and we speculate that, for important evolutionary reasons, our emotions fluctuate within a genetically determined range to which they will—even if altered dramatically by an unusual event—eventually return (Tellegen *et al.* 1988; Brickman *et al.* 1978). It is possible (and worthwhile) to increase the amount of positive emotion in our lives, but we can boost our hedonics only so high. Further, when people fluctuate within a relatively 'down' range of positive emotion yet live in a society like the US that promotes an upbeat disposition, they can feel discouraged and even defective. Fortunately, positive emotion is not the sole determinant of happiness, and our most liberating goal is to offer a broader conception of happiness than mere hedonics (Seligman 2002).

A second route to happiness involves the pursuit of 'gratification'. The key characteristic of a gratification is that it engages us fully. It absorbs us. Individuals may find gratification in participating in a great conversation, fixing a bike, reading a good book, teaching a child, playing the guitar, or accomplishing a difficult task at work. We can take shortcuts to pleasures (e.g. eating ice cream, masturbating, having a massage, or using drugs), but no shortcuts exist to gratification. We must involve ourselves fully, and the pursuit of gratifications requires us to draw on character strengths such as creativity, social intelligence, sense of humor, perseverance, and an appreciation of beauty and excellence.

Although gratifications are activities that may be enjoyable, they are not necessarily accompanied by positive emotions. We may say afterwards that the concert was 'fun', but what we mean is that during it, we were one with music, undistracted by thought or emotion. Indeed, the pursuit of a gratification may be, at times, unpleasant. Consider, for example, the gratification that comes from training for an endurance event such as a marathon. At any given point during the grueling event, a runner may be discouraged or exhausted or even in physical pain, yet he or she may describe the overall experience as intensely gratifying.

Csikszentmihalyi (1990) describes the experience of complete absorption described above as 'flow'. According to Csikszentmihalyi's framework, flow occurs when a person is performing a task that challenges him or her, but not so much that it is frustrating. Flow as described in Csikszentmihalyi's writings consists in many of the qualities described above—a person in flow feels as if time has stopped, and is completely focused on the task at hand. It seems that there are positive benefits of being in such a state on a regular basis; People who experience high levels of flow are more motivated and creative in both work and leisure activities than people who are not experiencing flow (Csikszentmihalyi and LeFevre 1989). There is no evidence to indicate that there are limits in our ability to experience flow. In fact, Csikszentmihalyi (2000) argues that the more flow one experiences, the more 'psychological capital' is accrued and the more flow one will be able to experience in the future.

Finding flow in gratifications need not involve anything larger than the self. Although the pursuit of gratifications involves deploying our strengths, a third route to happiness comes from using these strengths to belong to and in the service of something larger than ourselves—something such as knowledge, goodness, family, community, politics, justice, or a higher spiritual power. The third route gives life meaning. It satisfies a longing for purpose in life and is the antidote to a 'fidgeting until we die' syndrome.

There has been less research on meaning than the previous two constituents, but we do know that the ways in which people find meaning are extremely varied from individual to individual (Baumeister and Vohs 2002). It also appears that spirituality is correlated with high global happiness ratings as well as positive coping in response to negative events such as divorce and bereavement (Myers 2000). In other words, it seems that meaning—even if it is not well understood—can lead to substantial positive benefits.

It appears, then, that pleasure, engagement, and meaning are distinct types of happiness, all of which have positive benefits. However, one might rightfully wonder whether each of these is equally important in determining one's global happiness or life-satisfaction ratings, or if one would have more 'weight' in a regression equation that predicts happiness. Peterson, Park, and Seligman (2003, unpublished) developed reliable measures for all three routes to happiness and demonstrated that people differ in their tendency to rely on one rather than another. We call a tendency to pursue happiness by boosting positive emotion, the 'pleasant life', the tendency to pursue happiness via gratifications, the 'good life', and the tendency to pursue happiness via using our strengths toward something larger than ourselves, the 'meaningful life'. Importantly, the pursuit of engagement and the pursuit of meaning make much larger contributions to life satisfaction than does the pursuit of pleasure.

A person who uses all three routes to happiness leads the 'full life', and recent empirical evidence suggests that those who lead the full life have much greater life satisfaction than individuals who use only one or two routes. In fact, they report more life satisfaction than would be predicted based on the impact of each of the three lives alone; the predicted life satisfaction of the three lives combined appears to be greater than the sum of its parts (Peterson, Park, and Seligman 2003, unpublished).

Increasing happiness

From our previous discussion, we know that happiness is a good thing, and that there seem to be three different kinds of happiness, each with an empirical basis. But the question remains, can we actually increase happiness? Several research groups have attempted to address this question and it appears that, across the board, the answer is 'yes'.[1]

Early work attempting to increase happiness found that, if you give individuals exercises that mimic the habits of people who are already happy, they become happier compared to a control group and maintain the increase for over a year (Fordyce 1977, 1983).

More recently, researchers have tested the efficacy of gratitude-related exercises (i.e. writing down things for which you are thankful on a regular basis). Emmons and McCullough (2003) found that individuals who wrote down five things for which they were grateful on a weekly basis for 10 weeks were significantly more satisfied with their lives and more optimistic about the next week, and they experienced more positive affect than individuals who wrote about neutral or mildly annoying life events. In a similar study, Lyubomirsky *et al.* (in press) compared participants who 'counted their blessings' once a week to a no-intervention

---

[1] Note that the research designs used in the studies described below are exactly parallel to the random-assignment, placebo-controlled experiments that are the bulwark of the medication and psychotherapy outcome literature.

control and found that the 'blessings' group was significantly happier after 6 weeks of the exercise. Interestingly, a group of participants who performed the 'blessings' exercise three times a week did not appear to benefit from the exercise (Lyubomirsky *et al.*, in press). In contrast, our own data (discussed in greater detail below) indicate that doing a similar 'blessings' exercise on a daily basis is quite effective in making people happier.

Lyubomirsky *et al.* (in press) tested the efficacy of an 'acts of kindness' exercise that involved performing five acts of kindness either in 1 day or across a week. While the 'acts of kindness' exercise did result in higher levels of happiness than a no-intervention control, they found a larger effect in the 'all in 1 day' condition than in the 'across the week' condition. This indicates that there is perhaps a dose-dependency curve for the happiness-increasing effects of this exercise; that is, more acts of kindness concentrated at once lead to a larger effect on increasing happiness.

Our own lab has designed and tested interventions to nurture each of the three routes to happiness (pleasure, gratification, and meaning). Positive emotions are increased, and the 'pleasant life' is promoted, by exercises that increase gratitude, that increase savoring, that build optimism, and that challenge discouraging beliefs about the past. Interventions that increase the 'good life' identify participants' signature strengths and use them more often and in creative new ways. 'Meaningful life' interventions aim toward participants' identifying and connecting with something larger than themselves by using their signature strengths.

We are in the process of testing the efficacy of these interventions by randomly assigning individuals to interventions or to a placebo control and measuring their level of happiness and depression before, immediately after the intervention, 1 week later, 1 month later, and 3 months later. Early results demonstrate that: (1) it is possible to boost individuals' levels of happiness; and (2) these effects do not fade immediately after the intervention (as is the case with the placebo).

The 'good things in life' exercise (similar to the 'blessings' exercise described above) provides an example of an efficacious intervention. Designed to increase positive emotion about the past, this exercise requires individuals to record every day for a week three good things that happened to them each day and why those good things occurred. After completing this exercise, individuals were happier and less depressed when compared to a placebo-control group, and this change was sustained through the 3 month follow-up (Seligman, Peterson, and Steen 2004, unpublished).

Seligman, Peterson, and Steen (2004, unpublished) also found that another exercise, the 'gratitude visit', was effective in making people happier and less depressed for up to 1 month compared to a placebo group, who returned to their baseline happiness scores within a week. For this exercise, participants were asked to write a letter of gratitude to someone and read it aloud to the person. Similarly, participants showed significant improvements after taking the VIA—a questionnaire that provides feedback about one's top five strengths—and being asked to use one of their signature strengths each day. Like the 'good things' exercise, this exercise's effects also lasted through the 3 month follow-up (Seligman, Peterson, and Steen 2004, unpublished).

In summary, it appears that individuals can be made happier via positive exercises, and that the increases brought about by these exercises can produce lasting effects. This brings us to another question: what are the implications of making people happier? Who is most likely to benefit from a targeted increase in happiness? These questions are the topic of the next section.

Future directions: alleviating and preventing distress?

While the primary focus of this chapter thus far has been on the importance of increasing happiness in its own right, we also believe that positive psychology has something to offer in the way of buffering individuals against distress. As we saw in previous sections of this chapter, there are many positive benefits that come along with experiencing high amounts of pleasure, engagement, and meaning, one of which is enhanced coping. It seems that positive emotions have the ability to buffer individuals against stress and depression, and for this reason we believe that positive interventions have potential as interventions for treating and preventing depression.

In a pilot study testing this hypothesis, Parks and Seligman (2004, unpublished) randomly assigned 33 mild–moderately depressed participants to either a 6-week positive intervention ($n = 13$) or a no-treatment control group ($n = 20$). At the end of the intervention, participants who received the intervention were significantly less depressed than control participants: there was a 6 point difference between groups in their scores on the Beck Depression Inventory, and there was a moderate–large ($d = 0.66$) effect size for this difference. In addition, intervention participants experienced notable, but not statistically significant, increases in happiness and decreases in anxiety.

While we are still in the process of replicating these findings with a larger sample size, we believe that these findings are quite promising. A positive intervention was able to decrease mild–moderate depressive symptoms—a risk factor for developing major depression—and we expect that the effects of the workshop will prevent depressive episodes down the road. However, we are still in the process of empirically testing that hypothesis.

In short, we believe that, in addition to being useful for increasing happiness, positive interventions may be useful tools for prevention and treatment. While we intend to continue research in this area, we also hope that other researchers will see if these interventions are effective against mental disorders, particularly those that are resistant to the treatments currently in place.

Concluding remarks

Our research places us among a growing number of positive psychologists who are committed to understanding and cultivating those factors that nurture human flourishing, and we are encouraged that the field of positive psychology seems to be thriving as well. Researchers who were studying positive strengths, emotions, and institutions long before the term 'positive psychology' was coined are receiving increased recognition and support for their work, while young researchers world-wide can apply for research and intellectual support via positive psychology research awards and conferences.

One reason for optimism that the field of positive psychology may make substantial gains in the next several years is that it does not start from square one. Rather, it draws on the proven methodologies that advanced the understanding and treatment of the mental illnesses. When it is no longer necessary to make distinctions between 'positive psychology' and 'psychology as usual', the field as a whole will be more representative of the human experience. Our goal is an integrated, balanced field that integrates research on positive states and traits with research on suffering and pathology. We are committed to a psychology that concerns itself with repairing weakness as well as nurturing strengths, a psychology that concerns itself with remedying deficits as well as promoting excellence, a psychology that concerns itself with reducing that which diminishes life as well as building that which makes life worth living. We are committed to a balanced psychology.

References

American Psychiatric Association (APA) (1994). *Diagnostic and statistical manual of mental disorders*, 4th edn. APA, Washington, DC.

Baumeister, R.F. and Vohs, K.D. (2002). The pursuit of meaningfulness in life. In *The handbook of positive psychology* (ed. C.R. Snyder and S.J. Lopez), pp. 608–18. Oxford University Press, New York.

Brickman, P., Coates, D., and Janoff-Bulman, R. (1978). Lottery winners and accident victims: Is happiness relative? *J. Pers. Soc. Psychol.* **36** (8), 917–27.

Csikszentmihalyi, M. (1990). *Flow.* Harper Collins, New York.

Csikszentmihalyi, M. (2000). The contribution of flow to positive psychology. In *The science of optimism and hope* (ed. J.E. Gillham), pp. 387–98. Templeton Foundation Press, Philadelphia.

Csikszentmihalyi, M. and LeFevre, J. (1989). Optimal experience in work and leisure. *J. Pers. Soc. Psychol.* **56** (5), 815–22.

Diener, E. and Lucas, R.E. (2000). Subjective emotional well-being. In *Handbook of emotions* (ed. M. Lewis and J. M. Haviland-Jones), pp. 325–37. Guilford Press, New York.

Emmons, R.A. and McCullough, M.E. (2003). Counting blessings versus burdens: an experimental investigation of gratitude and subject well-being in daily life. *J. Pers. Soc. Psychol.* **84** (2), 377–89.

Fordyce, M.W. (1977). Development of a program to increase personal happiness. *J. Couns. Psychol.* **24** (6), 511–21.

Fordyce, M.W. (1983). A program to increase happiness: further studies. *J. Couns. Psychol.* **30** (4), 483–98.

Fredrickson, B.L. (1998). What good are positive emotions? *Rev. Gen. Psychol.* **2** (3), 300–19.

Fredrickson, B.L. (2000). Cultivating positive emotions to optimize health and well-being. *Prevention and Treatment* **3**, Article 0001a.

Fredrickson, B.L. (2001). The role of positive emotions in positive psychology: the broaden and build theory of positive emotions. *Am. Psychol.* **56** (3), 218–26.

Fredrickson, B.L. and Joiner, T. (2002). Positive emotions trigger upward spirals toward emotional well-being. *Psychol. Sci.* **13** (2), 172–5.

Fredrickson, B.L. and Levenson, R.W. (1998). Positive emotions speed recovery from the cardiovascular sequelae of negative emotions. *Cogn. Emotion* **12** (2), 191–220.

Fredrickson, B.L., Tugade, M.M., Waugh, C.E., and Larkin, G.R. (2003). What good are positive emotions in crises? A prospective study of resilience and emotions following the terrorist attacks on the United States on September 11th, 2001. *J. Pers. Soc. Psychol.* **84** (2), 365–76.

Kessler, R.C., McGonagle, K.A., Zhao, S., Nelson, C.B., Hughes, M., Eshelman, S., Wittchen, H., and Kendler, K.S. (1994). Lifetime and 12-month prevalence of DSM-III-R psychiatric disorders in the United States. *Arch. Gen. Psychiatry* **51**, 8–19.

Lyubomirsky, S., Sheldon, K.M., and Schkade, D. (in press). Pursuing happiness: the architecture of sustainable change. *Rev. Gen. Psychol.*

Myers, D.G. (2000). Hope and happiness. In *The science of optimism and hope* (ed. J.E. Gillham), pp. 323–36. Templeton Foundation Press, Philadelphia.

Peterson, C., Maier, S., and Seligman, M.E.P. (1993). *Learned helplessness: a theory for the age of personal control.* Oxford University Press, New York.

Tellegen, A., Lykken, D.T., Bouchard, T.J., Wilcox, K.J., Segal, N.L., and Rich, S. (1988). Personality similarity in twins reared apart and together. *J. Pers. Soc. Psychol.* **54** (6), 1031–9.

Tugade, M.M. and Fredrickson, B.L. (2004). Resilient individuals use positive emotions to bounce back from negative emotional experiences. *J. Pers. Soc. Psychol.* **86** (2), 320–33.

World Health Organization (WHO) (1977). *International classification of diseases*, 9th revision. WHO, Geneva.

Daniel Kahneman is Eugene Higgins Professor of Psychology and Professor of Public Affairs at Princeton University, USA. He won the 2002 Nobel Prize in Economic Sciences for his work on the psychology of judgement and choice. His main current interest is the development of measures of well-being that could help policy makers.

Jason Riis is a postdoctoral research associate at the Center for Health and Wellbeing, Princeton University. He is investigating several aspects of well-being: methods of measurement; international differences; mechanisms of adaptation; and the influences of health, wealth, and consumption. He also studies the psychology of choice and preference.

Chapter 11

Living, and thinking about it: two perspectives on life

Daniel Kahneman and Jason Riis

Introduction

It is a common assumption of everyday conversation that people can provide accurate answers to questions about their feelings, both past (e.g. 'How was your vacation?') and current (e.g. 'Does this hurt?'). Although the distinction is mostly ignored, the two kinds of questions are vastly different. Introspective evaluations of past episodes depend on two achievements that are not required for reports of immediate experience: accurate retrieval of feelings and reasonable integration of experiences that are spread over time. The starting point for this chapter is that the retrieval and the temporal integration of emotional experiences are both prone to error, and that retrospective evaluations are therefore less authoritative than reports of current feelings. We first consider the dichotomy between introspection and retrospection from several perspectives, before discussing its implications for a particular question: how would we determine who is happier, the French or the Americans?

Two selves

An individual's life could be described—at impractical length—as a string of moments. A common estimate is that each of these moments of psychological present may last up to 3 seconds, suggesting that people experience some 20 000 moments in a waking day, and upwards of 500 million moments in a 70-year life. Each moment can be given a rich multidimensional description. An individual with a talent for introspection might be able to specify current goals and ongoing activities, the present state of physical comfort or discomfort, mental content, and many subtle aspects of subjective experience, of which valence is only one. What happens to these moments? The answer is straightforward: with very few exceptions, they simply disappear. The experiencing self that lives each of these moments barely has time to exist.

When we are asked 'how good was the vacation', it is not an experiencing self that answers, but a remembering and evaluating self, the self that keeps score and maintains records. Unlike the experiencing self, the remembering self is relatively stable and permanent. It is a basic fact of the human condition that memories are what we get to keep from our experience, and the only perspective that we can adopt as we think about our lives is therefore that of the remembering self. For an example of the biases that result from the dominance of the remembering self, consider a music lover who listens raptly to a long symphony on a disk that is scratched near the end, producing a shocking sound. Such incidents are often described by the statement that the bad ending 'ruined the whole experience'. But, in fact, the experience was not ruined, only the memory of it. The experience of the symphony was almost entirely good, and the bad end did not undo the pleasure of the preceding half hour. The confusion of experience with memory that makes us believe a past experience can be ruined is a compelling cognitive illusion. The remembering self is sometimes simply wrong.

The approach to well-being that we describe here emerged from an empirical study of the rules that govern the remembered scores of such episodes as brief emotional films (Fredrickson and Kahneman 1993), painful medical procedures (Redelmeier and Kahneman 1996), or annoyingly loud sounds (Schreiber and Kahneman 2000). The principal finding of this line of research was that episodes are scored by the value of a representative moment, which can be the feeling associated with its end or a weighted average of the ending moment and the most intense one—this has been called the peak/end rule. As implied by the peak/end rule, the evaluation of episodes is remarkably insensitive to their durations—this phenomenon has been called duration neglect.

The rules of evaluative memory can lead to bad choices. For example, subjects in a study by Kahneman *et al.* (1993) were exposed to two cold-pressor episodes and then given a choice of which of them to repeat on a third trial. In the 'short' episode, they held a hand in water at 14°C for 60 seconds, experiencing substantial pain. The 'long' episode lasted 90 seconds. The first 60 seconds were identical to the short episode; over the final 30 seconds the temperature was gradually raised to 15°C, still unpleasant but less so. From the point of view of the experiencing self, the long trial is clearly worse. For the remembering self, however, the peak/end rule implies that the added period of diminishing pain makes the memory of the long trial less aversive. The choice to repeat the inferior experience reflects the misguided preferences of the remembering self.

Constituents of well-being

The exclusive concern with the remembering self is a notable feature of well-being research. The vast body of literature devoted to subjective well-being is dominated by the questions, 'How satisfied are you with your life as a whole?' and 'How happy are you these days?' The happiness question explicitly requires the respondent to retrieve, integrate, and evaluate memories. The life satisfaction question involves evaluations that are even more remote from actual experience. In a different vein, researchers who prefer a eudaemonic conception of well-being also measure it by consulting stable aspects of the self-concept, such as purpose in life (Ryff and Keyes 1995), self-actualization (Ryan and Deci 2000), and optimism (Seligman 2002). The well-being of the experiencing self has been the object of much less research (but see Csikszentmihalyi and Larson 1984; Riis *et al.* 2005; Stone *et al.* 1999). In the remainder of this chapter we explore some conceptual and methodological issues that arise in measuring experienced well-being, and present some preliminary results and conclusions.

The hybrid concept of well-being that is illustrated in Fig. 11.1 distinguishes two components, which are labeled 'experienced well-being' and 'evaluated well-being'. Both are subjective, and both refer to a period of time that may be measured in days or months. The first component includes the statistics of the momentary affective states experienced during the reference period. The second component includes global subjective evaluations of one's life during the same period. The two components are not independent. Subjective evaluations are strongly influenced by emotional experiences—an individual who has recently experienced mostly negative affect is unlikely to describe herself as very happy or satisfied. Conversely, subjective evaluations of the state of one's life certainly have affective consequences, at least while the thoughts of these evaluations are active. The two components of well-being are therefore expected to be correlated, but they are distinct, empirically as well as conceptually.

We expect that, in many situations, measures of experienced well-being and of satisfaction with life or with some aspect of it (work, marriage) may be highly

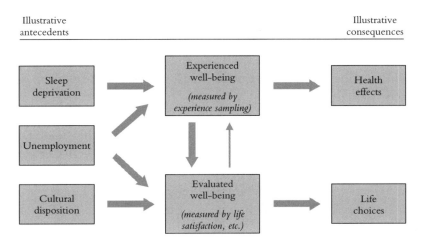

Fig. 11.1 Examples of different antecedents and consequences of experienced and evaluated well-being. Thickness of arrows indicates strength of the relationship. Not all relationships are shown.

correlated, and may share the same pattern of antecedents and consequences. Experienced well-being will be considered primary in such cases. When routine mood and general life satisfaction correspond closely, the conclusion that mood is the major determinant of life satisfaction is more plausible than the claim that people's routine moods are mainly determined by their general view of their lives.

Of course, the point of distinguishing experience from evaluation in well-being is that the two are not always in perfect correspondence. Figure 11.1 presents several hypotheses about factors that influence one of the constituents of well-being more than the other. For example, we suggest that a temporary state of anemia or sleep deprivation has a *direct* effect on affective experience, but only an indirect effect on self-reported happiness and life satisfaction. At the other extreme, we propose that the low level of life satisfaction among the French (who report much lower life satisfaction than, for example, Americans and Danes) could be due entirely to the rules that govern evaluations and their expressions in the French culture. We know of no evidence that the actual affective experience of the French is generally worse than that of the Americans or the Danes.[1] Finally, some life circumstances have a direct impact on both experience and evaluation; we list unemployment as a likely example.

The occasional dissociation of evaluated from experienced well-being is of modest interest on its own. The significance of the distinction between affective experience and life evaluations mainly depends on whether they have different consequences. As the relevant data have not been collected, we do not know the answer. However, it is a plausible speculation that the stress of negative affective experiences may have a direct cumulative effect on health (Marmot 2004; Sapolsky 1999). On the other hand, a decision to seek therapy or get a divorce could be a direct consequence of an evaluation of the state of one's life, related only indirectly to affective experience. In a study of vacations, Wirtz *et al.* (2003) found substantial discrepancies between respondents' recalled enjoyment of their vacation and their actual experienced enjoyment. It was the recalled enjoyment, however, and not the experienced enjoyment, that predicted people's desire to repeat the vacation experience.

The compound nature of well-being has been recognized in many discussions of the concept. For example, an authoritative review of the field (Diener and Seligman 2004, p. 1) begins with the following definition: 'Well-being, which we define as people's positive evaluations of their lives, includes positive emotions, engagement, satisfaction and meaning' (Seligman 2002). Diener *et al.* (1999, p. 277) had earlier defined well-being as 'a broad category of phenomena that

---

[1] A dissociation of the kind that we propose was found in the only published study looking directly at this type of question. Oishi (2002) reported that Japanese-American students report lower levels of well-being than white American students in retrospective (evaluative) reports, but equivalent levels in online (experiential) reports.

includes people's emotional responses, domain satisfactions, and global judgments of life satisfaction'. The lists of constituents of well-being in both definitions are related to the distinction drawn in Fig. 11.1: positive emotions belong to experienced utility; meaning and life satisfaction belong to evaluation. These representative definitions of well-being clearly imply the existence of components that are at least partly independent of each other. Nevertheless, the dominant practice in well-being research effectively ignores the issue, and continues to treat the determinants and consequences of happiness as if it were a unitary concept.

The simplifying assumption that well-being is unitary is implicitly invoked in many discussions of well-being. It plays a particularly important role in the common practice of using both national differences and individual differences in accounts of individual well-being (Diener and Seligman 2004; Diener and Suh 1999; Frey and Stutzer 2000; Helliwell 2003; Layard 2005). This practice can only be justified if the phrase 'happier than' has a similar meaning in the contexts of 'John is happier than Peter; he scores higher by 1 point on the happiness scale' and 'the Americans are happier than the French; they score higher by 1 point on the happiness scale'. Figure 11.1 represents the hypothesis that happiness may have quite different meanings in these contexts. Specifically, we agree that it is reasonable to infer from the difference in reported life satisfaction that John probably *experiences* more happiness than Peter, but we know of no evidence that would justify a similar inference about the Americans and the French.

The main operational conclusion from this discussion is that experienced and evaluated well-being should both be measured, and that the measures should be explicitly separated. Contrary to a position that one of us espoused earlier (Kahneman 1999), measures of evaluated well-being are not simply flawed indicators of objective happiness (experienced well-being). Evaluation and memory are important on their own, because they play a significant role in decisions, and because people care deeply about the narrative of their life. On the other hand, an exclusive focus on retrospective evaluations is untenable if these evaluations do not accurately reflect the quality of actual experience. Experienced well-being must be measured separately, because it cannot be inferred with adequate precision from global reports of happiness or life satisfaction. The logical foundations of this measurement are discussed next. Subsequent sections describe preliminary results from studies using this approach, and sketch ideas for future research.

Objective happiness and the logic of experienced utility

The approach to experienced well-being that we present here extends an earlier analysis of Bentham's notion of utility (Kahneman 1999, 2000; Kahneman *et al.* 1997). Bentham's concept was labeled *experienced utility*, to distinguish it from the usage now current in decision theory and in economics, where utility is interpreted as *decision utility*, and is inferred from choices and applied to explain choices. In Bentham's classic usage, utility refers to the hedonic dimension of experience (Bentham 1789/1948): each moment is characterized by the quality and intensity of pleasure or pain. The utility of more complex events is described by additional attributes, including intensity, duration, certainty, and fecundity. Bentham's concept of utility is a good starting point for our enquiry.

The basic element of experienced utility is *moment utility*, the true answer to the question: 'how good or bad is your experience *now*?' Meaningful outcomes, however, last more than a moment. A useful measure of experienced utility must be applicable to temporally extended outcomes, but it must be derived from measurements of moments according to some logical rules. One of these rules is temporal monotonicity: adding an extra period of pain to an episode of pain can only make things worse. As we mentioned earlier, judging episodes of pain by the peak and the end leads to violations of monotonicity, both in retrospective evaluations and in choices. The conclusion from that research was that the scores kept by the remembering self (they have been called 'remembered utilities') violate logic.

A formal analysis of experienced utility was offered by Kahneman *et al.* (1997), who examined the requirements for deriving an adequate measure of the *total utility* of an episode from reports of moment utility over its duration. They first identify three requirements that the measure of moment utility must satisfy: (1) reports of the *sign* of experience (positive or negative) can be trusted; (2) reports of the positive or negative *intensity* of the experience satisfy the conditions of ordinal measurement; and (3) reports of experience are interpersonally comparable. Next, two normative conditions identify factors that should *not* influence evaluations of the total utility experienced over a period of time: (4) separability: the order in which moment utilities are experienced is not relevant to total utility; and (5) time neutrality: all moments have equal weights. If these five conditions hold, then cumulative distributions of moment utility over time provide a measurement of total utility that is adequate for most research purposes. Except for the rare cases in which the cumulative distributions cross, the mean (or the median) of the distribution of moment utility is an ordinal measure of total utility that can be compared across situations, people, and populations (Kahneman 2000).[2]

---

[2] Additional conditions must be satisfied to achieve a cardinal measurement of utility in terms of time (Broome 1991; Kahneman *et al.* 1997), in which total utility is the temporal integral of the rescaled utility measure. This is the level of measurement that is assumed when states of health are specified in terms of QALYs (quality-adjusted life years), e.g. when the judgment is made that 10 years of survival on dialysis have the same utility as 8 years of survival in perfect health.

The formal analysis establishes, at least in principle, the feasibility of a measure of experienced well-being (objective happiness), as proposed in Fig. 11.1. Specifically, it shows how a global measure of the well-being of individuals or of populations can be obtained by using the experience-sampling method (ESM; Csikszentmihalyi 1994; Stone *et al.* 1999). Participants in ESM studies carry a hand-held computer that calls them at random moments during the waking day to answer questions about their current situation and their current feelings. These reports of momentary subjective state provide an ordinal measure of moment utility. The cumulative distribution of these moment utilities can be used to compare the well-being of populations that differ in their life circumstances. For example, we could assess the proportion of time that the rich and the poor spend at a utility level of 6, or below 3 on a 6-point scale. This concept of well-being has been called *objective happiness* (Kahneman 1999, 2000), because the global evaluation of happiness is constructed according to an objective rule, although it is ultimately based on subjective reports of the experience of moments. In terms of the distinction that was drawn earlier, objective happiness is moment-based, and draws only on immediate introspection; unlike other measures of well-being, it does not involve retrospection at all. The remainder of this chapter deals with theoretical and empirical aspects of objective happiness, and compares it with other approaches to the measurement of well-being. First, we consider some objections to both the feasibility and the appropriateness of this approach to well-being.

Is moment utility unidimensional?

An immediate difficulty is that the method, at least in its simplest form, depends on a single measure of moment utility as a building block. We must be willing to believe that people are capable of applying the scale coherently to headaches, pangs of guilt, beautiful music, and the pleasure of hoping for a better future—and that they are also capable of appropriately weighting such experiences whenever several of them occur at the same moment. This may be too much to ask from untrained introspection. It is worth noting, however, that the standard procedure in which people are asked to report a single value of happiness or life satisfaction implicitly involves similar assumptions about the ability to compare and combine heterogeneous experiences.

The best rationale for a single dimension of experienced utility may well be its relation to decision utility, and to the fundamental dichotomy of approach and avoidance. Intensities of positive and negative utilities may be difficult to compare across categories of experiences, but the basic distinction between good and bad moments—the sign of the utility measure—is easier to determine. Choices between approach and avoidance occur frequently in the lives of humans and other animals, and conflicting tendencies must be weighed and resolved. The ability to resolve such conflicts implies that a common currency is used to combine and compare the values of different experiences (Cabanac 1992; Montague and Berns 2002; Shizgal 1999). Cabanac boldly suggested that this currency is pleasure.

The idea of anchoring the concept of experienced utility in approach and avoidance suggests a measure: respondents can be asked to indicate whether they feel impatient for their current situation to end, or would prefer for it to continue (Kahneman *et al.* 2004*b*). This measure of well-being is far from perfect—it reflects expectations about the next state as well as a response to the current state. When aggregated over a long period of time, however, the average of the momentary preference for continuing or stopping is an appealing measure: it identifies well-being with the extent to which individuals live their lives in a state of wishing for the present to extend, as against wishing they were somewhere else—or not caring one way or the other. Notably, this measure applies equally well to eudaemonic and to hedonic states, to intense absorption ('flow') and to sensory pleasure, to boredom and to pain.

Although the present discussion assigns a privileged role to the elicitation of action tendencies, we do not recommend that moment utility be measured by a single question. Affective experience varies along many dimensions, and moments can be good or bad in many different ways: we can be angry, afraid, sad, bored, humiliated, wasting time; we can be proud, serene, involved, in control, pleased, purposeful, or affectionate. Each of these distinct feelings has its own pattern of causes and consequences, which are likely to be of interest to the student of well-being. A single summary measure of objective happiness is useful, but it is not all we want to know about the life of the experiencing self.

The preceding discussion has assumed that moment utility is measured by collecting introspective reports, but this restriction is not necessary. Appropriately validated physiological measures of moment utility could be used instead, and may have important advantages (Stone *et al.* 1999). The most promising physiological indicator of momentary affect is the prefrontal cortical asymmetry in the electroencephalogram (EEG), which has been extensively validated by Davidson and his team as a measure of the balance of positive and negative feelings, and of the relative strength of tendencies toward approach or avoidance (Davidson 1992, 2004; Sutton and Davidson 1997). A portable measuring instrument is not yet available, but is technically feasible. When success is achieved, Davidson's technique will be a candidate for a continuous and non-intrusive indicator of moment utility. As we show in a subsequent section, even the existing laboratory measure could be useful in validating and correcting verbal reports of experienced well-being.

The rationale for separability and time neutrality

Next, we discuss some common misunderstandings of the assumptions of separability and time neutrality. By separability, total utility is independent of the order in which moment utilities are experienced. This assertion appears to contradict the compelling intuition that the sequencing of experiences does matter. However, the source of the apparent conflict is a failure to distinguish moment utilities from the events that give rise to those utilities. The order of *events* certainly matters to the total utility of a sequence, but the separability assumption only asserts that the order of *utilities* does not. The order of events matters when their utilities are affected by it, which is a common occurrence. For example, the order in which a rich lunch and a strenuous tennis game are experienced surely makes a difference to total utility, because the tennis game will be much less enjoyable soon after lunch, but this is not a violation of separability in the present sense.

The order in which events occur also matters when their experienced utility is influenced by comparison processes. For example, we recognize that, if an individual wins the lottery twice in successive months, total utility will be higher if the first win is $1000 and the second $1 000 000 than if the order is reversed. Clearly, the joy of winning $1000 is attenuated in the context of the larger sum. A much larger amount—perhaps $100 000—might be required in that sequence to match the utility of an unexpected win of $1000 when it occurs first. Assuming for simplicity that the utility of the million win is not affected by order, separability implies that the sequences ($1000–$1 000 000) and ($1 000 000–$100 000) have the same total utility, because the constituent utilities in the two sequences are the same. This does not appear unreasonable. For another illustration of the separability of utilities, consider the life of a 60-year-old woman, of whom it is only known that she experienced very different levels of happiness in her thirties and in her forties. We see no compelling reason for the evaluation of her life as a whole to depend on the order of these experienced utilities. As these examples show, separability is quite plausible for sequences specified in terms of utilities.

The assumption of time-neutrality asserts that the contribution of a moment to the utility of a longer episode is determined only by the utility of that moment, not by its content. Note that time neutrality does *not* assert that all moments of life are equally significant or important. Privileged moments acquire special significance by affecting the utility of other moments. For example, the event of graduating from college is both anticipated for a long time and frequently recalled after it happens, thereby inducing experienced utility both early and late. Graduation may also bring about changes in the individual's activities, circumstances, and self-image, and influence experienced utility in these ways. But the utility of the moment of graduation itself is not privileged in the assessment of total utility. For the purpose of this assessment, it is just another moment.

Time neutrality is essential for the present approach, in which experienced well-being is measured by the temporal distribution of moment utility. The focus on the duration of experiences calls attention to the allocation of time as one of the more practical ways to improve experienced well-being. Competing views of well-being have one feature in common: in every conception some activities and mental states are considered better than others—pleasure is better than pain; intense absorption in contemplation or in activity is more valued than lethargic emptiness; intense communion with others is valued more than loneliness and hostility. Other things being equal, well-being is increased by spending more time in the good states and less time in bad or empty states. This formulation holds whether the good state is defined by positive affect or by intense engagement in a task or in a spiritual pursuit. Time is the ultimate finite resource of life, and finding ways to spend it well is a worthy objective both at the individual level and at the level of a social policy that is concerned with human well-being.

Measuring experienced well-being: the day reconstruction method

The application of experience sampling to measure and compare the well-being of populations is technically feasible but quite impractical, because the method places a high burden on respondents and severe constraints on recruitment and compliance. In an attempt to overcome this difficulty, Kahneman *et al.* (2004*a*, *b*) developed the day reconstruction method (DRM), which combines a time-budget study with a technique for recovering detailed information about specific experiences from the previous day. The DRM is intended to reproduce the information that would have been collected by measuring the experience immediately, as in the experience sampling method, and to do so more efficiently. The method allows a characterization of the experienced utility associated with the diverse settings and activities of people's lives, and it also provides a measure of how they allocate their time among settings, activities, and partners in social interactions.

The DRM employs a structured questionnaire that elicits a detailed description of a particular day in the respondent's life. The method is based on research indicating that accurate retrospective reports of affect can be achieved by encouraging retrieval of specific and relatively recent episodes (Robinson and Clore 2002). Respondents first revive memories of the *previous* day by constructing a short diary consisting of a sequence of episodes. Next the respondents describe each episode in detail by indicating: (1) when the episode began and ended; (2) what they were doing; (3) where they were; (4) persons with whom they were interacting; and (5) how they felt on multiple affect dimensions (in each case on a scale ranging from 0 (not at all) to 6 (very much)).

An illustrative study using the DRM was conducted, in which 1018 working women in Texas participated (Kahneman *et al.* 2004*b*). The central conclusions of the study are as follows.

1 Most people report themselves in at least a moderately good mood most of the time. Negative affect (anger, frustration, depression, feeling criticized) is reported only 34% of the time. However, people often describe themselves as

tired (76% of the time above 0) and also, at least to some degree, as 'impatient for it [the current episode] to end' (55% of the time above zero).

2 The overall distributions of ratings, as well as the diurnal rhythms (which are quite pronounced), observed with the DRM replicate the results of prior research using experience sampling. This finding is particularly important, because the DRM is explicitly intended to provide a more efficient substitute for experience sampling.

3 Any individual's affective state varies substantially in the course of the day, depending on the activity in which he or she is engaged, as well as on the people with whom he or she interacts. The mean level of enjoyment reported, on a 0–6 scale of 'enjoying myself', was 4.68 for episodes in which the respondents socialized with friends, 2.97 when they were commuting alone, and 2.15 when the only activity recorded was interacting with one's boss.

4 Some general aspects of the respondents' circumstances had a substantial effect. For example, the mean enjoyment at work was 2.88 for respondents who described their work situation as involving high time pressure, and 3.96 for those who reported low time pressure.

5 Individual differences were also large: the standard deviation of the mean level of enjoyment reported over the entire day was 1.58. Some individual characteristics of the respondents were strongly associated with their reports of mood. Respondents who reported that their general sleep quality was 'very good' had a mean enjoyment level of 4.05 over the entire day, which contrasts with the value of 2.80 for respondents whose sleep quality was said to be 'very bad'.

6 In contrast to the large effects of personality, immediate circumstances, and current activities, more general features of individuals' life situations had relatively small effects. For example, the correlation between income and mean enjoyment over the day was 0.05, which is significantly lower than the correlation of 0.20 between income and general life satisfaction in the same sample.

7 In accord with the model presented in Fig. 11.1, the respondents' global evaluations of their lives (life satisfaction, happiness) and their reported affect during a particular day (a measure of experienced well-being) were substantially correlated ($r = 0.38$), but far from identical.

8 Some circumstances of life and work had different effects on evaluation and on experience. For example, divorced women reported slightly lower life satisfaction than married women, as expected. Surprisingly, the divorced women also reported slightly better affect. And a high level of time pressure at work reduced enjoyment at work significantly more than it reduced job satisfaction.

The results of this study suggest that the DRM or some variant of it can be used as a tool to assess the experienced well-being of individuals and populations and to identify dissociations between experienced and evaluated well-being. In the next section we propose an application of the DRM to answer the question that we posed at the beginning of this chapter, 'Who is happier, the French or the Americans?' We show that measurements of experienced well-being are needed to answer this question. We also suggest that physiological measures are needed to remove possible effects of language and culture on verbal reports of both evaluated and experienced well-being.

A possible application: comparing well-being across countries

A significant strand of well-being research has been concerned with comparisons across countries. The studies of national differences have revealed a robustly consistent pattern: the highest levels of average life satisfaction are reported by northern European democracies, there is no correlation between gross domestic product (GDP) and happiness among relatively wealthy countries, the nations of the former Soviet Empire are very dissatisfied (perhaps historically), and those of South America are surprisingly happy. Some authors (e.g. Veenhoven 1996) conclude that the problem of national differences has been solved, because a few variables account for most of the variance (see also Helliwell 2003). But others have noted that a stubborn puzzle remains: the differences between countries are too large to be plausible, especially when compared to the small effects of life circumstances (Inglehart and Rabier 1986). To illustrate this puzzle, consider the Americans and the French. The distributions of life satisfaction in the US and in France differ by about half a standard deviation. For comparison, this is also the difference of life satisfaction between the employed and the unemployed in the US, and it is almost as large as the difference between US respondents whose household income exceeds $75 000 and others whose household income is between $10 000 and $20 000 (in 1995).[3] In this section we describe the implications of our analysis for research that compares the well-being of nations.

The large country effects that have been documented were observed in the answers to evaluation questions. In the present framework, the critical issue is whether the same country differences will also be found in measurements of experienced well-being. Indeed, we believe that comparisons of well-being across countries and cultures are commonly understood as referring to affect (i.e. experience). To appreciate this, imagine the following pattern of results: the population of country A reports low satisfaction with life but is consistently cheerful; the population of country B indicates high satisfaction with life but is

---

[3] The difference in satisfaction between the unemployed and employed is based on data from waves 2 and 3 of the World Values Survey (Inglehart *et al.* 2003). Only wave 3 data were used in comparing the income groups.

generally in a sad or angry mood. Which of the conflicting measurements would be considered more compelling? We surmise that, when the goal is to compare populations, most people (and most scholars) will find a measure of experience more compelling than a measure of satisfaction as an indicator of well-being.

Is it possible to infer from the large differences in evaluated well-being that experienced well-being is also much lower in France than in the USA? We doubt it, because the sheer size of the difference seems implausible: it is hard to believe that the experienced well-being of the average *employed* Frenchman really matches that of the average *unemployed* American. Further reasons for doubt are found by comparing country effects in life satisfaction and in self-reported health status (SRHS; for a more detailed analysis see Riis, Schwarz, and Kahneman, manuscript in preparation). Among the 18 wealthiest, Western OECD (Organization of Economic Cooperation and Development) nations for which data were available, the correlation between national averages of self-reported health and of life satisfaction is 0.85[4]. The high correlation between health and happiness in international data sets has been interpreted as an indication of the importance of (real) health as a determinant of happiness (Helliwell 2003). However, this interpretation runs into difficulties in the set of prosperous countries, because national differences in SRHS are completely uncorrelated ($r = -0.03$) with the most widely used objective measure of national health, adult life expectancy. For example, the French describe themselves as much less healthy than the Americans do, but they live 3 years longer.

It is fair to describe national differences in self-reported health, at least among developed countries, as 'reality-free'. Could the same be true as well of national differences in self-reported life satisfaction? If the French are pessimistic or grumpy in describing their (generally good) health, could they be equally grumpy in evaluating their good lives? Could the French be objectively happier than the Americans, in spite of being less satisfied? Is there a dimension of culturally determined positivity that accounts for national difference in both subjective health and life satisfaction? At this time we do not know the answers. We next sketch a research design that might help us find them.

The discussion of the design also provides an opportunity to consider a set of hypotheses that we currently consider plausible. Most importantly, we expect that

---

[4] The data are from waves 2 and 3 of the World Values Survey (Inglehart *et al.* 2003). Data from two other wealthy, Western OECD nations, Luxembourg and New Zealand, were not available. Adding wealthy but non-Western Japan and/or Western but less wealthy Portugal does not substantially change the results. Furthermore, the same pattern of results is observed when overall (evaluative) happiness is used instead of life satisfaction, and in data from the Eurobarometer (Reif and Marlier 2002*a, b*) which is a different survey of many of the same countries. We are indebted to Richard Suzman, who drew our attention to the strikingly similar pattern of the differences between Denmark and France in the two variables.

the variance in country averages of life satisfaction will turn out to be associated mainly with evaluation, and largely unrelated to real differences in experienced well-being. This hypothesis is consistent with results reported by Diener and his associates (2000) that imply that national differences in positivity are much smaller for ratings of specific aspects of life than for global evaluations. The same conclusion is supported by the finding, mentioned earlier, that Japanese-American students indicate lower levels of well-being than White American students in retrospective (evaluative) reports, but equivalent levels in online (experiential) reports (Oishi 2002).

A good starting point for resolving the question of national differences in experienced well-being would be to use the DRM in national samples of countries that differ in measures of life satisfaction. However, the results of such a study are unlikely to be accepted as conclusive. Although Oishi's (2002) result suggests that the reports of moment utility are less susceptible than reports of life satisfaction to the effects of cultural disposition, critics will surely question whether any affective self-description can be free of cultural influence (Wierzbicka 2004). In anticipation of such concerns, we propose a design that incorporates a physiological measure as a means of validating (and perhaps correcting) self-reports of moment utility.

Can a physiological index be used to measure objective happiness?

The background of the present proposal is the availability of a physiological index—the Davidson index of PFCA (prefrontal cortical asymmetry)—that is a valid predictor of affective responses and of approach/avoidance tendencies both within and between persons (for a review, see Davidson, Chapter 5, this volume). A recent study in an American sample also showed substantial correlations between the Davidson index and several measures of evaluated well-being (Urry et al. 2004). The advantage of this measure, of course, is that it is not susceptible to linguistic biases. It can therefore be used as an anchor for the calibration of verbal reports of both experienced and evaluated well-being.

The following data would be obtained from a French and an American sample.

1 Physiological indicators of affect, such as the Davidson index, measured under standard resting conditions. This measure has been interpreted as an indicator of a general propensity to experience positive or negative feelings (Davidson 2004).

2 Self-reports of moment utility under the same standard conditions, obtained concurrently with the physiological measure.

3 Physiological indicators of affect, measured while people are induced to focus on positive or negative features of various domains of their lives.

4 Self-reports of the intensity of the pain and of the pleasure that people experience in the same domains of life.

5 Self-reports of life satisfaction (i.e. evaluated well-being).

6 Amount of time spent in several activities (e.g. work and leisure), measured by the DRM.

7 Self-reports of the experienced utility of these activities, measured by the DRM.

8 From measures 6 and 7, a duration-weighted summary measure of experienced well-being (objective happiness).

Next, consider what we could learn from these data. We have formulated four hypotheses.

H1 The populations do not differ in the *disposition* to experience positive or negative moment utility, as indicated by measure 1, for which the issue of linguistic correspondence does not arise.

H2 The correlation between the physiological measures (1 and 3) and the self-report measures will be positive and moderately high (a) within subjects (for measures 3 and 4), and (b) within each of the two national groups, confirming that all self-report measures are sensitive to individual differences in affective disposition.

H3 The French will report lower life satisfaction (measure 5) than the Americans. The regression lines predicting life satisfaction from the physiological indicator will have similar slopes in the two countries, but the American intercept will be higher.

H4 The differences between the populations in measures of momentary experience (measures 2 and 7) will be significantly smaller, or null (cf. Diener *et al.* 2000).

If these hypotheses hold—including the strong version of H4—then the only source of difference between the two populations in (duration-weighted) experienced well-being is the amount of time they spend at leisure and at work. Because they have more leisure, the French would then be said to have higher well-being than the Americans. We highlight this hypothesis, not because we are particularly attached to the specific conclusion that the French are happier than the Americans, but to illustrate the possibility of significant dissociations between measures of experience and evaluation in the context of national differences.

We have discussed the simplest possible pattern of results, but more complex patterns are possible. In particular, consider the conjunction of H1 and the weak version of H4: in the standard laboratory conditions, the physiology shows no national differences but the French report lower experienced utility. This pattern of result would suggest that the French are more cautious and less positive than the Americans even when reporting momentary mood, not only when making global judgments. The physiological measure could then be used to control for this difference in the expression of experience, particularly in the event that the weak version of H4 held, and the French reported lower experienced utility of various activities. Even more complex patterns can be anticipated. For example, it would not be particularly surprising to find that the French draw less enjoyment from work and more enjoyment from leisure than the Americans do, because of a culture that emphasizes pleasure (for empirical work relating to this possibility see Rozin *et al.* 1999). But the opposite result could also make sense if leisure is more appreciated by the Americans because they have less of it.

All these hypotheses are readily testable with technology currently available, and it is rather surprising that they have not yet been examined. Furthermore, technical developments that are already foreseeable will eventually allow nearly continuous non-intrusive monitoring of physiological correlates of experienced utility, as a substitute and supplement to the DRM and experience sampling. This idea would have been in the domain of science fiction in the relatively recent past, but it could turn into reality in the relatively near future.

Concluding remarks

In this chapter we have urged that the experiencing self be given due regard in well-being research. The conjunction of a powerful structural bias in favor of the remembering self and of the challenging difficulties in the measurement of experience has caused an almost exclusive emphasis on evaluated well-being. We have argued that the experiences that make the moments of life worth living deserve to be studied. We have not proposed a general answer to the puzzling question raised by Fig. 11.1: when experienced well-being and evaluated well-being diverge, what is well-being? A comprehensive solution will involve both normative and empirical issues and will be hard to find. In some cases, including the one on which we focused, the answer seems clear. If cultural differences regulate self-satisfaction but do not affect experienced well-being, the differences in satisfaction are interesting and significant, but well-being is better indexed by the quality of experience. If measurements indicate that the French are grumpy about their lives, or the Japanese are humble about theirs, while experiencing more pleasure and less pain than the American unemployed, most of us will feel more pity for the latter group. In the context of politics within a single country, on the other hand, some variations in people's satisfaction with their lives could be much more significant than variations in actual experience. Social movements arise when dissatisfied people agree on a common attribution for the source of their dissatisfaction, and also agree on what they want to do about it. We acknowledge that the scheme of Fig. 11.1 makes the study of well-being even more complicated than it was, but suggest that the complexity is real and that it is useful to admit it.

In closing, we comment on the relationship between the present analysis and the eudaemonic approach to well-being that has become highly influential in recent years (e.g. Ryan and Deci 2000; Ryff 1989; Ryff and Singer 1998; Seligman 2002). The eudaemonic approach draws on the Aristotelian concept of the good life, and appeals to the widely shared intuition that there is more to life than a favorable balance of pleasure and pain. Authenticity, affection, participation, efficacy, and the full utilization of human capabilities are elements of the lay concept of well-being, along with vitality and good spirits (Ryff and Keyes 1995, Seligman 2002). Eudaemonia is usually construed in this literature as a global property of an individual's life, and measured by consulting the remembering–evaluating self. But, of course, eudaemonia (or its absence) also has manifestations in the life of the experiencing self. The thrill of 'flow', the joy of intimacy, the sense of engagement in purposeful action, all can be identified and reported as characteristics of a moment. At the other extreme of the dimension, boredom and futility are also attributes of experience.

Whether or not the subjective states related to eudaemonia are 'feelings'—this is sometimes disputed—they are readily available to introspection, and they have a valence. People know when they are engaged, they know when they are killing time, and they much prefer to be engaged. The frequency and duration of these aspects of experience should therefore be included in a comprehensive description of internal life, and distinguished from the individual's self-evaluations. Indeed, it appears quite likely that experienced eudaemonia and evaluated eudaemonia will be found to have different antecedents and different consequences.

Detailed study of the experiential aspects of eudaemonia along with other types of good and bad moments will both raise and help answer several significant questions. Some of these questions are normative: how should weights be assigned to good and bad experiences that differ in kind? For example, how do experienced mastery and experienced enjoyment trade off in the measurement of well-being? Other questions are empirical. Do the activities that are most closely associated with eudaemonia have effects that spread to the experiences of more mundane situations? Do eudaemonia and other pleasures share a physiological representation in a common approach system? What is the experiential significance of different allocations of time to activities? Do different populations—perhaps the French and the Americans—create for themselves good experiences that differ in kind? There is much to learn about the well-being of the experiencing self.

Acknowledgements

The research we describe was supported by the Hewlett Foundation, the Woodrow Wilson School of Public and International Affairs at Princeton, the National Science Foundation, and the National Institute of Aging. We are grateful to Alan Krueger, David Schkade, Norbert Schwarz, and Arthur Stone for the long-term collaboration that produced many of the ideas and findings reported here.

References

Bentham, J. (1948). *An introduction to the principle of morals and legislation*. Blackwell, Oxford. [Originally published in 1789.]

Broome, J. (1991). *Weighing goods*. Blackwell, Oxford.

Cabanac, M. (1992). Pleasure: The common currency. *J. Theor. Biol.* **155**, 173–200.

Csikszentmihalyi, M. (1994). *Flow: the psychology of optimal experience*. Harper Collins, New York.

Csikszentmihalyi, M. and Larson, R. (1984). *Being adolescent: conflict and growth in the teenage years*. Basic, New York.

Davidson, R.J. (1992). Anterior cerebral asymmetry and the nature of emotion. *Brain Cogn.* **6**, 245–68.

Davidson, R.J. (2004). Well-being and affective style: neural substrates and biobehavioural correlates. *Phil. Trans. R. Soc. Lond. B* **359**, 1395–411.

Diener, E. and Seligman, M.E.P. (2004). Toward an economy of well-being. *Psychol. Sci. Public Interest* **5** (1), 1–31.

Diener, E. and Suh, E. (1999). National differences in subjective well-being. In *Well-being: the foundations of hedonic psychology* (ed. E. Diener, N. Schwarz, and D. Kahneman), pp. 434–50. Russell Sage Foundation, New York.

Diener, E., Suh, E., Lucas, R.E., and Smith, H.L. (1999). Subjective well-being: three decades of progress. *Psychol. Bull.* **125** (2), 276–302.

Diener, E., Scollon, C.N., Oishi, S., Dzokoto, V., and Suh, E. (2000). Positivity and the construction of life satisfaction judgments: global happiness is not the sum of its parts. *J. Happiness Stud.* **1** (2), 159–76.

Fredrickson, B.L. and Kahneman, D. (1993). Duration neglect in retrospective evaluations of affective episodes. *J. Pers. Soc. Psychol.* **65**, 45–55.

Frey, B.S. and Stutzer, A. (2000). Happiness prospers in democracy. *J. Happiness Stud.* **1** (3), 79–102.

Helliwell, J.F. (2003). How's life? Combining individual and national variables to explain subjective well-being. *Econ. Modeling* **20** (2), 331–60.

Inglehart, R. and Rabier, J.R. (1986). Aspirations adapt to situations—but why are the Belgians so much happier than the French? In *Research on the quality of life* (ed. F.M. Andrews), pp. 1–56. Institute for Social Research, University of Michigan, Ann Arbor, Michigan.

Inglehart, R., *et al.* (2003). *World values surveys and European values surveys, 1981–1984, 1990–1993, and 1995–1997* [Computer file]. ICPSR version. Institute for Social Research, Ann Arbor, Michigan [producer], 1999. Inter-university Consortium for Political and Social Research, Ann Arbor, Michigan [distributor].

Kahneman, D. (1999). Objective happiness. In *Well-being: the foundations of hedonic psychology* (ed. E. Diener, N. Schwarz, and D. Kahneman), pp. 3–27. Russell Sage Foundation, New York.

Kahneman, D. (2000). Evaluation by moments: past and future. In *Choices, values and frames* (ed. D. Kahneman and A. Tversky), pp. 293–308. Cambridge University Press and the Russell Sage Foundation, New York.

Kahneman, D., Fredrickson, B.L., Schreiber, C.A., and Redelmeier, D.A. (1993). When more pain is preferred to less: adding a better end. *Psychol. Sci.* **4**, 401–5.

Kahneman, D., Wakker, P.P., and Sarin, R. (1997). Back to Bentham? Explorations of experienced utility. *Q. J. Econ.* **112**, 375–405.

Kahneman, D., Krueger, A.B., Schkade, D.A., Schwarz, N., and Stone, A.A. (2004a). Toward national well-being accounts. *Am. Econ. Rev.* **94** (2), 429–34.

Kahneman, D., Krueger, A.B., Schkade, D.A., Schwarz, N., and Stone, A.A. (2004b). A survey method for characterizing daily life experience: the day reconstruction method (DRM). *Science* **306** (5702), 1776–80.

Layard, R. (2005). *Happiness.* The Penguin Press, New York.

Marmot, M. (2004). *The status syndrome.* Bloomsbury, London.

Montague, P.R. and Berns, G.S. (2002). Neural economics and the biological substrates of valuation. *Neuron* **36** (2), 265–85.

Oishi, S. (2002). The experiencing and remembering of well-being: a cross-cultural analysis. *Pers. Soc. Psychol. Bull.* **28** (10), 1398–406.

Redelmeier, D. and Kahneman, D. (1996). Patients' memories of painful medical treatments: real-time and retrospective evaluations of two minimally invasive procedures. *Pain* **116**, 3–8.

Reif, K. and Marlier, E. (2002a). *Eurobarometer 44.3 OVR: employment, unemployment, and gender equality, February–April 1996,* Report No. 2443. Conducted by INRA (Europe), Brussels. ICPSR edition. Inter-university Consortium for Political and Social Research, Ann Arbor, Michigan [producer]. Cologne, Germany: Zentralarchiv fur Empirische Sozialforschung, Cologne, Germany/Inter-university Consortium for Political and Social Research, Ann Arbor, Michigan [distributors].

Reif, K. and Marlier, E. (2002b). *Eurobarometer 443: health care issues and public security, February–April 1996,* Report No. 2443. Conducted by INRA (Europe), Brussels. ICPSR edition. Inter-university Consortium for Political and Social Research, Ann Arbor, Michigan [producer]. Cologne, Germany: Zentralarchiv fur Empirische Sozialforschung, Cologne, Germany/Inter-university Consortium for Political and Social Research, Ann Arbor, Michigan [distributors].

Riis, J., Loewenstein, G., Baron, J., Jepson, C., Fagerlin, A., and Ubel, P.A. (2005). Ignorance of hedonic adaptation to hemo-dialysis: a study using ecological momentary assessment. *J. Exp. Psychol. Gen.* **134** (1), 3–9.

Robinson, M.D. and Clore, G.L. (2002). Episodic and semantic knowledge in emotional self-report: evidence for two judgment processes. *J. Pers. Soc. Psychol.* **83** (1), 198–215.

Rozin, P., Fischler, C., Imada, S., Sarubin, A., and Wrzesniewski, A. (1999). Attitudes to food and the role of food in life: comparisons of Flemish Belgium, France, Japan and the United States. *Appetite* **33**, 163–80.

Ryan, R.M. and Deci, E.L. (2000). On happiness and human potentials: a review of research on hedonic and eudaimonic well-being. *Annu. Rev. Psychol.* **52**, 141–66.

Ryff, C.D. (1989). Happiness is everything, or is it? Explorations on the meaning of psychological well-being. *J. Pers. Soc. Psychol.* **57** (6), 1069–81.

Ryff, C.D. and Keyes, C.L.M. (1995). Psychological well-being in adult life. *Curr. Dir. Psychol. Sci.* **69**, 719–27.

Ryff, C.D. and Singer, B. (1998). The contours of positive human health. *Psychol. Inquiry* **9** (1), 1–28.

Sapolsky, R.M. (1999). The physiology and pathophysiology of unhappiness. In *Well-being: the foundations of hedonic psychology* (ed. E. Diener, N. Schwarz, and D. Kahneman), pp. 453–69. Russell Sage Foundation, New York.

Schreiber, C.A. and Kahneman, D. (2000). Determinants of the remembered utility of aversive sounds. *J. Exp. Psychol. Gen.* **129**, 27–42.

Seligman, M.E.P. (2002). *Authentic happiness: using the new positive psychology to realize your potential for lasting fulfillment*. Free Press, New York.

Shizgal, P. (1999). On the neural computation of utility: implications from studies of brain stimulation reward. In *Well-being: the foundations of hedonic psychology* (ed. E. Diener, N. Schwarz, and D. Kahneman), pp. 502–26. Russell Sage Foundation, New York.

Stone, A.A., Shiffman, S.S., and DeVries, M.W. (1999). Ecological momentary assessment. In *Well-being: the foundations of hedonic psychology* (ed. E. Diener, N. Schwarz, and D. Kahneman), pp. 61–84. Russell Sage Foundation, New York.

Sutton, S.K. and Davidson, R.J. (1997). Prefrontal brain asymmetry: a biological substrate of the behavioral approach and inhibition systems. *Psychol. Sci.* **8** (3), 204–10.

Urry, H., Nitschke, J.B., Dolski, I., Jackson, D.C., Dalton, K.M., Mueller, C.J., Rosenkranz, M.A., Ryff, C.D., Singer, B.H., and Davidson, R.J. (2004). Making a life worth living: neural correlates of well-being. *Psychol. Sci.* **15** (6), 367–72.

Veenhoven, R. (1996). Developments in satisfaction research. *Soc. Indicators Res.* **37** (1), 1–46.

Wierzbicka, A. (2004). 'Happiness' in cross-linguistic and cross-cultural perspective. *Daedelus* **133** (2), 34–43.

Wirtz, D., Kruger, J., Scollon, C.N., and Diener, E. (2003). What to do on spring break? The role of predicted, on-line, and remembered experience in future choice. *Psychol. Sci.* **14** (5), 520–4.

Felicia Huppert is a neuropsychologist with degrees from the University of Sydney, University of California at San Diego, and Cambridge University. Her research combines psychology, epidemiology, and neuroscience, and focuses on the determinants of positive well-being across the life-course. She is chair of the European Network for Positive Psychology and director of CIRCA, the Cambridge Interdisciplinary Research Centre on Ageing.

Photo courtesy of Herbert Huppert.

Chapter 12*

Positive mental health in individuals and populations

Felicia A. Huppert

Introduction

Recent years have witnessed an exhilarating shift in the literature from an emphasis on disorder and dysfunction to a focus on well-being and positive health. Within psychology, the heralds of this revolution have included researchers such as Seligman (1991, 2002), Ryff (1989; Ryff and Singer 1998), and Diener (1984, 2000; Diener *et al.* 1999), building on the earlier work of investigators such as Bradburn (1969), Argyle (1987), and Csikszentmihalyi (Csikszentmihalyi and Csikszentmihalyi 1988). This perspective is also enshrined in the constitution of the World Health Organization (WHO 1948) where health is defined as 'a state of complete physical, mental and social well-being and not merely the absence of disease or infirmity'.

This chapter explores what we know about the factors that determine positive mental health and how this knowledge may be utilized to improve mental health in individuals and populations. Positive mental health is here defined as a combination of subjective well-being and of being fully functional (i.e. realizing or developing one's potential). This corresponds to Keyes's terminology, in which positive mental health comprises positive feelings and positive functioning (Keyes 2002a). Importantly, this definition integrates two distinct philosophical approaches to defining well-being: the hedonic perspective that emphasizes feelings of pleasure or happiness, and the eudaimonic perspective that emphasizes modes of thought and behaviour that provide engagement and fulfilment. The distinction between the hedonic and eudaimonic perspectives has been elegantly reviewed by Ryan and Deci (2001). It is assumed in this chapter that positive feelings alone are not sufficient for positive mental health since: (1) they do not necessarily lead to personal growth and fulfilment; (2) they may be transitory and achieved in unsustainable ways (e.g. through the use of drugs); and (3) there are occasions when positive mental health requires the experience of negative emotional states (e.g. when a close friend or family member has died). Conversely, it is assumed that positive mental health cannot be achieved solely by realizing one's potential or leading a good life, because these behaviours do not necessarily lead to happiness, contentment, or joy.

The chapter incorporates perspectives from a number of disciplines and research areas. These include the individual difference approach in psychology, a developmental or life-course perspective, cognitive approaches drawn from clinical psychology, and an epidemiological perspective on population characteristics and population-level change. It is argued that a scientific understanding of positive mental health is required not only to provide insight into why some people enjoy their life while others do not, but also to develop effective interventions so that more of us can thrive and flourish.

The chapter begins by examining individual differences in happiness or subjective well-being. Here the focus is on the concept of a set point for happiness, that is, an individual's basal level of happiness to which they tend always to return even if there are major positive or negative fluctuations. It is usually assumed that this set point is genetically determined and hence unalterable (e.g. Lykken 2000; Sheldon and Lyubomirsky 2004). However, the chapter challenges both the assumption that the set point is determined by our genes and the assumption that it cannot be changed. It is further suggested that some of the misapprehensions in the field arise from the reliance on cross-sectional research, a point also made by other investigators (e.g. Diener *et al.* 1999), and the failure to use a life-course approach. Finally, the chapter examines possibilities for interventions to increase flourishing, including changing the set point, changing external circumstances, and changing attitudes and behaviours. It concludes that, whereas most of the emphasis in the field has been on change at an individual level, there is great scope for societal or population-level change to enhance mental flourishing in most of our citizens.

Individual differences in happiness

People vary widely in their typical emotional tone, that is, whether they tend to feel generally positive or generally negative. As well as these marked differences between individuals, there is also variation within each individual in response to particular events and circumstances. Psychologists refer to these two phenomena as trait and state, respectively. Trait denotes the person's underlying emotional characteristics, sometimes referred to as temperament or affective style, and state denotes an individual's current mood or emotions, which may be momentary or longer term. This distinction is, of course, not entirely clear-cut, since a person's affective style can influence their state or the responses to questions about their state and, conversely, while a person's state cannot influence their affective style, it can influence responses to questions about affective style. Put simply, people with an underlying positive affective style tend to be in a positive state much of the time, whereas people with an underlying negative affective style tend more often to be in a negative state.

Affective style and the set point for happiness

The trait concept is closely linked to the notion of a set point for happiness. Some view the set point as simply the average of affective states over a given time period, but the set point is better understood as a baseline or starting point from which an individual deviates in response to circumstances. This raises two important questions: (1) how is the set point set? and (2) how set is the set point? These questions are important because, if we are interested in improving levels of happiness or subjective well-being, we need to know whether we are intervening at the level of a state or a trait, that is, whether we are making short-term changes or more fundamental and enduring changes.

Figure 12.1(a) is a schematic diagram representing an individual's set point or basal level of happiness. This is represented as the height of the liquid in a thermometer but, unlike a normal thermometer that measures ambient temperature, the set point for happiness is conceptualized as being relatively fixed. The factors that play a role in establishing the set point are discussed later.

The set point is not the only component of affective style that is characteristic of an individual. The frequency of variations around the set point is also characteristic, as is the amplitude or intensity of emotional responses, i.e. the magnitude of the change from baseline. Figure 12.1(b) depicts an individual with a relatively low basal level of happiness, whose moods frequently go up and down and show large deviations from their basal level, indicating someone who is highly emotionally reactive. Figure 12.1(c) depicts an individual with a high basal level of happiness, with relatively small and infrequent variations in mood, indicating someone who takes things in their stride. In both (b) and (c), the variations are shown as symmetrical around the individual's baseline, so the mean level of happiness remains constant over time. But we do not need to assume that variations are symmetrical. Figures 12.1(d, e) depict individuals whose fluctuations around their basal level are *asymmetric*. In 12.1(d) the fluctuations above the set point (positive direction) are more frequent than those below the set point (negative direction); hence the average level of happiness tends to increase over time. This could arise from improvements in circumstances, lifestyle, or attitudes increasing the person's overall level of well-being. The opposite is seen in Fig. 12.1(e) where the fluctuations below the set point are more frequent than the fluctuations above it, resulting in a decrease in the average level of happiness over time. This could arise from adverse changes in circumstances, lifestyle, or attitudes, leading to a deterioration in the person's overall level of well-being. A further parameter of affective responsiveness is duration (not depicted). For a given frequency and intensity of response, some individuals may remain in the new emotional state for a long period, while others

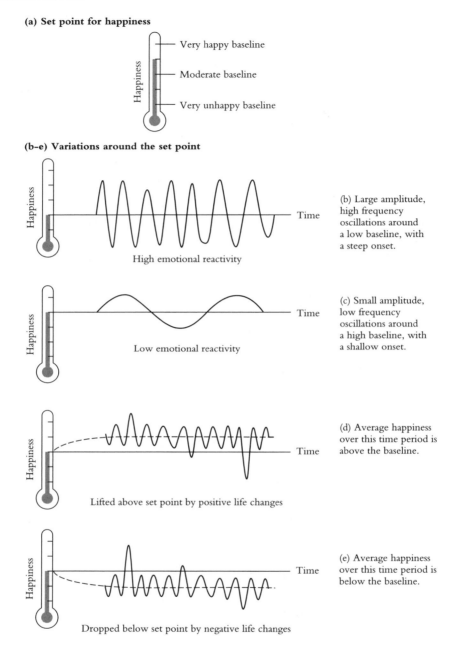

(a) Set point for happiness

— Very happy baseline

— Moderate baseline

— Very unhappy baseline

(b–e) Variations around the set point

(b) Large amplitude, high frequency oscillations around a low baseline, with a steep onset.

High emotional reactivity

(c) Small amplitude, low frequency oscillations around a high baseline, with a shallow onset.

Low emotional reactivity

(d) Average happiness over this time period is above the baseline.

Lifted above set point by positive life changes

(e) Average happiness over this time period is below the baseline.

Dropped below set point by negative life changes

Fig. 12.1

may return more quickly to baseline. Duration of negative affect (e.g. following a stressful event) can be regarded as a measure of resilience, i.e. how quickly a person 'bounces back' after an adverse experience (Fredrickson *et al.* 2000).

The various components of affective style outlined above have been described by Davidson (2002) as 'affective chronometry'. Earlier work in the literature on temperament was concerned with establishing the relationship between the different components of affective style. In the classic study by Wessman and Ricks (1966), 38 students were asked to report their highest, lowest, and average levels of mood every evening for 6 weeks. On the basis of their results, the authors concluded that average hedonic level and mood variability were independent of each other, and proposed four categories: happy–stable; happy–unstable; unhappy–stable; and unhappy–unstable. However, as Kardum (1999) points out in his extensive review of the area, the variability measure used by Wessman and Ricks and other investigators was based on the standard deviation of the mood measures, and does not differentiate between intensity and frequency. He concludes that attempts to measure intensity and frequency independently have not so far been successful. Therefore, we do not yet know the extent to which the components of affective style (set point, frequency, intensity, duration) are independent of one another. For example, are emotionally reactive individuals more likely to have a low set point for happiness or a high set point for happiness? Are individuals with a high set point more likely to be lifted by positive life changes than individuals with a low set point? To answer these questions, we need prospective longitudinal studies of representative samples of individuals, using measures that are capable of differentiating between these four parameters. While it is likely that some progress can be made by prospective studies using survey questionnaires, a better approach may be the use of experience sampling methods (ESM), where reports of current mood are made at random intervals throughout the day in response to the bleep of a wristwatch (e.g. Penner *et al.* 1994; Van Eck *et al.* 1996; Oishi *et al.* 2004). ESM can help provide more reliable on-line data about mood valence (positive or negative) and subjective intensity, but accurate data on frequency and duration as well as objective intensity are best obtained through physiological measures. These too should be examined prospectively, since response characteristics might vary over time and at different stages in an individual's life. These considerations point to a need for a programme of research that applies the methods of affective neuroscience in a prospective cohort study.

How is the set point set?

It is commonly assumed that our set point is determined by our genes (e.g. Lykken 2000; Sheldon and Lyubomirsky 2004). It is further assumed that, since our set point is under genetic control, it cannot be altered. However, both of these beliefs are erroneous. While genetic factors obviously play a part in all our physical and psychological characteristics, their role may be relatively minor compared to the effect of our environment. There is abundant evidence that early environment exerts a powerful influence on subsequent physical and mental health. For example, low birthweight (which can occur following serious nutritional problems during fetal development) results in a hugely increased risk of diseases such as cardiovascular disease and diabetes later in life (Barker 1995, 1998, and Chapter 3, this volume). Such outcomes appear to be independent of the individual's genetic makeup (genotype), so the same genotype can result in many different possible phenotypes (the individual's actual characteristics) depending upon his or her early environment.

From the perspective of mental health, one of the most important determinants of later mental health or mental disorder is early nurturing, particularly the closeness of the bond between mother and infant (Bowlby 1988). Assuming that basic physical needs are met, children whose social and emotional needs are also met, i.e. those who experience a warm and loving relationship, tend to grow up feeling secure and trusting, seeing the world as a benevolent place in which they can grow and develop. In contrast, children who experience little or inconsistent love, or those who experience childhood neglect or abuse tend to grow up feeling insecure and untrusting, seeing the world as a hostile place (e.g. Ainsworth 1989; Ainsworth and Bell 1970; Maccoby and Martin 1983; Shaw *et al.* 2004; Weiss and Schwarz 1996).

Of course, it could be argued that nurturing is itself under genetic control and that children who have been well nurtured become well-adjusted adults not because of the nurturing *per se* but because of the genes that influence both nurturing and later adjustment. This issue can be resolved in studies of non-human mammals where it is possible to experiment on the role of nurturing in later adjustment. Taking advantage of natural variations in rodent maternal behaviour that can be seen across generations, Meaney and colleagues took the offspring of anxious, emotionally reactive mothers and reared them with calm, stress-resistant mothers whose maternal behaviour is characterized by high levels of licking and grooming. When the offspring reached maturity, their behaviour, including reactivity to stress, resembled the behaviour of their adoptive mothers rather than that of their biological mothers (Meaney 2001). This supports the view that experiences that take place early in development may play a crucial role in adult mental health.

Exciting recent evidence is linking the behavioural effects of the very early environment with the chemistry and physiology of the developing nervous system. Some studies of rodents during their first 2 weeks of life show that being separated from their mother for several hours a day results in significantly increased levels of adrenal hormones (e.g. the stress hormone cortisol) in response to stressful stimuli (see review by Cirulli *et al.* 2003). It has also been shown that prolonged maternal separation leads to a lower density of sites for the neurotransmitter dopamine, which in turn leads to profound and lasting changes in the responsiveness of dopamine neurons to stress and psychostimulants, suggesting a neurobiological basis for individual differences in vulnerability to compulsive drug taking (Brake *et al.* 2004). Numerous studies have demonstrated that brief daily handling of laboratory rats and mice during the neonatal period improves both responsiveness to stress and cognitive ability. Handled animals show higher levels of glucocorticoid secretion immediately after shock exposure, and a more rapid return to basal levels; non-handled animals show a much slower rise and a higher peak in the post-shock secretion of these adrenal hormones, although basal levels do not differ between the two groups (Levine 1957; Meaney *et al.* 1996). These differences are long-lasting and can persist for the entire life of the animal. It has been suggested that the pattern of endocrine responsiveness shown by the handled animals is extremely adaptive; the speed and short duration of secretion enable the animal to respond rapidly to a challenging situation while avoiding prolonged exposure to adrenal steroids, which is known to exert toxic effects on the nervous system (McEwen 1998). Handled animals also show an increased number of glucocorticoid receptors in the hippocampus, a brain region strongly involved in learning and memory (Meaney *et al.* 1989).

In an impressive series of recent studies, Meaney and his associates have demonstrated numerous effects of maternal care on brain structure and function. Liu *et al.* (1997) showed that plasma levels of adrenal hormones were significantly correlated with the frequency of maternal licking and grooming, suggesting that maternal behaviour serves to 'programme' hypothalamic–pituitary–adrenal responses to stress in the offspring. Rats reared by mothers that exhibit high levels of licking and grooming also show increased synaptic density and increased cholinergic innervation in the hippocampus, and improved hippocampal-dependent spatial learning and memory compared with rats reared by mothers that show low levels of licking and grooming (Liu *et al.* 2000). There is also evidence that animals reared by mothers that show high rates of licking and grooming have improved survival of hippocampal neurons, which may have implications for late-life responses to stress and cognitive challenges (Weaver *et al.* 2002). Other studies have suggested that variations in maternal behaviour may be directly related to variations in receptor levels of the hormone oxytocin (Francis *et al.* 2000). A study in humans, taking advantage of a natural disaster (an ice storm in Canada), has shown that level of maternal stress during pregnancy (controlling for other possible confounders) was directly related to the general intellectual and language ability established when their babies were assessed at 2 years of age (Laplante *et al.* 2004). Taken together, all these studies provide compelling evidence for the influence of the early fetal and postnatal environment on endocrine and neurotransmitter function, as well as on emotional and cognitive function later in life.

Adverse early environments can clearly produce devastating long-term effects, but some recent studies have found that these can be partially or entirely reversed. Francis *et al.* (2002) showed that providing a stimulating environment during the peripubertal period completely reversed the effects of maternal separation on both endocrine and behavioural responses to stress, and suggested that this functional reversal indicated a compensation for, rather than reversal of the neural effects of early life adversity. Environmental enhancement also appears to eliminate hippocampal and cognitive differences between groups of rats with high versus low levels of maternal care (Bredy *et al.* 2003, 2004). These findings provide further support for the powerful effect of environmental factors both in setting enduring levels of emotional responsiveness and cognitive ability during the critical postnatal period and in providing opportunities for remediation at a later stage in the life-course. A deeper understanding is required of the content and timing of remediation and the extent to which the effects of early adversity can be compensated for by interventions at different stages in the life-course. Carefully designed studies will also enable us to establish conclusively how much these environmental variations affect basal levels of physiology and behaviour, and how much they affect reactivity parameters (frequency, magnitude, and duration of change).

The relative contribution of nature and nurture

The majority of our estimates of the degree to which a trait is inherited or is environmentally determined come from twin studies. The basic design is to compare identical twins (monozygotic (MZ) twins), who have 100% of their genes in common, with non-identical twins (dizygotic (DZ) twins; also known as fraternal twins), who share only 50% of their genes on average, usually restricting the latter group to those with the same sex. Both types of twin come into the world at the same time and are assumed to experience similar family environments (shared environment). Twins also have non-shared environments that are unique to them as individuals, e.g. their own friends and independent experiences. The simplest model on which heritability estimates are based assumes that the members of any twin pair have a similar shared environment (principally family environment) and hence, for characteristics on which identical twins are more similar than fraternal twins, the degree of similarity indicates the magnitude of the genetic contribution to the characteristic.

A fundamental error in this simple model is the assumption that the family environment is as similar for non-identical twins as it is for identical twins. This is certainly not the case. Pairs of fraternal twins (same sex) often have very different physical and behavioural characteristics, so their family and others are likely to treat them differently, and the fraternal twins themselves are likely to emphasize their differences rather than their similarities. Thus, even the shared environment of fraternal twins is not as similar as the shared environment of identical twins. In contrast, identical twins have identical genes plus (virtually) identical environments, so it is hardly surprising that they have very similar characteristics. Accordingly, interpreting this similarity as primarily genetic will overestimate the effect of the genetic contribution (see Maccoby 2000).

The best known twin study in the field of well-being was undertaken by Lykken and Tellegen on a large sample of twins in the Minnesota Twin Study (Lykken 2000; Lykken and Tellegen 1996). There are two unique features in their studies: (1) they estimated heritability both for a single (cross-sectional) measure of well-being and for a repeat (longitudinal) measure made between 4 and 10 years later; (2) they compared the heritability estimates in twins reared together and in twins reared apart. The measure they used was the 18 items of the well-being scale that is part of Tellegen's Multidimensional Personality Questionnaire (MPQ). Analysing data on the first (cross-sectional) measure of well-being, they report a correlation of 0.44 within MZ twin pairs ($n = 663$) and a correlation of 0.08 within DZ twin pairs ($n = 715$).* This indicates that MZ twins are much more similar in their level of well-being than DZ twins. But, for the reasons given above, these figures do not in themselves reveal the cause of the similarity (genes or shared environment). Lykken and Tellegen also analysed data from the small sample of twins reared apart. The comparable correlations for these twins are 0.53 for MZ twins ($n = 69$) and 0.13 for DZ twins ($n = 50$). Obtaining higher correlations for the twins reared apart compared to those reared together is counterintuitive, and probably reflects the larger errors of estimation due to small sample size. Comparing the correlations between MZ twins reared apart and reared together, Lykken (2000, Chapter 2, p. 39) states, 'These MZ correlations indicate that 40% to 50% of the variation across people in current happiness levels is associated with genetic difference. . . . This is the same as saying that the *heritability* of single measurements of subjective well-being is more than 40%.' A different conclusion is reached when Lykken examines the stability of the well-being measure over time. He reports that the retest correlation for individual twins over an average 9-year period is 0.55, indicating only moderate stability of the well-being measure. On the other hand, the correlation between the scores of one MZ twin at time 1 and their co-twin at time 2 was 0.54 ($n = 131$ pairs). The fact that the co-twins correlated with each other over this long interval almost as strongly as they correlated with themselves over that period leads Lykken in the same publication to say 'we can estimate that the heritability of the stable component of well-being, or the happiness *set-point*, is about 0.54/0.55 or close to 100%. Nearly 100% of the variation across people in the happiness set point seems to be due to individual differences in genetic makeup!' (p. 41). This figure increases Lykken and Tellegen's (1996) earlier estimation of 80%.

Is it really the case that our genes play such a huge role in determining our set point? It seems as if interpretation of the twin data boils down to what one chooses to emphasize. For MZ twins, one can either emphasize the fact that a twin's performance at time 2 is predicted equally well by their own performance at time 1 or their co-twin's performance at time 1. Or one can emphasize the fact that an individual's performance at time 1 is a rather poor predictor of their performance at time 2, and their co-twin's performance at time 1 is an equally

---

* Their publications do not specify whether the DZ twins are same sex but, in a personal communication, Lykken has said that they are.

poor predictor. As regards the comparison between MZ and DZ twins, the greater similarity between the MZ twins does not tell us about the relative contribution of genes and environment for the reasons already given. The only compelling data from Lykken and Tellegen's research is the evidence that the well-being scores of pairs of MZ twins reared apart are as similar as the scores of pairs of MZ twins reared together, implying that genes rather than environment play the dominant role. Lykken does, however, point out that these correlations are for twins reared in typical middle-class environments and the correlations might be lower if the environmental variability between twins was greater.

From the preceding sections, it appears that there is no simple answer to the question of whether nature or nurture plays the greater role in determining our characteristic level of happiness. But perhaps investigators have been asking the wrong question. Since genes are not deterministic, and are only expressed in response to environmental triggers (described in more detail below), perhaps what we should be looking for is the circumstances under which genes are or are not expressed, that is, gene–environment interactions.

A study by Caspi *et al.* (2003) provides clear evidence of the way in which genes and environments interact in producing mental health outcomes. In order to understand why stressful experiences lead to depression in some people but not others, they analysed data on the association between adverse life events and depression in a prospective longitudinal study of a representative birth cohort in New Zealand. Depressive symptoms and major depressive episodes during the past year were assessed when the 847 cohort members were aged 26, and the association of depression with adverse life events occurring after their 21st birthday was examined. DNA analysis focused on the serotonin transporter gene (5-HTT) because the serotonin system is known to play a role in depression and is a target of selective serotonin reuptake-inhibitor (SSRI) drugs such as Prozac that are widely used in treating depression (Tamminga *et al.* 1995). The serotonin transporter gene comes in two common versions: the long allele and the short allele. The short allele reduces the transcription efficacy of the 5-HTT gene promoter, resulting in decreased expression and function of serotonin. Previous studies had shown an association between the short allele and responses to stress. Monkeys with the short allele displayed decreased serotonergic function when they were reared in stressful conditions, but not when they were reared in normal conditions (Bennett *et al.* 2002). A neuroimaging study in humans showed that the stress response was moderated by variations in the 5-HTT gene; individuals with one or two copies of the short allele exhibited greater neuronal activity in the amygdala in response to fearful stimuli compared to individuals homozygous for the long allele (Hariri *et al.* 2002). This suggests that individual differences in emotional responsiveness may be in part genetically driven, although the heritability estimates are very low, around 4% of total variance (Lesch *et al.* 1996).

Caspi *et al.* (2003) reported that 69% of the birth cohort sample carried the short allele of the 5-HTT gene, 17% of the sample had a major depressive episode in the past year, and 15% of the sample had experienced four or more adverse life events between ages 21 and 26. Their analysis showed that having four or more life events predicted depression, but only among cohort members who were carriers of the short allele of the 5-HTT gene. This effect did not appear to be due to a direct influence of genotype on either life events or depression, since genotype was not related to the number of stressful life events, and the interaction between life events, depression, and genotype was found even amongst those with no prior history of depression. Additional analyses showed that childhood maltreatment predicted adult depression—but only among those who were carriers of the short allele of the 5-HTT gene. Thus, the short allele variant of the 5-HTT gene confers vulnerability to depression when there are appropriate environmental triggers, while the long allele acts as a resilience or protective factor. Although the sample size was small for a complex study of this type, the findings suggest that asking whether it is genes or environment that contribute more to our mental health is to ask the wrong question. Our genes may play the dominant role, but only if we have certain experiences; alternatively, our environment may play the dominant role but only if we have a particular genotype. A similar point has been made in a review using a more sophisticated approach to examining the role of parental influences on behaviour (Collins *et al.* 2000).

How set is the set point?

Questions of origin aside, it is important to know whether our characteristic level of happiness is fixed and immune to long-term change, or whether it can be altered. The widely held view that a person's basic level of happiness is fixed and unchanging stems from two different types of inference. One is based on evidence that, no matter how extreme a change may take place in our circumstances, within a short time we adapt to our new circumstances and return to our basal level of happiness (e.g. Brickman *et al.* 1978; Kahneman 1999). The other is the assumption that our basal level of happiness is primarily under genetic control and, being genetically determined, cannot be changed. Each of these assertions is examined in more detail below.

Probably the most striking illustration of how we adapt to profound changes in our circumstances and return to our basal level of happiness comes from the classic study of Brickman *et al.* (1978). They reported that the well-being scores of lottery winners soared initially, while the scores of accident victims who had become paraplegic plummeted, but that within a relatively short period both groups had returned to their original baseline. This phenomenon has been described as hedonic adaptation (Kahneman 1999). Economists and policy makers who had hoped that their policies would improve average levels of well-being have been greatly discouraged by the evidence of hedonic adaptation, since it implies that any beneficial change that is introduced will have only transitory effects.

Recent studies by Diener, Lucas, and colleagues demonstrate that enduring changes in well-being do sometimes occur. Using data from a 15-year longitudinal study of over 24 000 German adults, they found that individuals reacted strongly to unemployment but then moved back towards their baseline level of life satisfaction. On average, however, individuals did not completely return to their former levels even after they became re-employed (Lucas *et al.* 2004). They also found that, following marital transitions, individuals showed an initial strong reaction and then, on average, returned towards their basal level. But there were substantial individual differences; those who reacted very strongly initially were still far from their baseline years later (Lucas *et al.* 2003). This confirms earlier findings of Winter *et al.* (1999) that marriage and widowhood among older adults were still producing positive or negative effects, respectively, on subjective well-being for a long period after their occurrence. So, although levels of well-being are relatively stable over time and individuals tend to return towards their baseline remarkably quickly (Silver 1982: Suh *et al.* 1996), life events can have a long-lasting influence.

The second reason why it has often been assumed that the set point is fixed and unchangeable is the assumption that it is genetically determined. However, it is important to realize that, even if genes play a dominant role, it does not follow that the set point is predetermined. This is because genes do not of themselves make things happen. Genes need to be expressed (turned on or turned off) and it is environmental triggers that lead to genes being expressed. For example, a person might have a gene that increases his/her vulnerability to dietary salt, putting the person at high risk of cardiovascular disease. But, if that person has a diet low in salt, this vulnerability may never become manifest. Likewise, someone may have a genetic predisposition to depression under conditions of high stress (as seen in the Caspi *et al.* 2003 study), but if that person avoids highly stressful environments, this genetic propensity may not be triggered. Therefore, even if a trait can be shown to have a large genetic component, genes do not determine outcomes. It may be possible to adapt the environment in such a way that it reduces the risk of the adverse outcome, or increases the probability of an advantageous outcome.

The two lines of evidence presented above, show that our set point for happiness is less set than some have supposed. It appears that certain life events and some aspects of our external environment or circumstances can have long-term effects on our average level of happiness. The idea that the set point can drift considerably above or below its pre-established level was depicted in Fig. 12.1(d, e). However, the data suggest that the magnitude of changes in our set point during adult life are very modest if such changes occur at all. Yet most people yearn to be happier than they are, to feel greater contentment or experience more joy in their lives. Is this simply an impossible dream, or can it become a reality? In the next section we consider a different approach to increasing well-being in a sustainable way.

Towards sustainable increases in happiness

Thus far, the emphasis has been on the way in which individuals respond to the contexts and circumstances in which they find themselves, and to the life events they experience. An alternative to this passive approach is to consider the active role that people can play in shaping their own lives. This approach has been eloquently expounded by Sheldon and Lyubomirsky (2004) who emphasize the importance of intentional activities in determining the 'chronic happiness level'. They distinguish between three broad classes of intentional activities, each of which can be shown to be related to subjective well-being. These are: (1) overt behaviours such as taking regular exercise or showing kindness to others (Keltner and Bonanno 1997; Magen and Aharoni 1991); (2) cognitions such as interpreting events in a positive light or savouring the moment (Seligman 1991; Emmons and McCullough 2003); and (3) motivations such as striving towards valued goals or putting effort into worthwhile activities (Sheldon and Houser-Marko 2001). Other conceptualizations have also shown a link between intentional activities and subjective well-being (SWB). For instance, the seminal work of Csikszentmihalyi (1997; Csikszentmihalyi and Csikszentmihalyi 1988) demonstrates that optimal experience or 'flow' results when a person is fully engaged in a task in which his/her interests or skills are fully utilized. It has been found that optimal experience arises when the person's level of skill matches the level of challenge; subjective well-being is compromised if the challenge is either too difficult, leading to frustration, or too easy, leading to boredom or apathy. Extensive research on goal pursuit shows that enhanced well-being is associated with goals being intrinsic, i.e. self-generated (e.g. Kasser and Ryan 1996), with progress towards a valued goal (Sheldon and Kasser 1998), the pursuit of approach goals rather than avoidance goals (Elliot *et al.* 1997), and the pursuit of goals congruent with personal values (Brunstein *et al.* 1998: Sheldon and Elliot 1999). In addition, a large body of work shows that active participation in social activities and involvement in one's community is associated with high levels of happiness and life satisfaction (Putnam 2000; Helliwell 2003; Helliwell and Putnam, Chapter 17, this volume).

One of the most significant aspects of the work of Sheldon and Lyubomirsky (2004) is their insight into how to solve the problem of hedonic adaptation. Adaptation of any kind (sensory, physiological, hedonic) occurs when stimuli are constant, unchanging, or regularly repeated. Because intentional activities are under our control, and can be altered in their content and timing, adaptation does not occur. For example, although a person might show hedonic adaptation to (i.e. no longer derive pleasure from) a rigid exercise regime or routine contributions to a voluntary organization, the enjoyment can return if small changes are made to the sequence, location, or content of these activities.

There is another important reason why intentional activities do not suffer from the problem of hedonic adaptation, and this seems to have been overlooked in the literature. Intentional activities of the type described by Sheldon and Lyubomirsky (2004) typify eudaimonic well-being, not hedonic well-being; that is, they are not directly concerned with seeking pleasure or happiness, but rather with seeking fulfilment. These activities have positive effects on subjective well-being (SWB) not because feeling good is their aim, but because feeling good is a byproduct of being engaged and fully functional. There are, of course, intentional activities that are almost entirely hedonistic in nature, such as sunbathing or eating chocolate, where the pursuit of pleasure is virtually the sole aim of the activity. However, such activities have only transient effects on well-being compared with the long-lasting and more complex effects produced by eudaimonic activities. While many eudaimonic activities produce 'flow' or other types of hedonic well-being, there is no guarantee that all valued activities are associated with positive emotions. Someone might, for instance, work hard on behalf of a charity but derive little hedonic pleasure from the work. And, as stated in the introduction, this chapter takes the view that only a person whose life combines eudaimonic and hedonic well-being, i.e. fulfilment as well as enjoyment, can be regarded as having positive mental health.

Happiness and positive functioning: cause or consequence?

The preceding section implies that many eudaimonic activities result in hedonic well-being, but it would be a mistake to think that the causal direction always flows from positive functionings to positive feelings. As Ryan and Deci (2001) point out, people high in SWB may have attributional styles that are more self-enhancing and perhaps more enabling than those low in SWB, raising the possibility that happiness can lead to positive cognitions that in turn contribute to further happiness. Certainly, there is a great deal of observational data showing that characteristically happy people tend to construe the same experiences and life events more favourably than unhappy people (Lyubomirsky and Tucker 1998) and be less responsive to negative feedback (Lyubomirksy and Ross 1999). Diener (2000) also reported that happy people are typically more socially engaged and more productive. Observational studies, particularly cross-sectional ones, cannot, of course, establish the direction of causality, so it is not clear whether positive cognitions and behaviours are the consequence of happiness or its cause, or whether both are influenced by a third factor such as temperament.

Longitudinal research can go some way towards establishing causal relationships, but the most persuasive evidence comes from experimental studies. Research using mood induction techniques demonstrates unequivocally that positive mood states can enhance flexible and creative thinking. Fredrickson has shown that, compared with individuals in negative or neutral mood states, subjects in a positive mood state have a broader focus of attention ('see the bigger picture') and generate many more ideas in problem-solving tasks (Fredrickson and Branigan 2005). Further evidence comes from Levy and colleagues who have found that elderly subjects who have been primed (using subliminal presentation) with positive descriptions of ageing (and are hence likely to have positive affect) show higher self-esteem and perform better on cognitive tests than the comparison group who have been primed with negative descriptions of ageing (Levy 1996; Levy and Banaji 2002). Experimental social psychology is full of examples showing that positive emotional experiences have beneficial effects on the way that people perceive and interpret social behaviours and the way in which they initiate social interactions (e.g. Forgas 2001; Isen 1987). Thus it is clear from the experimental research on induced mood that happiness or other positive emotions can have a direct effect on cognitive appraisal and social relationships. Fredrickson's 'broaden-and-build' theory of positive emotions proposes that the frequent experience of positive affect broadens cognitive processes and builds enduring coping resources that lead to later resilience (Fredrickson 2001, 2004, and Chapter 8, this volume).

The experimental research has mainly been carried out in adults, and demonstrates that the benefits of positive emotions, at least over the short term, can be seen right across the adult lifespan. It is probable that the beneficial effects of positive emotions have an even greater impact during childhood and adolescence. Early studies of mother–infant bonding showed that infants with secure attachment were more confident in exploring their environment and responding to strangers than infants whose attachment was insecure (Ainsworth and Bell 1970). The research of Ainsworth and later investigators (e.g. Maccoby and Martin 1983) provides evidence that, even in infancy, positive emotions are associated with positive cognitive, social, and emotional behaviour that may provide a basis for resilience throughout life.

Overall, the available data indicate that the causal pathway between hedonic and eudaimonic well-being is bi-directional. Positive feelings can produce positive functionings and positive functionings can produce positive feelings.

Enhancing mental health and individual well-being

The state of complete mental health has been described as 'flourishing' (Keyes 2002a). According to Keyes (2002a, p. 262), 'flourishing individuals have enthusiasm for life and are actively and productively engaged with others and in social institutions.' To establish how many of us are in this fortunate state, Keyes analysed data from a nationally representative sample of US adults who participated in the Midlife in the United States (MIDUS) study (Keyes 2002b). He estimated that only around 17% of individuals are flourishing at this stage in life. The majority were described as moderately mentally healthy (54%), while the remainder either have mental illness (18%) or are 'languishing' but without mental illness (11%). The term 'languishing' refers to a condition in which a person's life seems empty or stagnant, 'a life of quiet despair' (Keyes 2002a, p. 210). Keyes has shown that 'languishers' are at greatly increased risk of depression and of physical disorders including cardiovascular disease (Keyes 2004). He also suggests that languishing may have very high prevalence among young people, who are showing increasing rates of depression and suicide, and who are seeking ways to fill the void of their lives. Sex, drugs, and alcohol are often used in this way, but these only deepen the void and make the person more dysfunctional. A schematic version of the mental health spectrum, from flourishing to mental disorder, is depicted in Fig. 12.2.

Vast amounts of time, energy, and expenditure have gone into trying to improve the mental state of those with mental disorders, and this is without doubt an important area of research and application. But is there scope for improving the mental health of the majority (65% by Keyes's calculations) who are not flourishing and yet do not have mental illness? In other words, is it possible to enhance the well-being of those who have only moderate mental health, and particularly of those who are languishing? There could be immense benefits to individuals, their families, and society in general to have more of us 'flourishing' more of the time. But how might this be accomplished? While there is no end of advice and well-intentioned self-help books available on how to improve our lives, very little serious research has been undertaken to establish the evidence base for the many interventions that have been proposed. There may be a strong evidence base for what works with mental disorder, but it is not necessarily the case that the interventions that help to shift an individual from a state of disorder to a state of no disorder will also be effective in shifting individuals from languishing to flourishing.

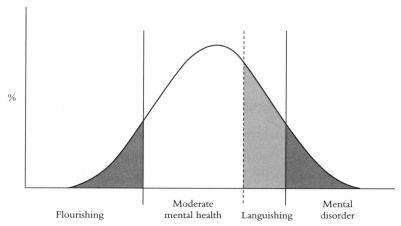

Fig. 12.2 The mental health spectrum. Thresholds for selected mental health categories are shown schematically.

One approach to evidence-based interventions to enhance well-being is to adapt methods that are solidly grounded in theory and for which there is good observational data on their strong association with measured well-being. Preceding sections of this chapter point to ways in which it may be possible to enhance well-being in ordinary people. Using the framework proposed by Ryan and Deci (2001), attempts could be made to modify relevant behaviours, cognitions, and motivation. Examples of behavioural modifications that were described earlier include exercising regularly, which produces major benefits for psychological well-being (see review by Biddle and Ekkekakis, Chapter 6, this volume), and showing kindness to others. Cognitive modifications include learning to look on the positive side of things (sometimes referred to in cognitive therapy as 'reframing') and appreciating the good things in life (having an optimistic outlook, or savouring the moment). Changes in motivation or goal pursuit include ensuring that goals are set appropriately so that progress can readily be made, or changing one's goal if reasonable progress is not possible; identifying goals that involve approach rather than avoidance, e.g. having as a goal not the giving up of smoking or fatty foods, but of becoming healthy. A further illustration is ensuring that goals are congruent with our deeply held values, such as undertaking socially responsible work with youth or disadvantaged groups rather than having a lucrative job doing less socially responsible work, e.g. advertising cigarettes or encouraging clients to get into serious debt.

Experimental evidence for the efficacy of individual-level interventions

While approaches such as these have a sound empirical basis, and may indeed enhance individual levels of well-being, the real criterion for whether an intervention should be recommended is its proven effectiveness in an adequately controlled trial and subsequent replications. Fortunately, some serious attempts have begun to establish the efficacy of potential interventions in enhancing well-being in ordinary people. To date, most of the interventions undertaken with adults have used small groups of subjects (mainly undergraduates) and examined short-term effects. One example is the work of Emmons and McCullough (2003) who reported that students who listed things they were grateful for ('counting their blessings') on a daily or weekly basis obtained higher scores on positive affect and other well-being measures than students who recorded hassles or more neutral events. In a similar vein, Seligman and colleagues instructed subjects to record every day for 1 week three good things that happened to them each day and why these good things occurred. Compared with a control group, individuals who completed this exercise were happier and less depressed even at the 3-month follow-up (Seligman, Peterson, and Steen 2004, unpublished). The finding that the benefit extended so far beyond the intervention exercise suggests that individuals learn to use these techniques in their daily life. Similar methods were adopted by Fredrickson and colleagues whose research showed the psychological benefits of actively finding meaning and long-term benefits in daily events, no matter whether those events were positive, neutral, or negative (Fredrickson *et al.* 2004).

Mindfulness training as a well-being intervention

These very specific techniques can be contrasted with a far broader approach that aims to increase well-being through enhancing our quality of consciousness. The best known of these approaches is the age-old Buddhist practice of mindfulness. Mindfulness is defined as the state of being attentive to and aware of what is taking place in the present. This includes awareness of the external environment, that is, being conscious of our sensory inputs (e.g. perceiving the unique qualities of everyday objects and events) as well as our internal environment, that is, observing our thoughts and feelings rather than letting them influence our behaviour in unconscious ways. Many years of observation of the effects of mindfulness training (e.g. Kabat-Zinn 1990) show that mindfulness facilitates a variety of well-being outcomes. These include increases in positive affect (e.g. becoming more peaceful and loving and experiencing more joy in life) and improvements in functioning (e.g. acting with more compassion, and developing more flexible and creative thinking styles). A major series of investigations by Brown and Ryan (2003) has linked individual differences in mindfulness to the self-regulation of behaviour. Using experience sampling methods in both students and a general adult sample, they established that there are stable individual differences in mindfulness, and that both the characteristic level of mindfulness and variations in the state of mindfulness are related to autonomous behaviour regulation and emotional well-being at the momentary level.

However, as noted before, while such observational studies provide a valuable starting point for the development of interventions, the most persuasive evidence of efficacy comes from randomized controlled trials. There have been several well-designed trials using training in mindfulness meditation to improve quality of life, but almost all have been carried out on clinical samples. For instance, mindfulness training has been shown to improve outcomes in patients with chronic pain (Kabat-Zinn *et al.* 1986), cancer (Brown and Ryan 2003; Speca *et al.* 2000), and other physical disorders (Kabat-Zinn *et al.* 1998), and in patients with anxiety disorders (Kabat-Zinn *et al.* 1992; Miller *et al.* 1995) and depression (Ma and Teasdale 2004; Teasdale *et al.* 2000). Only a few studies have been undertaken on nonclinical samples, mainly medical students. In several studies, subjects who had 8–10 weeks of mindfulness training showed significant decreases in psychological and physical symptoms compared to controls (Rosenzweig *et al.* 2003; Shapiro *et al.* 1998; Williams *et al.* 2001) and, in one study that tested subjects post-intervention, the benefits were maintained at the 3-month follow-up (Williams *et al.* 2001).

Surprisingly, only one of these studies looked at positive well-being outcomes; Shapiro *et al.* (1998) found increased scores on overall empathy and a measure of spiritual experiences.

It is possible that mindfulness, an intense awareness of what is taking place in the present, also plays a role in other activities that can boost our well-being. For example, during physical exercise we tend to be very focused on the present moment, and the beneficial effects of physical activity both in the short- and long-term are well known (see review by Biddle and Ekkekakis, Chapter 6, this volume). Likewise, the immediacy of the natural environment often hijacks our senses, making us focus on the immediate moment, and this may contribute to the powerful effects that nature can exert on our well-being (see Burns, Chapter 16, this volume, for a review). Mindfulness training makes us fully conscious, so that even the most mundane experience or object can absorb or fascinate us, and mindfulness enhances our sense of connection with others and with the wider universe.

An important study by Davidson and colleagues (2003) has extended this research by examining the relationship between neurophysiological measures and mindfulness. They report that: (1) expert meditators have patterns of brain activation at rest that resemble the extreme of positive affect (much higher activation of the left compared with the right frontal cortex); and (2) training ordinary people in mindfulness techniques produces positive brain activation patterns and is associated with a substantial boost in immune function 6 months after training, compared to the control group (Davidson *et al.* 2003).

Future direction for individual-level interventions

Although most of the well-being interventions described above have been used on an experimental basis to date, and follow-up has been relatively short-term, the results are encouraging. They suggest that such techniques could begin to be applied more widely following further evaluation. The gold standard for evaluating research in this field is to have large-scale randomized controlled trials undertaken on representative samples of the population, with long-term follow-up to establish the sustainability of changes. Future studies should also give greater consideration to individual differences. They should examine the magnitude and duration of change in relation to differences in pre-intervention levels of happiness and mood variation. Research also needs to begin considering whether 'one size fits all' in well-being interventions or whether interventions should be tailored to the relevant characteristics of the individual, such as their personality and life stage (Henry 2004) and their sociocultural context (Delle Fava and Massimini, Chapter 15, this volume).

As noted earlier, the power and beauty of interventions that nurture positive emotions, attitudes, and behaviours is that they can potentially benefit the normal majority of people whose lives may not be as happy or fulfilled as they might wish. This is in contrast with the usual approach of restricting interventions to the small minority of the population who already have a problem or are at high risk of developing a problem.

Enhancing well-being at the population level

The preceding sections highlighted the importance of people taking responsibility for their own health and recognizing that there are actions that are within their control that can make their lives more enjoyable and meaningful. It is equally important to take an epidemiological perspective and recognize that all aspects of our health and behaviour are profoundly influenced by the society to which we belong. It has been shown that for many common disorders there are community-level as well as individual-level risk factors and protective factors. If a disease or disorder were entirely the result of individual risk factors, then we might expect the prevalence of the condition to be much the same in any population. On the other hand, if community-level characteristics play an important role, then we might expect that the prevalence of the condition would be related to the average level of the risk factors in the population. We might expect, for example, that common physical health problems such as cardiovascular disease would be more common in populations that have high levels of the known risk factors, e.g. high blood pressure associated with smoking, sedentary lifestyle, and diets high in fat and salt. This has indeed been shown to be the case for cardiovascular disease (e.g. Puska *et al.* 1998). The renowned epidemiologist Geoffrey Rose (1992) demonstrated that the prevalence of many common diseases in a population or subgroup is directly related to the population mean of the underlying risk factors. He proposed accordingly that the most effective way to reduce the prevalence of the disorder was to shift the mean of the population in a beneficial direction. Shifting the mean of the risk factors in the population has been responsible for the substantial drop in cardiovascular disease in many developed countries. The large reduction in prevalence has been driven mainly by population-level interventions, such as making smoking less socially acceptable and promoting exercise and healthy diet, which have had the effect of lowering the mean levels of cholesterol and blood pressure (Puska *et al.* 1998). While treatments with drugs, surgery, or behavioural techniques like stress reduction have been extremely effective at controlling heart disease at an individual level, the population approach ensures that fewer people develop heart disease in the first place.

A striking example of the relationship between the population mean and the prevalence of disorder can be seen in the association between mean alcohol consumption and the prevalence of heavy or problem drinking. Based on data from over 32 000 adults in the Health Survey for England, Colhoun *et al.* (1997) showed that, across all the regions in England, the mean alcohol consumption (excluding heavy or problem drinkers) is strongly correlated with the prevalence of heavy drinking and problem drinking in that region. They concluded that a small reduction in the mean consumption of the light or moderate drinkers (for example, by increasing the cost of alcohol or discouraging binge drinking) is likely to result in a substantial decrease in serious alcohol-related problems. This conclusion follows from the fact that there is essentially a normal distribution of alcohol consumption in the population and, for normal distributions, when there is a small shift in the mean, there is a large shift in the tails of the distribution. Rose (1992, p. 64) has suggested that, in a similar manner, deviant behaviours such as violence are 'simply the tail of the population's own distribution'. Therefore, reducing the mean level of violence or the mean of its risk factors in a population should reduce the prevalence of serious violence and related crime.

The population study that arguably has the most direct relevance to mental health was undertaken by Anderson, Huppert and Rose in 1993. They examined the scores on a mental health questionnaire (GHQ-30 of Goldberg 1978) in subpopulations of a nationally representative UK sample who participated in the Health and Lifestyle Survey (Cox et al. 1987). They found that the prevalence of clinically significant disorder was directly related to the mean number of symptoms in the subpopulation (excluding those with a clinically significant disorder). In other words, whether or not a person has a mental health problem is influenced not only by individual characteristics and experiences, but also by population-level factors.

The key test of the proposition that an individual's chance of having a mental health problem is related to the mean number of symptoms in the subpopulation is to establish whether the prevalence of such problems changes as the mean number of symptoms in the subpopulation changes. In a 7-year longitudinal follow-up of participants in the Health and Lifestyle Survey, Whittington and Huppert (1996) demonstrated that the change in the mean number of symptoms in various subpopulations (excluding those with disorder) was highly correlated with the prevalence of clinically significant disorder. Moreover, the relationship was linear; as the number of symptoms decreased by one point, the prevalence of disorder dropped by 7%. The Whittington and Huppert study was observational, but it implies that, if population-level interventions were developed to improve overall levels of mental health, they could potentially have a substantial effect on reducing the prevalence of common mental disorder as a direct result. As stated by Anderson et al. (1993, p. 475), 'Populations thus carry a collective responsibility for their own mental health and well-being. This implies that explanations for the differing prevalence rates of psychiatric morbidity must be sought in the characteristics of their parent populations; and control measures are unlikely to succeed if they do not involve population-wide changes.'

Strategies for population-level interventions

Most population health strategies are targeted at groups with a disorder or at high risk for a disorder, with the aim of reducing the likelihood of the condition or its severity, duration, or relapse rate. In contrast to these prevention strategies, population intervention or health promotion strategies are universal rather than targeted and are an appropriate technique to use when their benefits are regarded as desirable for everyone in the population. In the physical health domain, population intervention strategies include the compulsory wearing of seat belts in cars, fluoridation of drinking water, and advocating healthy behaviours (balanced diet, frequent exercise, sensible drinking, smoking cessation).

Mental health can be regarded as a suitable target for population-level interventions of the type described by Rose (1992) since its manifestations form a continuous distribution in the population (see Fig. 12.2). The clinical diagnosis of depression or anxiety disorder can be regarded as an arbitrary cut-point along a distribution of symptoms; that is, a point at which the symptoms have become so numerous and so disabling that the individual can no longer manage his/her normal activities. Individuals who do not quite meet diagnostic criteria for mental disorder may nevertheless be suffering in their daily life. They may be languishing rather than flourishing, and may benefit from appropriate intervention.

The approach to population intervention proposed here would not focus on the negative end of the distribution where high numbers of psychological symptoms occur; rather it would aim to shift the whole population in a positive direction. Figure 12.3 shows that, by reducing the mean number of psychological symptoms in the population, many more individuals would cross the threshold for flourishing.

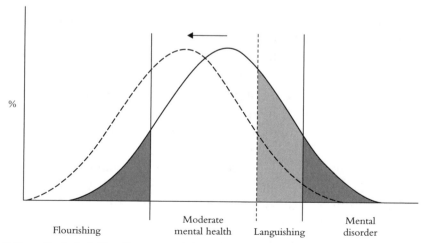

Fig. 12.3 The effect of shifting the mean of the mental health spectrum towards positive mental health. A higher percentage of individuals would cross the threshold for flourishing and the number in the categories of mental disorder and languishing would be substantially reduced.

Moreover, as a direct consequence of a small shift in the mean of symptoms or risk factors, there would be a substantial decrease in the number of people in the pathological tail of the distribution who meet the criteria for mental disorder or who are languishing. The principles and empirical findings underlying population intervention for mental and behavioural disorders have been presented in more detail in Huppert (2004).

Delivering population interventions

What are the practical steps that can be taken to shift the mental health of the population? It is proposed here that we can adopt the principles and practices that have proven success at the individual level, but make them more universally available. How can this be done? The key is the way in which the information is delivered. Educational or health promotion approaches can be offered through organizations such as schools, workplaces, health clinics, or community centres. The media also have a major role to play, both in providing reliable information and in modelling desirable patterns of behaviour. Evidence that the media can influence mental health in a negative way can be found in reviews by Phillips and Carstensen (1986) and Gould (2001) that show a very strong link between media presentations of actual or fictional suicides and the subsequent suicide rate. In Germany, Schmidtke and Hafner (1988) found that each of two showings of a television drama depicting a young man's fictional suicide on a railway track was followed by a significant increase in railway suicides by young men. In light of such influence, the media's power for good should also not be underestimated. For example, soap operas that currently tend to portray the seamy side of life, thereby normalizing or appearing to condone violent behaviour, eating disorders, drug taking, or teenage suicide, could as readily portray and thus promote values such as compassion, responsibility, and tolerance, as well as activities that encourage sustainable happiness and fulfilment. Portraying only positive attitudes and behaviours would certainly be counterproductive, but there is ample scope for a better balance in the portrayal of desirable and undesirable human characteristics.

Effective delivery of population interventions could adopt the principles of 'social marketing' (Kotler and Zoltman 1971; Donovan and Henley 2004). This approach recognizes that the same marketing principles that are used to sell products to consumers can also be used to 'sell' ideas, attitudes, and behaviours. However, social marketing differs from other areas of marketing since the objective is not to benefit the marketer but to benefit the target group or the general society. Social marketing has evolved from a reliance on public service announcements to a more sophisticated approach that draws on successful techniques used by commercial marketers. The techniques have been used extensively in international health programmes (e.g. promoting contraceptive use) and are being used more frequently in other health areas (e.g. drug abuse, heart disease, inoculation), as well as for environmental issues (e.g. energy efficiency). A recent programme in Australia entitled Freedom from Fear proved very successful in reducing male violence towards intimate partners (Donovan and Henley 2004).

The potential of the internet and other communication devices (mobile/cell phones) is only beginning to be glimpsed. In a pioneering study in Australia, Christensen *et al.* (2004) report the successful delivery of cognitive behaviour therapy via the internet in a randomized controlled trial of 525 participants in a longitudinal survey, whose depressive symptoms had increased. The availability of helpful information and on-line techniques to enhance well-being could readily be extended to the general public. A crucial proviso is that the websites involved would need to carry some form of guarantee for the reliability of the information provided, the efficacy of interventions, and a genuine assurance of confidentiality.

The overwhelming majority of population interventions undertaken so far to improve mental health and behaviour, have been administered to children. To illustrate the way in which such interventions operate, we briefly review some of these programmes.

Child-focused intervention programmes

There are two principal reasons why most population interventions have focused on children. One is the recognition of the fundamental importance of the early years for later health and adaptation and the evidence for a sensitive period with respect to the effects of stress on the developing nervous system and behavioural outcomes (Dawson *et al.* 2000). The other is that it is relatively easy to gain access to children and their parents. The universality of schooling in developed countries makes schools the ideal place for the delivery of mental health interventions. This has been recognized in a number of programmes, some of which teach cognitive and emotional skills, while others emphasize interpersonal relationships. In one large programme from Virginia, adolescents are taught life skills by high school students in interactive classroom sessions. The programme is called Going for the Goal and involves adolescents learning how to identify positive life goals, focus on the process of goal attainment rather than the outcome, identify behaviours that facilitate or compromise goal attainment, and find or create social support (Danish 1996). The programme has been administered to tens of thousands of adolescents in numerous cities in the US.

While potentially very effective, such programmes have not always been evaluated against matched control groups. However, a well-designed intervention that did use controls is the Penn Resiliency Program, an exciting school-based initiative that aims to prevent depression in adolescents. It involves 12 2-hour sessions of cognitive behaviour therapy and its effectiveness compared with controls has been demonstrated (Freres *et al.* 2002). An excellent review of universal interventions for the prevention of mental health disorders in school-age children has been carried out by Greenberg *et al.* (1999). They identified 14 universal preventive interventions that had undergone a quasi-experimental or randomized evaluation and had been found to produce positive outcomes in either symptom reduction or reduction of risk factors. A recent special issue of the *American Psychologist* (Weissberg *et al.* 2003) presents the most up-to-date information on prevention of mental health and behavioural problems in children, and includes examples of successful universal interventions.

While most interventions have been applied directly to children, others have involved changing parental behaviour to improve children's well-being (e.g. Kumpfer and Alvarado 2003). An impressive series of linked programmes in Australia has focused on parental education (Sanders 1999; Sanders *et al.* 2000, 2002). Known as Triple P (Positive Parenting Program), this programme is administered in a variety of ways, including both community-based and media interventions. Sanders and his colleagues examined the effects of a 12-episode television series, 'Families', on disruptive child behaviour and family relationships. Compared to the control group, parents in the television viewing condition reported significantly lower levels of disruptive child behaviour and higher levels of perceived parenting competence and, as measured by a 36-item child behaviour inventory, the prevalence of disruptive behaviour dropped from around 43% to 14%. The authors also report that all post-intervention effects were maintained at 6-month follow-up.

The Triple P intervention has now been extended so that all parents of pre-school children in a participating region are eligible to take part. A shortened version of the programme is delivered through community health services. It involves an 8-week parenting skills training programme consisting of group sessions followed by telephone consultations. Many thousands of families in Western Australia have participated in the programme. Programme evaluation has documented an increase in the use of positive parenting strategies as opposed to harsh, punitive strategies, a decrease in parental reports of the number and intensity of child behaviour problems, a decrease in parental depression, anxiety, and stress, and an increase in marital or relationship satisfaction. It has been estimated that, if carried out at a universal level and if all eligible families enrolled, Triple P would reduce the total proportion of pre-school children with significant behaviour problems by around 40% (Lewis 1996).

Future directions for population interventions

Authors of universal intervention studies such as those described above usually acknowledge the value of promoting competence and positive mental health among all children, but the measured outcomes usually focus on reductions in psychopathology rather than positive benefits for the whole population. Testing the applicability of the Rose (1992) model described earlier would be a valuable exercise. That is, such programmes need to ascertain the magnitude of the mean change in the population, and the extent to which this increases the number of individuals in the positive tail of the distribution (flourishing) and decreases the number in the negative tail of the distribution (disorder or languishing).

Future population interventions need to broaden the participant base to include the general adult population and they need to adopt a life-course approach. Evidence described earlier makes it plain that providing a solid foundation for well-being later in life requires good nurturing in the earliest years of life. More could be done to educate prospective parents about optimal fetal and postnatal development. The ubiquity of antenatal check-ups and childbirth classes provides an ideal opportunity for universal intervention. The importance of well-being could be given more weight at all stages of education, from pre-school to tertiary, with schools being encouraged to take social, emotional, physical, and creative development as seriously as academic achievement (Gardner 1993, 2000). Workplaces could play a more proactive role in enhancing the well-being of their employees, and not just their profit margins. It is surprising how few employers appear to recognize that a happy workforce is a more productive workforce; meta-analysis of the extensive research in this field shows that job satisfaction is associated with lower absenteeism, higher staff retention, and better work performance (Judge *et al.* 2001). Positive work environments are those in which employees can experience a good level of engagement and autonomy, challenge and the opportunity to demonstrate their competence, and positive relationships with others in the workplace. All organizations need to recognize that negative stereotypes (e.g. of older people, women, or ethnic minorities) have a damaging effect on the emotional well-being, social interactions, and capabilities of those groups, whereas positive attitudes enable them to thrive and utilize their potential. A new generation of evidence-based population interventions has enormous potential to benefit a wide range of individuals and society at large.

This chapter has emphasized two broad approaches to mental health interventions: (1) those that rely on an individual seeking information about ways of improving their well-being and taking responsibility for improvement; and (2) those that are made available to all members of a population or subpopulation (i.e. universal interventions). The latter are provided by responsible bodies (government or private organizations), but it is still up to the individual whether or not they take advantage of the methods on offer. A third approach is to identify the societal factors that influence our levels of well-being, and to make changes at the level of social policy in order to create environments in which more of us can flourish. The social policy approach has been persuasively articulated by Layard (2003, 2005), and by Shah and Marks of the new economics foundation (nef) in their *Well-being manifesto* which is reproduced in Chapter 20, this volume.

Conclusion

Positive mental health, or mental well-being, describes a condition in which individuals are fulfilling their potential and enjoying their life. Mental well-being varies widely across individuals, and only a relatively small percentage of the population can be described as flourishing. This chapter has examined some of the factors that influence our level of well-being and some interventions that may enhance it. There are large individual differences in characteristic levels of subjective well-being, including the set point for happiness. There are also large and stable individual differences in the magnitude, frequency, and durations of fluctuations around the set point. While genetic factors play a part in establishing our set point and other parameters of emotional reactivity, evidence is reviewed that suggests that the early environment makes a very large contribution to our characteristic level of well-being. External circumstances and life events can change our level of subjective well-being, but these changes are usually fairly short-lived. However, actions that are under our control, including positive behaviours, cognitions, and motivation, can have profound and long-lasting effects on our enjoyment of life and effective functioning. Evidence from well-designed experimental research is accumulating in support of this view, but longer-term studies on representative population samples are needed. Mindfulness training, i.e. being attentive to one's ongoing sensations, thoughts, and emotions, provides a particularly promising avenue of enquiry in light of the states of clear consciousness, love, joy, and contentment experienced by its practitioners.

While it is important to recognize that individuals can take actions to enhance their well-being, it is equally important to recognize the influence of population-level factors on well-being. It is argued that there is a strong case to be made for enhancing well-being among the normal majority, combining individual- and population-level approaches. Early interventions with children and adolescents are particularly important and could make a major difference to outcomes later in life including higher levels of social and occupational functioning, creativity, resilience, and enjoyment of life. Moreover, the evidence suggests that improving average levels of mental health in the population will have the direct result of reducing the prevalence of mental disorder. Therefore, a positive mental health agenda represents a win–win situation in which individuals can flourish and society is the beneficiary.

Acknowledgements

I am grateful to Eve Bowtell and George Ploubidis for assistance with researching sections of this chapter and to Nick Baylis, Eve Bowtell, George Burns, and Barry Keverne for their helpful comments on the draft manuscript. My special thanks go to Caroline Wright and Julian Huppert for creating the figures.

References

Ainsworth, M.D.S. (1989). Attachments beyond infancy. *Am. Psychol.* **44**, 709–16.

Ainsworth, M.D. and Bell, S.M. (1970). Attachment, exploration, and separation: illustrated by the behaviour of one-year-olds in a strange situation. *Child Dev.* **41**, 49–67.

Anderson, J., Huppert, F.A., and Rose, G. (1993). Normality, deviance and minor psychiatric morbidity in the community. *Psychol. Med.* **23**, 475–85.

Argyle, M. (1987). *The psychology of happiness.* Routledge, London.

Barker, D.J.P. (1995). Fetal origins of coronary heart disease. *Br. Med. J.* **311**, 171–4.

Barker, D.J.P. (1998). *Mothers, babies and health in later life,* 2nd edn. Churchill Livingstone, Edinburgh.

Bennett, A.J., Lesch, K.P., Heils, A., Long, J.C., Lorenz, J.G., Shoaf, S.E., Champoux, M., Suomi, S.J., Linnoila, M.V., and Higley, J.D. (2002). Early experience and serotonin transporter gene variation interact to influence primate CNS function. *Mol. Psychiatry* **7** (1), 118–22.

Bowlby, J. (1988). *A secure base: parent–child attachment and healthy human development.* Basic Books, New York.

Bradburn, N.M. (1969). *The structure of psychological well-being.* Aldine, Chicago.

Brake, W.G., Zhang, T.Y., Diorio, J., Meaney, M.J., and Gratton, A. (2004). Influence of early postnatal rearing conditions on mesocorticolimbic dopamine and behavioural responses to psychostimulants and stressors in adult rats. *Eur. J. Neurosci.* **19** (7), 1863–74.

Bredy, T.W., Grant, R.J., Champagne, D.L., and Meaney, M.J. (2003). Maternal care influences neuronal survival in the hippocampus of the rat. *Eur. J. Neurosci.* **18** (10), 2903–9.

Bredy, T.W., Zhang, T.Y., Grant, R.J., Diorio, J., and Meaney, M.J. (2004). Peripubertal environmental enrichment reverses the effects of maternal care on hippocampal development and glutamate receptor subunit expression. *Eur. J. Neurosci.* **20** (5), 1355–62.

Brickman, P., Coates, D., and Janoff-Bulman, R. (1978). Lottery winners and accident victims: is happiness relative? *J. Pers. Soc. Psychol.* **36** (8), 917–27.

Brown, K.W. and Ryan, R.M. (2003). The benefits of being present: mindfulness and its role in psychological well-being. *J. Pers. Soc. Psychol.* **84** (4), 822–48.

Brunstein, J.C., Schultheiss, O.C., and Grassman, R. (1998). Personal goals and emotional well-being: the moderating role of motive dispositions. *J. Pers. Soc. Psychol.* **75**, 494–508.

Caspi, A., Sugden, K., Moffitt, T.E., Taylor, A., Craig, I.W., Harrington, H.L., McClay, J., Mill, J., Martin, J., Braithwaite, A., and Poulton, R. (2003). Influence of life stress on depression: moderation by a polymorphism in the 5-HTT gene. *Science* **301**, 386–9.

Christensen, H., Griffiths, K.M., and Jorm, A.F. (2004). Delivering interventions for depression by using the internet: randomised controlled trial. *Br. Med. J.* **328** (7434), 265.

Cirulli, F., Berry, A., and Alleva, E. (2003). Early disruption of the mother–infant relationship: effects on brain plasticity and implications for psychopathology. *Neurosci. Biobehav. Rev.* **27** (1–2), 73–82.

Colhoun, H., Ben-Shlomo, Y., Dong, W., Bost, L., and Marmot, M. (1997). Ecological analysis of collectivity of alcohol consumpton in England: importance of average drinker. *Br. Med. J.* **314**, 1164–8.

Collins, W.A., Maccoby, E.E., Steinberg, L., Hetherington, E.M., and Bornstein, M.H. (2000). Contemporary research on parenting. The case for nature and nurture. *Am. Psychol.* **55** (2), 218–32.

Cox, B.D., Blaxter, M., Buckle, A.L.J., Fenner, N.P., Golding, J.F., Gore, M., Huppert, F.A., Nickson, J., Roth, M., Stark, J., Wadsworth, M.E.J., and Whichelow, M. (1987). *The Health and Lifestyle Survey: preliminary report of a nationwide survey of the physical and mental health, attitudes and lifestyle of a random sample of 9003 British adults.* The Health Promotion Research Trust, London.

Csikszentmihalyi, M. (1997). *Activity, experience and personal growth.* Human Kinetics, Champaign, Illinois.

Csikszentmihalyi, M. and Czikszentmihalyi, I.S. (1988). *Optimal experience: psychological studies of flow in consciousness.* Cambridge University Press, New York.

Danish, S.J. (1996). Going for the goal: a life skills program for adolescents. In *Primary prevention works* (ed. G.W. Albee and T.P. Gullotta), pp. 291–312. Sage Publications, Newbury Park, California.

Davidson, R.J. (2002). Anxiety and affective style: role of prefrontal cortex and amygdala. *Biol. Psychiatry* **51**, 68–80.

Davidson, R.J., Kabat-Zinn, J., Schumacher, J., Rosenkrantz, M., Muller, D., Santorelli, S.F., *et al.* (2003). Alterations in brain and immune function produced by mindfulness meditation. *Psychosom. Med.* **65**, 564–70.

Dawson, G., Ashman, S.B., and Carver, L.J. (2000). The role of early experience in shaping behavioural and brain development and its implications for social policy. *Dev. Psychopathol.* **12**, 695–712.

Diener, E. (1984). Subjective well-being. *Psychol. Bull.* **95**, 542–75.

Diener, E. (2000). Explaining differences in societal levels of happiness: relative standards, need fulfilment, culture and evaluation theory. *J. Happiness Stud.* **1** (1), 41–78.

Diener, E., Suh, E.M., Lucas, R.E., and Smith, H.L. (1999). Subjective well-being: three decades of progress. *Psychol. Bull.* **125**, 276–302.

Donovan, R. and Henley, N. (2004). *Social marketing: principles and practice.* IP Communications, Victoria, Australia.

Elliot, A.J., Sheldon, K.M., and Church, M.A. (1997). Avoidance personal goals and subjective well-being. *Pers. Soc. Psychol. Bull.* **23**, 915–27.

Emmons, R.A. and McCullough, M.E. (2003). Counting blessings versus burdens: an experimental investigation of gratitude and subjective well-being in daily life. *J. Pers. Soc. Psychol.* **84** (2), 377–89.

Forgas, J.P. (2001). *The handbook of affect and social cognition.* Erlbaum, Mahwah, New Jersey.

Francis, D.D., Champagne, F.C., and Meaney, M.J. (2000). Variations in maternal behaviour are associated with differences in oxytocin receptor levels in the rat. *J. Neuroendocrinol.* **12** (12), 1145–8.

Francis, D.D., Diorio, J., Plotsky, P.M., and Meaney, M.J. (2002). Environmental enrichment reverses the effects of maternal separation on stress reactivity. *J. Neurosci.* **22** (18), 7840–3.

Fredrickson, B.L. (2001). The role of positive emotions in positive psychology: the broaden-and-build theory of positive emotions. *Am. Psychol.* **56**, 218–26.

Fredrickson, B.L. (2004). Gratitude, like other positive emotions, broadens and builds. In *The psychology of gratitude* (ed. R.A. Emmons and M.E. McCullough), pp. 145–66. Oxford University Press, New York.

Fredrickson, B.L. and Branigan, C. (2005). Positive emotions broaden the scope of attention and thought–action repertoires. *Cogn. Emotion* **19** (3), 313–32.

Fredrickson, B.L., Mancuso, R.A., Branigan, C., and Tugade, M. (2000). The undoing effect of positive emotions. *Motivation and Emotion* **24**, 237–58.

Fredrickson, B.L., Brown, S., Cohn, M.A., Conway, A., Crosby, C., McGivern, M., and Mikels, J. (2004). Finding positive meaning and experiencing positive emotions builds resilience. In *The functional significance of positive emotions* (S. Brown and K.J. Johnson, Chairs), Symposium presented at the annual meeting of the Society for Personality and Social Psychology, Austin, Texas.

Freres, D.R., Gillham, J.E., Reivich, K., and Shatte, A.J. (2002). Preventing depressive symptoms in middle school students: the Penn Resiliency Program. *Int. J. Emerg. Ment. Health* **4**, 31–40.

Gardner, H. (1993). *Multiple intelligences: the theory in practice*. Basic, New York.

Gardner, H. (2000). *Intelligence reframed: multiple intelligences for the 21st century*. Basic, New York.

Goldberg, D.P. (1978). *Manual of the General Health Questionnaire*. NFER-Nelson, Windsor.

Gould, M.S. (2001). Suicide and the media. *Ann. NY Acad. Sci.* **932**, 200–21.

Greenberg, M.T., Domitrovich, C., and Bumbarger, B. (1999). *Preventing mental disorders in school-age children: a review of the effectiveness of prevention programs*, Report for Center for Mental Health Services. US Department of Health and Human Services, Washington, DC.

Hariri, A.R., Mattay, V.S., Tessitore, A., Kolachana, B., Fera, F., Goldman, D., Egan, M.F., and Weinberger, D.R. (2002). Serotonin transporter genetic variation and the response of the human amygdala. *Science* **297** (5580), 400–3.

Helliwell, J.F. (2003). How's life? Combing individual and national variations to explain subjective well-being. *Econ. Modelling* **20**, 331–60.

Henry, J. (2004). The relationship between personal development challenge, personality type and the perceived effectiveness of different interventions. Paper presented at the 2nd European Positive Psychology Conference, Verbania Pallanza, Italy.

Huppert, F.A. (2004). A population approach to positive psychology: the potential for population interventions to promote well-being and prevent disorder. In *Positive psychology in practice* (ed. P.A. Linley and S. Joseph), pp. 693–709. John Wiley and Sons Inc, Hoboken, New Jersey.

Isen, A.M. (1987). Positive affect, cognitive prcesses and social behaviour. *Adv. Exp. Soc. Psychol.* **20**, 203–53.

Judge, T.A., Thoresen, C.J., Bono, J.E., and Patton, G.K. (2001). The job satisfaction–job performance relationship: a qualitative and quantitative review. *Psychol. Bull.* **127** (3), 376–407.

Kabat-Zinn, J. (1990). *Full catastrophe living. Using the wisdom of your body and mind to face stress, pain and illness.* Bantam Doubleday Dell Publishing, New York.

Kabat-Zinn, J., Lipworth, L., Burney, R., and Sellers, W. (1986). Four-year follow-up of a meditation-based stress reduction program for the self-regulation of chronic pain: treatment outcomes and compliance. *Clin. J. Pain* **2**, 159–73.

Kabat-Zinn, J., Massion, A.O., Kristeller, J., Peterson, L.G., Fletcher, K.E., Pbert, L., Lenderking, W.R., and Santorelli, S.F. (1992). Effectiveness of a meditation-based stress reduction program in the treatment of anxiety disorders. *Am. J. Psychiatry* **19**, 936–43.

Kabat-Zinn, J., Wheeler, E., Light, T., Skillings, A., Scharf, M.J., Cropley, T.G., Hosmer, D., and Bernhard, J.D. (1998). Influence of a mindfulness meditation-based stress reduction intervention on rates of skin clearing in patients with moderate to severe psoriasis undergoing phototherapy (UVB) and photochemotherapy (PUVA). *Psychosom. Med.* **60** (5), 625–32.

Kahneman, D. (1999). Objective happiness. In *Well-being: the foundations of hedonic psychology* (ed. D. Kahneman, E. Diener, and N. Schwartz), pp. 3–27. Russell Sage Foundation, New York.

Kardum, I. (1999). Affect intensity and frequency: their relation to mean level and variability of positive and negative affect and Eysenck's personality traits. *Pers. Indiv. Diff.* **26**, 33–47.

Kasser, T. and Ryan, R.M. (1996). Further examining the American dream: differential correlates of intrinsic and extrinsic goals. *Pers. Soc. Psychol. Bull.* **22**, 280–7.

Keltner, D. and Bonanno, G.A. (1997). A study of laughter and dissociation: distinct correlates of laughter and smiling during bereavement. *J. Pers. Soc. Psychol.* **73**, 687–702.

Keyes, C.L.M. (2002*a*). Promoting a life worth living: human development from the vantage points of mental illness and mental health. In *Promoting positive child, adolescent and family development: a handbook of program and policy innovations* (ed. R.M. Lerner, F. Jacobs, and D. Wertlieb), Vol. 4 (15), pp. 257–74. Sage, Newbury Park, California.

Keyes, C.L.M. (2002*b*). The mental health continuum: from languishing to flourishing in life. *J. Health Soc. Behav.* **43** (2), 207–22.

Keyes, C.L.M. (2004). The nexus of cardiovascular disease and depression revisited: the complete mental health perspective and the moderating role of age and gender. *Aging Ment. Health* **8** (3), 266–74.

Kotler, P. and Zaltman, G. (1971). Social marketing: an approach to planned social change. *J. Marketing* **35** (3), 3–12.

Kumpfer, K.L. and Alvarado, R. (2003). Family-strengthening approaches for the prevention of youth problem behaviors. *Am. Psychol.* **58** (6–7), 457–65.

Laplante, D.P., Barr, R.G., Brunet, A., Galbaud du Fort, G., Meaney, M.J., Saucier, J.F., Zelazo, P.R., and King, S. (2004). Stress during pregnancy affects general intellectual and language functioning in human toddlers. *Pediatr. Res.* **56** (3), 400–10.

Layard, R. (2003). Happiness—Has social science a clue? Lionel Robbins Memorial Lectures 2002/3. London School of Economics and Political Science.

Layard, R. (2005). *Happiness: Lessons from a New Science.* Penguin, New York.

Lesch, K.P., Bengel, D., Heils, A., Sabol, S.Z., Greenberg, B.D., Petri, S., Benjamin, J., Muller, C.R., Hamer, D.H., and Murphy, D.L. (1996). Association of anxiety-related traits with a polymorphism in the serotonin transporter gene regulatory region. *Science* **274** (5292), 1527–31.

Levine, S. (1957). Infantile experience and resistance to physiological stress. *Science* **126**, 405–6.

Levy, B. (1996). Improving memory in old age through implicit self-stereotyping. *J. Pers. Soc. Psychol.* **71** (6), 1092–107.

Levy, B.R. and Banaji, M.R. (2002). Implicit ageism. In *Ageism: stereotypes and prejudice against older persons* (ed. T. Nelson), pp. 49–75. MIT Press, Cambridge, Massachusetts.

Lewis, J. (1996). *Triple P—Positive Parenting Program: evidence based action in child health promotion.* Report of the Eastern Perth Public and Community Health Unit, Western Australia.

Liu, D., Diorio, J., Tannenbaum, B., Caldji, C., Francis, D., Freedman, A., Sharma, S., Pearson, D., Plotsky, P.M., and Meaney, M.J. (1997). Maternal care, hippocampal glucocorticoid receptors, and hypothalamic–pituitary–adrenal responses to stress. *Science* **277** (5332), 1659–62.

Liu, D., Diorio, J., Day, J.C., Francis, D.D., and Meaney, M.J. (2000). Maternal care, hippocampal synaptogenesis and cognitive development in rats. *Nat. Neurosci.* **3** (8), 799–806.

Lucas, R.E., Clark, A.E., Georgellis, Y., and Diener, E. (2003). Reexamining adaptation and the set point model of happiness: reactions to changes in marital status. *J. Pers. Soc. Psychol.* **84** (3), 527–39.

Lucas, R.E., Clark, A.E., Georgellis, Y., and Diener, E. (2004). Unemployment alters the set point for life satisfaction. *Psychol. Sci.* **15** (1), 8–13.

Lykken, D. (2000). *Happiness: the nature and nurture of joy and contentment.* St Martin's Press, New York.

Lykken, D. and Tellegen, A. (1996). Happiness is a stochastic phenomenon. *Psychol. Sci.* **7** (3), 186–9.

Lyubomirsky, S. and Ross, L. (1999). Changes in attractiveness of elected, rejected and precluded alternatives: a comparison of happy and unhappy individuals. *J. Pers. Soc. Psychol.* **76**, 988–1007.

Lyubomirsky, S. and Tucker, K.L. (1998). Implications of individual differences in subjective happiness for perceiving, interpreting and thinking about life events. *Motivation and Emotion* **22**, 155–86.

Ma, S.H. and Teasdale, J.D. (2004). Mindfulness-based cognitive therapy for depression: replication and exploration of differential relapse prevention effects. *J. Consult. Clin. Psychol.* **72** (1), 31–40.

Maccoby, E.E. (2000). Parenting and its effects on children: on reading and misreading behavior genetics. *Annu. Rev. Psychol.* **51**, 1–27.

Maccoby, E.E. and Martin, J.A. (1983). Socialization in the context of the family: parent–child interaction. In *Handbook of child psychology*, 4th edn (ed. P.H. Mussen), Vol. 4. *Socialization, personality, and social development* (vol. ed. E.M. Hetherington), pp. 1–101. Wiley, New York.

Magen, Z. and Aharoni, R. (1991). Adolescents' contributing toward others: relationship to positive experiences and transpersonal commitment. *J. Hum. Psychol.* **31**, 126–43.

McEwen, B.S. (1998). Protective and damaging effects of stress mediators. *New Engl. J. Med.* **338**, 171–9.

Meaney, M.J. (2001). Maternal care, gene expression, and the transmission of individual differences in stress reactivity across generations. *Annu. Rev. Neurosci.* **21**, 1161–92.

Meaney, M.J., Aitken, D.H., Sharma, S., Viau, V., and Sarrieau, A. (1989). Postnatal handling increases hippocampal type II glucocorticoid receptors and enhances adrenocorticoid negative feedback efficacy in the rat. *Neuroendocrinology* **50**, 597–604.

Meaney, M.J., Diorio, J., Francis, D., Widdowson, J., LaPlante, P., Caldji, C., Sharma, S., Seckl, J.R., and Plotsky, P.M. (1996). Early environmental regulation of forebrain glucocorticoid receptor gene expression: implications for adrenocortical responses to stress. *Dev. Neurosci.* **18**, 49–72.

Miller, J.J., Fletcher, K., and Kabat-Zinn, J. (1995). Three-year follow-up and clinical implications of a mindfulness meditation-based stress reduction intervention in the treatment of anxiety disorders. *Gen. Hosp. Psychiatry* **17**, 192–200.

Oishi, S., Diener, E., Napa Scollon, C., and Biswas-Diener, R. (2004). Cross-situational consistency of affective experiences across cultures. *J. Pers. Soc. Psychol.* **86** (3), 460–72.

Penner, L.A., Shiffman, S., Paty, J.A., and Fritzsche, B.A. (1994). Individual differences in intraperson variability in mood. *J. Pers. Soc. Psychol.* **66** (4), 712–21.

Phillips, D.P. and Carstensen, L.L. (1986). Clustering of teenage suicides after television news stories about suicide. *New Engl. J. Med.* **315** (11), 685–9.

Puska, P., Vartiainen, E., Tuomilehto, J., Salomaa, V., and Nissinen, A. (1998). Changes in premature deaths in Finland: successful long-term prevention of cardiovascular diseases. *Bull. WHO* **76**, 419–25.

Putnam, R.D. (2000). *Bowling alone*. Simon and Schuster, New York.

Rose, G. (1992). *The strategy of preventive medicine*. Oxford University Press, Oxford.

Rosenzweig, S., Reibel, D.K., Greeson, J.M., Brainard, G.C., and Hojat, M. (2003). Mindfulness-based stress reduction lowers psychological distress in medical students. *Teach. Learn. Med.* **15** (2), 88–92.

Ryan, R.M. and Deci, E.L. (2001). On happiness and human potentials: a review of research on hedonic and eudaimonic well-being. *Annu. Rev. Psychol.* **52**, 141–66.

Ryff, C.D. (1989). Happiness is everything, or is it? Explorations on the meaning of psychological well-being. *J. Pers. Soc. Psychol.* **57**, 1069–81.

Ryff, C.D. and Singer, B.H. (1998). The contours of positive human health. *Psychol. Inq.* **9** (1), 1–28.

Sanders, M.R. (1999). Triple P–Positive parenting program: towards an empirically validated multilevel parenting and family support strategy for the prevention of behaviour and emotional problems in children. *Clin. Child Fam. Psychol. Rev.* **2**, 71–90.

Sanders, M.R., Montgomery, D.T., and Brechman-Toussaint, M.L. (2000). The mass media and the prevention of child behavior problems: the evaluation of a television series to promote positive outcomes for parents and their children. *J. Child Psychol. Psychiatry* **41**, 939–48.

Sanders, M.R., Turner, K.M., and Markie-Dadds, C. (2002). The development and dissemination of the Triple P–Positive Parenting Program: a multilevel, evidence-based system of parenting and family support. *Prev. Sci.* **3**, 173–89.

Schmidtke, A. and Hafner, H. (1988). The Werther effect after television films: new evidence from an old hypothesis. *Psychol. Med.* **18**, 665–76.

Seligman, M. (1991). *Learned optimism.* Knopf, New York.

Seligman, M. (2002). *Authentic happiness.* Free Press, New York.

Shapiro, S.L., Schwartz, G.E., and Bonner, G. (1998). Effects of mindfulness-based stress reduction on medical and premedical students. *J. Behav. Med.* **21**, 581–99.

Shaw, B.A., Krause, N., Chatters, L.M., Connell, C.M., and Ingersoll-Dayton, B. (2004). Emotional support from parents early in life, aging, and health. *Psychol. Aging* **19** (1), 4–12.

Sheldon, K.M. and Elliot, A.J. (1999). Goal striving, need satisfaction, and longitudinal well-being: the self-concordance model. *J. Pers. Soc. Psychol.* **76**, 482–97.

Sheldon, K.M. and Houser-Marko, L. (2001). Self-concordance, goal-attainment and the pursuit of happiness: can there be an upward spiral? *J. Pers. Soc. Psychol.* **80**, 325–39.

Sheldon, K.M. and Kasser, T. (1998). Pursuing personal goals: skills enable progress but not all progress is beneficial. *Pers. Soc. Psychol. Bull.* **24**, 1319–31.

Sheldon, K.M. and Lyubomirsky, S. (2004). Achieving sustainable new happiness: prospects, practices and prescriptions. In *Positive psychology in practice* (ed. P.A. Linley and S. Joseph), pp. 127–45. John Wiley and Sons Inc, Hoboken, New Jersey.

Silver, R.L. (1982). Coping with an undesirable life event: a study of early reactions to physical disability. Unpublished PhD dissertation. Northwestern University, Evanston, Illinois.

Speca, M., Carlson, L.E., Goodey, E., and Angen, M. (2000). A randomised, wait-list controlled clinical trial: the effect of a mindfulness meditation-based stress reduction program on mood and symptoms of stress in cancer outpatients. *Psychosom. Med.* **62**, 613–22.

Suh, E., Diener, R., and Fujita, F. (1996). Events and subjective well-being: only recent events matter. *J. Pers. Soc. Psychol.* **70**, 1091–102.

Tamminga, C.A., Nemeroff, C.B., Blakely, R.D., Brady, L., Carter, C.S., Davis, K.L., Dingledine, R., Gorman, J.M., Grigoriadis, D.E., Henderson, D.C., Innis, R.B., Killen, J., Laughren, T.P., McDonald, W.M., Murphy, G.M. Jr, Paul, S.M., Rudorfer, M.V., Sausville, E., Schatzberg, A.F., Scolnick, E.M., and Suppes, T. (1995). Developing novel treatments for mood disorders: accelerating discovery. *Biol. Psychiatry* **52** (6), 589–609.

Teasdale, J.D., Segal, Z.V., Williams, J.M.G., Ridgeway, V.A., Soulsby, J.M., and Lau, M.A. (2000). Prevention of relapse/recurrence in major depression by mindfulness-based cognitive therapy. *J. Consult. Clin. Psychol.* **68**, 615–23.

Van Eck, M.M., Nicholson, N.A., Berkhof, H., and Sulon, J. (1996). Individual differences in cortisol responses to a laboratory speech task and their relationship to responses to stressful daily events. *Biol. Psychol.* **43** (1), 69–84.

Weaver, I.C., Grant, R.J., and Meaney, M.J. (2002). Maternal behavior regulates long-term hippocampal expression of BAX and apoptosis in the offspring. *J. Neurochem.* **82** (4), 998–1002.

Weiss, L.H. and Schwarz, J.C. (1996). The relationship between parenting types and older adolescents' personality, academic achievement, adjustment, and substance use. *Child Dev.* **67** (5), 2101–14.

Weissberg, R.P., Kumpfer, K.L., and Seligman, M.E.P. (2003). Prevention that works for children and youth: an introduction. *Am. Psychol.* **58**(6/7), 425–32.

Wessman, A.E. and Ricks, D.F. (1966). *Mood and personality.* Holt, Rinehart and Winston, New York.

Whittington, J.E. and Huppert, F.A. (1996). Changes in the prevalence of psychiatric disorder in a community are related to changes in the mean level of psychiatric symptoms. *Psychol. Med.* **26**, 1253–60.

Williams, K.A., Kolar, M.M., Reger, B.E., and Pearson, J.C. (2001). Evaluation of a wellness-based mindfulness stress reduction intervention: a controlled trial. *Am. J. Health Promot.* **15** (6), 422–32.

Winter, L., Lawton, M.P., Casten, R.J., and Sando, R.L. (1999). The relationship between external events and affect states in older people. *Int. J. Hum. Dev. Aging* **50**, 1–12.

World Health Organization (WHO) (1948). *Preamble to the constitution of the World Health Organization.* [This definition remains unchanged.] WHO, Geneva.

Part 4

Cultural perspectives

Susan Verducci, PhD, is a research associate at the Stanford Center on Adolescence and an assistant professor of humanities at San Jose State University. A philosopher of education by training, she also co-directs a summer program that brings high school students to Stanford to study philosophy.

Howard Gardner is Hobbs Professor of Cognition and Education at the Harvard Graduate School of Education. He is the author of many books and articles, including, most recently, *Changing minds, Intelligence reframed*, and *The disciplined mind*. With Mihaly Csikszentmihalyi and William Damon, he directs the GoodWork Project, which is described in this chapter.

Photo by Jay Gardner (2002)

Chapter 13

Good work: its nature, its nurture

Susan Verducci and Howard Gardner

Introduction

The subject of work has been a mainstay in both economics and sociology. However, relatively little research has come out of psychology. Research in the psychology of work generally focuses on issues such as productivity, effectiveness, or the relationship between work characteristics and well-being, when well-being is measured in terms of job satisfaction and physical and mental health (Whitehall studies I and II). The characteristics of these latter studies include social relationships at work, the degree of control one has over one's work, and the balance of demand and reward. An understanding of what determines an individual's quality of work and how a worker derives meaning from that work has been neglected.

In 1995 three American psychologists, Howard Gardner, William Damon, and Mihaly Csikszentmihalyi, launched an empirical investigation of work that focuses on these issues. Now a decade later, these investigators, with their respective research teams at Harvard, Stanford, and Claremont Graduate University, are deeply engaged in what has come to be called the GoodWork Project (hereafter GWP). Guiding the GWP is a search for individuals and institutions that exemplify 'good' work—work that is at once excellent in technical quality and ethical. Of course, work can be good simply in a technical sense, or good in only an ethical sense, but our project is particularly interested in persons and institutions that manage to realize these dual connotations of 'good'. The GWP stands out from other studies then in that we seek to examine the psychological states of good work and their determinants.

In a volume that foregrounds the field of positive psychology, we also hypothesize that good work also 'feels good'—it harbors meaning for its practitioner who is likely to feel fully engaged by his or her daily undertakings. Although there is some evidence that good work evokes flow, or feelings of well-being associated with being engaged in a creative unfolding of something larger than one's self (Csikszentmihalyi 2003), we have not investigated this correlation systematically.

Still speculative is the extent to which this meaning is generated from the performance of excellent work or from feelings associated with acting ethically.

The pursuit of good work in our time

Even in the best of times, it is not easy to achieve good work. It proves especially challenging to do so in the US at the present time—when material conditions are changing very quickly, our sense of time and space is being radically altered by technology, market forces are powerful, and few if any counterforces of comparable power exist. Indeed, we believe that the enormous power of market forces, including the profit motive, constitutes the signature factor influencing the prevalence of good work in our time. We worry that, at times, the market's potentially destructive power can eclipse its benefits.

Good work, although certainly present in the US, rarely gains the sort of attention given to compromised or bad work. The havoc that the executives at Enron created in the lives of their workers, stockholders, and in the field of business more generally speaks to the worst sort of work. Stories of these executives filled the front pages of American and international newspapers for over a year. Journalism itself has much to be ashamed of even when technically excellent. The prioritizing of bottom-line issues has many newsrooms focusing on attracting readers and viewers rather than reporting news to these people. Further, a growing number of American journalists have been passing off fiction for fact in such venerable publications as *The New Republic* and *The New York Times*. Work in the field of philanthropy, a field that by nature should be beyond reproach, also has its share of dubious practices— work that is not illegal, but is ethically compromised. Questionable compensation packages and loans and gifts to foundation staff and board members have begun to cause talk of increasing regulation on Capitol Hill. It is all too easy to chronicle examples of bad work in most professions these days. There are fewer chronicled examples of work that is good in the dual sense of the word.

Further, the US finds itself in a situation where government regulations, ethics courses in professional training, and standards that are not enforced are proving inadequate in ameliorating the problems that compromised work creates. The pervasive presence of greed, the lack of professional morality, and the lack of focus on positive role models all take their toll on the behavior of current workers and, one can imagine, future workers. We fear a time when compromised work becomes the norm and good work the rare exception. Although this dystopic situation may not yet be the case, there is growing concern that the 'cheating culture' that David Callahan (2004) documents in America and Sue Cameron (2002) documents in the British elite will inevitably lead to more compromised and bad work. Particularly chilling are polls and studies that indicate that cheating in schools has become routine. The consequences of such a situation can be disastrous because healthy professions and healthy societies require individuals with character and competence working within institutions that facilitate good work.

The professions: a realm for conceptualizing good work

With this scenario in mind, GWP investigators elected to focus primarily on work in the professions. Consistent with common usage, we construe professions as collections of individuals who perform a service deemed important by the larger society. In return for performing this service in ways that are expert and disinterested, the professional is accorded a certain degree of autonomy and status from the rest of the society (Abbott 1983; Freidson 1994).

While it is useful to delineate the features of a profession, we should note that professionals and professional behavior are not coterminous. Alas, many professionals do not behave as good professionals; and sometimes a profession as a whole (such as accounting in the US since 2000) comes into disrepute because of widespread unprofessional behavior. On the happier side, any individual may choose to carry out work in a professional manner. It is quite possible that the individual who sets type for this manuscript or who drives the truck that brings the book to the bookstore carried out his or her work in ways that are excellent and ethical; and it also possible that some of the professional contributors to or readers of this essay may violate the basic codes of their chosen profession.

In conceptualizing professional realms, we have found it useful to delineate three components: the individual practitioner; the professional domain; and the professional field. The term *domain*, an epistemological one, refers to the ideas and codes that constitute the knowledge base of a profession. Thus, the Hippocratic oath delineates the domain of medicine. The term *field*, a sociological one, refers to the current roles and institutions with which the profession currently practices. In the US 50 years ago, most physicians were solo practitioners, whereas today most are members of large health maintenance organization; in the UK, the field is virtually coterminous with the National Health Service.

While there is no formula for producing good work, our research indicates that good work is most likely to come about when the principal stake-holders are in agreement about the importance of the work being carried out and the way in which it is currently being pursued. When current practitioners, the long-standing precepts of the domain, and the current institutions of the field are 'singing from the same hymn sheet', good work is more likely to occur.

Our study

In the GWP we seek to understand good work with the ultimate aim of promoting it. Accordingly, we examine the experiences of new and veteran professionals in a range of occupational domains. The domains surveyed so far include journalism, genetics, law, higher education, philanthropy, theater, and business—all but last two would generally be considered professions. Over the last decade, we conducted in-depth, primarily face-to-face, semi-structured interviews with approximately 1000 people. The workers range in age from adolescents setting forth on a career to veteran practitioners to 'trustees'—seasoned workers no longer actively practicing but deeply concerned about the health of their chosen calling. We interviewed our subjects about their missions, guiding principles, formative influences, greatest challenges, the strategies devised to meet these challenges, the changes taking place in their profession, and allied issues. In some cases we also posed specific dilemmas and administered an inventory of values. While much of our own work remains to be analyzed and studied, we have already published several books and many articles on our initial findings. These are cited and a number can be directly accessed at our website <www.goodworkproject.org>.

We located our subjects in a number of ways. In this chapter, all of the business workers cited individually and in the aggregate were nominated by experts in the field for doing excellent and ethical work. Those in higher education were nominated by a random sample of co-workers after their colleges and universities were nominated as 'good schools' by a range of experts. In the other fields we studied, we did not have such formal nominating processes. In some, we asked experts to tell us who were the people doing good work in the field and then took advantage of the snowball effect; these workers led us to other workers, who then led us to others. We also looked for subjects who had been recognized publicly for their work. We believe that most of our subjects are indeed good workers but, in any particular case, we cannot know for sure. Hence, we see our study as illuminating the phenomenon of good work, even if some of our informants turn out to violate one or more of its tenets. Subjects had the choice of anonymity or identification, and we adhere to their preferences here.

Our subjects came primarily from the US, but related investigations of good work are underway in Scandinavia and, we hope, in other parts of the world as well. Given that increasing globalization and Westernization are powerful trends, we hope that our findings will prove helpful to those concerned about similar issues in other countries.

We hypothesized that formative as well as environmental influences would prove important to the ways in which workers carried out their work. In what follows we delineate the factors that appear to contribute to the making of good workers or good work across various fields. We group these factors under three headings: the person-related or individual resources described by our subjects; the social resources encountered at work or in the wider society; and the various opportunities or barriers that are part-and-parcel of the particular occupational realm in which they have chosen to work. While in many cases the broad lines of support are not surprising, the particular trends encountered by specific persons working in specific fields are notable. Indeed, we believe that the power of our approach derives from its potential for uncovering the distinctive features that contribute to good work across several occupational domains at a particular historical moment.

Individual resources

We found that workers, and in particular good workers, brought a history of positive values, influences, and supports to their professional work. Not surprisingly, early experiences served to shape their career choices, aspirations, values, beliefs, and approaches to work. Mentioned across all the fields we studied, although present to differing extents in each, were formative influences such as family and religion.

Family

The importance of family came up in numerous ways in the stories that workers told. Families served as formative sources of positive values, good habits, and professional direction. In the case of many workers, family continues to influence their work by lending perspective. Young workers report struggling for a balance between work and family.

In our study of journalism, 32% of our subjects cited the families they grew up in when asked to discuss positive influences on their work. This influence was also seen clearly in the business people we studied. Referring to his mother and extended family, Mike Hackworth, Chairman and former CEO of Cirrus Logic Inc, said, 'Well she was an incredibly dedicated person . . . . Her whole life was around the kids and she had a set of values which she just stayed with throughout her entire life, and the consistency of that has had an effect on all of us, all the siblings, without a doubt.' Earlier in the interview, he noted that, '[T]he ethics of the farming community is about as fundamental as you get—and I've thought about that and I said, you know, a lot of the values that I must have, it's got to emanate from there. The fact that I had a lot of support, I'll use that term . . . a lot of attention, let me put it that way. A lot of attention from that [first] year through year four . . . had to have an extraordinarily positive impact on me.'

Family was not only a source of positive values; it was a place where good habits developed that were later used in their work. Veteran journalist, Cliff Radel, connected an important strategy he used at work to his father:

> I have a rule that I apply when I write columns that my dad would always apply. I call it that 'now wait a minute rule.' . . . we would be telling stories about what happened at school . . . He would say, 'now wait a minute'. He would turn the argument inside out. We would look at it from the other aspect of the other person. Or, from some totally different angle, some totally unbiased angle. He would get both sides of the story and turn things inside out.

Family provided career direction for workers as well. In a phase of our study that focused on young professionals, geneticists talked about how visiting their parents or older siblings at their labs influenced their professional direction. (Fischman *et al.*

2004, p. 73) Young journalists also frequently said that their motivating forces were supportive parents. A well-known veteran journalist told us that in high school 'I was invited to join the newspaper staff, and I didn't want to because I wanted to cheerlead . . . my mother informed me that I would be joining the newspaper staff.'

These workers, in turn, are concerned about their own families. Some subjects saw their families as lending balance and perspective to their work and have set clear boundaries for themselves.

> So the contract that I've always had with myself was this: I'll work as hard as I can work on the job, then I'm also going to go home and have my family life and my personal life. And if that's not good enough for the company, they have a choice. The choice is: they can ask me to leave. But I'm never going to win the award for the person who put in the longest hours on the job. And I can't let my ego feel bad about that, because I've chosen a different yardstick that I'm going to run my life by.

While priorities were clear with veteran workers, young workers across fields struggled to prioritize and maintain balance. Newly practicing geneticists work inordinately long hours in laboratories under pressure of fierce competition. Scientists working in academia feel the same pressures. A female professor spoke to us about how her struggles with infertility resulted from doing exactly what is required to excel in her field—getting grants and working in the lab. As reported by the geneticists, long hours in the lab have taken a toll on her family life. Indeed some subjects reported that they were considering leaving their chosen work precisely because it was impossible to achieve a desired balance in their lives.

Religious/spiritual values

Workers also drew upon the values and resources provided by their religions. Nowhere was this stronger than in our interviews in the field of business. Sixteen of our 17 workers nominated for the excellence and ethics of their work spoke of being inspired by religious values. Anita Roddick, founder of the Body Shop, trains her employees to do community service and described how five or six hundred staff travel to Albania and Romania each summer to educate children through play. Raised a Christian, she spoke of being influenced by the Bible and by what the nuns taught her. Having spent time in a kibbutz, she was moved by the Jewish sense of community. She also spoke of being inspired by priests working with the poor in the fields of El Salvador. In the wake of these values and her passion for social justice, she created an activist and service-oriented company.

Overwhelmingly, our business workers spoke about doing God's work through their companies. An active member in the Mormon Church, Richard Jacobson, told us, 'If my faith is such that you and I share a relationship in that sort of eternal, cosmic nature of things, then that teaches me that I should treat you in a way that I would want to be treated and so that has an effect on the way that I conduct my business.' Similarly, another businessman cited the Jewish faith and the notion of 'unity' as catalyst to his beliefs about work and doing good: 'It's a unification of the economic process with the ethical.' It's not one, and then the other, but all at once.

To varying degrees, religious values provided strength and professional guidance in higher education as well as business. From good workers who came from religious colleges, such as Mount St Mary's and Xavier Jesuit University, to colleges like Swarthmore founded on the Quaker tradition, it was not surprising that religious and spiritual values were part of how these workers conceived of themselves and their work. It was more surprising to find similar attitudes at other institutions, including a small liberal arts college in Minnesota. Only at Carleton College was a chaplain nominated by the general population as a good worker. It was also here that the clearest example of how religious values play out in the professional lives of workers came to light. Tammy Metcalf-Filzen, the women's basketball coach, told us,

> My faith is really important to me . . . . I believe I'm here for a purpose and that God wants me to use the gifts that I have to reach other people. It's not anything that I'm very verbal about here on this campus, but I hope that people see that the way I live my life is consistent with the Christian standards that I set for myself or that I believe are set for me. I believe I've been called to do this, to teach and to coach as kind of, maybe, a form of a ministry or a form of mission work. My parents were missionaries, so I grew up as a missionary kid . . . .

This nominated good worker connects these values directly to her work as a teacher and coach. At an out-of-state basketball game, her team worked in a soup kitchen preparing a Thanksgiving meal after a morning of basketball practice. On a trip she organized to Thailand, Metcalf-Filzen described basketball as a 'bridge' to learning about women's issues in Thailand.

> Prostitution is just crazy over there. And we spent six nights in a remote village, Kurin Village, in the foothills of the Himalayas in a . . . 'compound' isn't the right word, but like a hostel where there's a woman from India who is training young women in trades so that they don't go into prostitution . . . So, basketball was kind of the bridge.

She was not just 'missionary' in her approach to her professional work, she was exemplary in its technical deliverance. When we talked, she had recently led her team to win its division's championship, an unlikely feat for a small liberal arts college ranked fourth in its institutional category by *US News and World Report*.

Although present in all the domains we studied, the impact of religious values varied markedly across the professions. The topic came up much less frequently with geneticists, journalists, and philanthropists. In a survey of values, geneticists ranked religious ones near the bottom. Professional philanthropists, those who work in foundations, also did not speak much of religion as a source of strength and direction. Yet, some donors talked directly about the topic. Sir John Templeton even used his immense fortune to set up a charitable foundation whose mission is to explore the boundary between theology and science. His religious values permeate his work not only as a businessman, but also as a philanthropist.

We speculate that certain factors—like family values and practices—may be important across domains, while others may prove more fungible. There will always be people for whom religion and spirituality are not important. In addition to these people, we found individuals for whom the work itself was spiritual. According to Gregory Boyd, Artistic Director of the Alley Theatre in Houston, Texas, 'Acting is about being in the room with the audience and breathing the same air as the audience. Those are the relationships that have [to do with] organized religion as well. There are very few opportunities in our daily life where we sit in rows, agree to be silent, and pay attention to something else that's not us.'

In addition to traits of the individual, qualities specific to each field may factor into the differences we saw regarding religion. Particular domains may provide their own philosophy as an alternative to religion. Genetics provides one such example. The laws and procedures of science aim for truth, and its procedures require a certain philosophy pertaining to confirmation and validation of the truth. This stance is considered by many to be a philosophy that replaces religion. Indeed, Nietzsche meant that rationalism and science had become the dominant *Weltanschauung* when he declared, 'God is dead' (Nietzsche, 2001/1882, section 126). Many scientists speak in quasireligious terms about the beauty that they encounter in the natural or physical world and about the deep, virtually spiritual satisfaction they derive from discovering the basic laws of the universe.

Finally, there may be an interaction between individuals and the institutions they choose to work in. Journalism and science might attract workers who are naturally more skeptical. Theater or philanthropy might attract non-religious persons seeking other forms of spirituality.

Social resources

Resources from workers' pasts were supplemented by the relationships that they share with others in their professions. The workers we interviewed relied on two complementary sources: horizontal and collaborative networks of peers and colleagues; and vertical support from authority figures and mentors who inspire them to do their work well.

Collaboration—peers and colleagues

All fields we studied require that workers cooperate to varying degrees with their peers and colleagues. Some, like theater and higher education, require intense collaboration. These realms feature formal and structural requirements for collaboration related to the goals and means of the endeavor. Even a 'one-person' theatrical project requires an actor to collaborate with a director, designers, technical theater workers, agents, and marketers. According to GWP researcher, Paula Marshall,

> The essence of theater is in the relationships of those who work on stage and off; it is marked by their ability to share their knowledge, to communicate effectively with each other, and to

pursue their common goals with originality and boldness. Creating a strong ensemble is demanding work. It requires that those involved take risks, have the courage to fail, trust their colleagues, and defend and mediate their artistic visions with one another . . . . [Marshall, 2002, p. 18]

She quotes artistic director George Wolfe,

A good collaboration is one in which there are very big egos in the room, very talented people, gifted people, people with vision. People who are possessive of their vision and insistent on seeing it materialize in the performance. People who are tenacious, and people who have a great deal of stamina. Those are, I would say, the major requirements. Then you put all those people in a room and let them mix it up and fight and make decisions and enjoy one another and respect each other. But without the first you'll end up with sort of a low common denominator, where everyone can agree—that isn't a good collaboration. [Marshall, 2002, p. 23]

Institutions of higher education also require intensive collaboration but for a different reason: they are self-governing. Committees are formed and groups of people come together to run departments, schools, and the institution as a whole. Without being asked directly, 66% of our subjects at Swarthmore described the process of decision-making as coming to consensus. Whereas consensus was seen as the death of collaboration for director Wolfe and many other artists, it is honored at this Quaker institution. Carr Everbach, a professor in the engineering department, said,

We operate most of the time, almost all the time, on the view that we only make big decisions after a lot of consensus. And after everyone has a chance to—either everyone has a chance to put his or her oar in to be heard, and then to acclimate him or herself to the will of the majority, or to dissent from it respectively but not angrily or with defiance. Let's say everyone eventually is either resigned or encouraging of decisions we make.

Another sort of collaboration occurs in theater and higher education: collaboration with those outside the workers' particular project or institution. For faculty to proceed in their respective disciplines, whether it be genetics or English, they collaborate with others through journals, professional associations, and book writing. The creation of knowledge, regardless of discipline, is tested in collaboration with others. Likewise, theater artists collaborate and gauge the success of their work by the standards of other artists and the audience. The artistic director of a well-known off-Broadway theater remarked, 'When people talk about how much they've enjoyed the play, that's what gauges me. When people come up and start talking about how the play affected them, then that's when I feel I've done a good job. It's when I'm doing what I'm supposed to be doing.'

We also studied the field of organized philanthropy, looking in particular at workers in foundations whose assets exceed US $500 000 000. Within this group, collaboration ranked highest as a recommended strategy for good work. Given organized philanthropy's absolute reliance on the work of nonprofit organizations to accomplish its own mission, it is logical that collaboration was highly valued.

Said one president of a large foundation, 'The good foundation recognizes that standing alone it rarely accomplishes anything . . . . When you think about it, it facilitates, it abets, it brokers, in encourages, it supports, and it celebrates the work of others.' Without nonprofits, organized philanthropy loses its principal rationale and its principal partners. Eighty-three per cent of the subjects we interviewed in philanthropy described some form of collaboration as critical.

Even though collaboration was cited as a primary characteristic of good work, philanthropists and grant-makers are hard-pressed to collaborate well with the nonprofits they fund. An enormous power differential exists in most philanthropic collaborations because foundations have money that nonprofits require to survive. At times this imbalance can stymie a nonprofit's ability to implement its own good work. To satisfy what grantors see as important, nonprofits shift from their own missions. The foundation-driven initiative, currently popular, is a glaring example of what can color any type of collaboration between funders and grant-seekers.

Traditional domains such as journalism and law have not required the intense collaboration described above. Nor do they foreground the same co-dependent relationship that permeates philanthropy. Although there may be teams of people working with journalists and lawyers, we usually think of these workers as single actors in their domains. Despite the substantial teams amassed in the notorious O.J. Simpson trial, Marcia Clark was seen as *the* prosecutor and Johnnie Cochran as *the* defense lawyer. In spite of the camera people, editors, and the like backing her up, it is Diane Sawyer whom we see as *the* journalist. Even geneticists who work together in labs generally do so under the guidance of a single principal investigator, the structure being more hierarchical than that of theater and higher education. As fields like philanthropy, journalism, and genetics continue to evolve, it will be instructive to observe the evolving importance of horizontal collegiality.

Authority figures and mentors

Good workers bring strong vertical supports to their work in the form of authority figures and mentors. In fields such as genetics, law, and higher education, mentors tend to come as part of a worker's professional training. Apprenticeships are mandatory in these fields and training includes college and graduate school. In the case of genetics, vertical support can span a series of postdoctoral positions. Mentors included faculty advisors, senior scientists, and professors. Business, philanthropy, theater, and journalism, on the other hand, do not rely on professional training: one can become a worker—even a good worker—in these fields without any formal training whatsoever. Although not dismissive of training, the majority of these workers got their most important training and mentoring on-the-job.

Of the areas studied, philanthropy offers the least formal training. Although there are numerous training programs in nonprofit management, there are few programs for people interested in philanthropy. In response to the growing need for training, a cottage industry of short-term workshops has arisen to fill the void. None of the workers we interviewed found their mentors in such locales.

What field-specific mentoring does occur in philanthropy happens on the job. 'You have to learn from the ground up because there's no training, there's no apprenticeship', Ray Bacchetti, formerly of the William and Flora Hewlett Foundation, reported, there's just, 'Come in, you gotta give away ten million dollars. And make sure you don't make a mistake.'

There are many reasons for the lack of professional training and mentoring. Organized philanthropy is a relatively new undertaking, just over 100 years old. There are no accrediting bodies and few substantive legal regulations. The accepted body of literature on the field is small, and people still question whether philanthropy is a profession at all. Somewhat surprisingly, we found that the majority of donors and grant-makers we interviewed disdained the 'professionalization' of philanthropy. Most working 'professionals' were either related to the donor or came from successful careers in academia or grant-seeking nonprofits. Many cycled into philanthropy from these other arenas and then cycled back after a number of years. Donors, on the other hand, hailed primarily from the business sector and found their mentoring there. It is not surprising, then, that the people we interviewed were much more likely to identify mentors from outside the field than inside.

Although most journalists we interviewed had professional training, few found mentors there. Nor did many find mentors in the field. Despite the clearly defined and widely understood values and codes of professional behavior, the field does not possess well-regulated training and supervisory systems.

In contrast to both philanthropy and journalism, the field of genetics provides highly elaborate structures for training and mentoring. Fischman *et al.* (2004, pp. 144–5) write,

> [A]pprenticeships for aspiring scientists are essentially mandatory, lengthy, and fraught with uncertainties. Attending and excelling in college and graduate school are no longer sufficient. Nowadays, young scientists often do several postdoctoral stints. In the course of ten or more years of tertiary education, these students acquire a great deal of knowledge, skill, and networking links. They often have the opportunity to work with several distinguished mentors. Yet only after a decade of training are they finally given the chance to set up their own labs, devise their own research program, secure grants, hire assistants and launch the careers for which they have been preparing.

All the workers we talked with drew upon mentors and authority figures to aid their work or wished they had them to draw upon. Where the lucky ones found them depended upon not only the training required by the domain, but also on the structures of the actual workplace itself. When training associates required, mentors tend to come from the pools of people with whom the trainee associates with regularly during this period. When training is on-the-job, mentors may come from both inside and outside the field. On the whole, mentoring is more fitful in these latter domains.

To be sure, not all mentors and colleagues were exemplary professionals. Several workers told us they also drew on examples of compromised work and bad workers as part of compensatory efforts to keep their own compasses pointing in the right direction. We characterize this use of bad and compromised work as one form of 'inoculation'. Even if inclined toward good work, workers needed booster shots from encounters with other good workers or antibodies to counter negative lessons from scandals. The cases of the journalist who wrote fiction and plagiarized in *The New York Times* and executives at Enron not only help workers within these two fields understand ethical and legal limits; these exemplars have the potential to remind *all* workers of the limits in their own work. On a more positive note, workers may be reinforced in their inclinations toward good work when they learn of the achievements of a broadcast journalist like Edward R. Murrow, who covered the most important stories at mid-century and risked his career to denounce the antidemocratic actions of Senator Joseph McCarthy, or the achievements of a physician like Albert Schweitzer, who gave up a career as a world class organist to minister to the poorest inhabitants of the Belgian Congo.

The market and its restraints

In our comprehensive study, as noted above, we found that good work is most likely to emerge when all of the interest groups involved in a sector have the same goals (Gardner *et al.* 2001). Studied in the late 1990s, the realm of genetics emerged as well-aligned within the US, that is, the individual scientists, the values of science, the current institutions of the field, the shareholders of for-profit biotech companies, and the stake-holders of the wide society basically desired the same thing: longer and healthier lives. Although there may have been differences regarding the means to these ends and conflicts pertaining to property rights about genetic discoveries, so long as the scientists were advancing toward these ends, there were few constraints on their behavior. In the late 1990s, virtually the only obstacles to good work were scientists' own behaviors and values.

In contrast, journalism in America in the late 1990s emerged as a sector that was poorly aligned. Most journalists wanted to pursue stories in a careful and deliberate manner, to verify their sources, to focus on stories that they themselves considered important and in need of scrupulous documentation. But they found themselves in a milieu that exerted enormous pressures to report quickly, to cut corners, to focus on the dramatic and the horrific ('If it bleeds, it leads'), and to avoid stories that were complex, difficult to unravel, and might possibly embarrass the owners or the advertisers. As a result, these journalists felt unhappy, frustrated, and unable to pursue good work, and many of them were looking for a way to escape the profession as currently practiced.

Many factors contribute to this discouraging state of affairs, and no party is immune from criticism. According to our analysis, economic markets have exerted the most powerful influence on the news media. During the time when news reporting was not seen as a leading profit center to owners and they were content with smaller profit margins, journalists were given more resources and more leeway in their reporting. But, as the news media become increasingly owned by large, and at times multinational conglomerates, the news has become seen (and relied upon) as a source of profits each year. The journalists we talked with perceive important consequences of this fact: newsrooms now cut more corners and focus on gaining readers and viewers. A side-effect of this latter consequence has been the glamorization of news content. These journalists see these negative effects of market forces as exceedingly difficult to counter. As we document in the books, *Good work* and *Making good*, good work is all-too-often sacrificed in the process of making money.

Of course, in any capitalistic society, markets will be powerful—and markets are neither moral nor immoral in and of themselves. We think that the current era is distinguished by the lack of forces that can counter or temper today's markets. With strong family, religious, and communal values on the wane in many parts of the society and with the muting of governmental regulation, the line between a putative profession like journalism and the routine practice of business is erased. As famed editor Harold Evans has remarked, 'The problem many organizations face is not to stay in business but to stay in journalism' (quoted in Gardner *et al.* 2001, p. 131). Market forces have taken a large toll on what the journalists we interviewed perceived to be their field's mission. Joe Birch, a broadcast journalist from Tennessee, summed up the sentiments of many:

> Because the temptation, see, is to get an audience by having all these lurid stories and some celebrities. And say to yourself, 'Well see, look at our ratings. Isn't that wonderful?' And that is an abdication of our responsibility. While news can be entertaining, that's not our job, to be entertainers. Our job is to be informers . . . And that's a tremendous challenge today because these forces of infotainment . . . are crashing through the door. And the ratings are imperative, you have to have them or you don't survive, generally. Or if you do survive, you survive at a very meager level.

The incessant push for greater market share combined with a retreat from in-depth coverage of serious stories are two primary sources of pessimism. When asked about change in the field, a full 64% of the journalists we interviewed noted the increasing importance of complying with the business goals of the industry and 63% perceived a decline in values and ethics within the field. Only 24% felt positive about the changes that have been happening in the field.

The sorts of pressures described by journalists can be discerned across the board in the fields we studied. Starting closest to home, it is no surprise that American institutions of higher education feel sharp pressure from the market. Public institutions are particularly hard-pressed as state governments become less willing to fund the bulk of their operations at the very time that operating costs are rising dramatically. To be sure, some positive outcomes *may* result from the pressure. But it is not the positives that the workers we interviewed focused on; they spoke primarily of the negatives.

Examples abound of how pressures to make money for the university, or in private industry for that matter, can adversely affect the work that scientists do. Many of the young geneticists express concerns about having data 'scooped' by another researcher or lab and feel pressure to release information before others working on the same topics release theirs. The pressure to be first makes them wary about data accuracy, i.e. inappropriate characterization of data or overstatement of findings. Graduate students in particular feel uncertain about their role in speaking out against such possible ethical breaches. (Fischman *et al.* 2004, p. 99)

The pressures wrought by commodification and marketization can be found in all the fields we studied with the possible exception of philanthropy. Philanthropy is relatively insulated from what typically affects other fields. Of course, assets are invested in the market and when it is down, there may be less money to give and perhaps people will be laid off. Also, the 'model of the market' affects foundations when they seek to fund only the most visible persons or the 'sexiest' projects or when they spend a significant amount of money on publicizing their 'good works'. But market forces mostly skip over foundations to land heavily on the nonprofits they fund. Less money interest earned means less money given away. The foundation itself survives.

Turning to the sphere of values, our veteran informants viewed the market as fostering destructive competitive practices, a dog-eat-dog mentality, greed, and a resulting dilution of their traditional missions and standards. Moreover, our study of young workers documented that these aspiring workers would like to do good work but they feel that they cannot afford to do so at the present time. They want to succeed and so many of their competitors are cutting corners that they feel that they must compromise for the time being. When they achieve success, they told us, *then* they will revert to admirable professional behavior. Of course, this reasoning embodies a classical ethical fallacy: the end justifies the means.

In our view, the structure of the field offers the strongest protection against the unmitigated and destructive pressures of the market. Certain professions, like law and medicine, stand out in terms of their relatively strong and well-defined structures. They each possess accepted bodies of knowledge and procedures for transmitting the knowledge, as well as codes of ethical standards that have teeth. It is important to note that these are relatively old professions. It is easy to spot compromised work in professions that have existed for some time and whose principal values are widely known and widely shared. In these strong domains, we find consensual processes of training, recognized mentors, and established procedures in place for censuring or ostracizing those whose work violates norms of the domain, with the ultimate sanction a process of disbarment or loss of license.

Universities are also ancient institutions. However, the professional desiderata of professors and other teachers are less well defined and, as we have seen, currently most universities have relatively few protections against the market. Science, a more recently emerging domain, has strong mentoring and supervisory systems. As long as it remains outside of the commercial realm, or its goals fit comfortably with those of investors, then scientists have relatively smooth sailing to carry out their research. Like science, journalism also has strong values and ethical principles, but it lacks the licensure and training structures of the more venerable professions. In theater it is not clear there are any ethical principles that individuals must abide by. Perhaps the arts are more like business, where people can choose to behave as if they are professionals, but there are few extrinsic rewards for doing so and few punishments for not doing so as long as one does not break the law. Also like business, theater is closely tied to the market in terms of accountability, with few external incentives to do good in the ethical sense.

As noted, organized philanthropy lacks salient domain features. It is young, not yet organized, and without an accepted body of knowledge to transmit to new practitioners. Moreover, it is only loosely affected by external forces, including the government and the market. The head of a New York foundation captured philanthropy's isolation, '[T]he basic problem in philanthropy is that the people who give away the money are in no way accountable to any quantifiable success. They can do a really lousy job, they can have counterproductive effects and nobody's going to stop them as long as their intentions are good.'

If the constitution of the domain is the strongest protector against compromised work, it is not the only one. Fifty years ago, it was said that the problem with American newspapers was that they were owned by families; if only they were publicly owned, the argument went, these media outlet would be more reliable and more accountable. Precisely the opposite has occurred. It is generally agreed that the best American newspapers—for example, *The Washington Post, The Wall Street Journal*, and *The New York Times*—are among the few that are still largely under family control. When a family's name and reputation is on the line, that fact may in itself encourage—though obviously it does not ensure—good work.

Finally, we should point out that, in a democratic society, any individual or group of individuals has the power to begin a new institution that embodies—or reembodies—the core values of a field. That is what has happened time and again in the news media, and it is happening as well in professions, ranging from medicine to accounting. Anthropologist Margaret Mead provides the classic formula: 'Never doubt that a small group of committed people can change the world. Indeed it is the only thing that ever has.'

Towards a culture of good work

How can a milieu of good work be achieved, particularly when a profession or occupational realm has weak structural features? As we have shown above, certain variables appear to affect specific aspects of good work. For example, family and religious/spiritual values are linked to ethical work, while social resources appear to impact excellence. In our project, we also speak of the four Ms that help to propagate good work. These were initially designed to address individuals, but they can be applied as well to institutions and even entire professions. The Ms seek answers to the following questions: (1) What is the Mission of our field? (2) What are the positive and negative Models that we must keep in mind? (3) When we look into the Mirror as individual professionals, are we proud or embarrassed by what we see? and (4) When we hold up the Mirror to our profession—or, indeed, our society—as a whole, are we proud or embarrassed by what we see? And, if the latter, what are we prepared to do about it? We suggest that if individuals and institutions examine the four Ms, they would be more likely to be on the road to doing good work.

Going beyond exhortation, we have ourselves undertaken more applied projects. We have partnered with the Committee of Concerned Journalists to bring educational workshops to print, broadcast, and television newsrooms all over the US. This curriculum has had enormous success in raising consciousness about issues of good work and the strategies by which it can be achieved. We have spoken across Europe and Asia about GoodWork, as well as to numerous professional and quasiprofessional associations in the US. We are also in the process of creating a GoodWork Toolkit to be used in classrooms from kindergarten through secondary school.

We recognize that good work is more demanding than average or compromised work. We take heart from the fact that professionals in most domains do have a clear vision of what constitutes good work and how it might be achieved, even as they recognize the challenges to carrying out such work in the current milieu. As we have noted, good work is most readily carried out when workers bring a host of individual and social resources to their occupations, as well as when the various interest groups are well-aligned, that is, when the long-term professional mission, the individual practitioner, the current institutions and gatekeepers, the share-holders (in the case of for-profit enterprises), and the society's stake-holders all want the same thing from current work.

In contrast, when wide disparities exist within and across these interest groups, good work is elusive. Nonetheless, even under less favorable conditions, some individuals and institutions succeed in carrying out good work and these exemplars serve to inspire others. Moreover, some individuals are actually stimulated by misalignment and proceed to 'right' the institutions in which they work or to devise new ones that realize their ideals.

Conclusion

Though it began as a research endeavor, the principal goal of the GoodWork Project is to help to bring about the scenario of more good work across the occupational landscape. Occupations will always feel pressures of one type or another and, at the time of powerful market forces, these pressures can be decisive. The forces cannot be ignored; they must be negotiated, not succumbed to. Individuals, institutions, and professions that actively negotiate these forces, while adhering to the central and irreplaceable values of the domain, are most likely to survive, to thrive, and to carry out work that is excellent and ethical.

Acknowledgements

We owe a debt of gratitude to a number of foundations and individuals for their generous support of the GoodWork Project. Initial grants were provided by the William and Flora Hewlett Foundation, the Ross Charitable Foundation, and the Carnegie Corporation of New York. These grants were later supplemented by the Ford Foundation, the John Templeton Foundation, the Christian Johnson Endeavor Foundation, the Atlantic Philanthropies, the Bauman Foundation, and the Pew Charitable Trusts. Louise and Claude Rosenberg and Jeffrey Epstein have also supported our project in a number of its phases. For helpful suggestions on an early draft of this essay, we thank William Damon and the editors of this volume.

References

Abbott, A. (1983). Professional ethics. *Am. J. Sociol.* **88** (5), 855–85.

Callahan, D. (2004). *The cheating culture: why more Americans are doing wrong to get ahead.* Harcourt Press, Orlando, Florida.

Cameron, S. (2002). *The cheating classes: how Britain's elite abuse their power.* Simon and Schuster, London.

Csikszentmihalyi, M. (2003). *Good business: leadership, flow and the making of meaning.* Viking Penguin, New York.

Fischman, W., Solomon, B., Greenspan, D., and Gardner, H. (2004). *Making good: how young people cope with moral dilemmas at work.* Harvard University Press, Cambridge, Massachusetts.

Freidson, E. (1994). *Professionalism reborn: theory, prophecy, and policy.* Polity Press, Cambridge, England.

Gardner, H., Csikszentmihalyi, M., and Damon, W. (2001). *Good work: when excellence and ethics meet.* Basic Books, New York.

Marshall, P. (2002). Goodwork in Theater: Report to the Hewlett Foundation. Unpublished.

Nietzsche, F. (2001/1882). *The gay science.* Cambridge University Press, Cambridge.

Whitehall Studies I and II. <http://www.workhealth.org/projects/pwhitepub.html.>

Robert J. Sternberg is IBM Professor of Psychology and Education and director of the Center for the Psychology of Abilities, Competencies, and Expertise at Yale University. His PhD is from Stanford University and he holds five honorary doctorates. His main research interests are in human intelligence, creativity, wisdom, thinking styles, and leadership.

Elena L. Grigorenko is associate professor of child studies and psychology at Yale and associate professor of psychology at Moscow State University. She has received awards for her work from five different divisions of the American Psychological Association, and in 2004 won the APA Distinguished Award for an Early Career Contribution to Developmental Psychology.

Chapter 14*

Intelligence and culture: how culture shapes what intelligence means, and the implications for a science of well-being

Robert J. Sternberg and Elena L. Grigorenko

Introduction

The field of intelligence is relatively old. It has made some mistakes. In particular, its practitioners have often assumed that what applies to one culture applies to another. It is important that the much newer field of positive psychology does not repeat these mistakes: that in attempting to understand well-being, it understands intelligence in its multicultural context. Moreover, it is important that the field of positive psychology understands how intelligence, broadly defined, is mostly an attempt to use one's cognitive skills to achieve a state of well-being within one's own cultural context. Intelligence is always displayed in a cultural context. The acontextual study of intelligence imposes a (usually Western) investigator's view of the world on the rest of the world. Can research provide an understanding of intelligence that is not so culturally constrained? Can it help us to understand the role of intelligence in well-being? We address these questions in this chapter.

The chapter is divided into four parts. In the first section, we introduce our main ideas. In the second section, we briefly present the theory of successful intelligence, which underlies our work (Sternberg 1985, 1990, 1997, 1999). In the third section, we discuss cultural studies relevant to these ideas. Some of these studies ask people to behave intelligently, whereas others query people as to their conceptions of what it means to behave intelligently. In the final section, we draw some conclusions. Our own personal experiences motivated our interest in the interface between culture and intelligence. Three experiences were particularly instrumental.

The first experience occurred during our work in Jamaica in the mid-1990s. R.J.S. was sitting in a school listening to a lesson. The school was situated in one big room, such that each 'classroom' was merely a section of that room. There were no partitions between class groupings. Each teacher thus had to talk over the voices of the other teachers. R.J.S. was seated towards the edge of one of the class groups and realized that he could hardly hear the teacher whose class he was supposed to be observing. Indeed, he could better hear the teacher of another class that was proximal to the class he attended, and realized that many of the other children who were not near to the teacher of their own group had the same problem. How could the children maximally profit from instruction that they could scarcely hear? How could their achievement equal that of the children who were better situated in the classroom? And how could they possibly equal the performance of Western children who actually had their own walled classroom in which to listen to the teacher?

The second experience occurred in India in late 1995 (Sternberg and Grigorenko 1999). We were carrying out research in a school. It was 113° F (45°C) in the shade. The stench of surrounding litter, excrement, and assorted waste was overwhelming. E.L.G. was asking a child to solve a linear syllogism (e.g. one relating the heights of three children to each other). Upon hearing E.L.G. present a problem, R.J.S. thought to himself that she had made a mistake: it seemed that the problem she had just presented was indeterminate and had no solution. However, the young child to whom she presented the problem proceeded successfully to solve it. R.J.S. had made a mistake in trying to solve a very simple problem that a young child could solve. He realized that the kinds of teaching and testing conditions that apply in most of the developed world, however defective they may be, scarcely compare with those in the developing world. Anyone can be affected by such conditions, the uninitiated, like R.J.S., more than others. How often is any kind of test given in the developing world in conditions even approaching these?

The third experience occurred while we were doing research in Tanzania at the turn of the century. This experience truly gave new meaning to the concept of bad conditions for testing. The building in which we were testing collapsed at the time of testing! How could children possibly perform at a maximal level when they could not even count on the structural integrity of the building in which they were working? The testing had to be abandoned at that time because of the loss of the building.

These experiences suggested to us that intelligence, considered outside its cultural context, is in large measure a mythological construct. There are some aspects of intelligence that transcend cultures, namely, the mental processes underlying intelligence and the mental representations upon which they act. For example, individuals in all cultures need to recognize and define problems, formulate strategies to solve these problems, monitor and evaluate these strategies, and so forth. The nature of the problems may differ, but there are always problems, regardless of where or when one lives. One's skill in solving these life problems contributes to one's well-being, but the operations that one performs to solve problems gain expression in performance differently from one culture to another. As soon as one assesses performance, one is assessing mental processes and representations in a cultural context. How do these contexts manifest themselves?

Most psychological research is executed within a single culture, but we believe that single-cultural studies, in some respects, do an injustice to psychological research. In particular they: (1) introduce limited and often narrow definitions of psychological phenomena and problems; (2) engender risks of unwarranted assumptions about the phenomena under investigation; (3) raise questions about cultural generalizability of findings; (4) engender risks of cultural imperialism: the belief that one's own culture and its assumptions are somehow superior to other cultures and their assumptions; and (5) represent missed opportunities to collaborate and develop psychology and psychological understanding around the world.

Some investigators have realized the importance of cultural context (for reviews of relevant literature see Laboratory of Comparative Human Cognition 1982; Serpell 2000; see also Greenfield 1997). For example, Berry (1974) reviewed concepts of intelligence across a wide variety of cultural contexts. Carraher *et al.* (1985; see also Ceci and Roazzi 1994; Nuñes 1994) studied a group of children in whom intelligence as adaptation to the environment was especially relevant. This was a group of Brazilian street children, who are under great contextual pressure to form a successful street business. If they do not, so-called 'death squads' may murder children who, unable to earn money, resort to robbing stores (or who are suspected of resorting to robbing stores). They found that the same children who were able to do the mathematics needed to run their street business were often little able or unable to do school mathematics. In fact, the more abstract and removed from real-world contexts the problems were in their form of presentation, the worse the children did on the problems. These results suggest that differences in context can have a powerful effect on performance.

Such differences are not limited to Brazilian street children. Lave (1988) showed that Berkeley, California housewives who could successfully do the mathematics needed for comparison shopping in the supermarket were unable to do the same mathematics when they were placed in a classroom and given isomorphic problems presented in an abstract form. In other words, their problem was not at the level of mental processes but at the level of applying the processes in specific environmental contexts.

The theory of successful intelligence provides a way of understanding these and other results.

The theory of successful intelligence

The nature of successful intelligence

In the theory of successful intelligence, intelligence is defined as one's ability to achieve success in life in terms of one's personal standards, within one's sociocultural context. The field of intelligence has, at times, tended to 'put the cart before the horse', defining the construct conceptually on the basis of how it is operationalized rather than vice versa. This practice has resulted in tests that stress the academic aspect of intelligence, as one might expect, given the origins of modern intelligence testing in the work of Binet and Simon (1916) in designing an instrument that would distinguish children who would succeed from those who would fail in school. However, the construct of intelligence needs to serve a broader purpose, accounting for the bases of self-defined success throughout one's life.

The use of societal criteria of success (e.g. school grades, personal income) can obscure the fact that these measures of performance often do not capture people's personal notions of success. Some people choose to concentrate on extracurricular activities such as athletics or music, and pay less attention to grades in school; others may choose occupations that are personally meaningful to them but that will never yield the income that they could gain by doing work that is less personally meaningful. In the theory of successful intelligence, the conceptualization of intelligence is individually determined but always occurs within a sociocultural context. Although the processes of intelligence may be common across such contexts, what constitutes success is not. Being a successful member of the clergy of a particular religion may be highly rewarded in one society, but viewed as a worthless pursuit in another culture.

In the theory, one's ability to achieve success depends on the capitalization of one's strengths and correction or compensation for one's weaknesses. Theories of intelligence typically specify some relatively fixed set of abilities, whether this be one general factor and several specific factors (Spearman 1904), seven multiple factors (Thurstone 1938), eight multiple intelligences (Gardner 1983, 1999), or 150 separate intellectual abilities (Guilford 1982). Such a way of looking at intelligence may be useful in establishing a common set of skills to be tested. People achieve success, even within a given occupation, in many different ways. For example, successful teachers and researchers achieve success through many different blendings of skills rather than through any single formula that works for all of them.

The theory states that a balancing of abilities is achieved so as to adapt to, shape, and select environments. Definitions of intelligence traditionally have emphasized the role of adaptation to the environment (Intelligence and its Measurement 1921; Sternberg and Detterman 1986). But intelligence involves not only modifying oneself to suit the environment (adaptation), but also modifying the environment to suit oneself (shaping) and sometimes finding a new environment that is a better match to one's skills, values, or desires (selection).

Not all people have equal opportunities to adapt to, shape, and select environments. In general, people of higher socio-economic standing tend to have more opportunities and people of lower socio-economic standing have fewer. The economy or political situation of the society can also be factors. Other variables that may affect such opportunities are education (especially literacy), political party, race, religion, and so forth. For example, someone with a college education typically has many more career options than does someone who has dropped out of high school to support a family. Thus, how and how well an individual adapts to, shapes, and selects environments must always be viewed in terms of the opportunities available to them.

Finally, success is attained through a balance of analytical, creative, and practical abilities. Analytical abilities are those primarily measured by traditional ability tests. Success in life requires one not only to analyse one's own ideas as well as those of others, but also to generate ideas and to persuade other people of their value. This necessity occurs in the world of work, for example, when a subordinate tries to convince a superior of the value of his or her plan, in the world of personal relationships when a child attempts to convince a parent to do what he or she wants or when a spouse tries to convince the other spouse to do things in his or her preferred way, and in the school when a student writes an essay arguing for a point of view.

The theory would interpret the studies described earlier as showing the importance of context in understanding human intelligence. For street children, knowing how to do the mathematics needed to run a street business is a matter of survival; knowing how to solve similar or even identical problems in the classroom is not. The children have adapted to the exigencies of their own environments. The processes needed for solving problems may be largely the same in the classroom and the street contexts, but the different contexts elicit different behaviour, just as we may behave very differently in school from the way we do at work, or at work from the way we do at home.

Cultural studies

In a series of studies in a variety of cultures, we have investigated some of our notions about intelligence and how they might apply in diverse contexts. As explained later in this section, they may apply quite differently, depending on where they need to be applied.

Children may develop contextually important skills at the expense of academic ones

Investigations of intelligence conducted in settings outside the developed world can often yield a picture of intelligence that is quite at variance with the picture one would obtain from studies conducted only in the developed world. In a study in 1996 in Usenge, Kenya, near the town of Kisumu, we were interested in school-aged children's ability to adapt to their indigenous environment.

We devised a test of practical intelligence for adaptation to the environment (see Sternberg and Grigorenko 1997; Sternberg et al. 2001). The test of practical intelligence measured children's informal tacit knowledge of natural herbal medicines that the villagers believe can be used to fight various types of infections. More than 95% of the children suffer from parasitic illnesses. Children in the villages use their knowledge of these medicines at an average frequency of once a week in medicating themselves and others. Thus, tests of how to use these medicines constitute effective measures of one aspect of practical intelligence as defined by the villagers, as well as their life circumstances in their environmental contexts. Their well-being hinges upon being able to self-medicate. Those who cannot suffer to a greater degree the consequences of the illnesses. Middle-class Westerners might find it quite a challenge to thrive or even survive in these contexts or, for that matter, in the contexts of urban ghettos often not distant from their comfortable homes.

We measured the Kenyan children's ability to identify the natural herbal medicines, where they come from, what they are used for, and how they are dosed. Based on work that we had carried out elsewhere, we expected that scores on this test would not correlate with scores on conventional tests of intelligence (Sternberg *et al.* 2000). To test this hypothesis, we also administered to the 85 children the 'Raven coloured progressive matrices test' (Raven *et al.* 1992), which is a measure of fluid or abstract-reasoning-based abilities, as well as the 'Mill Hill vocabulary scale' (Raven *et al.* 1992), which is a measure of crystallized or formal knowledge-based abilities. In addition, we gave the children a comparable test of vocabulary in their own Dholuo language. The Dholuo language is spoken in the home; English is spoken in the schools.

We found no significant correlation between the test of indigenous tacit knowledge and scores on the fluid-ability tests. But, to our surprise, we found statistically significant correlations of the tacit-knowledge tests with the tests of crystallized abilities. The correlations, however, were negative. In other words, the higher the children scored on the test of tacit knowledge, the lower they scored, on average, on the tests of crystallized abilities. Tests of fluid abilities also showed correlations with practical intelligence in the negative direction.

These surprising results can be interpreted in various ways but, based on the ethnographic observations of the anthropologists on the team, P. Wenzel Geissler and Ruth Prince, we concluded that a plausible scenario takes into account the expectations of families for their children. Many children drop out of school before graduation, for financial or other reasons. Moreover, many families in the village do not particularly value formal Western schooling. There is no reason why they should, since the children of many families will, for the most part, spend their lives farming or engaged in other occupations that make little or no use of Western schooling. Few, if any, will go to universities. These families emphasize teaching their children the indigenous informal knowledge that will lead to successful adaptation to the environments in which they will really live. Children who spend their time learning the indigenous practical knowledge of the community generally do not invest heavily in doing well in school, whereas children who do well in school generally do not invest as heavily in learning the indigenous knowledge: hence the negative correlations. In some cases, they do not learn the indigenous knowledge because no one wants to take them on as apprentices to teach them. They may therefore be perceived as the 'losers' in the village.

The Kenya study suggests that the identification of a general factor of human intelligence may tell us more about how abilities interact with patterns of schooling and especially Western patterns of schooling than it does about the structure of human abilities. In Western schooling, children typically study a variety of subjects from an early age and thus develop skills in a variety of areas. This kind of schooling prepares children to take a standard test of intelligence. Such a test typically measures skills in a variety of areas. Intelligence tests often measure skills that children were expected to acquire a few years before taking the intelligence test but, as Rogoff (1990, 2003) and others have noted, this pattern of schooling is not universal and has not even been common for much of the history of humankind.

Throughout history and in many places still, schooling, especially for boys, takes the form of apprenticeships in which children learn a craft from an early age. The children learn what they will need to know to succeed in a trade, but not a lot more. They are not simultaneously engaged in tasks that require the development of the particular blend of skills measured by conventional intelligence tests. Hence it is less likely that one would observe a general factor in their scores, much as we discovered in Kenya.

The context-specificity of intellectual performance does not apply only to countries far removed from North American or Europe. One can find the same on these continents, as we did in our studies of Eskimo children in southwestern Alaska.

Children may have substantial practical skills that go unrecognized in academic tests

We found related although certainly not identical results in a study of Yup'ik Eskimo children in southwestern Alaska (Grigorenko *et al.* 2004). We were particularly interested in these children because their teachers thought them, for the most part, to be quite lacking in the basic intelligence needed for success in school. However, many of the children had tremendous practical knowledge that few, if any, of the teachers had, such as how to travel from one village to another in the winter on a dogsled in the absence of landmarks that would have been recognizable to the teachers (or to us).

We assessed the importance of academic and practical intelligence in rural and urban Alaskan communities. A total of 261 high-school children were rated for practical skills by adults or peers in the study: 69 in grade 9; 69 in grade 10; 45 in grade 11; and 37 in grade 12. Out of these children, 145 were females and 116 were males, and they were from seven different communities—six rural and one relatively urban. We measured academic intelligence with conventional measures of fluid and crystallized intelligence. We measured practical intelligence with a test of tacit (informally learned) knowledge as acquired in rural Alaskan Yup'ik communities.

The urban children generally outperformed the rural children on a measure of crystallized intelligence, but the rural children generally outperformed the urban children on the measure of Yup'ik tacit knowledge. The test of tacit knowledge was superior to the tests of academic intelligence in predicting the practical and, particularly, the hunting skills of the rural children (for whom the test was created) but not those of the urban ones. Thus, in terms of the skills that mattered most to the children's everyday lives, the test of practical intelligence was distinctly preferable.

Practical intellectual skills may be better predictors of health than academic ones

In their study, Grigorenko and Sternberg (2001) tested 511 Russian schoolchildren (ranging in age from 8 to 17 years) as well as 490 mothers and 328 fathers of these children. They used entirely distinct measures of analytical, creative, and practical intelligence.

Fluid analytical intelligence was measured by two subtests of a test of non-verbal intelligence. The 'test of g: culture fair, level II' (Cattell and Cattell 1973) is a test of fluid intelligence designed to reduce, as much as possible, the influence of verbal comprehension, culture, and educational level, although no test completely eliminates such influences. In the first subtest, 'series', individuals were presented with an incomplete, progressive series of figures. The participants' task was to select, from among the choices provided, the answer that best continued the series. In the 'matrices' subtest, the task was to complete the matrix presented at the left of each row.

The test of crystallized intelligence was adapted from existing traditional tests of analogies and synonyms or antonyms used in Russia. Grigorenko and Sternberg (2001) used adaptations of Russian rather than American tests because the vocabulary used in Russia differs from that used in the USA. The first part of the test included 20 verbal analogies (internal-consistency reliability, 0.83). An example is 'circle ball = square?: (i) quadrangular; (ii) figure; (iii) rectangular; (iv) solid; (v) cube'. The second part included 30 pairs of words, and the participants' task was to specify whether the words in the pair were synonyms or antonyms (internal-consistency reliability, 0.74). Examples are 'latent–hidden' and 'systematic–chaotic'.

The measure of creative intelligence also comprised two parts. The first part asked the participants to describe the world through the eyes of insects. The second part asked participants to describe who might live and what might happen on a planet called 'Priumliava'. No additional information on the nature of the planet was specified. Each part of the test was scored in three different ways to yield three different scores. The first score was for originality (novelty); the second was for the amount of development in the plot (quality); and the third was for creative use of prior knowledge in these relatively novel kinds of task (sophistication). The mean interstory reliabilities were 0.69, 0.75, and 0.75 for the three respective scores, all of which were statistically significant at the $p < 0.001$ level.

The measure of practical intelligence was self-report and also comprised two parts. The first part was designed as a 20-item, self-report instrument, assessing practical skills in the social domain (e.g. effective and successful communication with other people), in the family domain (e.g. how to fix household items, how to run the family budget), and in the domain of effective resolution of sudden problems (e.g. organizing something that has become chaotic). For the subscales, internal-consistency estimates varied from 0.50 to 0.77. In this study, only the total practical intelligence self-report scale was used (Cronbach's alpha, 0.71). The second part had four vignettes, based on themes that appeared in popular Russian magazines in the context of discussion of adaptive skills in the current society. The four themes were, respectively, how to maintain the value of one's savings, what to do when one makes a purchase and discovers that the item one has purchased is broken, how to locate medical assistance in a time of need, and how to manage a salary bonus one has received for outstanding work. Each vignette was accompanied by five choices and participants had to select the best one. Obviously, there is no one 'right' answer in this type of situation. Hence, Grigorenko and Sternberg used the most frequently chosen response as the keyed answer. To the extent that this response was suboptimal, this suboptimality would work against us in subsequent analyses relating scores on this test to other predictor and criterion measures.

Clear-cut analytical, creative, and practical factors emerged for the tests. Thus, with a sample of a different nationality (Russian), a different set of tests and a different method of analysis (exploratory rather than confirmatory analysis) supported the theory of successful intelligence.

In this same study, the analytical, creative, and practical tests that we employed were used to predict mental and physical health among the Russian adults. Mental health was measured by widely used paper–and–pencil tests of depression and anxiety, and physical health was measured by self-report. The best predictor of mental and physical health was the practical intelligence measure (or, because the data are correlational, it may be that health predicts practical intelligence, although the connection here is less clear). Analytical intelligence came second and creative intelligence came third. All three contributed to prediction, however. Thus, we again concluded that a theory of intelligence encompassing all three elements provides better prediction of success in life than does a theory comprising just the analytical element.

The results in Russia emphasized the importance of studying health-related outcomes as one measure of successful adaptation to the environment. Health-related variables can affect one's ability to achieve one's goals in life, or even to perform well on tests, as we found in Jamaica.

Physical health may moderate performance on assessments

In interpreting results, whether from developed or developing cultures, it is always important to take into account the physical health of the participants one is testing. In a study that we carried out in Jamaica (Sternberg *et al.* 1997), we found that Jamaican school children who suffered from parasitic illnesses (for the most part, whipworm or Ascaris) performed more poorly on higher-level cognitive tests (such as of working memory and reasoning) than did children who did not suffer from these illnesses, even after controlling for socio-economic status.

Thus, many children were poor achievers not because they lacked abilities, but because they lacked good health. If you are moderately to seriously ill, you probably find it more difficult to concentrate on what you read or what you hear than if you are healthy. Children in developing countries are ill much and even most of the time. They simply cannot devote the same attentional and learning resources to schoolwork as do healthy children.

Do conventional tests, such as those of working memory or of reasoning, measure all of the skills possessed by children in developing countries? Work that we have done in Tanzania suggests that they do not.

Dynamic testing may reveal cognitive skills not revealed by static testing

A study that we conducted in Tanzania (see Sternberg and Grigorenko 1997, 2002; Sternberg *et al.* 2002) demonstrates the risks of giving tests, scoring them, and interpreting the results as measures of some latent intellectual ability or abilities. We administered to 358 schoolchildren between the ages of 11 and 13 years near Bagamoyo, Tanzania tests, including a form-board classification test, a linear syllogisms test, and a 20 questions test, which measure the kinds of skills required in conventional tests of intelligence. Of course, we obtained scores that we could analyse and evaluate, ranking the children in terms of their supposed general or other abilities. However, we administered the tests dynamically rather than statically (Vygotsky 1978; Brown and French 1979; Brown and Ferrara 1985; Lidz 1991; Haywood and Tzuriel 1992; Guthke 1993; Grigorenko and Sternberg 1998; Sternberg and Grigorenko 2002).

Dynamic testing is like conventional static testing in that individuals are tested and inferences about their abilities are made. But dynamic tests differ in that children are given some kind of feedback to help them to improve their scores. Vygotsky (1978) suggested that children's ability to profit from guided instruction that they received during a testing session could serve as a measure of the children's zone of proximal development, or the difference between their developed abilities and their latent capacities. In other words, testing and instruction are treated as being of one piece rather than as being distinct processes. This integration makes sense in terms of traditional definitions of intelligence such as the ability to learn

(Intelligence and its Measurement 1921; Sternberg and Detterman 1986). What a dynamic test does is to measure directly processes of learning in the context of testing, rather than measuring these processes indirectly as the product of past learning. Such measurement is especially important when not all children have had equal opportunities to learn in the past.

In the assessments, children were first given static ability tests. Experimental-group children were then given a brief period of instruction in which they were able to learn skills that would potentially enable them to improve their scores. Control-group children were not given such instruction. Then they were all tested again. Because the instruction for each test lasted for only approximately 5–10 minutes, one would not expect dramatic gains. However, on average, the gains in the experimental group were statistically significant. The experimental group also showed significantly greater gains than did the control group. More importantly, scores of the experimental-group children on the pre-test showed only weak although significant correlations with scores on the post-test. These correlations, at about the 0.3 level, suggested that, when tests are administered statically to children in developing countries, the results may be rather unstable and easily subject to influences of training. The reason for this could be that the children are not accustomed to taking Western-style tests, and so profit quickly even from small amounts of instruction as to what is expected from them. By contrast, the correlations for the control group were at the 0.8 level, as would be expected when one merely administers a pre-test and a post-test without an experimental intervention.

Of course, the more important question is not whether the scores changed or even correlated with each other, but rather how they correlated with other cognitive measures. In other words, which test was a better predictor of transfer to other cognitive performance, the pre-test score or the post-test score? We found the post-test score to be the better predictor in the experimental group.

In the Jamaica study described earlier, we had failed to find effects of anti-parasitic medication, albendazole, on cognitive functioning. Might this have been because the testing was static rather than dynamic? Static testing tends to emphasize skills developed in the past. Children who suffer from parasitic illnesses often do not have the same opportunities to profit from instruction that healthy children have. Dynamic testing emphasizes skills developed at the time of test. Indeed, the skills or knowledge are specifically taught at the time of the test. Would dynamic testing show effects of medication (in this case, praziquantel for schistosomiasis) not shown by static testing?

The answer was yes. Over time, treated children showed a distinct advantage over children who received a placebo, and were closer after time had passed to the control (uninfected) group than were the placebo-treated children. In other words, dynamic testing showed both hidden skills and hidden gains not shown on static tests.

New 'intermediate tests' of cognitive skills reveal new aspects of cognitive performance

In cultural research, we may want to assess school-related skills that are intermediate between abilities and achievement. Traditional tests of cognitive abilities are quite far removed from school performance. Achievement tests are a form of school performance.

In our work in Zambia, we devised such an intermediate test (Grigorenko, Jarvin, Kaani, Kapungulya, Kwiatkowski, and Sternberg, manuscript in preparation). Children in school and outside it continually need to be able to follow instructions. Often they are not successful in their endeavours because they do not follow instructions as to how to realize these endeavours. Following complex instructions is thus important for the children's success.

The Zambia cognitive assessment instrument (Z-CAI) measures working memory, reasoning, and comprehension skills in the oral, written, and pictorial domains. The Z-CAI was designed to: measure children's ability to follow oral, written, and pictorial instructions that become increasingly complex; be simple to implement, so that teachers could be easily trained to administer the instrument; be sensitive specifically to any improvement in cognitive functioning that was a result of improved health status; and be psychometrically sound (valid and reliable) in Zambia.

We found that children tested on the Z-CAI who were treated for parasitic illnesses outperformed children who were not treated relative to baseline performance.

Intelligence may be different things in different cultures

Intelligence may be conceived in different ways in different cultures (see reviews in Berry 1997; Sternberg and Kaufman 1998). Yang and Sternberg (1997a) reviewed Chinese philosophical conceptions of intelligence. The Confucian perspective emphasizes the characteristic of benevolence and of doing what is right. As in the Western notion, the intelligent person expends a great deal of effort in learning, enjoys learning, and persists in life-long learning with a great deal of enthusiasm. The Taoist tradition, in contrast, emphasizes the importance of humility, freedom from conventional standards of judgement, and full knowledge of oneself as well as of external conditions.

The difference between Eastern and Western conceptions of intelligence may persist even to the present day. Yang and Sternberg (1997b) studied contemporary Taiwanese Chinese conceptions of intelligence and found five factors underlying these conceptions: (1) a general cognitive factor, much like the g-factor in conventional Western tests; (2) interpersonal intelligence (i.e. social competence); (3) intrapersonal intelligence; (4) intellectual self-assertion: knowing when to show that you are smart; and (5) intellectual self-effacement: knowing when not to show that you are smart. In a related study but with different results, Chen (1994) found three factors underlying Chinese conceptualizations of intelligence: non-verbal reasoning ability; verbal reasoning ability; and rote memory. The difference may be a result of different subpopulations of Chinese, differences in methodology, or differences in when the studies were done.

The factors uncovered in Taiwan differ substantially from those identified in US citizens' conceptions of intelligence by Sternberg *et al.* (1981)—(1) practical problem-solving; (2) verbal ability; and (3) social competence—although, in both cases, people's implicit theories of intelligence seem to go quite far beyond what conventional psychometric intelligence tests measure. Of course, comparing the Chen (1994) study with the Sternberg *et al.* (1981) study simultaneously varies both language and culture.

Studies in Africa, in fact, provide yet another window on the substantial differences. Ruzgis and Grigorenko (1994) argued that, in Africa, conceptions of intelligence revolve largely around skills that help to facilitate and maintain harmonious and stable intergroup relations; intragroup relations are probably equally important and at times more important. For example, Serpell (1974, 1996) found that Chewa adults in Zambia emphasize social responsibilities, cooperativeness, and obedience as important to intelligence; intelligent children are expected to be respectful of adults. Kenyan parents also emphasize responsible participation in family and social life as important aspects of intelligence (Super and Harkness 1982, 1986, 1993). In Zimbabwe, the word for intelligence, *ngware*, actually means to be prudent and cautious, particularly in social relationships. Among the Baoule, service to the family and community and politeness towards, and respect for, elders are seen as key to intelligence (Dasen 1984).

It is difficult to separate linguistic differences from conceptual differences in cross-cultural notions of intelligence. In our own research, we use converging operations to achieve some separation. That is, we use different and diverse empirical operations to ascertain notions of intelligence. So we may ask in one study that people identify aspects of competence, in another study that they identify competent people, in a third study that they characterize the meaning of 'intelligence', and so forth.

The emphasis on the social aspects of intelligence is not limited to African cultures. Notions of intelligence in many Asian cultures also emphasize the social aspect of intelligence more than does the conventional Western or intelligence quotient-based notion (Lutz 1985; Poole 1985; White 1985; Azuma and Kashiwagi 1987).

It should be noted that neither African nor Asian notions emphasize exclusively social notions of intelligence. These conceptions of intelligence focus much more on social skills than do conventional US conceptions of intelligence, while simultaneously recognizing the importance of cognitive aspects of intelligence. In a study of Kenyan conceptions of intelligence (Grigorenko *et al.* 2001), it was found that there are four distinct terms constituting conceptions of intelligence among rural Kenyans—*rieko* (knowledge and skills), *luoro* (respect), *winjo* (comprehension of how to handle real-life problems), and *paro* (initiative)—with only the first directly referring to knowledge-based skills (including but not limited to the academic).

It is important to realize, again, that there is no one overall US conception of intelligence. Indeed, Okagaki and Sternberg (1993) found that different ethnic groups in San Jose, California had rather different conceptions of what it means to be intelligent. For example, Latino parents of schoolchildren tended to emphasize the importance of social-competence skills in their conceptions of intelligence, whereas Asian parents tended rather heavily to emphasize the importance of cognitive skills. 'White' parents also emphasized cognitive skills more. Teachers, representing the dominant culture, emphasized cognitive skills more than social-competence skills. The rank order of children of various groups' performance (including subgroups within the Latino and Asian groups) could be perfectly predicted by the extent to which their parents shared the teachers' conception of intelligence. In other words, teachers tended to reward those children who were socialized into a view of intelligence that happened to correspond to the teachers' own. However, social aspects of intelligence, broadly defined, may be as important as or even more important than cognitive aspects of intelligence in later life. Some, however, prefer to study intelligence not in its social aspect, but in its cognitive one.

Conclusion

When cultural context is taken into account: (1) individuals are better recognized for and are better able to make use of their talents; (2) schools teach and assess children better; and (3) society uses rather than wastes the talents of its members. We can pretend to measure intelligence across cultures simply by translating Western tests and giving them to individuals in a variety of cultures. But such measurement is only pretence. Individuals in other cultures often do not do well on our tests, nor would we do well on theirs. The processes of intelligence are universal, but their manifestations are not.

Intelligence can be used to maximize well-being, but it also can be used to destroy it, as Hitler, Stalin, Amin, and many other leaders have shown. By understanding cross-cultural meanings of intelligence and of well-being, we can seek to match intelligence to the attainment of well-being, rather than to its destruction.

Acknowledgements

Our work on culture and intelligence has been supported primarily by the Partnership for Child Development, centred at Imperial College, University of London.

References

Azuma, H. and Kashiwagi, K. (1987). Descriptions for an intelligent person: a Japanese study. *Jpn. Psychol. Res.* **29**, 17–26.

Berry, J.W. (1974). Radical cultural relativism and the concept of intelligence. In *Culture and cognition: readings in cross-cultural psychology* (ed. J.W. Berry and P.R. Dasen), pp. 225–9. Methuen, London.

Berry, J.W. (1997). Immigration, acculturation and adaptation. *Appl. Psychol. Int. Rev.* **46**, 5–68.

Binet, A. and Simon, T. (1916). *The development of intelligence in children.* Williams and Wilkins, Baltimore, Maryland. [Originally published in 1905.]

Brown, A.L. and Ferrara, R.A. (1985). Diagnosing zones of proximal development. In *Culture, communication, and cognition: Vygotskian perspectives* (ed. J.V. Wertsch), pp. 273–305. Cambridge University Press, New York.

Brown, A.L. and French, A.L. (1979). The zone of potential development: implications for intelligence testing in the year 2000. In *Human intelligence: perspectives on its theory and measurement* (ed. R.J. Sternberg and D.K. Detterman), pp. 217–35. Ablex, Norwood, New Jersey.

Carraher, T.N., Carraher, D., and Schliemann, A.D. (1985). Mathematics in the streets and in schools. *Br. J. Dev. Psychol.* **3**, 21–9.

Cattell, R.B. and Cattell, A.K. (1973). *Measuring intelligence with the culture fair tests.* Institute for Personality and Ability Testing, Champaign, Illinois.

Ceci, S.J. and Roazzi, A. (1994). The effects of context on cognition: postcards from Brazil. In *Mind in context: interactionist perspectives on human intelligence* (ed. R.J. Sternberg and R.K. Wagner), pp. 74–101. Cambridge University Press, New York.

Chen, M.J. (1994). Chinese and Australian concepts of intelligence. *Psychol. Dev. Soc.* **6**, 101–17.

Dasen, P. (1984). The cross-cultural study of intelligence: Piaget and the Baoule. *Int. J. Psychol.* **19**, 407–34.

Gardner, H. (1983). *Frames of mind: the theory of multiple intelligences.* Basic Books, New York.

Gardner, H. (1999). *Intelligence reframed: multiple intelligences for the 21st century.* Basic Books, New York.

Greenfield, P.M. (1997). You can't take it with you: why abilities assessments don't cross cultures. *Am. Psychol.* **52**, 1115–24.

Grigorenko, E.L. and Sternberg, R.J. (1998). Dynamic testing. *Psychol. Bull.* **124**, 75–111.

Grigorenko, E.L. and Sternberg, R.J. (2001). Analytical, creative, and practical intelligence as predictors of self-reported adaptive functioning: a case study in Russia. *Intelligence* **29**, 57–73.

Grigorenko, E.L., Geissler, P.W., Prince, R., Okatcha, F., Nokes, C., Kenny, D.A., Bundy, D.A., and Sternberg, R.J. (2001). The organisation of Luo conceptions of intelligence: a study of implicit theories in a Kenyan village. *Int. J. Behav. Dev.* **25**, 367–78.

Grigorenko, E.L., Meier, E., Lipka, J., Mohatt, G., Yanez, E., and Sternberg, R.J. (2004). Academic and practical intelligence: A case study of the Yup'ik in Alaska. *Learn. Indiv. Diff.* **14**, 183–207.

Guilford, J.P. (1982). Cognitive psychology's ambiguities: some suggested remedies. *Psychol. Rev.* **89**, 48–59.

Guthke, J. (1993). Current trends in theories and assessment of intelligence. In *Learning potential assessment* (ed. J.H.M. Hamers, K. Sijtsma, and A.J.J.M. Ruijssenaars), pp. 13–20. Swets and Zeitlinger, Amsterdam.

Haywood, H.C. and Tzuriel, D. (1992). *Interactive assessment.* Springer, New York.

Intelligence and its measurement 1921: a symposium (1921). *J. Educ. Psychol.* **12**, 123–47, 195–216, 271–5.

Laboratory of Comparative Human Cognition (1982). Culture and intelligence. In *Handbook of human intelligence* (ed. R.J. Sternberg), pp. 642–719. Cambridge University Press, New York.

Lave, J. (1988). *Cognition in practice.* Cambridge University Press, New York.

Lidz, C.S. (1991). *Practitioner's guide to dynamic assessment.* Guilford, New York.

Lutz, C. (1985). Ethnopsychology compared to what? Explaining behaviour and consciousness among the Ifaluk. In *Person, self, and experience: exploring Pacific ethnopsychologies* (ed. G.M. White and J. Kirkpatrick), pp. 35–79. University of California Press, Berkeley, California.

Nuñes, T. (1994). Street intelligence. In *Encyclopedia of human intelligence*, Vol. 2 (ed. R.J. Sternberg), pp. 1045–9. Macmillan, New York.

Okagaki, L. and Sternberg, R.J. (1993). Parental beliefs and children's school performance. *Child Dev.* **64**, 36–56.

Poole, F.J.P. (1985). Coming into social being: cultural images of infants in Bimin–Kuskusmin folk psychology. In *Person, self, and experience: exploring Pacific ethnopsychologies* (ed. G.M. White and J. Kirkpatrick), pp. 183–244. University of California Press, Berkeley, California.

Raven, J.C., Court, J.H., and Raven, J. (1992). *Manual for Raven's progressive matrices and Mill Hill vocabulary scales.* Oxford Psychologists Press, Oxford.

Rogoff, B. (1990). *Apprenticeship in thinking. Cognitive development in social context.* Oxford University Press, New York.

Rogoff, B. (2003). *The cultural nature of human development.* Oxford University Press, London.

Ruzgis, P.M. and Grigorenko, E.L. (1994). Cultural meaning systems, intelligence and personality. In *Personality and intelligence* (ed. R.J. Sternberg and P. Ruzgis), pp. 248–70. Cambridge University Press, New York.

Serpell, R. (1974). Aspects of intelligence in a developing country. *Afr. Soc. Res.* **17**, 576–96.

Serpell, R. (1996). Cultural models of childhood in indigenous socialization and formal schooling in Zambia. In *Images of childhood* (ed. C.P. Hwang and M.E. Lamb), pp. 129–42. Lawrence Erlbaum, Mahwah, New Jersey.

Serpell, R. (2000). Intelligence and culture. In *Handbook of intelligence* (ed. R.J. Sternberg), pp. 549–80. Cambridge University Press, New York.

Spearman, C. (1904). The proof and measurement of association between two things. *Am. J. Psychol.* **15**, 72–101.

Sternberg, R.J. (1985). *Beyond IQ: a triarchic theory of human intelligence.* Cambridge University Press, New York.

Sternberg, R.J. (1990). *Metaphors of mind: conceptions of the nature of intelligence.* Cambridge University Press, New York.

Sternberg, R.J. (1997). *Successful intelligence.* Plume, New York.

Sternberg, R.J. (1999). The theory of successful intelligence. *Rev. Gen. Psychol.* **3**, 292–316.

Sternberg, R.J. and Detterman, D.K. (1986). *What is intelligence?* Plume, Norwood, New Jersey.

Sternberg, R.J. and Grigorenko, E.L. (1997). The cognitive costs of physical and mental ill health: applying the psychology of the developed world to the problems of the developing world. *Eye on Psi Chi* **2**, 20–7.

Sternberg, R.J. and Grigorenko, E.L. (1999). A smelly 113° in the shade, or, why we do field research. *APS Observer* **12**, 1, 10–11, 20–21.

Sternberg, R.J. and Grigorenko, E.L. (2002). Just because we 'know' it's true doesn't mean it's really true: a case study in Kenya. *Psychol. Sci. Agenda* **15**, 8–10.

Sternberg, R.J. and Kaufman, J.C. (1998). Human abilities. *Annu. Rev. Psychol.* **49**, 479–502.

Sternberg, R.J., Conway, B.E., Ketron, J.L., and Bernstein, M. (1981). People's conceptions of intelligence. *J. Pers. Soc. Psychol.* **41**, 37–55.

Sternberg, R.J., Powell, C., McGrane, P.A., and McGregor, S. (1997). Effects of a parasitic infection on cognitive functioning. *J. Exp. Psychol. Appl.* **3**, 67–76.

Sternberg, R.J., Forsythe, G.B., Hedlund, J., Horvath, J., Snook, S., Williams, W.M., Wagner, R.K., and Grigorenko, E.L. (2000). *Practical intelligence in everyday life*. Cambridge University Press, New York.

Sternberg, R.J., Nokes, K., Geissler, P.W., Prince, R., Okatcha, F., Bundy, D.A., and Grigorenko, E.L. (2001). The relationship between academic and practical intelligence: a case study in Kenya. *Intelligence* **29**, 401–18.

Sternberg, R.J., Grigorenko, E.L., Ngrosho, D., Tantufuye, E., Mbise, A., Nokes, C., Jukes, M., and Bundy, D.A. (2002). Assessing intellectual potential in rural Tanzanian school children. *Intelligence* **30**, 141–62.

Super, C.M. and Harkness, S. (1982). The development of affect in infancy and early childhood. In *Cultural perspectives on child development* (ed. D. Wagnet and H. Stevenson), pp. 1–19. Freeman, San Francisco.

Super, C.M. and Harkness, S. (1986). The developmental niche: a conceptualization at the interface of child and culture. *Int. J. Behav. Dev.* **9**, 545–69.

Super, C.M. and Harkness, S. (1993). The developmental niche: a conceptualization at the interface of child and culture. In *Life-span development: a diversity reader* (ed. R.A. Pierce and M.A. Black), pp. 61–77. Kendall/Hunt Publishing Co, Dubuque, Iowa.

Thurstone, L.L. (1938). *Primary mental abilities*. University of Chicago Press, Chicago.

Vygotsky, L.S. (1978). *Mind in society: the development of higher psychological processes*. Harvard University Press, Cambridge, Massachusetts.

White, G.M. (1985). Premises and purposes in a Solomon Islands ethnopsychology. In *Person, self, and experience: exploring Pacific ethnopsychologies* (ed. G.M. White and J. Kirkpatrick), pp. 328–366. University of California Press, Berkeley, California.

Yang, S. and Sternberg, R.J. (1997a). Conceptions of intelligence in ancient Chinese philosophy. *J. Theor. Phil. Psychol.* **17**, 101–19.

Yang, S. and Sternberg, R.J. (1997b). Taiwanese Chinese people's conceptions of intelligence. *Intelligence* **25**, 21–36.

Antonella Delle Fave, MD, is professor of psychology at the Medical School of the University of Milan, Italy. Her studies focus on optimal experience and its long-term developmental role in promoting personal growth and social integration. She has investigated these topics among over 5000 participants in 22 different cultures.

Fausto Massimini, MD, is professor of psychology at the Medical School of the University of Milan, Italy. His research interests concern the evolution mechanisms promoting social change, and the interactions between culture, biology, and individual behavior. These processes were investigated in societies undergoing cultural transitions, such as the Kapauku of Papua New Guinea and Native American tribes.

Chapter 15

The relevance of subjective well-being to social policies: optimal experience and tailored intervention

Antonella Delle Fave and Fausto Massimini

Introduction

Researchers and professionals in the domains of psychology, health, and social sciences are paying increasing attention to subjective well-being, to the formalization of models of healthy behavior throughout the life span, and to their implications for intervention and social policies. This approach entails a theoretical and empirical shift, from the prominent use of interventions to treat the outcomes of problems to the development of strategies aimed at preventing their onset.

The psychological investigation of subjective well-being allows researchers to explore individual resources, strengths, and potentials. This information can be used to design intervention programs addressed to all categories of individuals and groups in any society. As several studies have shown (Bronstein *et al.* 2003), an approach emphasizing resources and abilities instead of weaknesses and deficits can promote individual development and well-being in various domains: education; physical and mental health; work; and social integration. In particular, this approach can be used to design projects addressed to people who have to cope with disadvantages, be they related to health conditions or to social maladjustment. In most cultures, negative stereotypes often surround people who—for various reasons—differ from the average citizen or do not belong to socially accepted groups and communities. The investigation of their perceived resources and quality of life can help change this attitude and promote both their subjective well-being and their social integration.

Studies on subjective well-being derive from two main perspectives: hedonism and eudaimonism (Ryan and Deci 2001). The former emphasizes human search for pleasure (Kahneman *et al.* 1999). The latter stems from Aristotle's concept of eudaimonia as the fulfillment of one's true nature, which includes both self-actualization and commitment to socially shared goals (Ryff and Keyes 1995; Waterman 1993). The framework adopted in this chapter to explore subjective experience and well-being belongs to the eudaimonic approach. It is centered on two constructs: psychological selection (Csikszentmihalyi and Massimini 1985) and optimal experience (Csikszentmihalyi 1975), described in the following section.

Through this framework, we will highlight the relevance to social policies of the subjective experience reported by two disadvantaged categories of citizens: people with physical disablement and street children. The former face the daily constraints due to impairments at the biological level, while the latter have to cope with maladjustment at the social and developmental levels.

Two main reasons drove us to investigate subjective well-being among disabled people. The first one is related to their number. In 2002, at the opening session of the International Day of people with disabilities (December 3), the UN General Secretary stated that there are about 500 million disabled people around the world, accounting for 10% of the whole population. This percentage is increasing every year due to several factors. All over the world, road accidents, the powerful weapons of modern wars, and the devastating consequences of environmental pollution are frequent and tragic causes of disablement. In Western countries, inadequate nutrition habits and sedentary lifestyle increase the risk of disabling cardiovascular pathologies. People enjoy a longer life expectancy than in the past, and therefore they are exposed to impairments related to aging. The technological advancements in medicine allow people with life-threatening pathologies to survive or to live longer.

The second reason for us to discuss subjective well-being and disablement is related to the attention that international agencies have recently devoted to this issue. The promotion of well-being and social integration of people with disabilities is among the prominent goals of UN agencies. As concerns Europe in particular, 2003 was declared the Year of Disabled Citizens. The 'Community Action Program to combat discrimination 2001–2006' was launched to develop an integrated system of policies allowing disabled people to practically achieve the status of full citizens within daily activities, through the removal of architectural, occupational, educational, and social barriers. In all these initiatives, the investigation of the disabled person's subjective perspective and evaluation of daily life is explicitly considered as the starting point for designing intervention.

As for street children, their condition represents an unprecedented phenomenon in the history of human cultures. It is mostly related to the massive worldwide urbanization that—especially in developing countries—is taking place as a consequence of poverty, ethnic conflicts, uncontrolled modernization, and disruption of traditional cultures. The total number of street children in the world is not easy to estimate, since they do not attend schools and institutions that usually provide data for surveys and statistics. It ranges from 50 to 150 million according to different agencies. Country- and city-level surveys conducted by local organizations are more reliable. However, 'regardless of the statistics, even one child on the streets is too many if their rights are being violated' (UNICEF 2004).

Street children represent a great challenge for international agencies. They are excluded from public educational and health care programs; they are exposed to physical and sexual abuse, to malnutrition, to severe pathologies such as AIDS; they are exploited as illegal laborers; and they are often killed during police actions. The public image of these children is overwhelmingly negative, and intervention projects addressed to them often consist in short-term programs mostly focused on their social, economic, and health problems (Human Right Watch 2004). Only limited attention is paid to their psychological needs and to their potentials and developmental resources. From this perspective, the investigation of the perceived needs and daily experience of these children could be of paramount importance in order to design tailored intervention aimed at the promotion of their well-being, personal growth, and social participation.

Subjective experience and social context

The impact of culture and social norms on human behavior and on the individual's quality of life can hardly be overestimated (Berry *et al.* 1997). Day by day, people acquire cultural information from their social environment by means of various forms of learning (Flinn 1997; Tomasello *et al.* 1993), and they subsequently replicate and transmit it. This process is partially regulated by culture itself (Durcham 1991). For example, in socially stratified communities people have access to a limited variety of jobs, according to the class or caste they belong to. In cultures characterized by gender stratification, women are usually restrained from getting involved in activities, interests, and professions that are considered men's domains.

These cultural constraints can be used as objective indicators to evaluate quality of life and well-being, but they are not sufficient to grasp the real-life conditions of an individual. If we exclude the most extreme conditions, each person has a more or less wide extent of autonomy and freedom in facing challenges and in discovering opportunities and stimuli in their daily activities, in interpreting life events, and in setting self-selected goals (Diener 2000; Veenhoven 2002).

Psychological selection and optimal experience

Far from simply being recipients and vehicles of cultural information, human beings actively take part in the process of cultural transmission and change. A process of active selection takes place at the psychological level, promoting the differential reproduction of cultural information units. A great number of cross-cultural studies have been conducted in order to detect the basic criterion that guides this process of *psychological selection* (Csikszentmihalyi and Massimini 1985; Massimini *et al.* 1987, 1988). Results show the paramount role played by the quality of experience that individuals associate with the activities, social contexts, situations—i.e. cultural information—to which they are exposed. In particular, individuals preferentially select for long-term cultivation situations and activities connected with optimal experience (Csikszentmihalyi 1975; Csikszentmihalyi and Csikszentmihalyi 1988). Optimal experience is characterized by concentration, focused attention, positive mood, clear goals, unselfconsciousness, and intrinsic motivation. People describe themselves as active and deeply involved in the task at hand, excited and relaxed at the same time. They perceive high challenges in the activity, and adequate personal skills in facing these challenges. They report engagement, arousal, enjoyment, and autonomy.

Optimal experience results from a complex and positive balance among the emotional, cognitive, and motivational components of the psychic system (Delle Fave and Bassi 2000). It should not be mistaken for peak experiences that are extreme and exceptional states of ecstasy and transcendence. On the contrary, several studies show that optimal experience is a rather common psychological state. Within a wide cross-cultural investigation involving over 5000 participants, optimal experience was reported by 85% of the people. Participants associated it with most activities of daily life, provided that they were structured and complex enough to foster active engagement, creativity, and involvement. In contrast, repetitive tasks and passive entertainment were very rarely cited (Massimini and Delle Fave 2000).

Besides its positive and rewarding features, optimal experience is also intrinsically dynamic. One of its core components, namely, the perception of high environmental challenges, promotes the increase of related skills and the subsequent search for increasingly higher challenges. This virtuous circle gradually enhances individual competencies and complexity in behavior in the domains associated with optimal experience. From this perspective, optimal experience plays a key role in shaping psychological selection and the development of the life theme (Csikszentmihalyi and Beattie 1979) that comprises the goals and values that each individual preferentially cultivates throughout life.

The activities associated with optimal experience are usually available in the daily environment; therefore, they shed light on the features of the culture to which individuals belong. However, human ability to set innovative goals, to selectively cultivate specific activities, and to transmit them to other members of the community directly affects culture. Each individual therefore, through the process of psychological selection, represents a prominent agent of cultural transmission and change (Bhattarai and Delle Fave 2002).

Well-being, agency, and social change

As stressed by several scholars, individuals and social groups are living systems, thus showing a tendency toward growth in complexity and information (Khalil and Boulding 1996). From this perspective, in order to yield positive outcomes and to foster authentic development, the cultivation of activities associated with optimal experiences has to promote complexity at the individual level, as well as a constructive information exchange at the social level (Delle Fave and Massimini 2000). However, psychological selection does not necessarily lead to development and adaptation. The ultimate result depends upon the type of activities that individuals associate with optimal experiences.

Some activities may lack the complexity that allows for a progressive enhancement of challenges and skills. Drug intake, for example, in the initial stages of addiction gives rise to states of consciousness that are subjectively perceived as similar to optimal experiences. They have been defined *mimetic optimal experiences* (Delle Fave and Massimini 2003). However, these states are only transitory and artificially induced. Moreover, in the long term, drug intake and related behaviors do not support refinement of any ability or competence, leading instead to health impairment and psychosocial maladjustment.

Similarly, some antisocial behaviors, such as stealing, can be associated with subjective experiences of thrill, concentration, and high engagement, and they often require a considerable amount of knowledge and technical abilities (Massimini and Delle Fave 2000). However, their cultivation often results in marginalization. Individuals engaged in these activities often circumscribe their interpersonal relations to small selected groups, such as gangs, and face isolation, prosecution, and restrictions in the opportunities for growth and social integration.

These examples show that the long-term consequences of optimal experiences can deeply differ according to the associated activities. This raises a crucial issue as concerns the distinction between well-being and social welfare, and the evaluation of what is good for the person and for the community. In our opinion, a major contribution to this issue comes from the works of the economist Amartya Sen (Sen 1987, 1992). He defines well-being through the concepts of functionings and capabilities. Functionings represent the basic constituents of well-being; they consist in what a person accomplishes, in terms of activities and development of personal identity. Capabilities comprise the set of functionings potentially available to each person in the environment; they reflect the extent of freedom each person enjoys in her social context, that is, the possibility to choose among different functionings. Throughout their life, individuals select and actualize some of these functionings, according to their capability set. This process is in many respects similar to psychological selection.

Sen's definition of well-being includes the satisfaction of basic survival needs, as well as the achievement of other functionings that can vary widely among individuals and in different cultures. However, Sen identifies an additional factor underlying human behavior and promoting the setting and pursuit of goals: agency (Sen 1992). Well-being and agency are two distinct constructs: the former is the attainment of individual functionings; the latter is the active pursuit of goals that are relevant and meaningful for the person within a broader perspective. This perspective takes into account the relationship of the individual with the social context and its values, and the needs of other people. As a consequence, agency achievements differ from well-being achievements. People can actively pursue goals that they consider important and valuable, though not necessarily related to their personal well-being. People can prominently invest their resources in activities that are valuable for the community, but that undermine some of their potential functionings (for example, free time or the availability of comforts in daily life).

Sen (1987) advocates an ethical approach to economics, deeply rooted in the Aristotelian conception of eudaimonia as the pursuit of human virtue. Agency, with its ethical implications, is a central construct in this approach (Nussbaum 1993; Sen 1992). In our opinion, a psychological analysis of behavior that takes into account the potential of humans to commit themselves to goals and values transcending individual well-being can fruitfully include the constructs of agency and optimal experience. They allow researchers to explore the relationship between perceived quality of life, goal setting, and social action.

As stated in the introduction, the relevance of these constructs to the development of policies addressed to people with disabilities and street children will be discussed in the following pages.

The subjective experience of disablement

The construct of health as a multifaceted result of cultural, biological, and individual factors has often been emphasized in the bio–psycho–social approach to disease and disability (Engel 1982; WHO 1946). In 1974 the European Council defined rehabilitation as 'a whole set of measures aimed at establishing and maintaining the most satisfactory relation possible between the individual and his environment after appearance of a handicap or an injury or illness resulting in a handicap' (resolution of June 27). The European Parliament, in its resolution of 11 March 1981, stressed the need to promote at the community level the economic, social, and vocational integration of disabled people. More recently, the Lisbon European Council (March 2000) and the Nice European Council (December 2000) emphasized the necessity to offer 'appropriate solutions reflecting disabled people's own perspective and experience'.

In the revised *International classification of impairments, disabilities and handicaps* (ICIDH-2), disablement is conceptualized as a variation of human functioning within a dynamic interaction between individual and environmental features (Thuriaux 1995). The last version of the ICIDH-2, the *International classification of functioning* (ICF; WHO 2001), has gone one step further in interpreting disablement within the bio–psycho–social perspective. The main focus is shifted from the features of disease to the components of health: *body functions, activities*, and *participation* are the basic elements of individual life, and they show wide variations among individuals. Disablement can be identified through changes in functioning within these three domains, which also include the cultural context the person lives in (Üstün *et al.* 2001).

This approach also sheds light on the personal and environmental factors that, in addition to biological conditions, affect individual functioning as a whole. Variations in psychological features, family support and relationships, material resources, job and educational opportunities, collective representations of disease, and social policies have a relevant impact on the degree of autonomy, participation, and self-actualization that individuals can achieve during their lives (Cousineau *et al.* 2003; Simeonsson *et al.* 2000).

Facing challenges and discovering opportunities

The subjective perception of functioning in daily activities and contexts influences the quality of life of disabled people. One of the first studies on this topic was conducted by Brickman *et al.* (1978). Their findings showed that paraplegic and quadriplegic victims of recent accidents rated their presently perceived level of happiness significantly lower than a control group did. However, their ratings of pleasure in performing ordinary daily activities did not differ from the control evaluations. More recently, several studies have focused on the perceived positive consequences of disability and chronic disease. They include improved interpersonal relationships, positive changes in personality, the discovery of new interests, and a revision in the hierarchy of life values (for a review see Sodergren and Hyland 2000). As concerns disability, Albrecht and Devlieger (1999) have highlighted the relationship between the perception of a satisfying quality of life and a good personal and relational balance.

We investigated psychological selection and optimal experience among disabled people, using both single administration questionnaires, and an on-line sampling procedure, the experience sampling method (ESM). The findings confirmed that people with disabilities can successfully achieve developmental goals, social integration, and a good quality of life despite severe biological constraints. This proved to be true both of people with congenital disabilities and people with impairments acquired later in life (Delle Fave and Massimini 2004*a*). Results highlighted the role of optimal experience in fostering individual's physical and social functioning, as concerns both congenital and acquired disabilities. In particular, the investigation of acquired disabilities allowed us to identify a specific process labeled *transformation of optimal experience* (Delle Fave 1996). The onset of physical impairments can make activities previously associated with optimal experiences unavailable. The disabled person is thus forced to find alternative sources of enjoyment and engagement, and new opportunities for action and skill development. These studies also showed that each individual develops a personal approach to disability, related challenges, and associated experience. The findings emphasized the need for paying more attention to the subjective perception of disablement, and to the resources that can be used to increase the developmental opportunities available to disabled people.

Some results can better clarify these concepts. In a study conducted among 36 people who were born blind (Delle Fave and Maletto 1992), all the participants reported optimal experiences in their life, mostly associating it with the use of media (35% of the answers) and with work (25.5%). Media included activities such as reading in Braille, listening to music, listening to radio news, watching (*sic*) TV news and films. Work mostly comprised handicrafts, traditionally representing the prominent learning and job opportunities offered to congenitally blind persons. Other activities, such as sports, hobbies, studying, social and family interactions, and religion were also quoted, but each of them accounted for a low percentage of answers. These results highlighted the successful integration of the participants in the social context through the implementation of alternative sensorial paths, namely, tactile and auditory channels, that took the place of sight. Reading in Braille and acquiring information from media represented both relevant daily opportunities for optimal experience, and a basic link between the individual and the social environment. The association of work with optimal experiences fostered both the implementation of skills and competencies, and participants' active role in the productive life.

As concerns acquired disability, we investigated the quality of experience of 45 people who had faced an illness or trauma during adolescence or adulthood. Among them, 12 were blind and 33 paraplegic or quadriplegic. Most participants in both groups (11 and 30, respectively) reported optimal experiences in their present life. Blind people most frequently quoted the use of media (28% of the answers; prominently comprising reading in Braille) followed by work (13%). Other activities were also quoted, however, accounting for remarkably lower percentages of answers (Delle Fave 2001). People with spinal lesions mostly associated with optimal experience structured leisure activities (sport and hobbies, 30% of the answers), followed by health care, particularly physiotherapy exercises (15%) and work (14%). We also investigated the activities associated with optimal experience before the illness or injury. Leisure and work were retrospectively reported to be the most frequent ones in both groups; leisure primarily included sport practice.

The prominent association of optimal experience with the use of media among blind people confirmed the findings obtained from the congenitally blind participants. However, it also implied the development of new skills, through the process of transformation of optimal experience. This skill acquisition was much more complex and demanding in the condition of acquired blindness, because the participants had to learn again to read, using fingers instead of eyes. They had to develop a new attention pattern in watching TV, learning to draw as much information as possible only through the auditory channel. In fact, they did not report the use of media as a prominent context for optimal experiences before blindness. Disability brought about radical changes in the life and opportunities for action of these people. They had to withdraw from sports (the prevailing leisure activity before the loss of sight) and to look for new jobs. Differently from congenitally blind people, who were often involved in handicrafts and manual tasks, most of these participants selected jobs requiring intellectual skills they had acquired before the onset of blindness, becoming, for example, teachers and musicians. This allowed them to exploit already existing resources, while laboriously developing the new manual skills necessary to cope with blindness.

People with spinal lesions also discovered new sources of engagement and enjoyment, particularly the development of new motor skills in sport practice (Delle Fave 1996). Similarly, the optimal experience associated with physiotherapy exercises fostered the optimization of the residual sensorimotor skills in the injured areas, and the improvement of vicarious motor abilities and sensorial channels, thus increasing participants' chances of integration in the active life (Delle Fave and Massimini 2004a).

However, these studies also highlighted some of the difficulties that people with disablement subjectively perceived in daily life, which are often overlooked in the planning of interventions. In a study conducted with ESM, 35 disabled participants aged 39 on average carried an electronic device, sending random acoustic signals during daytime for a period of 1 week (Delle Fave et al. 2003a). At each signal receipt, they filled out a self-report, describing the ongoing external context and the associated quality of experience. Within the experience fluctuations reported during daily activities, participants most frequently associated optimal experience with sports and creative hobbies (28.6%), social relations (25.3%), and work (18.8%). However, despite the relevance of interpersonal relations as opportunities for optimal experiences, these participants spent a very large portion of their time in solitude. In over 48% of the ESM forms they reported being alone (Ferrario et al. 2004). This result is rather peculiar, considering that in other ESM studies involving young and middle-aged adults solitude accounted for 30% of the reports on average (Csikszentmihalyi and Schneider 2000; Delle Fave and Massimini 2004b; Larson and Richards 1994). Moreover, disabled participants associated solitude with below average levels of affect and intrinsic motivation, confirming the results obtained in the other studies.

These findings highlight the need for intervention programs addressed to disabled people that take into account the subjective perception of daily opportunities and constraints, in order to effectively match the individual needs and potentials with the environmental resources. In most Western countries, there is a growing effort to remove architectural barriers and to guarantee the accessibility of private and public places to all citizens, overcoming limitations related to sensory or mobility impairments. More attention is also paid to the implementation of job and educational opportunities available to people with disabilities. However, less visible communication barriers that can undermine both the quality of subjective experience and the social integration process also need to be addressed.

Optimal experiences and disability in cross-cultural perspective

Taking into account the interplay between individual and environment, the constructs of health and well-being should also clearly include an evaluation of the actual chances individuals are offered to exploit their own resources and to develop their skills according to the environmental challenges. This information is essential to design intervention programs overcoming the compensation-focused perspective, which is based on medical treatment and on the satisfaction of survival and material needs only.

Paradoxically, in countries with limited material and economic resources, people with disabilities can be exposed to more meaningful opportunities for resource mobilization than in affluent societies. This finding emerged from a comparative study on optimal experience and psychological selection conducted among adolescents with motor disabilities in Nepal and in Italy (Delle Fave *et al.* 2003*b*). In Nepal, technological resources are scarce and services for disabled citizens are hardly available. There is a great need for low-cost rehabilitation. A community-based rehabilitation (CBR) project was launched in 1986 in order to develop programs for disabled people in the area of Bhaktapur, Kathmandu Valley. The project, initially run by volunteers, today has grown into an organization providing a variety of need-based services to over 1000 children (Save the Children, Norway 2000). The main resources of the CBR project are families and community members, who are informed through awareness campaigns and provided with technical abilities in the domain of physiotherapy and rehabilitation. Parents, friends, and teachers actively support disabled children, and they contribute to their development and social integration.

We investigated the occurrence of optimal experience among 20 disabled adolescents and young people who had joined the Bhaktapur CBR programs. All the participants associated it with the opportunities for action offered by their daily context. Over 46% of the answers referred to work, followed by the use of media (mainly reading and listening to the radio, 21%) and study (18%). Work included both traditional activities, such as woodcarving and tailoring, and modern activities, such as using a computer and printing. Some of these activities were part of the vocational training organized within CBR projects (Lombardi and Delle Fave 2002). Again, the association of optimal experience with work facilitated both participants' improvement of competencies and skills, and their active role in the productive life.

Moreover, some participants stressed the developmental and almost therapeutic role of facing daily challenges, related to the lack of facilities typical of a developing country. Kumar, an 18-year-old-boy with lower limb impairment due to poliomyelitis, described the most positive experience in his life as follows, '. . . I went to the temple of Changu Narayan, outside the city, on the top of a hill . . . When I returned to the bus station, I discovered that the last evening bus had already left, and that I had to come back home by foot . . . This was the greatest satisfaction in my life: I walked downhill . . . and I eventually reached home alone!'

Bishnu, aged 15, cannot use his upper limbs and hands because of cerebral palsy (Delle Fave and Massimini 2004a). However, he was trained to use his feet, and he learned to do almost everything: writing, painting, playing. He has some difficulties in walking, but his hometown Bhaktapur is an ancient city, with narrow and roughly paved streets, and wheelchairs would be of little help. Bishnu lives in the central area of the city where cars are not admitted; each day some friends go to school with him, and the neighborhood is ready to provide him with any kind of support. At school, he sits on a special wood chair, where he can straighten his legs and put books and pencils. After removing shoes and socks, he performs all sort of classwork by means of his feet.

What would the daily life of Kumar and Bishnu look like, if they lived in a Western country? How would they experience the events they described to us? After his visit to the temple, Kumar would probably be able to choose among various alternatives: a wheelchair; a mobile phone to call a taxi; a car belonging to family members or friends. Bishnu would have at his disposal a computer especially designed for people who cannot perform precise hand movements. He would not be expected to lie barefoot on a special chair in the classroom, and to walk long distances.

Would Kumar and Bishnu function better in a Western country? Would they enjoy a higher quality of life? Would they feel more integrated in the community? It is impossible to give a definitive answer, since they would probably face different problems and enjoy different facilities, such as those reported by the adolescents we interviewed in Italy. Annalisa, a girl aged 17, who had to undergo several operations during childhood and adolescence because of drop-gait, reported that she faces the following challenges: 'Now I can walk without problems. However, I have to use special shoes; therefore I cannot wear short dresses, I always wear trousers . . . I cannot drive the motorbike, and therefore I do not have many friends. My best friends are my cousins, who live in my grandmother's village. I visit them every summer . . . they do not treat me as a disabled person.' To quote Giacomo, a 15-year-old boy with spastic tetraparesis, 'I have been treated with physiotherapy and speech-therapy. At home, my parents equipped my bedroom with many technological aids. Also this electronic wheelchair is very easy to use . . . I love computers and mathematics. I like developing programs and games. My parents and the doctors, however, say that it is better for me to study humanities and social sciences; they maintain that I will have more chances to find a job . . .'

These quotations clearly exemplify the great variation in the opportunities offered to people with disabilities by the social context. However, they also suggest that the assessment of the quality of life should be primarily based on subjective evaluations. People with disablement should be asked to express their views, as concerns daily challenges, future goals, social inclusion, and interpersonal relations. They should be involved as consultants to point out aspects of the social environment to be implemented, removed, or changed in order to empower their status as citizens. They should be asked about perceived resources and quality of experience in daily activities, in order to provide them with job and educational opportunities that promote optimal experiences, skill development, and personal growth. First-hand knowledge entitles disabled people to be very effective advisors to the rehabilitation team, to other people with disabilities, and to the whole community.

Street children: stereotypes and potentials

The phenomenon of children taking to the streets has spread to most countries. Living in socio-economic systems that are often both turbulent and insensitive, a street child develops varied strategies of survival. The swelling number of street and homeless children in developing countries has created a whole new set of issues that needs to be addressed. Child abuse, illiteracy, malnutrition, disease, delinquency, and substance abuse are realities showing no signs of being under control, in spite of prevention efforts undertaken by governmental and non-governmental bodies (Inciardi and Surratt 1998; Jutkowitz *et al.* 1997; Le Roux 1996).

There is a common misunderstanding concerning street children: they are globally perceived as minors with no place to live and without family ties. The United Nations defines street children as boys and girls who habitually stay and live in the street, without any kind of protection and supervision from responsible adults (International Catholic Children's Bureau 1985). However, in fact, most of the children found in the streets all over the world do have parents daily interacting with them and caring for them to some extent. Lusk (1992) proposes four different typologies of these children.

1 Child workers who join their family home at night and in several cases manage to combine a job and school attendance. Recent studies indicate that this group accounts for over 90% of the children who work in the streets (Williams 1993).

2 Child workers who live with their family in the street and help parents earn a living through begging, occasional jobs, or more structured occupations, mainly occurring in the street environment.

3 Children who have only loose ties with their family, and begin to show deviant behaviors, such as petty crimes and dropping out of school.

4 Street children in the proper sense, who spend their lives in the streets, far from their families or without relations with them. These children are often workers; they do not attend school and are frequently members of gangs involved in illegal activities, such as theft, pickpocketing, and drug trafficking.

There are several reasons for the ambiguity in the characterization of street children. According to Aptekar (1994), the first reason is a political one: international agencies tend to inflate the number of street children, including categories that do not strictly fit the definition, in order to promote donations and financial investments. Secondly, street children are usually compared to 'the children from a fictitious ideal middle-class family in the developed world' (Aptekar and Abebe 1997, p. 478). Such an ethnocentric comparison is misleading in that, at least in the countries where the phenomenon is widespread, most of the other children belong to low-income families and actively contribute to the family's daily survival. They should be the actual comparison group in studies on street children.

Another problem concerns the hostile public attitude toward street children. Aptekar and Abebe (1997) identify three kinds of hostility that influence the strategies used to 'rehabilitate' street children and to neutralize their deviant behaviors. Firstly, street children are perceived as criminals. Secondly, they are considered a public danger, and as a consequence they are treated harshly by ordinary people and abused and severely punished in detention centers. The third hostility is related to the 'peripatetic' way of life of street children. Most cultures are sedentary, and attach a great value to the house as a private place where one can hide inappropriate or shameful behaviors. Street children lack this area of privacy, and they display all their good and bad aspects in the public space, including belongings that are normally concealed.

Most studies on street children and homeless youth emphasize their deviant behaviors, their weaknesses, and their poor physical and mental health (Rossi 1990; Robertson 1992; Toro 1999; Unger *et al.* 1998; van Wormer 2003). More recently, however, a growing number of studies point out that the common view of street children as deprived or emotionally and culturally impaired is a stereotype. Street children vary as do their peers, and could become real resources for cultural change and development, instead of being regarded as problems (de Oliveira *et al.* 1992). Homeless adolescents can develop coping styles that allow them to become psychologically adjusted to street life (Votta and Manion 2003; Rew and Horner 2003). However, without being offered basic education and opportunities for personal growth, they are exposed to risks at two different levels. First, their intellectual and psychological development is seriously limited. Second, marginalization prevents them from becoming active members of their community and agents of cultural change and empowerment.

Optimal experience and antisocial activities

The investigation of optimal experience among street children could provide some hints for the development of projects centered on individual resources and subjective well-being. At present, we can only offer findings obtained in exploratory studies (in preparation). Optimal experience was investigated among 11 Italian adolescents raised in disrupted and/or harmful family environments. All the participants had dropped out of school and spent most of their time in the street, going round in informal peer groups, listening to music, or fighting with other gangs. They were enrolled in an intervention program developed by educationalists who regularly met them in the streets. When asked about the occurrence of optimal experience in their life, only 6 (55%) identified this psychological state, mostly associating it with peer interactions and sexual activities. Even more striking results were obtained with a group of teenagers living in the streets of Nairobi, Kenya: only 2 participants out of 13 reported optimal experiences in their life, in association with play and praying.

Besides perceiving limited opportunities for engagement and skill development in street life, participants prominently associated optimal experience with free peer interaction. Previous cross-cultural studies on adolescents showed the two-sided effect of this activity on the quality of experience (Verma and Larson 2003). Spending free time with peers provides fun, positive affect, and pleasure; it fosters the development of social roles, but it is often associated with low engagement and low mobilization of personal skills. As concerns street youth, it often means idling around without specific goals and aims.

These findings suggest that street life, combined with an often problematic family background, does not provide children with meaningful occasions for optimal experiences. Most street children live and grow up in a harsh and hostile environment, characterized by uncertainty, insecurity, and the necessity to satisfy pressing survival needs. In less extreme circumstances, they spend most of their time in informal group interactions, without relevant opportunities for skill cultivation.

This situation exposes them to the risk of getting involved in deviant and criminal activities, which are often structured and engaging enough to require skills, control of the situation, and focused attention, thus becoming opportunities for optimal experiences. The positive psychological impact of these activities was clearly described by some of the participants in a study conducted among homeless people (Delle Fave et al. 1991). Sergio, a 24-year-old young man with a long history of homelessness and imprisonments for stealing, reported, 'I have optimal experiences when I steal. I always liked stealing. I am completely concentrated, I feel great . . . With a gun in my hand I feel I am the boss of the city, and I can have anything I want. While I am stealing, I do not feel anxious, I am calm, I know what I am doing and what will happen.' Giovanna, 22, provided a similar description of the optimal experience associated with stealing, 'I have fun, because I put myself to the test. I steal for fun, not because I need it, I am rich! I get very absorbed in the task, you know, in those situations you have to use your brain well . . . concentration is necessary, I don't like to play it by ear.' The shortcomings of these activities at the social and developmental levels are obvious; nevertheless, their psychological appeal should be taken into account in rehabilitation programs.

Intervention programs: from institutionalization to participation

Street children are often considered an isolated phenomenon, to be faced with *ad hoc*, circumscribed 'rehabilitation' programs. The fact that they are the living consequences of massive urbanization and socio-economic inequalities is often ignored (Rizzini 1996). As a consequence, in many projects children are institutionalized and provided with a safe environment and all material facilities that guarantee a good quality of daily life. However, in comparison with their peers and families, this sometimes makes them members of a privileged group. They become strangers in their own culture.

Moreover, to seclude street children into centers with strict discipline, rules, and obligations does not represent an adequate solution to their problems. Norberto Bobbio, an Italian philosopher of law, stated 'The orphans of rights are strangers in the city of duties' (quoted by Rigoldi, *Il Corriere della Sera*, April 2004). We cannot expect individuals who have suffered from daily exploitation and abuse to become suddenly obedient and trusting, and devote themselves to the prosperity of the society that had previously marginalized them. We cannot ask these children to quickly and permanently reverse their conflictual perception of society, just because some adults show them interest and offer them shelter and food. We cannot assume that they will automatically find optimal experiences and new opportunities for action in a context that radically differs from their usual environment. The investigation of optimal experience among adolescents living in institutions can help shed light on this problem. In a comparative study, we examined female adolescents living in two-parent families and girls entrusted to institutions for the custody of minors, because of family problems and individual maladjustment (Delle Fave and Massimini 2000). Institutionalized teenagers reported optimal experience in a lower percentage, compared with the girls raised in intact families (80% and 96%, respectively). Moreover, they prominently associated optimal experience with passive entertainment, such as watching TV soap-operas, and with free peer interactions. On the contrary, girls living at home mostly reported complex and creative tasks, such as hobbies, sports, and studying.

Taking into account the problems of institutionalization, and the frequent drop-outs of street children from residential centers, in several countries interventions addressed to children are presently combined with initiatives supporting families and communities.

Children are invited to join NFE (non-formal education) courses, which often take place directly in the streets (Rossi Doria 1999; Vithal 2003). These courses provide basic notions in the domains of literacy, mathematics, history, and humanities, but they also foster children's awareness of social values and human rights. Children involved in these courses can subsequently decide to join residential communities, and at this point they are usually enrolled in public school curricula. Additional training programs are customized according to individuals' interests;

each child can freely choose among various activities, including traditional local skills, arts, and handicraft. Efforts are made to facilitate family reunion and the reintegration of the children in their original social context, whenever possible. At the community level, advocacy programs are run to raise awareness on child rights, and literacy courses for adults are organized as well.

The theoretical and empirical work of the Brazilian pedagogist Paulo Freire has given a major impulse to this approach. Freire stressed the importance of promoting adult education as a means towards emancipation from poverty and social participation (Freire 1970, 1998, 1999). His model has been successfully applied in several countries, such as the Philippines (Too and Floresca-Cawagas 1997) and African-American communities in the United States (Potts 2003).

We gathered evidence of the positive outcomes of this community-centered approach in three different countries. In Brazil, a pilot study on psychological selection and life theme was conducted among adolescents living in centers run by the association 'Caminhos de Vida', settled in Goiania, state of Goias (Municipality of Goiania 1999). In Kenya we got in touch with Kivuli, a center for street children located in a slum area of Nairobi (Shorter and Onyancha 1999). In Nepal, we were involved in a cooperation project with CWIN (Child Workers in Nepal), an organization established in Kathmandu and addressed to working and street children (CWIN 1998, 2001).

In all these centers, educators perform the most crucial part of their job in the streets, to foster a spontaneous and free contact with the children. The intervention is primarily based on NFE courses. Teaching strategies are used to promote children's autonomous discovery of their own learning potentials. Moreover, these projects are deeply rooted in the local social system, and they include the active involvement of families and communities. This affects society as a whole, because, as selective transmitters of information, individuals influence the cultural evolution trend. Such a combined action of various individuals can foster social empowerment in the long run.

Investigating the experience of street children: a tool for tailored intervention

All over the world, there is a large amount of good work being done with street children. However, it is fragmented; there is a lack of networking and dissemination of information. Governmental and non-governmental bodies develop programs and plans at the national and regional level, but there is little recorded research on their impact.

Moreover, while careful evaluations are essential to demonstrate effectiveness, it is often difficult to plan a long-term follow-up. Intervention programs such as those illustrated above require great efforts. They have to provide support to children in various domains, from health care to education, from vocational training to reintegration into family and community. Children often run away from residential centers, despite the relative amount of freedom they enjoy there, and they become even more suspicious and refractory (Wilder 2002). Some children return home, but the benefits of intervention can be erased by negative attitudes of parents and community members, family poverty, and lack of educational or job opportunities. Other children cannot rejoin their parents because of family poverty or relational problems, and they have to be provided with shelter, education, and income alternatives. In some cases, counseling and psychological support are needed, but they are not always available. Because of their high costs in terms of financial and human resources, these programs can actually help only a relatively small number of children (Bhattarai and Delle Fave 2002).

The Declaration of Child Rights (1989) provides a legal basis for initiating activities aimed at promoting children's well-being and development. However, in order to design effective need-based and 'respectful' interventions (Panter-Brick 2002), more attention should be paid to the perspective of the children themselves, and to their subjective experience. The perspective of significant others in children's lives should also be taken into account (Snow et al. 1994; Bar-On 1998). The limited though relevant fieldwork studies highlight the importance of actively involving children in the research design, in order to get authentic insights into their real life (Baker et al. 1996; van Beers 1996). The effectiveness of this strategy was assessed through the development of instruments, such as the Children's Perspective Protocol, used in studies conducted in Bangladesh, Ethiopia, the Philippines, and Central America to get subjective reports of working and street life (Woodhead 1999).

In our opinion, the prominent weakness of most interventions with street children is the lack of systematic and longitudinal assessment of the children's quality of experience and perceived opportunities for development in the daily life. The approach centered on optimal experience—among others—could be used to explore children's perceived challenges, interests, and sources of engagement, thus enabling educationalists to design person-centered programs, and to reduce the frequent drop-out from centers.

The perceived lack of optimal experiences could represent relevant information as well. As illustrated in the previous pages, often street youth do not find opportunities for optimal experiences in ordinary daily life, and they can be exposed to the thrill and excitement provided by illegal activities. As most studies with teenagers have shown, disengagement in daily life has negative implications at the educational and developmental levels (Larson 2000). Communities and centers should help these children discover socially meaningful sources of optimal experiences, while at the same time supporting individual predisposition and intrinsic motivation, in order to promote the cultivation of long-term relevant skills and competencies. To achieve this aim, an environment is needed that fosters the autonomous search for goals, encouraging the selection of challenging and socially adaptive opportunities for action (Levy 1996). The individual's engagement in self-selected daily activities is a basic prerequisite to promote both subjective well-being, and a successful social integration (Arieli *et al.* 1990).

Policies for people or policies with people?

As previously highlighted, individuals play a central role in the process of cultural transmission and cultural change. However, people should be supported in finding meaningful and socially relevant challenges, especially under conditions such as disability or social maladjustment where they cope with restricted opportunities for action in daily life. From childhood, citizens should be exposed to opportunities for engagement, enjoyment, and optimal experiences in socially useful activities. They should be taught to appreciate the development potential embedded in agency and cooperation towards community objectives. Intrinsic motivation and the autonomous search for meanings and goals (Oles 1991; Levy 1996) should be sustained. Even under circumstances such as disability and marginalization that can obstruct and even distort the development path, the individual effort and ability to find a personalized way toward complexity have to be primarily supported (Delle Fave 2001).

The collaboration between researchers, intervention agencies, end-users, and ordinary people should be emphasized as the major tool to achieve successful results in social policies. This collaboration has already proved to be useful in various domains. For example, the need for improving life conditions in traditional settlements and slum areas has promoted a synergy between architects and local populations (Davidson and Serageldin 1995; Doshi 1990; Oppenheimer Dean and Hursley 2002). Thanks to their first-hand expertise, village and slum inhabitants have suggested creative and low-cost solutions to build houses, furniture, ventilation and heating systems, and water supplies. The Barefoot College and the Vastu-Shilpa Foundation in India, and the Rural Studio in the United States offer practical examples of how to combine cultural and aesthetic needs with financial and material restrictions. Above all, these projects prove the importance of recognizing marginalized people as fundamental resources for the community, and to develop projects authentically tailored to individual and community needs.

However, though it seems self-evident that tailored interventions could work better than standardized ones, no systematic studies have yet been conducted on this topic. As pointed out in the previous pages, only recently has attention been devoted to the relevance of subjective experience in designing rehabilitation programs. Professionals and policy makers are becoming increasingly aware of the importance of actively involving end-users in intervention programs. However, the psychological outcomes of the interventions have to be empirically assessed. Moreover, long-term follow-up studies are needed. Many centers provide education and vocational training to street children, but very rarely is their effectiveness in promoting social integration evaluated in the long run. A growing number of private and public companies offer job opportunities to people with disabilities, but the quality of experience associated with these activities and their long-term impact on individual well-being and socialization are only rarely investigated.

Moreover, intervention models should be compared across cultures, also taking into account social factors such as gender, family structure, and education, whose influence on daily experience and opportunities for development can vary widely.

As clearly stated in the 1948 Declaration of Human Rights, one of the goals of every society should be to provide all citizens with equal rights and equal access to the environmental resources, or set of capabilities (Sen 1992). At the individual level, this would broaden the range of opportunities for optimal experiences and skill development available to each person. At the social level, the individual tendency to pursue self-selected goals and personal well-being could be channeled to foster at the same time co-operation and community empowerment (Delle Fave and Massimini 2002). Following the long tradition initiated by Aristotle with his studies on virtues and ethics, we strongly believe in the human potential to match the pursuit of optimal experiences and personal well-being with agency and the active contribution to the improvement of society.

Conclusion

In this chapter we have emphasized the importance of investigating subjective well-being and the perceived quality of life in order to build person-centered interventions and social policies. Among the various approaches to the study of well-being, we have focused on the investigation of optimal experience, a state of deep engagement, involvement, and enjoyment reported by people of the most diverse cultures, socio-economic status, ages, and educational levels in association with daily activities. Several studies showed the positive impact of optimal experience on personal growth and development. These aspects can be exploited to design interventions aimed at promoting individual well-being and community empowerment. In particular, they could be used as guidelines to develop programs addressed to people who live in disadvantaged conditions, thus deserving specific attention as concerns social policies.

We specifically referred to two disadvantaged categories of citizens: people with physical disabilities and street children. The former have to cope with the constraints derived from biological impairments; the latter with the disadvantages related to social maladjustment and marginalization. International agencies and local governments are devoting increasing efforts to promote the quality of life of disabled people, whose number is increasing in most countries. The phenomenon of street children represents a widespread and tragic problem for both industrialized and developing societies. These children are deprived of all basic human rights, their physical and psychological development is seriously jeopardized, and they are at risk for criminality and marginalization.

Empirical findings gathered among people with disabilities and street children suggest that their subjective evaluation of the opportunities for optimal experience in the daily life could be effectively used as a tool to build person-centered interventions, aimed at fostering both the unfolding of individual potentials and the improvement of society.

References

Albrecht, G.L. and Devlieger, P.J. (1999). The disability paradox: high quality of life against all odds. *Soc. Sci. Med.* **48** (8), 977–88.

Aptekar, L. (1994). Street children in the developing world: a review of their conditions. *Cross-cult. Res.* **28**, 195–224.

Aptekar, L. and Abebe, B. (1997). Conflict in the neighborhood. Street and working children in the public space. *Childhood* **4**, 477–90.

Arieli, M., Beker, J., and Kashti, Y. (1990). Residential group care as a socializing environment: toward a broader perspective. *Child Youth Serv.* **13**, 45–58.

Baker, R., Panter-Brick, C., and Todd, A. (1996). Methods used in research with street children in Nepal. *Childhood* **3**, 171–93.

Bar-On, A. (1998). So what's so wrong with being a street child? *Child Youth Care Forum* **3**, 201–22.

Berry, J.W., Poortinga, Y.H., Pandey, J., Dasen, P.R., Saraswathi, T.S., Segall, M.H., and Kagitçibasi, C. (Eds.) (1997). *Handbook of cross-cultural psychology.* Allyn and Bacon, Boston.

Bhattarai, K. and Delle Fave, A. (2002). Children at work, children in the street: present programs and future plans. In *In pursuit of a sustainable modernisation: culture and policies in Nepal* (ed. A. Delle Fave and M.B. Pun), pp. 145–65. Arcipelago Edizioni, Milan.

Brickman, P., Coates, D., and Janoff-Bulman, R. (1978). Lottery winner and accident victims: is happiness relative? *J. Pers. Soc. Psychol.* **36**, 917–27.

Bronstein, M.H., Davidson, L., Keyes, C.L.M., and Moore, K.A., The Centre for Child Well-Being (Eds.) (2003). *Well-being. Positive development across the life course.* Lawrence Erlbaum Associates, Mahwah, New Jersey.

Cousineau, N., MsDowell, I., Hotz, S., and Hébert, P. (2003). Measuring chronic patients' feelings of being a burden to their caregivers. Development and preliminary validation of a scale. *Med. Care* **41**, 110–18.

Csikszentmihalyi, M. (1975). *Beyond boredom and anxiety.* Jossey-Bass, San Francisco.

Csikszentmihalyi, M. and Beattie, O. (1979). Life themes: a theoretical and empirical exploration of their origins and effects. *J. Hum. Psychol.* **19**, 677–93.

Csikszentmihalyi, M. and Csikszentmihalyi, I. (Eds.) (1988). *Optimal experience. Psychological studies of flow in consciousness.* Cambridge University Press, New York.

Csikszentmihalyi, M. and Massimini, F. (1985). On the psychological selection of bio-cultural information. *New Ideas Psychol.* **3**, 115–38.

Csikszentmihalyi, M. and Schneider, B. (2000). *Becoming adult.* Basic Books, New York.

CWIN (1998). *State of the rights of the child in Nepal.* Indreni Offset Press, Kathmandu, CWIN/Redd Barna Nepal.

CWIN (2001). The State of the Rights of the Child in Nepal. Summary of the National Report. <http://www.cwin-nepal.org>. Downloaded March 2004.

Davidson, C. and Serageldin, I. (1995). *Architecture beyond architecture*. Academy Editions, London.

Delle Fave, A. (1996). Il processo di 'trasformazione di Flow' in un campione di soggetti medullolesi [The process of flow transformation in a sample of people with medullary lesions]. In *La selezione psicologica umana* (ed. F. Massimini, P. Inghilleri, and A. Delle Fave), pp. 615–34. Cooperativa Libraria IULM, Milan.

Delle Fave, A. (2001). Deficiência, reabilitação e desenvolvimento do individuo: questões psicológicas e trans-culturais. *Paideia. Cadern. Psicolog. Educ.* **21**, 35–46.

Delle Fave, A. and Maletto, C. (1992). Processi di attenzione e qualità dell'esperienza soggettiva [Attention and the quality of subjective experience]. In *Vedere con la mente* (ed. D. Galati), pp. 321–53. Franco Angeli, Milan.

Delle Fave, A. and Massimini, F. (2000). Living at home or in institution: adolescents' optimal experience and life theme. *Paideia. Cadern. Psicolog. Educ.* **19**, 55–66.

Delle Fave, A. and Massimini, F. (2002). Cultural change and human behaviour: evolution or development? In *In pursuit of a sustainable modernisation: culture and policies in Nepal* (ed. A. Delle Fave and M.B. Pun), pp. 13–34. Arcipelago Edizioni, Milan.

Delle Fave, A. and Massimini, F. (2003). Drug addiction: the paradox of mimetic optimal experience. In *European Positive Psychology Proceedings 2002* (ed. J. Henry), pp. 31–8. British Psychological Society, Leicester.

Delle Fave, A. and Massimini, F. (2004a). Bringing subjectivity into focus: optimal experiences, life themes and person-centred rehabilitation. In *Positive psychology in practice* (ed. P.A. Linley and S. Joseph), pp. 581–97. Wiley and Sons, London.

Delle Fave, A. and Massimini, F. (2004b). Parenthood and the quality of experience in daily life: a longitudinal study. *Soc. Indic. Res.* **67**, 75–106.

Delle Fave, A., Massimini, F., and Maletto, C. (1991). Barboni [Homeless people]. *Psicolog. Contemp.* **104**, 55–63.

Delle Fave, A., Bassi, M., Lombardi, M., and Ferrario, E. (2003a). The positive consequences of disability: implications for intervention. *European psychotherapy, special edition abstracts VIII European Conference on Traumatic Stress*, abstract 248. Berlin, May 22–25.

Delle Fave, A., Lombardi, M., and Massimini F. (2003b). Disability and development: individual and cultural issues. In *Catching the future: women and men in global psychology* (ed. L. Loewenstein, D. Trent, and R. Roth.), pp.104–13. Pabst Science Publishers, Lengerich, Germany.

De Oliveira, W., Baizerman, M., and Pellet, L. (1992). Street children in Brazil and their helpers: comparative views on aspirations and the future. *Int. Soc. Work* **35**, 163–76.

Diener, E. (2000). Subjective well-being. *Am. Psychol.* **55**, 34–43.

Doshi, B.V. (1990). Planning for a community: Vidyadhar Nagar. *Int. Soc. Sci. J.* **42**, 387–94.

Durham, W.H. (1991). *Coevolution. Genes, culture and human diversity*. Stanford University Press. Stanford CA.

Engel, G.L. (1982). The biopsychosocial model and medical education. *N. Engl. J. Med.* **306**, 802–5.

Ferrario, E., Bassi, M., Lombardi, M., and Delle Fave, A. (2004). Disability and development: subjective experience and daily functioning. *2nd European Conference on Positive Psychology, book of abstracts*, pp. 195–6. Verbania Pallanza, Italy, July 5–8.

Flinn, M.V. (1997). Culture and the evolution of social learning. *Evol. Hum. Behav.* **18**, 23–67.

Freire, P. (1970). *Pedagogy of the oppressed*. Herder and Herder, New York.

Freire, P. (1998). Cultural action and conscientization. *Harvard Educ. Rev.* **68**, 499–521.

Freire, P. (1999). Education and community involvement. In *Critical education in the new information age* (ed. M. Castells, R. Flecha, P. Freire, H.A. Giroux, D. Macedo, and P. Willis), pp. 83–91. Rowman and Littlefield, Lanham, Maryland.

Human Right Watch (2004). <http://www.hrw.org>. Downloaded September 2, 2004.

Inciardi, J.A. and Surratt, H.L. (1998). Children in the streets of Brazil: drug use, crime, violence and HIV risk. *Subst. Use Misuse* **7**, 1461–80.

International Catholic Children's Bureau (1985). *Forum on street children and youth.* International Catholic Children's Bureau, Grand Bassam, Ivory Coast.

Jutkowitz, J.M., Spielmannm H., Koeler, U., Lohani, J., and Pande, A. (1997). Drug use in Nepal: the view from street children. *Subst. Use Misuse* **7–8**, 987–1004.

Kahneman, D., Diener, E., and Schwarz, N. (Eds.) (1999). *Well-being: the foundations of hedonic psychology.* Russell Sage Foundation, New York.

Khalil, E.L. and Boulding, K.E. (Eds.) (1996). *Evolution, order and complexity.* Routledge, New York.

Larson, R.W. (2000). Toward a psychology of positive youth development. *Am. Psychol.* **55**, 170–83.

Larson, R. and Richards, M.H. (1994). *Divergent realities.* Basic Books, New York.

Le Roux, J. (1996). The worldwide phenomenon of street children: conceptual analysis. *Adolescence* **124**, 965–71.

Levy, Z. (1996). Conceptual foundations of developmentally oriented residential education: a holistic framework for group care that works. *Resident. Treat. Child. Youth* **13**, 69–83.

Lombardi, M. and Delle Fave A. (2002). Disability and rehabilitation in Nepal: an example from Bhaktapur CBR. In *In pursuit of a sustainable modernisation: culture and policies in Nepal* (ed. A. Delle Fave and M.B. Pun), pp. 125–43. Arcipelago Edizioni, Milan.

Lusk, M. (1992). Street children of Rio de Janeiro. *Int. Soc. Work* **35**, 293–305.

Massimini, F. and Delle Fave, A. (2000). Individual development in a bio-cultural perspective. *Am. Psychol.* **55**, 24–33.

Massimini, F., Csikszentmihalyi, M., and Carli, M. (1987). Optimal experience: a tool for psychiatric rehabilitation. *J. Nerv. Ment. Dis.* **175**, 545–9.

Massimini F., Csikszentmihalyi M., and Delle Fave A. (1988). Flow and biocultural evolution. In *Optimal experience. Psychological studies of flow in consciousness* (ed. M. Csikszentmihalyi and I. Csikszentmihalyi), pp. 60–81. Cambridge University Press, New York.

Municipality of Goiania (1999). *Programa Brasil criança cidada. Situaçao de rua.* Fumdec, Goiania.

Nussbaum, M. (1993). Non relative virtues: an Aristotelian approach. In *The quality of life* (ed. M. Nussbaum, and A. Sen), pp. 242–69. United Nation University and WIDER, Helsinki.

Oles, T.P. (1991). Resident's perceptions of change: a basis for after care. *Resident. Treat. Child. Youth* **9**, 29–37.

Oppenheimer Dean, A. and Hursley, T. (2002). *Rural Studio: Samuel Mockbee and an architecture of decency.* Princeton Architectural Press, Princeton, New Jersey.

Panter-Brick, C. (2002). Street children, human rights, and public health: a critique and future directions. *Annu. Rev. Anthropol.* **31**, 147–71.

Potts, R.G. (2003). Emancipatory education versus school-based prevention in African American communities. *Am. J. Community Psychol.* **31**, 173–83.

Rew, L. and Horner, S.H. (2003). Personal strengths of homeless adolescents living in a high-risk environment. *Adv. Nursing Sci.* **26**, 90–101.

Rizzini, I. (1996). Street children. An excluded generation in Latin America. *Childhood* **3**, 215–33.

Robertson, J.M. (1992). Homeless and runaway youth. In *Homelessness: a national perspective* (ed. M.J. Robertson and M.D. Greenblatt), pp. 287–97. Plenum, New York.

Rossi, P.H. (1990). The old homeless and the new homeless in historical perspective. *Am. Psychol.* **45**, 954–9.

Rossi-Doria, M. (1999). *Di mestiere faccio il maestro* [My job is teaching]. L'Ancora, Naples, Italy.

Ryan, R.M. and Deci, E.L. (2001). On happiness and human potentials: a review of research on hedonic and eudaimonic well-being. *Annu. Rev. Psychol.* **52**, 141–66.

Ryff, C.D. and Keyes, C.L.M. (1995). The structure of psychological well-being revisited. *J. Pers. Soc. Psychol.* **69**, 719–27.

Save the Children, Norway (2000). *Meeting the rights of the children with disabilities. Mid-term review reports for SCN-CBR partners.* SCN, Oslo.

Sen, A. (1987). *On ethics and economics.* Oxford University Press, Oxford.

Sen, A. (1992). *Inequality reexamined.* Oxford University Press, Oxford.

Shorter, A. and Onyancha, E. (1999). *Street children in Africa—a Nairobi case study.* Paulines Publications Africa, Nairobi.

Simeonsson, R.J., Lollar, D., Hollowell, J., and Adams, M. (2000). Revision of the International Classification of Impairments, Disabilities, and Handicaps. Developmental issues. *J. Clin. Epidemiol.* **53**, 113–24.

Snow, D.A., Anderson, L., and Koegel, P. (1994). Distorting tendencies in research on the homeless. *Am. Behav. Sci.* **37**, 461–75.

Sodergren, S.C. and Hyland, M.E. (2000). What are the positive consequences of illness? *Psychol. Health* **15**, 85–97.

Thuriaux, M.C. (1995). The ICIDH: evolution, status and prospects. *Disabil. Rehabil.* **17**, 112–18.

Tomasello, M., Kruger, A.C., and Ratner, H.H. (1993). Cultural learning. *Behav. Brain Sci.* **16**, 495–552.

Too, S.H. and Floresca-Cawagas, V. (1997). Towards a people-centred education: possibilities and struggles in the Philippines. *Int. Rev. Educ.* **43**, 527–45.

Toro, P.A. (1999). Advances in research on homelessness: an overview of the special issue. *J. Community Psychol.* **27**, 115–17.

Unger, J., Kipke, M., Simon, T., Johnson, C., Montgomery, S., and Iversen, E. (1998). Stress and social support among homeless youth. *J. Adolesc. Res.* **13**, 134–57.

UNICEF (2004). Street children FAQ. <http://www.unicef.org>. Downloaded September 3, 2004.

Üstün, T.B., Chatterji, S., Bickenbach, J.E., Trotter II, R.T., Room, R., Rehm, J., and Saxena, S. (Eds.) (2001). *Disability and culture. Universalism and diversity.* Hogrefe and Huber Publishers, Göttingen, Germany.

Van Beers, H. (1996). A plea for a child-centered approach in research with street children. *Childhood* **2**, 195–201.

Van Wormer, R. (2003). Homeless youth seeking assistance: a research-based study from Duluth, Minnesota. *Child Youth Care Forum* **32**, 89–103.

Veenhoven, R. (2002). Why social policy needs subjective indicators? *Soc. Indic. Res.* **58**, 33–45.

Verma, S. and Larson, R. (Eds.) (2003). *Examining adolescent leisure time across cultures: developmental opportunities and risks,* New Directions in Child and Adolescent Development Series. Jossey Bass, San Francisco.

Vithal, R. (2003). Teachers and 'street children': on becoming a teacher of mathematics. *J. Math. Teach. Educ.* **6**, 165–83.

Votta, E. and Manion, I.G. (2003). Factors in the psychological adjustment of homeless adolescent males: the role of coping style. *J. Am. Acad. Child Adolesc. Psychiatry* **42**, 778–85.

Waterman, A.S. (1993). Two conceptions of happiness: contrasts of personal expressiveness (eudaimonia) and hedonic enjoyment. *J. Pers. Soc. Psychol.* **64**, 271–360.

Wilder, L.K. (2002). The homeless are people, too: including homeless students in educational programming. In *Educating all learners: refocusing the comprehensive support model* (ed. F.E. Obiakor and P.A. Grant), pp. 68–83. Charles C. Thomas Publisher, Springfield, Illinois.

Williams, C. (1993). Who are the 'street children'? A hierarchy of street use and appropriate responses. *Child Abuse Neglect* **17**, 831–41.

Woodhead, M. (1999). Combating child labour. Listen to what the children say. *Childhood* **6**, 27–49.

World Health Organization (WHO) (1946). *Basic documents* (15th edn). WHO, Geneva.

World Health Organization (WHO) (2001). *International classification of functioning, disability, and health: ICF.* WHO, Geneva.

George W. Burns is a clinical psychologist, author, therapist trainer, and director of the Milton H. Erickson Institute Western Australia. He has authored numerous articles, several book chapters, and five books, including *Nature-guided therapy, Standing without shoes* (with Helen Street and the Dalai Lama), *101 Healing stories* and *101 Healing stories for kids and teens*. He is also Adjunct Senior Lecturer, School of Psychology, Edith Caran University.

Chapter 16

Naturally happy, naturally healthy: the role of the natural environment in well-being

George W. Burns

Introduction

Does contact with the world of nature have beneficial effects on our well-being? If so, what are those effects? Can they enhance our psychological, or even, physical health? What is the evolutionary connection between well-being and nature? Is it relevant in modern, urbanized, high technology societies? How might it be utilized effectively for our well-being? And is it an effective enhancer of well-being for everyone?

These are some of the questions examined in this chapter, which begins by exploring what some researchers are calling a 'mismatch' in the relationship between humans and their natural environment—a factor that may contribute to a lack of well-being both emotionally and physically. From there I will describe a psychoevolutionary perspective of the development of our species before examining the shifting concepts of health and well-being as they relate to the ecology. After drawing a parallel between the correlates of happiness and the benefits derived from interactions with nature, I provide a more detailed examination of those physical, psychological, social, and spiritual benefits before offering implications for planning, therapy, and a lifestyle of well-being. Throughout the discussion, I draw on a range of research from environmental psychology, ecopsychology, clinical psychology, positive psychology, architecture, social geography, anthropology, environmental science, psychobiology, evolutionary medicine, and traditional medicines. Interspersed through the chapter I provide a case example of Mary to illustrate how the research and theories can be applied at a practical, clinical level, as well as what might be the limitations of such applications. Different people respond to different therapeutic interventions—a factor one needs to be cognizant of when using nature-guided processes either for the enhancement of well-being or as therapeutic interventions.

A 'mismatch'

The Western world is currently experiencing a number of interesting paradoxes. We have more wealth and possessions than our predecessors and yet we have rocketing rates of diagnosable depression and the resultant consumption of mood-altering medications. We have more advanced medical technology than has ever been known in the history of the planet and yet we suffer increasingly from conditions such as lifestyle illnesses—more so than our Neanderthal ancestors, say some (Bowden *et al.* 1990). We have bigger and better homes and, along with these, an increasing incidence of broken marriages. We have greater levels of knowledge and scientific understanding; we have explored almost every corner of our planet and beyond, yet people still suffer from poverty, war, and unhappiness. We live in burgeoning highly urbanized areas with the modern conveniences that city-based society can access yet we have exponential increments in social problems such as crime rates and mental illness.

Many researchers and authors are referring to this as a 'mismatch', arguing that industrialization and technology have outstripped our biological evolution, resulting in a negative effect on our personal well-being, interpersonal relationships, and the relationship with our environment. They even go so far as to claim that it may be a significant source of emotional discontent and physical disease. Bateson (1980) lamented that we, as people, have broken the unifying bond of biosphere and humanity, disrupting what he calls 'the pattern which connects'. Ornstein and Ehrlich (1989, p. 9) say, 'There is now a mismatch between the human mind and the world people inhabit.' While such authors define the detachment of humans from their natural environment and evolutionary history, they also point to the consequences of that dissonance on our physical and psychological well-being. Garrett (1994) and Nesse and Williams (1996) see it as one of the major causes of disease. At a physical level Nesse and Williams (1996, p. 9) claim that, because we live in an environment so alien from the one in which we evolved, 'Natural selection has not had time to revise our bodies for coping with fatty diets, automobiles, drugs, artificial lights and central heating. From this mismatch between our design and our environment arises much, perhaps most, preventable disease.' In Chapter 1, this volume, Nesse has furthered this argument asserting, 'Our modern world is vastly different from the environments in which we evolved. Much, even most, chronic disease results from this mismatch; atherosclerosis, diabetes, hypertension and the complications of smoking and alcohol are rare in hunter–gatherers even at older ages.' He goes beyond just the physical consequences, also arguing that there are neurological, cognitive, and emotional consequences. 'An evolutionary view of well-being shifts most people's perspectives substantially. It is disturbing to recognize that negative emotions exist because they have been useful in certain situations, and that positive emotions can be maladaptive . . . Much easier to accept is recognition that our brains were designed to function in a far different environment, with the resulting mismatch accounting for many malfunctions.'

As Nesse implies, if we have not made the adaptation at the physical level, it seems equally logical that we are unlikely to have made it at a psychological or emotional level either. Supporting this argument, Roszak (1996) lays the responsibility fairly and squarely at the feet of Western psychology and psychiatry for following the Cartesian split of mind and matter, and emphasizing the individual and the individual's inner psyche at the neglect of people's interactive relationship with the people, animals, and plants of their ecology. He cites the example of Freud who did not pull any punches in regards to his views about nature. 'Nature', declared Freud, 'is eternally remote. She destroys us—coldly, cruelly, and relentlessly' (cited in Roszak 1996, p. 22). Freud steered the search for emotional well-being in an inner direction, seeking to root out the causal factors behind an individual's unhappiness, and thus laying a course that the science and therapy of psychology, predominantly, would continue to follow for more than a century. Roszak (1992, p. 44) concedes that, 'Freud toiled under the influence of one of the most common placed images of our language: the spatial metaphor that locates the psyche "within" and the real world "outside".' This dichotomy, Roszak (1992, p. 44) concluded, resulted in 'a psychotherapy that separated person and planet'.

An evolutionary adaptation

Examining what many researchers have theorized as an evolutionary adaptation to, and connectedness with, the natural world may help us understand this mismatch between person and planet (Appleton 1996; Kellert and Wilson 1993; Nesse and Williams 1996; Nesse, Chapter 1, this volume). Pigram (1993, p. 402) illustrates this view when he says that we have 'a genetically coded pre-disposition to respond positively to natural-environment content' and that 'The implication is that everyday, unthreatening natural environments tend to promote more complete recuperation from stress than do urban settings'. This evolutionary view is perhaps most succinctly and lyrically expressed by biochemist and microbiologist Reanney (1994, p. 35) when, using the example of an atom of iron that resides in our bloodstream, he stated, 'It was smelted into being in the fiery furnace that was the brilliant core of a giant star; it was flung across space by the violence of a super-nova when that star exploded into an apocalypse that had the brilliance of a million suns; it congealed into the rocks of just-born planet; it was rubbled into soil by wind and water and the action of microbes; it was taken up and made flesh by plant; and it now lives in a red cell, circulating the rivers of your blood, helping you breathe and keeping your consciousness afire, here, now'.

Bowden *et al.* (1990) pick up the evolutionary story after our ancestors swung from the trees, began to explore the open savannah, and physically became more recognizable as a human species. They define four phases of evolution that both reflect and define the way we have related and do relate with the ecosystems of our planet.

1 *The hunter–gatherer phase.* The hunter–gatherer phase was characterized by a process of natural selection in which the most adapted and best environmentally suited individuals were able to survive and reproduce, passing on advantageous genes from generation to generation. It was during this, the longest of the four phases, that we see the shift from a hunched ape-like skeleton to a more erect human-like body on the charts so common in museums and encyclopaedias. What is harder to portray on such visual charts is that, along with this physiological adaptation, also came an adaptation in intelligence, emotional reactions, behaviour, and survival skills. Those individuals who best understood and adjusted to the demands and cycles of the environment were the most successful in overcoming inherent danger, utilizing natural resources, and surviving.

2 *The agricultural phase.* Coming with the origins of agriculture the second phase began ten to twelve thousand years ago. People became less nomadic and more settled, finding there were advantages to staying in the most fertile and productive areas where they could manage crop production to feed themselves and their animals on a more reliable basis. Humans still needed a close affinity with their ecology to understand and utilize factors such as the patterns of nature, the changes of season, and the growth cycles of different plants. Greater utilization of nature led to a greater sense of control, alterations of social relationships, and the opportunity to develop cultural attitudes, interests, and rituals.

3 *The urbanization phase.* Urbanization characterizes the third phase and began with the early cities of Mesopotamia and south-western Asia around five to nine thousand years ago. Being in close proximity with a sizeable group of other people was more convenient, increased security, and led to the satisfaction of social and cultural needs. Specialized trades developed such as the cobbler who could stay inside and make multiple pairs of shoes from the hides he purchased in contrast to the hunter–gatherer who would have stalked an animal through the forests observing its behaviour patterns, butchered the meal himself, tanned the hide, and made his own shoes. Urbanization and specialization contributed to a further reduction of contact with nature.

4 *The high-energy phase.* The final, most recent, and shortest of these phases, has been described as the high-energy phase, which spans only the last 150 to 200 years. Predominantly a Western or developed world phase, this has merged into the current stage of high technology in which televisions and home computers have found their ways into even the most isolated regions of the globe. Every country on the planet now has television, with Bhutan becoming the last nation in the world to establish its own broadcasting station in 1999. Even in remote Himalayan villages one can find Internet cafés. News is broadcast into homes every minute of the day with less need to discuss what is happening in the community. People can learn, work, bank, and shop from computers in climate-controlled homes without stepping outside to relate to the environment or other people in person.

If we consider that each human generation spans a period of 25 years, then it has been more than 100 000 generations since the beginning of the hunter–gatherer phase, allowing our ancestors considerable time to adapt both physically and emotionally to the different environment as they shifted from trees to savannah. From the beginnings of the agricultural phase there has been a period of only 500 generations for humans to adapt to urbanization. Comparatively, the high-energy phase has been attained in just five to eight generations, while the current high-technology period (the arrival of television and home computers) is something that many adults can remember occurring within their own life span.

Intriguingly, research by Bowden *et al.* (1990) and Nesse (Chapter 1, this volume) shows an interesting paradox. While the remains of our ancestral hunter–gatherers confirm that they suffered significant, traumatic injuries (such as fractured bones, presumably from their hunting practices) and had shorter lifespans than we can anticipate today, they generally lived a healthy, disease-free life. Being nomadic and living in small groups, they escaped the infectious diseases that spread rapidly in areas of high-density living (Garrett 1994). They did not carry new strains of influenza around the globe in a matter of hours aboard an international aircraft or spread it around office buildings through air-conditioning systems. They did not suffer from nutritional or lifestyle problems, which are common today. Cardiovascular disease, cancer, dental caries, duodenal ulcers, and diverticulitis were indeed much rarer. Bowden *et al.*'s (1990, p. 41) amazing conclusion is that 'hunters and gatherers were less likely to be sick than the modern city dwellers'.

Renewing contact with nature

As a clinical psychologist, my interest is in how this research and these theories can be applied to individuals seeking to enhance and improve their well-being. Let me illustrate this with a case example.

At the age of 50, Mary was one of those clients who has a therapist scratching his or her head, asking, 'Where do I begin?' Assessment on the Beck Depression Inventory suggested a 'potentially serious' level of depression, with significantly depressed affect and high levels of both self-denigration and self-criticism. She reported feelings of indecisiveness, introversion, and irritability along with fatigue and psychomotor retardation. There was a high level of guilt and some passive suicidal ideation.

From the history there appeared to be some good reasons for her to feel depressed. Feeling depressed, like grief and anxiety, can serve a functional purpose, at times, that needs to be respected and acknowledged. Mary reported being sexually abused as a child, her marriage had been going through a difficult period, and her children, who had received growth hormone treatment, were at risk of developing the fatal Creutzfeldt–Jakob Disease from contaminated batches. In the 4 years prior to seeking psychological assistance Mary experienced a disproportionate number of traumas and losses, including two motor vehicle accidents and the deaths of several close friends and family members. She was there for them all, like a universal mother.

Mary's problems were complex, long-standing, and real. The therapeutic question was 'What sort of interventions would best help Mary improve her state of affect, develop skills to cope with both her current and future difficulties, and improve her sense of well-being?' Since levels of stress have been shown to rapidly reduce after exposure to a transparency of a nature scene (Ulrich 1981; Ulrich *et al.* 1991*a*) and, as contact with nature is through our senses, Mary was asked to complete the Sensual Awareness Inventory (SAI; see Fig. 16.1) to help assess those individual items of sensual pleasure that might be therapeutically employed to help alter her state of affect. What was obvious from the SAI was the mismatch between what contributed to her sense of well-being and what she was actually doing. She was not doing the things that enhanced her comfort and pleasure, such as gardening. Instead she was doing things that exacerbated stress, such as tending to the unresolved and, in many ways, unresolvable family issues.

Among the items Mary listed on the SAI was the enjoyment of the sight and smell of roses. It may not have been therapeutically practical to tackle all of Mary's 'issues' but it did seem a pragmatic step to help broaden and build (Fredrickson 2000 and Chapter 8, this volume) at least one area of positive emotional well-being. Consequently, we spoke about roses, exploring ways in which Mary could engage in a task of sensate-focusing on the pleasurable experience of roses. She

SENSORY AWARENESS INVENTORY

Under each heading please list 10–20 items or activities from which you get pleasure, enjoyment or comfort.

SIGHT	SOUND	SMELL	TASTE	TOUCH	ACTIVITY

Fig. 16.1 Sensory Awareness Inventory (SAI).

wrote to me the day after our consultation:

> I thought I would write to you. I bet that it is a surprise to tell what I have decided to do. I am very uncomfortable focusing on doing something nice for myself each day so I have decided since my memories have given me most pain I would grant myself one good memory a day and write it down.
>
> Yesterday I was so embarrassed. I got such a shock that I wished myself anywhere but there [a reference to her suicidal ideation]. But it was so funny when I was on the train later I was sitting there visualising making my husband rose petal sandwiches, marinating his meat in rose perfume, rose petals in the stew, lighting rose candles, sprinkling rose petals all around the room, some other ideas that I won't write down. I started to giggle, people next to and opposite me moved and put a lot of space between us, all were giving me funny glances. By the time I got off the train I was laughing out loud while walking down the street. I found a freedom of spirit that hasn't been there for a long time. Thanks for letting me find my own way through it. The rose petals are a great idea.

The day before writing this letter Mary had expressed thoughts of suicide. Within a brief time of simply thinking about contact with nature, she had been able to shift her thoughts and feelings from issues of self-destruction to ones of pleasure, enjoyment, and laughter. Her sense of empowerment in doing this is reflected in her expression of gratitude about being permitted to find her own way through the issues.

A timeline of concepts about well-being

In Fig. 16.2 I have attempted to portray a rather potted history of some of the changes in the relationship between concepts of well-being and the natural environment. By and large, this history has been one of interaction and interconnectedness between the person and the environment. Health, according to shamans or traditional healers, is a holistic concept, referring to physical wellness, as well as to an emotional and spiritual well-being. It does not include boundaries as defined as the lines I have drawn to illustrate this in Fig. 16.2. There are no clear distinctions between the person, the ecology, and the cosmology for it is the oneness or connectedness that matters in these traditions more than the differences. Disrupting the balance of the relationship between these factors is the cause of ill health, whether physical, mental, or spiritual. Restoring the oneness, righting the balance, or re-establishing open communication with the forces of the universe is what restores health and well-being.

Well-being is often associated with natural sacred sites to which are attributed healing, therapeutic powers. The process of attachment to sacred places in natural landscapes, claim Mazumdar and Mazumdar (1993), is evidenced in the way in which Indian Hindus and Tibetan Buddhists venerate the Himalayas as the abode of deities and ancestors, the Shinto and Buddhists of Japan revere Mount Fuji, the Hindu Balinese sanctify Mount Agung, or Indian Hindus relate to the river they call, *Ma Ganga* (Mother Ganges).

Often it is not enough simply for a site to be deemed sacred. For health, healing, or well-being, the person must also interact with that site or aspect of the natural world. A shaman may direct a person to engage in interactive rituals like journeying

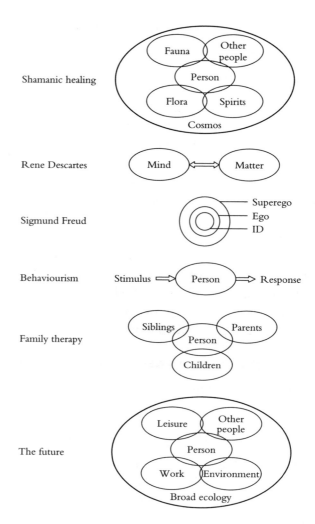

Fig. 16.2 A timeline of concepts about well-being.

to a particular natural site—a revered tree, cave, or thermal pool—and communicating with that place through the offering of sacrifices, meditation, or prayers, thus establishing and maintaining the vital connectedness between human and nature, so that the two are interacting in an ongoing, healthy dialogue. Discussing this process, Roszak (1992, pp. 79–80) states,

> One bargains with nature, apologises for intruding upon it, begs pardon of the animals one hunts and kills, tries to make good the losses one has brought about, offers sacrifices and compensation. Sanity is just such a matter of balance and reciprocity between the human and nothuman. The very idea that the two can be segregated, that the human world should or even *can* be treated as autonomously self-contained, would be the very height of madness for a traditional psychiatry. The connection between the two is not simply a matter of survival but of moral and spiritual well-being.

The interconnectedness of person, spirit, and universe is well summarized by Dr Tenzin Choedhak, a former chief medical officer at the Tibetan Medical Institute in Dharamsala, India, and personal physician to the 14th Dalai Lama who said, 'Based on the holistic Buddhist concept of mind-body, Tibetan medicine is a system of psycho-cosmo-physical healing' (Clifford 1984). The holistic philosophy he describes, whilst not limited to Tibetan medicine (even the Western term, 'medicine', can limit the broader conceptualization of such approaches), illustrates how traditional healing sees well-being as a balance between the mind, the body, and the universe. In fact, most if not all traditions of healing and well-being have taken this big picture perspective.

A narrowing of this perspective came in the philosophy of René Descartes. With the splitting of mind and matter, Descartes set science on a path that allowed for the discovery of antibiotics, the Salk vaccine, and a host of modern effective medications. Western medicine is the only major healing discipline to assume this reductionistic approach of which our culture—and now many others—has become so accepting. It has allowed for the microscopic study of the human body at a cellular level and opened up new healing approaches not possible in the broader cosmological view to healing. Consequently, it is not so much a question of either a reductionist *or* wholistic approach as perhaps an acknowledgment of the value of reductionist medicine *and* the importance of a healthy relationship with our environment.

Sigmund Freud, as founder of the modern science of psychology, narrowed the perspective even further by funnelling attention not just to the mind but the inner workings of the mind. He theorized the id, ego, and superego and, as quoted above, saw these as very segregated from an ecology that he considered so negatively. Then the behavioural scientists proposed that a solely inward direction of focus on mental well-being was too limiting and in response explored how specific aspects of the environment (stimuli) impact on an individual and how that individual responds. Following them, an awareness of the importance of the interaction of elements within systems was developed in the family therapy models, again broadening the basis of our understanding and treatment by including the relationships people have with significant others in their life and the sociocultural contexts within which they live.

I want to propose: that well-being is best conceived in an approach that acknowledges all the information we have available, both from the research laboratories and traditional wisdom—that it is not one *or* the other; that it is not as simplistic as assuming well-being exists either in the global cosmos (as tends to be the 'new age' approach) or in reductionistic Western medicine (as tends to be the 'scientific' approach). Sitting and smelling the roses may not remove a cancerous tumour that could be better treated surgically or chemically and examining a cancerous cell while failing to observe environmental factors, such as increasing levels of air pollution or chemicals in the food chain, may not prevent the growing incidence of such disorders.

Benefits of human–nature interactions

There is nothing novel in the idea that contact with nature has distinct benefits for human well-being. Back in 1882 Thoreau (2000) wrote of the importance of living close to nature in order to know one's true self. One hundred and twenty years later,

Nakamura and Csikszentmihalyi (2003, pp. 84–5) seem to echo a similar sentiment when they say, '. . . humans are socioculturally and historically situated actors whose experience is constituted jointly by environment and person. A person's goals influence transactions with the environment—but only through transactions with the environment will a self be realized.' Leopold (1949), a pioneering ecologist, linked the health of humans with the health of the land. Suzuki (1990) has made a lifetime commitment to studying the benefits of nature and arguing for the protection of the natural environment—for the well-being of humans. Indigenous cultures across the globe have historically seen our link with nature as an essential connectedness with health and well-being. The breaking or severing of this link is often considered the root cause of disease and unhappiness (Achterberg 1985; Eliade 1989; Roseman 1991; Mazumdar and Mazumdar 1993).

According to many researchers, it seems that humans are often at their best— physically, emotionally, socially, and spiritually—when engaged in non-threatening interactions with nature. Kaplan and Kaplan (1996) describe nature as being conducive to what they refer to as 'effective functioning'—a concept not dissimilar to Csikszentimihalyi's 'flow' (1990) or Nakamura and Csikszentmihalyi's 'vital engagement', which the latter describe as 'one feature of optimal development' (Nakamura and Csikszentmihalyi 2003, p. 100). Based on 2 decades of study, Kaplan and Kaplan point out that human well-being is linked to a preference for nature-dominated landscapes that include flowers and trees over human-constructed environments such as car parks and concrete buildings. They describe four interactive variables related to effective functioning and argue that these variables are most likely to occur in the natural environment. First, in nature there is a sense of 'being away' from the demands and stresses of one's day-to-day environment such as when vacationing at the beach, walking through a national park, or spending time in a home garden. Second, in nature there is a feeling of extent, of being part of an overall larger context, thus altering the perspective that we may have on other issues. Third, nature can be a rich source of fascination, stimulating our senses with a multiplicity of varying, intriguing, and pleasurable stimuli. Finally, they describe a variable of oneness that engenders a feeling of being within a supportive and harmonious environment. These variables, inherent in the natural ecology, contribute to optimal human functioning and such optimal levels of functioning have restorative qualities (Kaplan and Kaplan 1996).

Before examining the qualities of optimal health and well-being derived through human–nature interactions, let me preface my following discussion with a qualification. More than 130 years ago Ernst Haeckel, struggling with the increasing reductionist orientation of biology, considered that the subject of biology— including the subject of human beings—could only be truly understood and appreciated in the context of the relationships between organisms and their environment. From the Greek word *oikos* he coined the term 'ecology'. *Oikos* means household or living place and through it Haeckel sought to communicate the idea of interconnectedness between person and place as an essential aspect of our well-being. It is commonplace in Western scientific language and concepts not only to define the person as an island without reference to the surrounding sea but also to define the inner 'parts' of that person as though they were autonomous entities: the physical; psychological; social; and spiritual. For convenience, I follow this common literature classification but, like Haeckel, want to emphasize that the boundaries are probably more diffuse than defined, more interconnected than independent.

Nature and health

Contact with nature has long been demonstrated to promote healthy patterns of behaviour and thus serves a preventative role against the onset of health problems. Russell and Mehrabian (1976) found that showing subjects views of pleasant natural scenes promoted more health-oriented behaviours by increasing pleasurable emotional states and reducing the desire to engage in unhealthy behaviours such as smoking and drinking. Measuring alterations in indices such as heart rate, skin conductance, blood pressure, and muscle tension, Ulrich *et al.* (1991*b*) reported that exposure to nature scenes had positive physiological benefits, hypothesizing a direct beneficial effect on parasympathetic nervous system.

As well as a preventative effect, it seems that nature also has healing effects—and this may result from an action as simple as looking out a window on to a view of trees or gardens. Kaplan (1995) who claims restorative benefits of such contact with nature—both physical and emotional—has sought to offer an integrative framework for studying and employing this process. Stone and Irvine (1994) have demonstrated how access to a window view can reduce boredom and enhance perception of creative tasks. A decade earlier Ulrich (1984) was discovering that hospital patients with a natural landscape view (compared to those who looked out on to a brick wall) were discharged more quickly, used less major pain-killing medication, and were rated more cooperative by the hospital staff. Comparable studies done with prison populations have demonstrated similar preventative and healing benefits from natural views. Inmates in cells that have views on to natural landscapes make fewer calls on the medical services, suffer fewer stress symptoms, experience fewer digestive illnesses, and report fewer headaches (Moore 1982; West 1986). Other researchers claim that benefits from enhanced contact with nature are to be found not solely with physical illnesses but also in the treatment of psychiatric patients (Levitt 1991).

As well as this seemingly direct health benefit of human–nature interactions, we can further hypothesize an indirect benefit. It is well established that chronic psychological conditions such as anxiety and depression have a deleterious effect on physical health. Examining all the published investigations into the links between anxiety and chronic heart disease over a 17-year period, Kubzansky *et al.* (1998) found an overwhelming consensus that long-term states of anxiety increase the risk of heart disease and premature death. This, they say, can happen in three ways. First, behaviours directed toward the short-term relief of stress (such as cigarette smoking, increased alcohol consumption, poor or fatty diets, and lack of exercise) are unhealthy in the long run. Second, hypertension associated with anxiety escalates risk factors for the whole cardiovascular system and, third, severe anxiety in itself is enough to trigger fatal coronary events like strokes or heart attacks. Much the same, say Glassman and Shapiro (1998), holds true for health risks resulting from severe and sustained depression.

Fortunately, the other side of the coin also seems to be validated: positive and optimistic people have a generally higher level of physical health, suffering less severe illness in the first place and, if they do become ill, have better recovery rates (Fredrickson and Levenson 1998). A growing body of research in the field of cancer, for example, claims that patients can help—though not necessarily guarantee—the course of their recovery by the attitude they hold (Spiegel 1993). Happiness, positive attitudes, contentment, and emotional well-being all appear to put less stress on our body, permit it to function more healthily, and result in fewer physical breakdowns (Argyle 1997). Psychobiologists and psychoneuroimmunologists are claiming that states of mind and emotions—both positive and negative—not only affect the major bodily functions such as heart rate, respiration, and blood pressure but can also influence the more subtle physical structures of our being. Pert (1999), from her research into the relationship between neuropeptides and emotions, asserts that all the cells of the nervous system and endocrine system are functionally integrated by networks of peptides and their receptors. Through these systems, thoughts and feelings can thus be transduced to the body via neurotransmitters that communicate with the major organs as well as the genes, the basic building blocks of our body and being. If contact with nature helps facilitate positive thoughts and feelings, these are likely to be transduced into positive states of physical functioning. The bottom line thus seems to be that the happier a person is, the healthier they are likely to be, the better they will recover from illness, and the longer they are likely to live (Danner *et al.* 2001; Maruta *et al.* 2000; Ostir *et al.* 2000; Valliant 2002).

Mind, emotion, and milieu

I should point out that I do not want to over-romanticize the concept of nature. Just as exposure to sunlight can have health benefits so too can it cause skin cancer. The sunlight is not the relevant fact here so much as the way we interact with or relate with it that determines the benefits or detriments.

In nature there are wild bears, lions, and crocodiles that are only too willing, given the chance, to assert their dominance in the food chain. Nature can also swallow us up in floods, bury us in volcanic eruptions, or split our world apart in earthquakes. When I talk of the human–nature relationship, I refer to interactions with non-threatening environments that have a positive emotional or aesthetic value.

Fortunately, the aesthetic value that humans attribute to nature seems to be universal. Purcell *et al.* (1994) explored landscape preferences in culturally different subjects half a globe apart. University students in Australia and Italy were asked to rate natural or human constructed landscapes to indicate, first, their overall preference, second, their preference as a place to live and work, and, third, their preference as a place to holiday. Natural landscapes were the preferred choice across all three categories. McAndrew (1993, p. 254) claims that such preferences come about because natural settings are 'intrinsically satisfying' and that 'we find ourselves more relaxed and at ease in natural surroundings.' Fredrickson (2000, p. 11) states it simply when she says, 'certain nature scenes evoke contentment.'

The psychological benefits gained from human–nature interactions can be found at the cognitive, affective, and behavioural levels. With varying populations in varying circumstances, a number of researchers have explored how contact with nature affects cognitive functioning. Tennessen and Cimprich (1995), using an objective measure of attentional performance with college students at examination time, found those with a window view of a mostly or all natural scene showed significantly higher cognitive performance than those with a view of a mostly or entirely human-constructed scene. Cimprich (1993) studied post-operative attentional deficits displayed by women recuperating from surgery for breast cancer. Comparing a control group who received standard post-operative care with a treatment group who received the standard post-operative care plus nature-based activities (tending plants, sitting in a park, etc.) undertaken three times a week for a period of 3 months, Cimprich reported that the latter showed significant improvement in concentration as measured on standard neurocognitive instruments. Even when there are chronic difficulties such as with attention deficit disorders (ADD), progress can be made in focus of attention, completion of tasks, and the following of the instructions for children who participate in activities in natural, outdoor settings compared to children who do the same activities in human-constructed environments (Taylor *et al.* 2001).

In a large and comprehensive study, Greenway (1995), reported some of the psychological and lifestyle benefits gained from participation in a 2 week long wilderness course. With 1400 subjects, who did the course over a period of 2 decades, he accumulated extensive data from some 700 questionnaires, 700 interviews, and 52 studies that tracked participants over several years. Approximately 90% of respondents reported experiencing an increased sense of aliveness, well-being, and energy as a result of their wilderness experience. A similar percentage said that the course had allowed them to break old habits, such as the consumption of tobacco or alcohol, whilst 77% reported initiating a major life change—in either a personal relationship, employment, housing, or lifestyle—after returning from the wilderness. From the range of data available from this and the other research I have cited, the good news is that a person does not have to retreat to a wilderness experience or take up a 'hippy' lifestyle to gain the psychological benefits of nature contact. Even a view from a window may help. As Ulrich (1981, 1983; Ulrich *et al.* 1991*b*) has shown, brief exposure to slides of nature scenes can bring about rapid and positive psychophysiological changes in people with symptoms of stress.

Just as contact with pleasurable, non-threatening natural stimuli can provide positive emotional experiences so it seems that positive emotions can have us seeking out more pleasurable interactions with nature. Fredrickson (Chapter 8, this volume) says of this reciprocal and expansive process, 'experiences of positive emotions prompt individuals to engage with their environments and partake in activities, many of which were evolutionarily adaptive for the individual, its species, or both.' This enables the individual to 'exhibit the adaptive bias to approach and explore novel objects, people, or situations', thus tapping into long, historic, and healthy adaptational skills of seeking, exploring, and expanding on experiences that are likely to enhance well-being.

The social connection

One important matter to come out of the positive psychology movement has been the highlighting of the crucial role of relationships in the experience of happiness (Diener and Lucas 2001; Diener and Seligman 2002; Seligman 2002). Emmons (2003, p. 111) sees 'the ability to engage in close intimate relationships based on trust and affection' as 'the hallmark of psychosocial maturity and a key component in psychological growth', while Reis and Gable (2003, p. 129) assert, 'Relationships are an important, and perhaps the most important, source of life satisfaction and emotional well-being.' The correlation between the two appears to be interactive, having both cause and effect. Thus it seems like both a truism and a tautology to say that, when people are happy, they generally form better relationships and, when they are involved in close, supportive relationships, they report greater levels of happiness (Burns 2000).

If relationships are so important to happiness and well-being, does nature have an influence in enhancing relationships? If so, what sort of influence? And how can we make use of this in therapy, social activities, and community planning?

Kuo and Sullivan have been principal authors of a series of studies and resulting articles that examine the correlation between vegetated environments and social relationships for residents of urban public housing (Coley *et al.* 1997). Giving their subjects a choice of access to common outdoor areas that were either treed or treeless, these researchers initially found something that was not entirely new: residents showed a significant preference for the areas with trees. However, they took their research a step further and discovered that spending time in the treed public spaces served as a significant predictor of community bonds. Those residents who preferred and frequented the treed areas spoke more to other people, communicated better, were more likely to know their neighbours by name, and reported feeling a greater sense of community (Taylor *et al.* 1998). As well as enhancing such positive social interactions, contact with nature also appears to negate negative social behaviours. Families who have a window view of nature have a lower incidence of domestic violence, whether that aggression is directed toward an adult partner or a child (Kuo and Sullivan 2001).

Beyond the self

Science has long avoided the subject of spirituality, finding the concept too difficult to define and the processes even more difficult to study in its model of knowing, yet some say there are 'compelling' empirical and theoretical reasons to ensure it is included in any thorough account of human well-being (Emmons 1999; Piedmont 1999). In fact, a sense of spirituality rates as one of the correlates of a life well-lived. This relationship between happiness and taking a 'big picture' view of life is born out in research across gender, age, religion, and nationality (Meyers 2000). From his studies, Emmons comes to the conclusion that spiritual strivings are clearly correlated with higher levels of subjective well-being, particularly in regard to greater positive affect and higher satisfaction with both life and marriage (Emmons 2003; Emmons *et al.* 1998). Drawing together the results from such data, Burns and Street (2003, p. 197) claim that 'numerous researchers have found that those of us with strong spiritual beliefs are happier and better protected against depression than those who have no particular sense of spirituality. Similarly it seems that people cope better with major adversity in their life and major physical illness if they have a sense of established spirituality.'

From time immemorial nature has shaped how humans perceive the world. We have deified it, projecting our images of a creator into the sun, moon, or a sacred place. Our ancestors turned animals, trees, and mountains into gods and spirits. As mentioned earlier, many religions have venerated mountains, rivers, and other natural phenomena as abodes of their deities. Across cultures we see the defining of natural places as sacred sites attributed with qualities of power and healing.

If we have a 'creation', we have assumed there must be a 'Creator' who needs to be worshipped. Nature has thus been the source of many religious beliefs, the representation of what is bigger and more powerful than us. This is not surprising as nature is a painting of the universe in which the human species is but one small part. Attempting to understand the processes of nature, and to form a relationship with them, is one way of establishing at least a perceived sense of control.

Contributing to the sense of spirituality often reported in nature are several factors. First, people commonly describe a sense of awe when standing at the base of a powerful waterfall, gazing up at a towering tree, or sitting quietly on a clear evening looking at a canopy of sparkling stars. This awe relates to a something-bigger-than-me experience.

Second, spirituality tends to be enhanced by the experience of connectedness that has historically been associated with nature contact. If we take an example from ancient Taoism, humans are seen to affect the cosmos and, in turn, be affected by it. As the universe is perceived as a holistic unit, Taoist beliefs encourage actions and interactions designed to maintain harmony between fellow beings, the world of nature, and the universe as a whole. While these views are common across many cultures and belief systems, the Taoist concepts of connectedness are well summed up in the words of Tsui-Po (1994, p. 54), 'Human beings are one very delicate part of the whole picture. Humanity is considered to be a miniscule part of nature, and the key to understanding ourselves and the world we live in is through the understanding of nature.'

Third, people often describe transformational experiences from nature contact. I can think of no better illustration than the self-reported case of professor emeritus Lin Jensen who retired to his dream property of 20 acres where he planned to divide his time between gardening, bird-watching, and writing but an unfortunate back injury saw him confined to bed, morose and depressed, for a month and half before an incident about which he eloquently wrote (the case is described more fully in Burns 1998, pp. 219–23). A small flock of dark hooded juncos and white-crowned sparrows briefly landed in the winter-bared limbs of a tree outside his window, and then were gone. At that moment said Jensen (1997, pp. 5–6), 'I recognized that joy had come. And before this recognition, before the joy had been framed as a separate thought and thus set apart as an experience, I knew that I had simply been joy, the whole of me, without reservation, without anything set aside, so that I had not felt joy in addition to or in spite of my circumstances but rather that the whole of my circumstances had simply become joy.' He later concluded, 'The window birds are a memory now. Beyond the window, I see the garden buried beneath snow. Down under, the garden lies fallow and waits, as I lie here fallow and wait, and everywhere, in all places, at all times, and in all beings, joy lies fallow and waits.'

Seeing nature as a significant factor in our spiritual well-being offers many advantages: it adds to inner reflection and contemplation (Fredrickson and Anderson 1999), contributes to personal growth (Burns 1995, 1998, 1999), builds a sense of spiritual well-being (Heintzman 2000), and develops feelings of wholeness or belonging (Williams and Harvey 2001).

Nature-guided experiential learnings for clinical application

Knowing what we know about the role of the natural environment in human well-being, how can this be applied at an individual level to enhance recovery and well-being in day-to-day life and with clinical cases? While Mary (whose case was in part described previously) was receiving therapy, she announced that she and her husband were planning to take a holiday to central Australia and visit Ayer's Rock

(Uluru). Here was an opportunity to further her treatment through experiential nature-guided exercises. To help facilitate her relationship with the local ecology of Uluru, she was asked to take some quiet time to herself, experiencing the environment through each sense modality.

Again, in her prolific writing style, she put pen to paper about her experience. She described how her husband and others had started to climb Uluru at 5.30 a.m. so as to be at the summit for the sunrise, three-quarters of an hour later. Mary chose not to make the climb. She wrote,

> Instead, I walked around the base. I was on my own. I could neither see nor hear anyone else and in the expectant stillness of the dawn, it was like the beginning of time. The Rock seemed to be alive and it was as if it was whispering the secrets of the past, promising much for the future. I felt blessed to have had that time just there alone in such a spiritual place. Time stood still. Later I learnt the traditional owners prefer visitors not to climb Uluru. I am glad I chose not to climb. The experience of standing alone in the vastness was truly a spiritual experience. The healing is complete.

Her account clearly illustrates the personal journey she experienced as she walked around the base of Uluru. She established a relationship with nature, as intimate as if it was whispering secrets to her. These activities facilitated a therapeutic change more rapid and more effective than I think she would have gained from sitting in a consulting room lengthily exploring all the traumas of her past. She was able to look forward to a promise of the future and sense that her own healing was complete. With her letter she enclosed a photograph of a golden sunrise emerging behind the dark silhouette of Uluru.

Some concluding thoughts

Non-nature lovers

A question arises as to whether interacting with the natural environment is beneficial for everyone's well-being. Are there not some people who do not appear to relate to nature? People who are terrified of venturing into a forest for fear of snakes, spiders, or other creepy crawlies, just as there are people who are fearful of being in shopping centres or enclosed spaces? Will nature-based activities or therapeutic exercises help such people? These questions raise several issues. First, they highlight the mismatch that was discussed at the beginning of this chapter (Bateson 1980; Garrett 1994; Nesse 2000, Nesse and Williams 1996; Ornstein and Ehrlich 1989). Though we have had a long-established connectedness with the natural environment through our evolutionary history, the most recent high-technology phase has seen us becoming more and more detached from nature. This does not mean that natural environments are not pleasant or aesthetic to most people but that their day-to-day life has become detached from that context. Most of us arise in the morning putting our feet on to a carpeted floor before slipping them into shoes, walking across a concrete or asphalt surface to our car that we drive in air-conditioned comfort to an air-conditioned office before returning home at the end of the day. Our urbanized society, and our behaviours within that society, minimize the contact that we have with nature, giving little chance to feel the earth beneath our feet or the warmth of the sun on our skin. Yet, when people holiday or take recreational time at the weekends, they often drive into the forested hills, picnic beside a stream, or sunbathe on a beach, seeking out nature as a source of relaxation and pleasure.

Second, to have contact with nature, a person does not need to retreat into a wilderness experience. Nature contact can be as simple as stopping to smell a rose. Dr Sydney Rosen, a New York psychiatrist with an urban-dwelling clientele, asked the Phoenix-based, innovative psychotherapist and hypnotherapist, Dr Milton H. Erickson, about the environmentally based stories and exercises that Erickson often provided his patients. He enquired, 'Do you find that city people can also get some-thing from stories about flowers and gardens and such, even though they might not have had much experience with those things?' To this Erickson replied, 'I have sent more than one depressed man to go and dig and plant a flower garden for someone.' Referring to a mountain that Erickson often recommended that his clients climb, Rosen pressed the issue, 'I am trying to find the New York equivalent of climbing Squaw Peak.' Erickson responded that, even in New York, it was possible to 'find a nice crooked tree in Central Park with a squirrel in it' (Zeig 1980, pp. 250–2).

Third, when asked about what brings pleasure and enjoyment, people most com-monly describe nature-based experiences. In the 1970s while working with behaviour therapy programmes for a prison population, I discovered that prisoners, from maximum security prisons as well as open prisons set among hundreds of hectares of farmland and forest, rated nature-based experiences as pleasurable rein-forcers that contributed to their relaxation, happiness, and well-being (Burns 1974).

For more than 20 years I have been offering the SAI (see above) to relevant clients in private practice. It is an interesting observation, from the estimated 2000 or so of these that I have administered, that the majority of sensory pleasures and activities that people relate as being enjoyable are nature-based: watching waves crash on a shore; hearing the rhythmic swish of waves across sand; smelling the salty air at the beach; tasting the salt on one's lips; or feeling the cool dampness of wet sand beneath their feet.

Very rarely do 'canned' types of entertainment rate on these scales. It is surpris-ingly unusual for adolescents to list playing computer games and on only three occasions have adults listed watching TV. Though these responses may in part be a product of the nature of the inventory, they also indicate a high association with pleasurable nature-based sensations. This leads to an interesting paradox, for the Australian Bureau of Statistics (1998) figures on how Australians use their time show that people in my country spend the most number of daily hours sleeping, working, and watching television, in that order. While there may not be much choice about the time one spends sleeping or working, there is more choice about recreational time. The interesting thing is that watching television does not usually rate high on people's list of pleasurable activities yet it appears that this is what they devote most time to after sleep and work. In other words it seems that we do the things we *do not* enjoy doing but devote very little time to those nature-based experiences that *do* give us pleasure.

A fourth factor is that many researchers, theorists, and writers in the area claim that our connectedness with nature is universal (Appleton 1996; Kellert and Wilson 1993; Nesse, Chapter 1, this volume; Nesse and Williams 1996; Pigram 1993). Historically, they say, we have developed an evolutionary adaptation to and connection with the natural world. Pigram (1993, p. 402) even asserts that it is a 'genetically coded pre-disposition' to respond positively to the natural environment. In other words it seems not so much a question of whether people do or do not relate to the natural environment but more a question of what aspects of the environment they relate to in a most pleasurable way, and how that interaction can be effectively utilized for enhancing well-being.

The animal dimension

In discussing the benefits of interactions with nature I have not spoken directly about human interactions with animals as that is another field of study in itself. However, I want to recognize the research and work being done in this area and its contribution to human well-being. As previously mentioned, traditional societies have long seen our health and well-being as related to the connectedness with the total ecology and that includes flora, fauna, and other human beings as well as the spirit world. This relationship is such that in some indigenous cultures people apologize to animals for taking their lives, or thank the gods for providing them as food.

For dolphin researcher, Horace Dobbs, the discovery of the relationship between wildlife and emotion came as unexpected surprise. Whilst studying a friendly wild dolphin off the coast of Wales, he was asked to take a middle-aged, depressed man on one of his trips. The dolphin seemed to single out this man from the other people who had joined it in the water and, on their return to the shore, Dobbs observed a change in the man's affect, subsequently going on to study the relationship between dolphin contact and the alleviation of depression (Dobbs 1987, 1988).

More recently wildlife tourism, and its proclaimed benefits for human well-being, has expanded to an economically valuable growth industry (Burns and Howard 2003; Burns and Sofield 2001; Duffus and Deardon 1993; Moulton and Sanderson 1999; Orams 2002; Reynolds and Braithwaite 2001).

One interesting area of human–animal interaction research is in the promotion of empathy and pro-social behaviour in children. Thompson and Gullone (2003) consider that empathy and pro-social behaviour are essential building blocks in the development of psychological and social well-being among children. They review a range of literature that proposes direct contact with animals as an optimal method for promoting the development of empathy and helping children develop the necessary pro-social skills for a successful adult life (Ascione 1997; Ascione and Weber 1996; Dillman 1999; Fawcett and Gullone 2001; George 1999; Paul 2000; Rathman 1999; Vidovic *et al.* 1999).

Comparative studies

There is both scope and need for further investigation of these areas such as through randomized controlled studies of nature-guided therapy (and other nature-based therapeutic interventions) compared with other forms of therapy such as cognitive behavioural therapy or pharmacological approaches. It is important to bear in mind with such research that studies often give a good indication of what works for the majority of people under controlled experimental conditions but do not necessarily show what works best for the individual at any given point in time. Perhaps further investigations need to examine both quantitative and qualitative data. Different people respond to different therapeutic interventions and this needs to be borne in mind when using nature-guided processes either for the enhancement of well-being or as therapeutic interventions. The fact that there are some 400 different models of psychotherapy is a clear indication that there is no one-size-fits-all approach to offering psychotherapeutic interventions.

Some implications

As a clinical psychologist, my professional interest in the role of the natural environment in well-being has predominantly focused on its therapeutic benefits, whereas its applications are much broader than just the clinical perspective. Indeed, much of the research that I have cited comes not from the area of clinical or therapeutic psychology but rather from areas of architecture, social geography, environmental psychology, and ecopsychology, encouraging us to see the resources of nature in a broader perspective. Let me mention just some of these.

First, the relation of nature to our well-being has implications in planning and developing our current and future society structures (Kaplan and Kaplan 1997; Kaplan *et al.* 1998). If housing developments with window views and easy access to treed parklands correlate with a stronger sense of community and reduced levels of domestic violence, then providing such areas in housing developments is likely to lead to healthier communities with resulting reductions in crime rates and mental health issues such as depression. If viewing nature from a hospital window has significant benefits in healing, reduction of pain-killing medication, and quicker discharge from hospital, then there seem to be sound medical and economic reasons to be greening our medical facilities. If prisoners who have views of nature are less aggressive and require less from medical services, then planning such resources would seem to have good psychological, medical, and economic logic. If there is greater attention and higher cognitive performance for people who either have a view of, or are operating in, a natural environment as compared to a human-constructed environment, then there seems to be good reason for employers to take note of these results. If children with ADD show greater focus of attention and completion of tasks as well as display a better ability to follow instructions when engaged in activities in natural outdoor settings, the planners of educational facilities would be wise to develop such environments, and the educators would also be wise to plan activities in such environments.

Second, the implications of research and theories about the therapeutic benefits of nature for well-being are several-fold. Conventional therapy has long been directed to looking at the individual and the intrapsychic workings of that individual. This reductionist approach has been limiting, resulting in a 'psychotherapy that separated person and planet' (Roszak 1992, p. 44). By looking outside the literature on counselling and psychology to the multidisciplinary evidence available about the interaction between people and their environment, it may be possible to broaden the basis from which some approaches have commonly tackled psychological issues and start to build new effective strategies.

Interactions with nature may well enhance the efficacy and expediency of therapy. If stress can diminish within seconds of exposure to natural scenes (Ulrich 1981, 1983; Ulrich *et al.* 1991*a*), do we need to spend years exploring the inner workings of the psyche to find what may or may not have caused the stress? If therapeutic clients can learn to utilize the results of such nature-based research, are they not more empowered and in a better position to manage difficult life situations when they occur?

The pragmatic interventions of nature-based therapies do not necessarily require lengthy wilderness experiences as there may be effective mood-changing experiences in actions as simple as stopping to smell the cinnamon-like fragrance of a carnation, feed a bird in the backyard, or pause to watch a glistening dew drop on a spider's web. Learning such strategies helps people broaden their range of potential emotional responses and enables them to build on those new experiences as present and future strategies for well-being (Fredrickson, Chapter 8, this volume).

Finally, the findings I have sought to present have lifestyle implications for our day-to-day existence. Given the physical, psychological, social, and spiritual benefits that accrue from interacting with nature, there are simple strategies that can enhance our general sense of well-being and thus build resistance to the various challenges that we will encounter in life. Contact with nature enhances happiness and facilitates healing. People who are happy are healthier, enjoy better relationships, have a greater sense of well-being, and live longer. Often accessing these benefits is an easy and simple process: stepping out of the office at lunchtime to eat a sandwich on a park bench rather than behind the desk; exercising by cycling through a park or jogging along the beach rather than working out on an exercise machine in a back room; or watching a sunset with a picnic and wine at the end of a stressful day.

As well as having therapeutic applicability, the SAI can be used for the enhancement of pleasurable, enjoyable, day-to-day life experiences. A colleague illustrated this when she reported using the inventory on a recent family holiday. Having filled out her own SAI she asked her husband and three children to do the same. As a family, they based their seaside holiday around doing nature-based activities from their SAIs so as to provide maximum enjoyment.

Nature itself holds a multitude of stimuli and is therefore an invaluable resource in increasing pleasurable input. The stimuli in natural environments are softer, more pleasing, and have a better 'biological fit' than stimuli in human-made environments. Interacting with the magnitude and quality of natural stimuli makes it difficult to be depressed at the same time. In fact, it is proposed that natural environments can act as a reciprocal inhibitor of depression. Try to imagine how incompatible it must be to feel a sense of languishing while at the same time you watch a school of dolphins frolicking in the surf, gaze in awe at the kaleidoscopic display of a sunset, or cross-country ski over cotton-wool snow that softly decorates trees and mountains.

Could Mary have retained her suicidal ideation as she experimented with thoughts of roses, or watched the sunrise over Uluru? Perhaps, but the stimuli provided by nature are generally inconsistent with depression and thus likely to create a natural shift towards more positive, optimistic states.

Mary still has problems. Some of them won't go away. What has changed is the way in which she thinks and feels about them. She has discovered a vast reserve of natural stimuli that can readily assist her to create greater feelings of happiness and pleasure. Thinking of roses instantly brings a smile to her face. The poster-size photograph of an Uluru sunrise hanging on her kitchen wall puts her in touch with a sense of healing and spirituality. She knows the advantages of taking a walk by the river and is empowered in the knowledge that she can choose to be in touch with these facilitators of nature-guided well-being when she requires.

References

Achterberg, J. (1985). *Imagery in healing*. Shambala, Boston.

Appleton, J. (1996). *The experience of landscape*. John Wiley and Sons, New York.

Argyle, M. (1997). Is happiness a cause of health? *Psychol. Health* **12** (6), 769–81.

Ascione, F.R. (1997). Human education research: evaluating efforts to encourage children's kindness and caring toward animals. *Genet. Soc. Gen. Psychol. Monogr.* **123**, 59–77.

Ascione, F.R. and Weber, C.V. (1996). Children's attitudes about the humane treatment of animals and empathy: one-year follow-up of a school-based intervention. *Anthrozoos* **9**, 188–95.

Australian Bureau of Statistics (1998). *How Australians use their time, 1997.* Australian Bureau of Statistics, Canberra, Australia.

Bateson, G. (1980). *Mind and nature: anecessaryunity.* Bantam, New York.

Bowden, S., Dovers, S., and Shirlow, M. (1990). *Ourbiosphere under threat: ecological realities and Australia's opportunities.* Oxford University Press, Melbourne.

Burns, G.L. and Howard, P. (2003). When wildlife tourism goes wrong: a case study of stakeholders and management issues regarding Dingoes on Fraser Island, Australia. *TourismManage.* **24**, 699–712.

Burns, G.L. and Sofield, T. (2001). *The host community: social and cultural issues concerning wild life tourism,* Wildlife Tourism Research Report Series, no. 4. CRC Tourism, Brisbane.

Burns, G.W. (1974). Reinforcement preferences of a WA prison sample. *Department of Corrections Research Bulletin,* June 1974.

Burns, G.W. (1995). Psychoecotherapy: an hypnotic model. In *Contemporary International Hypnosis* (ed. G. Burrows and R. Stanley), pp. 279–84. John Wiley, London.

Burns, G.W. (1998). *Nature-guided therapy: brief integrative strategies for health and well-being.* Brunner/Mazel, Philadelphia.

Burns, G.W. (1999). Nature-guided therapy: a case example of ecopsychology in clinical practice. *Aust. J. Out door Educ.* **3** (2), 9–16.

Burns, G.W. (2000). When watching a sunset can help a relationship dawn anew: nature-guided therapy for couples and families. *Aust. NZJ. Fam. Ther.* **20** (4), 184–90.

Burns, G.W. and Street, H. (2003). *Standing without shoes: creating happiness, relieving depression, enhancing life.* Prentice Hall, Sydney.

Cimprich, B. (1993). Development of an intervention to restore attention in cancer patients. *Cancer Nurs.* **16** (2), 83–92.

Clifford, T. (1984). *Tibetan Buddhist medicine and psychiatry: the diamond healing.* Aquarian, Wellingborough.

Coley, R.I., Kuo, F.E., and Sullivan, W.C. (1997). Where does community grow: the social context created by nature in urban public housing. *Environ. Behav.* **29**, 468–94.

Csikszentmihalyi, M. (1990). *Flow: the psychology of optimal experience.* Harper Perennial, New York.

Danner, D., Snowdon, D., and Friesen, W. (2001). Positive emotions in early life and longevity: findings from the nun study. *J. Pers. Soc. Psychol.* **80**, 804–13.

Diener, E. and Lucas, R.E. (2000). Subjective emotional well-being. In *Handbook of emotions* (ed. M. Lewis and J.M. Hariland-Jones), pp. 325–37. Guilford Press, New York.

Diener, E. and Seligman, M.E.P. (2002). Very happy people. *Psychol. Sci.* **13**, 81–4.

Dillman, D. (1999). Kids and critters: an intervention to violence. In *Child abuse, domestic violence, and animal abuse: linking the circles of compassions for prevention and intervention* (ed. F.R. Ascione and P. Arkow), pp. 424–32. Purdue University Press, Lafayette, Indiana.

Dobbs, H. (1987). Dolphins—can they dispel the blues? *World Magazine.*

Dobbs, H. (1988). Dolphins and the blues: can dolphins help humans suffering from depression? *Caduceus* issue 4.

Duffus, D.A. and Deardon, P. (1993). Recreational use, valuation, and management, of killer whales (*Orchinusorca*) on Canada's Pacific Coast. *Environ. Conserv.* **20** (2), 149–56.

Eliade, M. (1989). *Shamanism: archaic techniques of ecstasy.* Arkana, London.

Emmons, R.A. (1999). *The psychology of ultimate concerns: motivation and spirituality inpersonality.* Guilford Press, New York.

Emmons, R.A. (2003). Personal goals, life meaning, and virtue: wellsprings of a positive life. In *Flourishing: positive psychology and the life well-lived* (ed. C.L.M. Keyes and J. Haidt), pp. 105–28. American Psychological Association, Washington, DC.

Emmons, R.A., Chueng, C., and Tehrani, K. (1998). Assessing spirituality through personal goals: implications for research on religion and subjective well-being. *Soc. Indic. Res.* **45**, 391–422.

Fawcett, N.R and Gullone, E. (2001). Cute and cuddly and a whole lot more? A call for empirical investigation into the therapeutic benefits of human–animal interactions for children. *Behav. Change* **18**, 124–133.

Fredrickson, B.L. (1998). What good are positive emotions? *Rev. Gen. Psychol.* **2**, 300–19.

Fredrickson, B.L. (2000). Cultivating positive emotions to optimize health and well-being. *Prevention and Treatment* 3. http://journals.apa.org/prevention/volume3/pre0030001a.html. Accessed 17 September 2003.

Fredrickson, B.L. and Levenson, R.W. (1998). Positive emotions speed recovery from the cardiovascular sequelae of negative emotions. *Cogn. Emotion* **12**, 191–220.

Fredrickson, L.M. and Anderson, D.H. (1999). A qualitative exploration of the wilderness experience as a source of spiritual inspiration. *J. Environ. Psychol.* **19** (1), 21–39.

Garrett, C. (1994). *The coming plague: newly emerging diseases in a world out of balance.* Virago, London.

George, H. (1999). The role of animals in the emotional and moral development of children. In *Child abuse, domestic violence, and animal abuse: linking the circles of compassion for prevention and intervention* (ed. F.R. Ascione and P. Arkow), pp. 380–92. Purdue University Press, Lafayette, Indiana.

Glassman, A. and Shapiro, P. (1998). Depression and the course of coronary artery disease. *Am. J. Psychiatry* **155**, 4–11.

Greenway, R. (1995). The wilderness effect and ecopsychology. In *Ecopsychology: restoring the earth, healing the mind* (ed. T. Roszak, M.E. Gomes, and A.D. Kanner), pp. 122–35. Sierra Book Club, San Francisco.

Heintzman, P. (2000). Leisure and spiritual well-being relationships: a qualitative study. *Soc. Leisure* **23** (1), 46–69.

Jensen, L. (1997). Window birds and maintenance. In *Bowing to the mountain* (ed. L. Jensen and E. Roberts), pp. 4–6. Sun Flower Ink, Carmel, California.

Kaplan, R. and Kaplan, S. (1996). *The experience of nature: a psychological perspective.* Ulrich's Bookstore, Ann Arbor, Michigan.

Kaplan, R., Kaplan, S., Ryan, R., and Ryan, R.L. (1998). *With people in mind: design and management for everyday nature.* Island Press, Washington, DC.

Kaplan, S. (1995). The restorative benefits of nature: toward an integrative framework. *J. Environ. Psychol.* **15** (3), 159–83.

Kaplan, S. and Kaplan, R. (1997). *Humanscape: environments for people.* Ulrich's Bookstore, Ann Arbor, Michigan.

Kellert, S.R. and Wilson, E.O. (Eds.). (1993). *The biophilia hypothesis.* Island Press, Washington, DC.

Kubzansky, L.D., Kawachi, I., Weisse, S.T., and Sparrow, D. (1998). Anxiety and coronary heart disease: a synthesis of epidemiological, psychological, and experimental evidence. *Ann. Behav. Med.* **20**, 47–58.

Kuo, F.E. and Sullivan, W.C. (2001). Aggression and violence in the inner city: effects of environment via mental fatigue. *Environ. Behav.* **33** (4), 543–71.

Leopold, A. (1949). *Sand County almanac.* Oxford University Press, New York.

Levitt, L. (1991). Recreation for the mentally ill. In *Benefits of leisure* (ed. B. Driver, P. Brown, and G. Peterson), pp. 161–77. Venture Publishing, State College, Pennsylvania.

Maruta, T., Colligan, R., Malinchoc, M., and Offord, K. (2000). Optimists vs. pessimists: survival rate among medical patients over a 30-year period. *Mayo Clin. Proc.* **75**, 140–3.

Mazumdar, S. and Mazumdar, S. (1993). Sacred space and place attachment. *J. Exp. Psychol.* **13**, 231–42.

McAndrew, F. T. (1993). *Environmental psychology.* Brooks/Cole, Pacific Grove, California.

Meyers, D. (2000). The funds, friends, and faith of happy people. *Am. Psychol.* **55**, 56–7.

Moore, E.O. (1982). A prison environment's effect on health care service demands. *J. Environ. Syst.* **11**, 17–34.

Moulton, M. and Sanderson, J. (1999). *Wildlife issues in a changing world,* 2nd edn. Lewis Publishers, London.

Nakamura, J. and Csikszentmihalyi, M. (2003). The construction of meaning through vital engagement. In *Flourishing: positive psychology and the life well-lived* (ed. C.L.M. Keyes and J. Haidt), pp. 88–104. American Psychological Association, Washington, DC.

Nesse, R.M. (2000). Is depression an adaptation? *Arch. Gen. Psychiatry* **57**, 14–20.

Nesse, R.M. and Williams, G.C. (1996). *Evolution and healing: the new science of Darwinian medicine.* Phoenix, London.

Orams, M. (2002). Feeding wildlife as a tourism attraction: a review of issues and impacts. *Tourism Manage.* **23** (2), 281–93.

Ornstein, R. and Ehrlich, P. (1989). *New world newmind: changing the way we think to save our future.* Methuen, London.

Ostir, G., Markides, K., Black, S., and Goodwin, J. (2000). Emotional well-being predicts subsequent functional independence and survival. *J. Am. Geriatr. Soc.* **48**, 473–8.

Paul, E.S. (2000). Love of pets and love of people. In *Companion animals and us: exploring the relationships between people and pets* (ed. A.L. Podberscek, E.S. Paul, and J.A. Serpell), pp. 168–86. Cambridge University Press, Cambridge.

Pert, C. (1999). *Molecules of emotion: why you feel the way you feel.* Simon and Schuster, New York.

Piedmont, R. (1999). Does spirituality represent the sixth factor of personality? Spiritual transcendence and the five-factor model. *J. Pers.* **67**, 985–1013.

Pigram, J.J. (1993). Human–nature relationships: leisure environments and natural settings. In *Behaviour and environment: psychological and geographical approaches* (ed. T. Gärling and R.G. Golledge), pp. 400–26. Elsevier Science Publishers, Amsterdam.

Purcell, A.T., Lamb, R.J., Perin, E.M., and Falchero, S. (1994). Preference or preferences for landscape. *J. Environ. Psychol.* **14**, 195–209.

Rathmann, C. (1999). Forget me not farm: teaching gentleness with gardens and animals to children from violent homes and communities. In *Child abuse, domestic violence, and animal abuse: linking the circles of compassion for prevention and intervention* (ed. F.R. Ascione and P. Arkow), pp. 393–409. Purdue University Press, Lafayette, Indiana.

Reanney, D. (1994). *Music of the mind: an adventure into consciousness.* Hill of Content, Melbourne.

Reis, H.T. and Gable, S.L. (2003). Toward a positive psychology of relationships. In *Flourishing: positive psychology and the life well-lived* (ed. C.L.M. Keyes and J. Haidt), pp. 129–59. American Psychological Association, Washington, DC.

Reynolds, P. and Braithewaite, D. (2001). Towards a framework for wildlife tourism. *Tourism Manage.* **22** (1), 17–26.

Roseman, M. (1991) *Healing sounds from the Malaysian rain forest: Temair music and medicine.* University of California Press, Berkeley, California.

Roszak, T. (1992) *The voice of the earth.* Simon and Schuster, New York.

Roszak, T. (1996) The nature of sanity. *Psychol. Today* **29**, 22–4.

Russell, J. and Mehrabian, A. (1976). Some behavioural effects of the physical environment. In *Experiencing the environment* (ed. S. Wapner, S. Cohen, and B. Kaplin), pp. 5–18. Plenum, New York.

Seligman, M.E.P. (2002). *Authentic happiness: using the new positive psychology to realize your potential for lasting fulfilment.* Random House, Sydney.

Spiegel, D. (1993). *Living beyond limits: new hope and help for facing life-threatening illness.* Vermilion, London.

Stone, N.J. and Irvine, J.M. (1994). Direct and indirect window access, task type and performance. *J. Environ. Psychol.* **14**, 57–63.

Suzuki, D. (1990). *Inventing the future: reflections on science, technology and nature.* Allen and Unwin, Sydney.

Taylor, A.F., Wiley, A., Kuo, F.E., and Sullivan, W.C. (1998). Growing up in the inner city: green spaces as places to grow. *Environ. Behav.* **30** (1), 3–27.

Taylor, A.F., Kuo, F.E., and Sullivan, W.C. (2001). Coping with ADD: the surprising connection to green play settings. *Environ. Behav.* **33** (1), 54–77.

Tennessen, C.M. and Cimprich, B. (1995). Views to nature: effects on attention. *J. Environ. Psychol.* **15** (1), 77–85.

Thompson, K. L. and Gullone, E. (2003). Promotion of empathy and prosocial behaviour in children through humane education. *Aust. Psychol.* **38** (3), 175–82.

Thoreau, H.D. (2000). *Walden and civil disobedience.* Houghton-Mifflin, Boston, Massachusetts.

Tsui-Po, P. (1994). *Healing secrets of ancient China.* Hill of Content, Melbourne.

Ulrich, R.S. (1981) Natural versus urban scenes: some psychophysiological effects. *Environment Behaviour* **13**, 523–56.

Ulrich, R.S. (1983). Aesthetic and affective response to natural environment. In *Human behavior and environment: advances in theory and research* (ed. I. Altman and J.F. Wohlwill), Vol. 6, pp. 85–125. Plenum Press, New York.

Ulrich, R.S. (1984). View through a window may influence recovery from surgery. *Science* **224**, 420–1.

Ulrich, R.S., Simons, R.F., Losito, B.D., Fiorito, E., Miles, M.A., and Zelson, M. (1991*a*). Stress recovery during exposure to natural and urban environments. *J. Environ. Psychol.* **11** (3), 201–30.

Ulrich, R.S., Dimberg, U., and Driver, B. (1991*b*). Psychophysiological indicators. In *Benefits of leisure* (ed. B. Driver, P. Brown, and G. Peterson), pp. 73–89. Venture Publishing, State College, Pennsylvania.

Vaillant, G. (2002). *Aging well.* Little, Brown, New York.

Vidovic, V.V., Stetic, V.V., and Bratko, D. (1999). Pet ownership, type of pet and socio-emotional development of school children. *Anthrozoos* **12**, 211–17.

West, M.J. (1986). Landscape views and stress response in the prison environment. L.M.A. thesis, Department of Landscape and Architecture, University of Washington, Seattle.

Williams, K. and Harvey, D. (2001). Transcendent experience in forest environment. *J. Environ. Psychol.* **21,3**, 249–60.

Zeig, J. (1980). *Teaching seminar with Milton H. Erickson, MD.* Brunner/Mazel, New York.

Part 5

Social and economic considerations

John F. Helliwell, usually based at the University of British Columbia, was Mackenzie King Visiting Professor of Canadian Studies at Harvard University 1991–94, Visiting Fellow of St Catherine's and Merton Colleges, Oxford (2001 and 2003), and Special Adviser at the Bank of Canada 2003–04. His most recent book is *Globalization and well-being* (UBC Press 2002).

Robert D. Putnam is Malkin Professor of Public Policy at Harvard University and a member of the National Academy of Sciences. He has written a dozen books in political science, including *Bowling alone*, a study of trends in social connectedness. He is working on strategies for civic renewal in the context of growing cultural diversity.

Chapter 17*

The social context of well-being

John F. Helliwell and Robert D. Putnam

Introduction

The purpose of this chapter is to survey the influence of social context on subjective well-being. We begin by defining key concepts, reviewing (briefly) the burgeoning literature on this topic, and indicating some of the most difficult methodological challenges in this field. We conclude by presenting original evidence from several countries, highlighting the close connection between well-being and social capital.

A prima facie case can be made that the ultimate 'dependent variable' in social science should be human well-being and, in particular, well-being as defined by the individual herself, or 'subjective well-being'. Whereas philosophers from Aristotle to John Stuart Mill have articulated this view, only in recent years have psychologists, economists, and others begun to demonstrate that subjective well-being can be measured with reliability and validity, using relatively simple self-rating questions about 'happiness' and 'life satisfaction'. Generally speaking, self-ratings of 'happiness' turn out to reflect relatively short-term, situation-dependent expressions of mood, whereas self-ratings of 'life satisfaction' appear to measure longer-term, more stable evaluations, but both produce broadly consistent findings (as indeed we confirm in our own research). This is not the place for a detailed review of the methodological issues involved in the measurement of subjective well-being, except to say that, for present purposes, a large body of literature has shown that responses to both sorts of question appear to reflect real differences across individuals that correspond with external reports on respondents (by friends, partners, and so on) and with observed behaviour (Wilson 1967; Diener *et al.* 1999; Diener 2000; Helliwell 2001; Donovan *et al.* 2003). Current research, though still preliminary, is beginning to establish biochemical correlations that reinforce this impression that measurements of subjective well-being are reasonably reliable and valid.

Among the most powerful predictors of subjective well-being, as reported in the literature, are genetic make-up and personality factors, such as optimism and self-esteem. Although not discounting such factors, we focus here instead on the social correlates of well-being. Another strong (and unsurprising) correlate of subjective well-being is physical health. Although the direction of causation underlying this correlation remains somewhat controversial, it seems quite likely that health is an important determinant of subjective well-being. In turn, a large and growing literature suggests that physical health itself is strongly conditioned by social factors, so it is plausible to conjecture that health constitutes one pathway through which social factors influence subjective well-being (Berkman and Glass 2000; House *et al.* 1982; Reed *et al.* 1983; Schoenbach *et al.* 1986; Seeman *et al.* 1993; Sugisawa *et al.* 1994; Farmer and Stucky-Ropp 1996; Kessler and Essex 1997; Roberts *et al.* 1997; Krumholz *et al.* 1998; Kawachi and Berkman 2000; Kawachi and Kennedy 1997; Ryff and Singer 2003). Although not exploring directly the putative impact of health on subjective well-being, our analysis here does include self-reports on physical health, as a way of estimating the possible *indirect* effects of social factors on subjective well-being. Our primary focus, however, is on the *direct* effects, holding physical health constant.

Which features of a person's social circumstances might be expected to affect her subjective well-being? One obvious answer is economic position or material well-being, as measured by wealth, income, or material possessions. Indeed, this factor seems so obviously important that, at least until recently, most economists have simply assumed that utility is, by definition, a product of material well-being. Much recent work, however, has questioned that assumption for, although at low levels of economic development income does indeed predict subjective well-being, at somewhat higher levels (say, above the median for Organization of Economic Cooperation and Development (OECD) countries) material well-being appears to have a quite modest effect (Diener and Oishi 2000; Diener and Biswas-Diener 2002; Frey and Stutzer 2002; Easterlin 2003). Money can buy you happiness, but not much, and, above a modest threshold, more money does not mean more happiness. Moreover, some evidence strongly suggests that, in fact, it is relative income, not absolute income, that matters most (Easterlin 1974, 1995, 2003; Kasser and Ryan 1993; Blanchflower and Oswald 2000). It is not my income itself that makes me happy, but rather a favourable comparison between my income and yours. This is one parsimonious explanation for the otherwise quite striking and startling fact that, although real per capita incomes have quadrupled in the past 50 years in most advanced economies, aggregate levels of subjective well-being have remained essentially unchanged (see Fig. 17.1 for the relevant evidence from Britain, a typical case).

Among other features of an individual's social location that have been shown in many studies to be predictive of subjective well-being are marital status, race, education, employment, and age (Glenn and Weaver 1985; Gove and Shin 1985; Gove *et al.* 1985; Coombs 1991; Clark and Oswald 1994; Clark *et al.* 2003). An

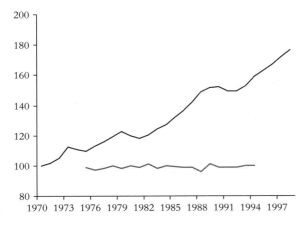

Fig. 17.1 Material and subjective well-being in Britain, 1973–97. Black line, gross domestic product per capita (1970, 100); grey line, life satisfaction (1973, 100). (From Donovan *et al.* (2003), p. 17, with permission.)

early review of the literature nearly 4 decades ago profiled the happy person as a 'young, healthy, well-educated, well-paid, extroverted, optimistic, worry-free, religious, married person with high self-esteem, job morale and modest aspirations, of either sex and of a wide range of intelligence' (Wilson 1967, p. 294, quoted by Diener *et al.* 1999). A more recent review of many subsequent studies in the US and Europe concluded that people who are married, White, better educated, employed, but not middle-aged, and have higher incomes are happier (Oswald 1997). This summary, although not identical with Wilson's initial findings 35 years earlier, is sufficiently similar to suggest that most of the key patterns seem to be relatively robust.

Marriage has universally been found to be a strong correlate of happiness, subject (like all these correlations) to some methodological cautions discussed below. Education has also been found to be a virtually universal correlate, although often its effects are substantially reduced or even absent when other variables are included. This suggests that education may be largely instrumental, acting mainly through its effects on human and social capital. Unemployment seems to be a strong negative predictor of happiness, substantially stronger than can be accounted for by the implied loss of income. Oswald's summary comment about age reflects the by now common finding that, controlling for marital status, the correlation between age and happiness is curvilinear—higher among the young and the elderly, lower among the middle-aged. Religiosity is often found to be associated with subjective well-being, although there is considerable debate about whether believing or belonging is more important, that is, whether what matters for subjective well-being is religious faith or rather participation in a religious community (Pollner 1989; Moberg and Taves 2000).

Our results will speak to all these social factors, but our distinctive focus here is on the contribution played by 'social capital'. It will thus be helpful to introduce this concept briefly.

Social capital

Physical capital generally refers to building and equipment (anything from a screwdriver to a power plant) used for production of goods and services. Several decades ago, economists started to think more explicitly of skills and education as another form of capital: human capital. More recently, social scientists in many countries have observed that social networks (and the associated norms of reciprocity and trust) can also have powerful effects on the level and efficiency of production and well-being, broadly defined, and they have used the term social capital to refer to these effects (Coleman 1993; Putnam 2000; OECD 2001; Woolcock 2001).

The core idea here is very simple: social networks have value. They have value to the people in the networks—'networking' is demonstrably a good career strategy, for example. But they also have 'externalities', that is, effects on bystanders. Dense social networks in a neighbourhood—barbecues or neighbourhood associations, etc.—can deter crime, for example, even benefiting neighbours who do not go to the barbecues or belong to the associations. Social capital can be embodied in bonds among family, friends, and neighbours, in the workplace, at church, in civic associations, perhaps even in Internet-based 'virtual communities'.

Although we do not, strictly speaking, include social trust within the core definition of social capital, norms of reciprocity and trustworthiness are a nearly universal concomitant of dense social networks. For that reason, social trust—that is, the belief that others around you can be trusted—is itself a strong empirical index of social capital at the aggregate level. High levels of social trust in settings of dense social networks often provide the crucial mechanism through which social capital affects aggregate outcomes. Indeed, so central is this relationship that some researchers include social trust within their definition of social capital.

Advocates of the 'social capital' lens have reported robust correlations in various countries between vibrant social networks and important social outcomes such as lower crime rates, improved child welfare, better public health, more effective government administration, reduced political corruption and tax evasion, and improved market performance, educational performance, etc. (Putnam *et al.* 1993; Verba *et al.* 1995; Knack and Keefer 1997; Sampson *et al.* 1997; Putnam 2000; Woolcock 2001). For example, several studies in Italy have shown that, controlling for all the other factors that might be thought to be relevant, places of higher social capital have more efficient financial and labour markets, exactly as the theory would predict (Putnam *et al.* 1993; Helliwell and Putnam 1995; Ichino and Maggi 2000; Cainelli and Rizzitiello 2003; Guiso *et al.* 2004).

Not all the externalities of social capital are positive. Some networks have been used to finance and conduct terrorism, for example. Just as physical and human capital—aircraft or knowledge of chemistry, for instance—can be used for bad purposes, so can social capital. Moreover, like physical and human capital, social capital comes in many forms, not all fungible (that is, useful for the same purposes). A dentist's drill and an oil-rigger's drill are not interchangeable, though both are physical capital.

Similarly, we need to distinguish among different types of social capital, such as the difference between 'bonding' social capital—these are links among people who are similar in ethnicity, age, social class, etc.—and 'bridging' social capital, which are links that cut across various lines of social cleavage. But the main point is that social networks can be a powerful asset, both for individuals and for communities.

How is social capital in the 'lean and mean' sense that we use it here—networks and norms of reciprocity and trust—related to subjective well-being? Empirical research on this issue is generally more limited and more recent, but such evidence as we have suggests that social connections, including marriage, of course, but not limited to that, are among the most robust correlates of subjective well-being. People who have close friends and confidants, friendly neighbours, and supportive co-workers are less likely to experience sadness, loneliness, low self-esteem, and problems with eating and sleeping. Indeed, a common finding from research on the correlates of life satisfaction is that subjective well-being is best predicted by the breadth and depth of one's social connections. In fact, people themselves report that good relationships with family members, friends, or romantic partners—far more than money or fame—are prerequisites for their happiness. Moreover, the 'happiness effects' of social capital in these various forms seem to be quite large, compared with the effects of material affluence. One preliminary study in the US found evidence that being married was, in round numbers, the happiness equivalent of quadrupling one's annual income, while monthly club meetings, monthly volunteering, and bi-weekly church attendance were each the happiness equivalent of a doubling of income. Because research on social capital is relatively recent, these findings have yet to be tested in other settings, and doing so is one of our purposes in this chapter.

Methodological cautions

Although for stylistic simplicity we have sometimes phrased the preceding literature review in terms of the 'causes' of subjective well-being, we want to emphasize four major methodological stumbling blocks that seriously complicate causal inference in this domain.

1 *Spuriousness.* Too often, analysis of the social correlates of subjective well-being has been based on simple bivariate analysis and, even when some other factors are included in the analysis, sample size has limited researchers' ability to control for all variables that might be causing spurious correlation. (The alleged effect of education on subjective well-being is one case in point, as the apparent importance of education has tended to diminish as other economic and social variables are taken into account.)

2 *Multilevel analysis.* Often, relevant hypotheses in this domain can be tested only by simultaneously examining variables at the individual and aggregate level. For example, to assess whether it is absolute or relative income that matters for happiness, we need to include both individual- and community-level measures of income in our analysis. Precisely analogous questions can be posed about the effects of social networks, education, ethnicity, and so on.

3 *Reverse causation and selection bias.* To the extent that a sunny disposition itself affects a person's location in the social structure, then correlations between social circumstance and subjective well-being might reflect the effects, not the causes of subjective well-being. In principle, this problem might even affect such 'hard' variables as income, but it seems even more threatening as regards social factors such as marital status and friendship patterns. It is especially apparent for the linkage between subjective well-being and subjective health status evaluations, both of which are likely to vary systematically with interpersonal differences in inherent optimism.

4 *Adaptation and the 'hedonic treadmill'.* If aspirations typically adjust quickly to changed circumstances (marriage, illness, income, etc.), then conventional cross-sectional data may overstate the permanency of social effects on happiness. For example, some studies report that, although lottery winners' happiness bounds upward initially, the 'high' is short-lived (Smith and Razzell 1973; Brickman *et al.* 1978). Conversely, severe physical trauma and permanent physical disability seem to have sharp negative effects on subjective well-being, but then gradually the victims become adjusted to their new circumstances, and their happiness tends to revert to the pre-trauma levels (Brickman *et al.* 1978).

The results we report here are based entirely on cross-sectional survey data. We are, therefore, precluded methodologically from addressing the second pair of issues just mentioned. Ultimately, longitudinal data and quasi-experimental methods will be necessary to resolve those uncertainties. However, the size of our samples and the abundance of measures of social context in our data do allow us to deal with the problems of spuriousness and the need for multilevel analysis. Therefore, at this stage of research, we present not confirmed causal claims, but a kind of *tour d'horizon* to highlight promising domains for future work.

Our results

This section brings together evidence on the determinants of life satisfaction, happiness, and self-assessed health status from several different national and international surveys of data on subjective well-being. Our primary focus is on the effects of social capital on alternative measures of well-being. We shall employ results from three different sources of survey data. The first source covers 49 countries, making use of data from the World Values Survey (WVS) and European Values Survey

(EVS). We shall mainly make use of a three-wave panel of roughly 88 000 observations used earlier in Helliwell (2003a). For some purposes we shall also add data from the 1999–2000 round of the EVS. The EVS data for 1999–2000 are not added in the first instance because the latest round of the EVS did not include the question on subjective health. They are, however, used in our comparisons of subjective well-being and suicide models, where the analysis is based on national average data, increasing the need to make the number of country waves as large as possible. The WVS and EVS samples average about 1000–1500 in each country wave, with the samples generally chosen to be nationally representative. The three waves were undertaken in about 1980, about 1991–92, and in 1995–97, respectively. The number of transition and developing countries increases from one wave to the next, and many of the industrial countries sampled in the first wave are absent for at least one of the two following waves. This changing sample from one wave to the next seriously limits our ability to analyse the dynamics of subjective well-being.

The second data source is the Social Capital Benchmark Survey in the US. This includes, for current estimates, about 29 000 observations drawn from a national random sample supplemented by samples from many participating communities. Although this means that the sample does not exactly match national characteristics, the very large size provides a great deal of power, and many tests suggest that the results can in most respects be treated as nationally representative. Moreover, like the Canadian data, the US Benchmark Survey has a much broader set of measurements of social capital than the WVS/EVS surveys, allowing us to explore more precisely the effects on subjective well-being of many different aspects of social context. The Canadian data are drawn from two national waves and two special oversamples (one of major urban centres and the other of forest industry communities) of a survey sponsored by the Social Sciences and Humanities Research Council of Canada. For our current analysis, the sample is about 7500. Tests reported elsewhere based on samples from the first Canadian ESC (equality, security, and community) survey waves (Soroka et al., in press) suggest that the parameter estimates from the national and oversample populations are fairly consistent. However, the much larger sample size and more balanced distribution of the US Benchmark Survey mean that it is somewhat more likely to be nationally representative.

We have already discussed different ways of measuring subjective well-being (SWB). Our survey evidence gives us some basis for comparing alternative measures. The two well-being measures that we are considering are life satisfaction measured on a 10-point scale, in the WVS and ESC samples, and happiness measured on a four-point scale in the WVS and Benchmark samples. Besides the definitional difference, we are also facing the discrepancy in the scale. To make the coefficients from survey linear estimation on the happiness equation more easily comparable with those from the 10-point life satisfaction equations, we multiplied the coefficients in the happiness equations by 2.5. An alternative way of dealing with the scale difference is to use survey-ordered probit estimation, which returns effects to the underlying latent index, thus making the estimates insensitive to the choice of scale. Our experiments that compared the 10-point life satisfaction and its collapsed four-point counterpart show almost exactly the same estimates. Because the survey-ordered probit results are almost identical in significance and relative size to the linear estimates using converted scales, and because the latter are easier to analyse, we report them in Table 17.1.

The first two columns of Table 17.1 show survey linear estimation results of life satisfaction and happiness equations from the WVS survey. The third column and the fourth columns are for the life satisfaction equation of the ESC survey and the happiness equation of the Benchmark survey, respectively.[1]

Comparison of the life satisfaction results with those of the happiness question shows specific differences within an overall context of substantial similarity. The differences are generally consistent with previous research suggesting that the life satisfaction question triggers answers that are more reflective of one's whole life experience than of one's current circumstances or mood. For example, there are striking differences in all three samples (global, US, and Canada) in the effects of age. Those aged over 65 have much higher life satisfaction than happiness scores. In the global sample, Scandinavians have higher measures of life satisfaction than of happiness, even though in both cases they show positive residuals. Those who report that God plays a very important role in their lives have higher reported measures of both life satisfaction and happiness, although the effect is larger and more significant for life satisfaction. The effects of trust show up more significantly (and are generally larger) in the equations for life satisfaction than in those for happiness.

In most other cases, the results from the two alternative measures of subjective well-being are fairly consistent. There are no variables where the results are generally stronger for happiness than for life satisfaction, but there are few if any variables in the equations that refer specifically to temporary circumstances that might be expected to have greater effects on happiness than on life satisfaction. One might think that this might be the case for unemployment, but even here the life satisfaction effects are stronger. In short, the 'life satisfaction' measure seems marginally better than the 'happiness' measure for our purposes of estimating the

---

[1] Electronic Appendix A (attached to the original publication of this chapter (*Phil. Trans. R. Soc. Lond. B* (2004) **359**, 1435–66; available at <www.journals.royalsoc.ac.uk>) shows results using the US and Canadian samples of the WVS data, for comparison with the results of the more recent national surveys. The appendix also shows results from happiness and life satisfaction equations estimated on comparable samples.

Table 17.1 Survey linear estimation of well-being and self-rated health equations[†]

Survey-wave	WVS 1–3	WVS 1–3	ESC 1 and 2	Benchmark	WVS 1–3	ESC 1 and 2	Benchmark
Nation(s)	World	World	Canada	USA	World	Canada	USA
Survey year(s)	1980s–mid-1990s	1980s–mid-1990s	2000–2003	2000	1980s–mid-1990s	2000–2003	2000
Dependent variable	Life satisfaction	Happiness	Life satisfaction	Happiness	Health status[¶]	Health status[¶]	Health status[¶]
r^2	0.25	0.23	0.16	0.19	0.21	0.04	0.16
Number of observations	83 520	83 520	7483	28 645	83 520	7483	28 766
Constant	2.5769**	4.3271**	5.1310**	5.7531**	2.6911**	3.1749**	2.5767**
National/community-level variables							
Per capita median income	−0.1399	−0.2387	−0.1319	−0.106	−0.0517	0.0589	−0.0313
Average membership	0.3006**	0.1644**	0.1198	−0.0022	0.1037**	−0.0615	0.1144
Average trust	0.3812	0.1168	0.1308	0.8432**	1.0362**	0.0883	0.7253**
Average importance of God/religion	0.8164*	0.6356**	0.8722	−0.1653	0.6430**	0.6703*	−0.0538
Governance quality	0.6918**	0.4304*	—	—	0.1434*	—	—
Individual-level variables							
Membership, 0–8 scale	0.0528**	0.0002	0.016	0.0274**	0.0161*	0.0056	0.0197**
Family‡	—	—	0.2795**	0.2108**	—	−0.008	0.0532**
Friends‡	—	—	0.5058**	0.5188**	—	0.0777	0.1186**
Neighbours‡	—	—	0.13	0.1276**	—	0.0960*	0.0626**
Trust, general‡	0.2192*	0.1403**	0.2475**	0.2117**	0.1428**	0.1026**	0.1508**
Trust in neighbours§	—	—	0.3307**	0.4248**	—	0.0488	0.2551**
Trust in police§	0.5519**	0.3962**	0.2519**	0.4050**	0.1330**	0.1074*	0.2292**
Importance of God/religion‡	0.4803**	0.2902**	0.1171	0.1166**	0.0583	−0.0011	−0.1069**
Freq. attend religious service‡	0.0982	0.1122*	0.1271	0.1206*	0.0009	0.0234	0.1949**

Table 17.1 (Continued)

Survey-wave	WVS 1–3	WVS 1–3	ESC 1 and 2	Benchmark	WVS 1–3	ESC 1 and 2	Benchmark
Dependent variable	Life satisfaction	Happiness	Life satisfaction	Happiness	Health status¶	Health status¶	Health status¶
Commute time to work, hours	—	—	—	−0.0827**	—	—	0.0690**
Self-reported health status¶	0.6458**	0.5382**	0.3335**	0.3512**	—	—	—
Male	−0.0251	−0.0768**	−0.1836**	−0.1124**	0.1096**	0.0532*	0.0018
Aged between 25–34 years	−0.1905**	−0.1987**	−0.2971**	−0.0504	−0.0813**	0.0151	−0.1209**
Aged between 35–44 years	−0.3316**	−0.3272**	−0.3329**	−0.0965**	−0.2326**	−0.0597	−0.2664**
Aged between 45–54 years	−0.2949**	−0.3444**	−0.3052**	−0.1178**	−0.3995**	−0.1432**	−0.4242**
Aged between 55–64 years	−0.113	−0.3130**	0.0473	0.0017	−0.5674**	−0.1403**	−0.5213**
Aged 65 years and up	0.1221	−0.1493*	0.4319**	0.0064	−0.7262**	−0.2413**	−0.5808**
Married	0.3656**	0.4816**	0.4276**	0.3281**	−0.0287	−0.0053	0.0307
Living with partner	0.2829**	0.3080**	0.4300**	0.1602**	−0.0273	0.0555	−0.026
Divorced	−0.2083**	−0.2076**	−0.2670**	0.0218	−0.0553*	−0.054	0.0091
Separated	−0.3658**	−0.3835**	−0.2401*	−0.1273	−0.1325**	0.0639	−0.1060*
Widowed	−0.1320**	−0.2169**	0.0433	0.0476	−0.1414**	−0.0094	−0.1124**
Unemployed	−0.6516**	−0.3642**	−0.8425**	−0.1684**	−0.0409	−0.0485	−0.1309**
High-school graduate equivalent	−0.1803**	−0.0452	−0.0144	0.0587	0.0705*	0.0565	0.2230**
Between high school and university	−0.1359*	0.0037	−0.1086	0.0645	0.1101**	0.1094**	0.2612**
University graduate equivalent	0.0356	0.0199	−0.1131	0.0229	0.1317**	0.1596**	0.4427**

† For comparison with life satisfaction, happiness has been rescaled into a 10-point scale from the four-point scale. The comparison group is female, whose educational attainment and total household income are in the lowest categories, are aged between 18–25 years, never married, employed or not in labour force. Income effects are shown in Fig. 17.2. *, Significance at the 0.05 level; **, significance at the 0.01 level.

‡ Index defined to be within the 0–1 range based on responses to the relevant survey questions.

§ Individual index minus national or community average of the index.

¶ The 1–5 range where 5 is the most healthy; please note that, unlike other surveys, Canadian ESC asked its respondents to report health status compared with others of their ages.

effects of relatively stable features of social context (and especially social capital) but, broadly speaking, our central results do not depend on the choice of indicators of subjective well-being.

The three right-hand columns of Table 17.1 show the estimation results for the correlates of self-assessed health status from our three survey sources: the WVS; the Canadian ESC Survey; and the US Benchmark Survey. To the greatest extent possible (as described in electronic Appendix A's table of definitions; see footnote 1, this chapter), the independent variables from the three surveys have been re-coded where necessary to make them as closely comparable as possible.

We shall discuss first the subjective well-being results, and then return to discuss the health results. In other words, the first part of this section compares the subjective well-being of people who feel equally healthy, but differ in other social characteristics. Of course, in our work as in many other studies, self-assessed health status is the single most important correlate of subjective well-being and, as the right-hand columns of Table 17.1 confirm, there are strong effects of social characteristics on subjective health status. Thus, in the second part of this section, we ask about the total effects of social context on subjective well-being, both direct (i.e. controlling for health) and indirect (through health). We recognize that the causal link between health and subjective well-being is not uncontroversial, although we tend to the view that well-being is mainly an effect not a cause in this relationship. In any event, that relationship itself is not the focus of our work here, and we present our findings in a form that will allow readers of either persuasion to see the independent effects of social context, which is our primary focus. Including self-reported health among the predictors of subjective well-being, as we do in the first part of this section, has the added advantage of tending to offset the effects of any 'positivity' or 'optimism' response bias, because such a response bias ought to affect both self-assessed health and subjective well-being.

Age and well-being

In all samples the cross-sectional age effect is lowest in the 35–44 or 45–54 years age group, and is usually highest for those over 65 years. It is important to note that this high subjective well-being of the elderly is a feature of equations in which physical health is already taken into account. If health is left out of the equation, the U-shaped pattern twists, and the low point appears later in life. Analysis of successive sets of Eurobarometer data show a persistent U shape linking age and happiness, even in the absence of health data, for annual surveys over the past 30 years. Our measurement of the effects of age on subjective health assessments will be considered later, but the basic point is simple: older people are (on average) less healthy and that substantially reduces their subjective well-being but, among equally healthy respondents, older people are more satisfied with their lives. The same pattern exists, but to a lesser extent, when health is not separately taken into account. It is also worth noting that the U shape would become shallower if marital status were not taken into account, because those who are married are happier than those who are single or widowed, and the last two conditions are more prevalent among the youngest and oldest of the survey respondents.

Income and well-being

In all samples, those with average or higher incomes show higher reported happiness than those at the bottom of the income distribution. Equations estimated with finer income gradations, as is done in the WVS global results, show diminishing returns to relative income above median levels, especially for those living in OECD countries. For the relatively poor, money can buy happiness, but for the relatively well-off, more money does not typically mean more happiness. Figure 17.2 compares the income effects in the different surveys for happiness, life satisfaction, and self-rated health.

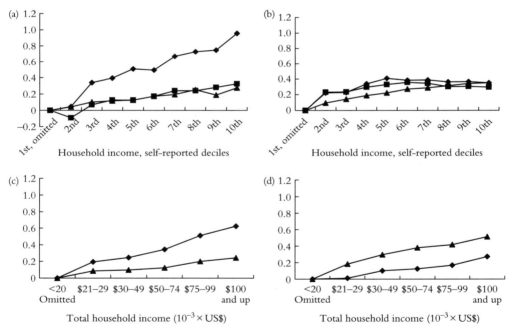

Fig. 17.2 Well-being, self-rated health status, and household income for (a) non-OECD nations, WVS survey; (b) OECD nations, WVS survey; (c) Canada, ESC survey; and (d) the USA, Benchmark survey. The y-axes show the coefficients of income controls from the survey linear regression of the 10-point life satisfaction (diamonds), 10-point happiness (originally in four-point scale; squares), and the five-point self-rated health status (triangles). Note that WVS asked its respondents to report in what decile they thought their total household income was, whereas ESC and Benchmark asked for estimates of income in dollars. Before the regression, we converted the income figures from ESC into US dollars by purchasing power. Unlike the other two surveys, the Canadian ESC asked its respondents to report health status in comparison with others of their ages.

Unemployment

As has been long established in the literature, individuals who are unemployed show significantly lower measures of subjective well-being (Clark 2003). Our current results show that this is true for the WVS global samples, for the Canadian and US WVS subsamples, for the ESC Canadian sample, and for the US Benchmark Survey of happiness. The effect is larger for the WVS global sample and the ESC Canadian sample than for the US Benchmark Survey. In all cases the SWB impact remains much larger than could be imputed to a present value calculation of the likely effects of current unemployment status on current and future incomes. Unemployment is thus likely to represent much more than a loss of income, perhaps reflecting the loss of workplace social capital as well as increases in family stress and individual loss of self-esteem.

Education

Education remains what might be referred to as an instrumental variable, being associated with higher levels of subjective well-being by simple correlations, but the effects tend to drop out (especially in equations in which health status is included) for higher levels of education in more fully specified models. Education improves health and thus indirectly improves subjective well-being but, net of that effect (and of the other factors in our analysis), education appears to have no direct impact on subjective well-being.

Gender

In the WVS/EVS global sample, overall life satisfaction is slightly higher among males than females (6.84 compared with 6.73 on the 10-point scale in the first three waves of the WVS), but this masks offsetting national differences. For example, in Scandinavia, Asia, and North America, life satisfaction is slightly higher among women than men, whereas the reverse is true, and to a larger extent, in the countries of the former Soviet Union. In more fully specified models, a gender effect sometimes arises and sometimes does not, depending on the specification of the model. One reason for the appearance of a negative male effect in some cases is that self-reported health status is worse among women than men in the WVS global sample (3.64 for women and 3.80 for men, where 5.0 is the best health status), and health status takes a strong coefficient in the life satisfaction and happiness equations. To get a more specific explanation for gender differences, we have estimated gender-specific equations. The general finding from these gender-specific equations is that the responses of males and females to different events and circumstances are strikingly similar, much more so, for example, than occurs when we model gender differences in the determinants of suicide. The only gender differences in happiness that are significant at the 1% level are that strong belief in God increases happiness more for females than males, that living in a country with a high quality of government increases happiness more for females than males, and that females are happier than males in Asia and some non-Asian developing countries. In short, unlike many other factors in our analysis, gender appears to have no strong and straightforward effect on subjective well-being.

Family-level social capital

At the family level, all samples show strong effects from family-level social capital, at least as measured by marital status. There are some differences across samples, however. In the WVS samples (including those for Canada and the US), the negative effects of divorce and being widowed are larger than in the US Benchmark Survey. In all samples, being married increases both life satisfaction and happiness, especially where the alternative is being separated or divorced. (The effect of cohabitation (as married) is generally positive, although not so strongly so as marriage. Contrary to what is sometimes believed, we find that marriage appears to increase subjective well-being equally among men and women.) The results from the new US and Canadian surveys add a new dimension to the results showing the importance of family, as those having frequent interactions with extended family members reveal systematically higher subjective well-being. Having a family enhances subjective well-being, and spending more time with one's family helps even more.

Faith and the church

In most samples, it is possible to establish separate positive linkages to subjective well-being from strong religious beliefs and from frequent church attendance. Comparing the ESC and Benchmark results, the effects of belief in God are almost equally strong in Canada and the US, but the linkages from church attendance to subjective well-being are much stronger in the US. There are also large differences between the two countries in the average levels of the variables themselves, with the prevalence of frequent church attendance and strong religious beliefs being more than twice as great in the US than it is in Canada. It is well known that, among advanced countries, religious observance is uniquely high in the US, and here we see that the impact of religious observance is also much higher in the US.

It has been suggested that church attendance creates community-level social capital (whether bridging or bonding depends on the divide under consideration), while belief in God provides alternative types of support for an individual's well-being. Support for this interpretation is provided by equations modelling the extent to which individuals think that others can be trusted. For example, if the WVS/EVS global sample is used to explain individual-level answers to the generalized trust question, conditional upon the national average value for that variable, those who attend church frequently, or who belong to more community organizations, are significantly more likely to think that others can be trusted, whereas those who have strong belief in God are significantly less likely to think that others can be trusted. In all cases, this is after taking into account national-level differences in these variables. This suggests that trust in God and trust in others are substitute modes of belief for individuals. By contrast, more frequent interactions with other people in both church and community settings tend to increase the extent to which those individuals think that others can be trusted and thereby to enhance their subjective well-being.

Friends and neighbours

The WVS data do not speak directly to the strength and nature of an individual's friendships. In the ESC and Benchmark surveys there are several relevant variables, most or all of which attest to the importance of such friendships as supports for subjective well-being. Frequent interactions with friends and neighbours are both associated with systematically higher assessments of subjective well-being. This is true in both the Canadian and US samples. In both countries, frequent interactions with friends are even more important (especially in the US) than those with neighbours and family, with family contact being slightly more important than that with neighbours. In short, informal social capital—what one of us (Putnam 2000, chapter 6) has previously termed 'schmoozing'—is strongly associated with higher subjective well-being.

Community involvement

All three of our survey samples have somewhat similar questions about the number of types of community organizations to which the respondent belongs. At the global level, civic participation matters for life satisfaction, as it does in some but not all of the smaller country samples. For happiness, there is no systematic global effect, although there is a strong effect in the US Benchmark data. Here, one might expect, as elsewhere, that there may be some two-way causality or joint influence from an excluded individual-level factor, such as extroversion or inherent optimism. However, studies at a more aggregate level, where individual-level personality differences should average out, tend to show undiminished effects. Another way of putting this is that participation is likely to have positive externalities, so that any effect at the individual level is likely to carry through to the aggregate level as long as the positive externalities are more than enough to offset the loss of personality-driven individual effects. We report our latest results on this when we discuss community-level effects more generally.

Trust

'Do you think that people can generally be trusted, or (alternatively) that you cannot be too careful in dealing with people?' This canonical question has been asked so many times over the past half-century as to become the standard assessment of trust levels. It has been much criticized and much analysed. Fortunately, many studies have found that results based on this broadly available measure tend to be confirmed by other ways of asking people about the trustworthiness of others. The radius of the question is ill-defined. Studies that link it to other questions with more specific objects of reference (friends, neighbours, police, clerks, and strangers) support the idea of using the measure to represent community-level trust. (For example, in the Benchmark Survey, a factor analysis of many different measures of social trust finds that the canonical question has the highest loading on the principal component and, over periods of 12 and 18 months, the canonical question has the highest test–retest reliability.) The level of community must extend in part to national boundaries, as cross-countries differences in average answers to this question predict differences in the frequency with which lost wallets are returned (Knack 2001). This fact, along with the fact that the question relates to the trustworthiness of others, and not to whether the individual is planning to act in a trusting manner, encourages us to use answers to this question as measures of trustworthiness. Thus, we would expect to find that those who believe themselves to live among others who can be trusted will tend to report higher subjective well-being. We do not treat these measures of trust as direct measures of social capital, but modelling of trust responses by us and by others suggests that trust levels are higher in communities that have higher social capital densities.

Individuals who report themselves as living in a high-trust environment report significantly higher levels of life satisfaction and of happiness, to extents that are roughly equal across the various surveys. The statistical strength of this relation is particularly great in the US Benchmark Survey.

Although answers to the general trust question are associated with high subjective well-being at the individual level, they by no means duplicate more specific assessments of the trustworthiness of others. The US Benchmark results are especially clear on this, as their large sample size permits the separate influence of several domains of trust to be established. The results show trust in police to be especially important in the Benchmark results, even when separate account is taken of trust in government (either local or national or, as in our equation, the average of these two measures), general trust, and trust in neighbours and in co-workers. The WVS results show the international data to have almost equally large SWB effects of trust in the police, even with aggregate measures of the quality of government taken into account. Other experiments with the Benchmark data show that the lower measures on SWB among Black Americans are related in large measure to their lower trust of the police. In short, feeling able to trust others—both those among whom one lives and works and those in authority—is strongly associated with higher subjective well-being.

Community-level effects

Supporting the many earlier findings that the subjective well-being effects of income relate mainly to relative income, the community or national level of income has an insignificant negative effect when it is added to the life satisfaction and happiness equations. This is what might be expected when incomes are measured in absolute form, as in the US and Canadian surveys. However, the same is also true in the WVS equations, where incomes are measured relative to the national average. However, in these equations, the effects of national income depend on what else is included in the equation, as noted in Helliwell (2003*a*).

If we are right to interpret individuals' answers to the trust question as their assessments of the average level of trust governing relations in their communities, we would expect to find a significant effect of community-level trust on well-being only in equations that do not control for the individual's own perception. In Table 17.1 we allow the individual's own perception to have an impact separate from those of others living in the same country or community. This is done by defining the individual-level trust assessments as differences between the individual's own assessment and those of others living in the same country or community. We then enter as a separate variable, an average trust assessment of others, where the average is across the three measures of trust and across individuals in the same country or community. Our results suggest the paramountcy of the individual's own perceptions, as the individual-level assessments are always significant, whereas the community-level trust measure, while positively signed for all surveys, is significant only for the US Benchmark Survey. (Note that in the WVS analyses, 'community-level' refers to the country as a whole, whereas in the ESC and Benchmark analyses 'community-level' refers to a level closer to the census district. We suspect that the latter is a less 'noisy' proxy for the actual environment within which each respondent lives and works.)

The situation is different for community engagement, because we might expect the satisfaction obtained by an individual from his or her own involvement to be affected by the extent to which others are similarly engaged. Some have argued that satisfaction from community activities is based on the relative degree of involvement (Nie *et al.* 1996). There is the opposite possibility, more likely in our view, that greater engagement by others would increase the satisfaction gained from our own engagement. A third and intermediate possibility is that the subjective well-being provided by one's own participation is neither increased nor decreased by the extent to which others are involved in the community. Which answer we would expect to find must depend on the model estimated. For example, we and others have found that high levels of community involvement are conducive to higher levels of trust. Thus, the total benefits of community-level participation would flow partly through their effects on trust, so that we would expect to find the estimated effects of community-level participation to be higher in models that did not include trust as a separate variable. Estimation of the WVS global life satisfaction does indeed support this notion, as the coefficients on both individual and national memberships rise when the trust variable is eliminated from the equation.

What do we find from the various bodies of data? As noted, for the global sample of WVS data, the community-level data are at the national level, whereas for the ESC and Benchmark surveys they are generally at the level of the census district. The latest results from the WVS are consistent with those reported in earlier papers: those nations with greater membership densities show higher average levels of life satisfaction, even after accounting for individual-level participation and estimated trust levels. For the ESC and Benchmark surveys, the community-level values of social capital variables generally neither add to nor subtract from what has already been found for the individual-level variables. If everyone in a community becomes more connected, the average level of subjective well-being would increase, but the channel appears, from these surveys, to be largely through the individual's own participation. There is no evidence here of the relative-participation effect matching the relative-income effect. The subjective well-being effects of income appear to flow entirely through relative incomes, so that community-wide increases in income are not accompanied by increases in measured life satisfaction. For measures of social capital, there appears to be no parallel negative effect from increased participation by others. On the contrary, the WVS estimates (but not the ESC and Benchmark samples) show that greater participation by others increases subjective well-being even for those whose own participation is not increased. Until this result is replicated among communities within nations, there remains a risk that it is capturing, in part, the effects of other important factors that differ among nations.

Social capital and health

The right-hand columns of Table 17.1 show equations for self-assessed health status, as measured on the same five-point scale used in all three surveys. These results indicate, echoing results from Berkman and Syme (1979), that there are strong links to physical health from family, friends, neighbours, and community involvement. Because our equations for subjective well-being included (and thus controlled for) each individual's self-reported health as an important determinant of subjective well-being, putting the results from the two side of Table 17.1 together allows some assessment of the total effects of social capital on well-being.

One of the biggest differences between the physical health and the subjective well-being equations relates to the effects of education, which are much larger on health than on subjective well-being (in well-being equations that include physical health). As shown in Fig. 17.2, the effects of income on self-reported health are smaller than for life satisfaction and happiness, except in the US, where the relationship is much steeper for self-reported health status than it is for happiness. All forms of social connectedness have strong positive effects on physical health, whereas strong religious beliefs do not. The age effects in the health equation are strong and almost linear, with each decade of age leading to a significant reduction in average health status. To find the overall effect of age on subjective well-being, it is necessary to combine the effects flowing through physical health with those estimated directly in the subjective well-being equation. This is done by multiplying the coefficient in the variable in question in the health equation by the health coefficient in the well-being equation, and adding this indirect effect to the direct effect of the same variable in the well-being equation. Using the linear estimation of the effects of living in a high-trust community as an example (based on the Benchmark Survey), the indirect effects flowing through health status increase the direct effects by approximately 30%.[2]

In other words, living in a high-trust community seems to improve health and thus indirectly to enhance subjective well-being, in addition to the even more powerful direct effect of a trustworthy community on subjective well-being. Generally speaking, most of our measures of social capital appear to have this 'turbocharged' effect on happiness and life satisfaction. There is a risk that the apparent effects of social capital on subjective health and on well-being can be at least partly traced to the fact that both are subjective measures. Because there are at least some international measures of measured health status, these data can be used with the national-level WVS data for social capital and subjective well-being to see if similar results can be found. Of the international variation of subjective well-being, 49% can be explained by differences in the World Health Organization (WHO) measure of healthy years of life expectancy, 72% by differences across countries in subjective measures of health status, and 76% when both measures are used. Of the variance explained by subjective health, slightly more than half is due to underlying differences in health outcomes as estimated by the WHO.

Suicide as an alternative measure of well-being

Although we have done no new suicide research specifically for this chapter, it is worthwhile bringing our earlier suicide results (Helliwell 2003*b*) into our account at this stage, for two reasons. First, it has been frequently said that the high subjective well-being and presumed high suicide rates in Scandinavian countries cast suspicion on the results linking social capital and well-being. Second, there have been many questions raised about the appropriateness of using subjective well-being data as 'true' measures of well-being. For example, the answers to

---

[2] The direct effects of 0.84 are increased by indirect effects of $0.254 = 0.725 \times 0.35$, where 0.725 is the trust effect in the health equation and 0.35 the effect of the health variable in the Benchmark happiness equation.

subjective well-being questions may reflect momentary circumstances, may mean different things to respondents of different ages, cultures, genders, and languages, and are sometimes thought to reflect too great an adaptation to current circumstances to be an acceptable means for comparing the quality of life among individuals or countries.

Because the suicide data are based on behaviour rather than subjective opinions and are collected on a population-wide basis, they provide a quite different way of measuring life satisfaction. It should be expected that the subjective well-being and suicide data might respond differently even when they are brought together for exactly the same countries and years, because the subjective well-being data are collected from a wide cross-section of the population, while the suicide data count final and often impulsive acts of individuals at the extreme lower end of the distribution from high hopes to hopelessness.

It is quite significant, therefore, that country-level measures of life satisfaction and suicide rates turn out to be explained by the same model, using the sample of 117 observations from 50 countries used previously to explain life satisfaction. Variables used included national average divorce and unemployment rates, the share of the population with a strong belief in God, two measures of social capital (extent of involvement with voluntary associations and the level of general trust), and external measures of each country's quality of government. The two equations give strikingly consistent results. Divorce and unemployment are associated with reduced life satisfaction and increased suicide (Kposowa 2000), trust and memberships are associated with increased life satisfaction and reduced suicide rates, and higher-quality government (Kaufman *et al.* 2003) is associated with increased life satisfaction (strongly) and reduced suicide (weakly). Sweden, which had previously been suggested as a puzzle because of very high subjective well-being combined with a reputedly high suicide rate, fits both equations exactly. Its better ranking on life satisfaction than on suicide reflects the different coefficients of two key variables in which Sweden differs from typical countries. Belief in God is more important in deterring suicide than in supporting life satisfaction, whereas the reverse is true for the quality of government. Sweden ranks very high on the quality of government and very low in belief in God.

The fact that the suicide data and the measures of life satisfaction show remarkably similar structures, especially with respect to the effects of social capital, thus represents a strong confirmation of the subjective well-being data. The coefficients are much larger for the suicide equation, but there is correspondingly greater international variation of suicide rates than of average measures of subjective well-being, so that the coefficients in the two equations are almost identical when compared with the standard deviations of the variables to be explained. In addition, the fact that the international differences in suicide rates are much larger than those for subjective well-being should reassure those who think that the international differences in average subjective well-being are implausibly large. This reassurance is all the greater because the suicide and subjective well-being data seem to be equally well explained by the same equation.

Perhaps the biggest difference between the suicide and subjective well-being data and results lies in the gender differences. Suicide is roughly four times more prevalent among males than females and the different rates are explained by differing models, while the gender differences in subjective well-being are far smaller, are sometimes of differing sign, and do not lead to large differences in equation structure and coefficients. This important issue of gender differences in suicide rates aside, these two independent analyses converge on one robust central finding: social context, and especially social capital, appears to have powerful effects on well-being. The suicide results also help to resolve the inevitable doubts about the direction of causation between social capital and subjective well-being in a cross-sectional setting. If subjective well-being and suicide rates are both correlated in closely comparable ways to differences in social capital and other aspects of the economic and social environment, this increases the likelihood that social capital has a causal role in both cases.

Conclusions

Our new evidence confirms that social capital is strongly linked to subjective well-being through many independent channels and in several different forms. Marriage and family, ties to friends and neighbours, workplace ties, civic engagement (both individually and collectively), trustworthiness, and trust all appear independently and robustly related to happiness and life satisfaction, both directly and through their impact on health. Moreover, the 'externalities' of social capital on subjective well-being (the effects of my social ties on your happiness) are neutral to positive, whereas the 'externalities' of material advantage (the effects of my income on your happiness) are negative, because in today's advanced societies, it is relative, not absolute, income that matters. In that sense, the impact of society-wide increases in affluence on subjective well-being is uncertain and modest at best, whereas the impact of society-wide increases in social capital on well-being would be unambiguously and strongly positive.

We emphasize again that the use of causal language in talking about the social context of subjective well-being (even as we have done for stylistic convenience) is premature, because of the possibility of selection effects, reverse causation, and adaptation effects. (Our previously reported suicide results certainly provide some evidence for the independent causal status of social capital.) The sort of cross-sectional survey research presented here cannot establish beyond doubt that (say) marriage and friendships enduringly foster happiness, rather than that happy people are simply more attractive mates. Nevertheless, the patterns we report here are sufficiently strong and pervasive to justify enhanced research to explore possible mechanisms linking social capital and subjective well-being, to look for contextual and interaction effects, and to seek instrumental variables and quasi-experimental settings that would provide more leverage on issues of causation.

Acknowledgements

We are especially grateful to Haifang Huang, Tom Sander, Elisabeth Jacobs, and Tami Buhr for their help with this research. Our revisions have been aided by suggestions from Felicia Huppert and Danny Kahneman.

References

Berkman, L. and Glass, T. (2000). Social integration, social networks, social support, and health. In *Social epidemiology* (ed. L. Berkman and I. Kawachi), pp. 137–73. Oxford University Press, Oxford.

Berkman, L. and Syme, L. (1979). Social networks, host resistance, and mortality: a nine-year follow-up study of Alameda County residents. *Am. J. Epidemiol.* **109**, 186–204.

Blanchflower, D.G. and Oswald, A.J. (2000). *Well-being over time in Britain and the USA*, NBER working papers 7487. National Bureau of Economic Research, Cambridge, Massachusetts.

Brickman, P., Coates, D., and Janoff-Bulman, R. (1978). Lottery winners and accident victims: is happiness relative? *J. Pers. Soc. Psychol.* **36**, 917–27.

Cainelli, G. and Rizzitiello, F. (2003). Social capital and local development in Italy: a note. *Ital. Politics Soc. Rev. Conf. Group Ital. Politics Soc.* **58**, 14–20.

Clark, A.E. (2003). Unemployment as a social norm: psychological evidence from panel data. *J. Labor Econ.* **21**, 323–51.

Clark, A.E. and Oswald, A.J. (1994). Unhappiness and unemployment. *Econ. J.* **104**, 648–59.

Clark, A.E., Diener, E., Georgellis, Y., and Lucas, R.E. (2003). Lags and leads in life satisfaction: a test of the baseline hypothesis. See http://www.delta.ens.fr/clark/BLINEaug03. pdf.

Coleman, J.S. (1993). Social capital in the creation of human capital. *Am. J. Sociol.* **94**, 95–120.

Coombs, R. (1991). Marital status and personal well-being: a literature review. *Fam. Relat.* **40**, 97–102.

Diener, E. (2000). Subjective well-being: the science of happiness, and a proposal for a national index. *Am. Psychol.* **55**, 34–43.

Diener, E. and Biswas-Diener, R. (2002). Will money increase subjective well-being? A literature review and guide to needed research. *Soc. Indicat. Res.* **57**, 119–69.

Diener, E. and Oishi, S. (2000). Money and happiness: income and subjective well-being across nations. In *Culture and subjective well-being* (ed. E. Diener and E.M. Suh), pp. 185–218. MIT Press, Cambridge, Massachusetts.

Diener, E., Suh, E.M., Lucas, R.E., and Smith, H.E. (1999). Subjective well-being: three decades of progress. *Psychol. Bull.* **125**, 276–302.

Donovan, N., Halpern, D., and Sargeant, R. (2003). *Life satisfaction: the state of knowledge and implications for government*. Cabinet Office Strategy Unit, London.

Easterlin, R.A. (1974). Does economic growth improve the human lot? Some empirical evidence. In *Nations and households in economic growth* (ed. P. A. David and M.W. Reder), pp. 89–125. Academic, New York.

Easterlin, R.A. (1995). Will raising the incomes of all increase the happiness of all? *J. Econ. Behav. Org.* **27**, 35–48.

Easterlin, R.A. (2003). Do aspirations adjust to the level of achievement? A look at the financial and family domains. Prepared for European Science Foundation Exploratory Workshop on Income, Interactions and Subjective Well- Being, Paris, France, 25–26 September. See <http://www.delta.ens.fr/swb/EasterlinParis.pdf>.

Farmer, J.E. and Stucky-Ropp, R. (1996). Family transactions and traumatic brain injury. In *Recovery after traumatic brain injury* (ed. B.P. Uzzell and H.H. Stonnington), pp. 275–88. Lawrence Erlbaum, Mahwah, New Jersey.

Frey, B.S. and Stutzer, A. (2002). What can economists learn from happiness research? *J. Econ. Lit.* **40**, 402–35.

Glenn, N.D. and Weaver, C.N. (1985). The changing relationship of marital status to reported happiness. *J. Marriage Fam.* **50**, 317–24.

Gove, W.R. and Shin, H. (1985). The psychological well-being of divorced and widowed men and women. *J. Fam. Iss.* **10**, 122–44.

Gove, W.R., Hughes, M., and Style, C.B. (1985). Does marriage have positive effects on the psychological well-being of the individual? *J. Health Soc. Behav.* **24**, 122–31.

Guiso, L., Sapienza, P., and Zingales, L. (2004). The role of social capital in financial development. *Am. Econ. Rev.* **94**, 526–56.

Helliwell, J.F. (2001). Social capital, the economy and well-being. In *The review of economic performance and social progress* (ed. K. Banting, A. Sharpe, and F. St-Hilaire), pp. 43–60. Institute for Research on Public Policy and Centre for the Study of Living Standards, Montreal and Ottawa.

Helliwell, J.F. (2003a). How's life? Combining individual and national variables to explain subjective well-being. *Econ. Model.* **20**, 331–360. [Previously issued as 2002 NBER working paper 9065. National Bureau of Economic Research, Cambridge, Massachusetts.]

Helliwell, J.F. (2003b). Well-being and social capital: does suicide pose a puzzle? Paper presented at the Conference on Social Capital and Well-Being, Harvard University, Weatherhead Center for International Affairs, November 7–9, 2003.

Helliwell, J.F. and Putnam, R.D. (1995). Economic growth and social capital in Italy. *Eastern Econ. J.* **21**, 295–307.

House, J.S., Robbins, C., and Metzner, H.L. (1982). The association of social relationships and activities with mortality: prospective evidence from the Tecumseh community health study. *Am. J. Epidemiol.* **116**, 123–40.

Ichino, A. and Maggi, G. (2000). Work environment and individual background: explaining regional shirking differentials in a large Italian firm. *Q. J. Econ.* **115**, 1057–90.

Kasser, T. and Ryan, R.M. (1993). The dark side of the American dream: correlates of financial success as a central life aspiration. *J. Pers. Soc. Psychol.* **65**, 410–22.

Kaufman, D., Kraay, A., and Mastruzzi, M. (2003). Governance matters III: governance indicators for 1996–2002, Social Science Research Network Working Paper Series. See <www.ssrn.com> and World Bank.

Kawachi, I. and Berkman, L. (2000). Social cohesion, social capital, and health. In *Social epidemiology* (ed. L. F. Berkman and I. Kawachi), pp. 174–90. Oxford University Press, Oxford.

Kawachi, I. and Kennedy, B. (1997). Health and social cohesion: why care about income inequality? *Br. Med. J.* **314**, 1037–41.

Kessler, R.C. and Essex, M. (1997). Marital status and depression: the importance of coping resources. *Soc. Forces* **61**, 484–507.

Knack, S. (2001). Trust, associational life and economic performance. In *The contribution of human and social capital to sustained economic growth and well-being* (ed. J. F. Helliwell), pp. 172–202. HDRC, Ottawa.

Knack, S. and Keefer, P. (1997). Does social capital have an economic payoff? A country investigation. *Q. J. Econ.* **112**, 1251–88.

Kposowa, A.J. (2000). Marital status and suicide in the national longitudinal mortality study. *Epidemiol. Community Health* **54**, 254–61.

Krumholz, H.M., Butler, J., Miller, J., Vaccarino, V., Williams, C., de Leon, C.F.M., Seeman, T.E., Kasl, S.V., and Berkman, L.F. (1998). The prognostic importance of emotional support for elderly patients hospitalized with heart failure. *Circulation* **97**, 958–64.

Moberg, D.O. and Taves, M.J. (2000). Church participation and adjustment in old age. In *Older people and their social world* (ed. A.M. Rose and W.A. Peterson), pp. 113–24. F.A. Davis, Philadelphia, Pennsylvania.

Nie, N.H., Junn, J., and Stehlik-Barry, K. (1996). *Education and democratic citizenship in America*. University of Chicago Press, Chicago.

OECD (2001). *The well-being of nations: the role of human and social capital*. OECD Centre for Educational Research and Innovation, Paris.

Oswald, A.J. (1997). Happiness and economic performance. *Econ. J.* **107**, 1815–31.

Pollner, M. (1989). Divine relations, social relations, and well-being. *J. Health Soc. Behav.* **30**, 92–104.

Putnam, R.D. (2000). *Bowling alone. The collapse and revival of American community*. Simon and Schuster, New York.

Putnam, R.D., Leonardi, R., and Nanetti, R. (1993). *Making democracy work: civic traditions in modern Italy*. Princeton University Press, Princeton, New Jersey.

Reed, D.M., LaCroix, A.Z., Karasek, R.A., Miller, D., and MacLean, C.A. (1983). Occupational strain and the incidence of coronary heart disease. *Am. J. Epidemiol.* **129**, 495–502.

Roberts, R.E., Kaplan, G.A., Shema, S.J., and Strawbridge, W.J. (1997). Prevalence and correlates of depression in an aging cohort: the Alameda County study. *J. Gerontol. Bull. Psychol. Sci. Soc. Sci.* **52**, 252–8.

Ryff, C.D. and Singer, B.H. (eds.) (2003). *Emotion, social relationships, and health*. Oxford University Press, Oxford.

Sampson, R.J., Raudenbush, S., and Earls, F. (1997). Neighborhoods and violent crime: a multilevel study of collective efficacy. *Science* **277**, 918–24.

Schoenbach, V.J., Kaplan, B.H., Fredman, L., and Kleinbaum, D.G. (1986). Social ties and mortality in Evans County, Georgia. *Am. J. Epidemiol.* **123**, 577–91.

Seeman, T.E., Mendes de Leon, C.F., Berkman, L.F., and Ostfeld, A.M. (1993). Risk factors for coronary heart disease among older men and women: a prospective study of community-dwelling elderly. *Am. J. Epidemiol.* **138**, 1037–49.

Smith, S. and Razzell, P. (1973). *The pools winners*. Caliban Books, London.

Soroka, S., Helliwell, J.F., and Johnston, R. (in press). Measuring and modelling trust. In *Diversity, social capital and the welfare state* (ed. F. Kay and R. Johnston). University of British Columbia Press, Vancouver.

Sugisawa, H., Liang, J., and Liu, X. (1994). Social networks, social support and mortality among older people in Japan. *J. Gerontol.* **49**, 3–13.

Verba, S., Brady, H., and Lehman–Schlozman, K. (1995). *Voice and equality: civic voluntarism and American politics*. Harvard University Press, Cambridge, Massachusetts.

Wilson, W. (1967). Correlates of avowed happiness. *Psychol. Bull.* **67**, 294–306.

Woolcock, M. (2001). The place of social capital in understanding social and economic outcomes. In *Proceedings of the OECD/HRDC Conference, Quebec, 19–21 March 2000: the contribution of human and social capital to sustained economic growth and well-being* (ed. J.F. Helliwell), pp. 65–88. HDRC, Ottawa.

Robert H. Frank is H.J. Louis Professor of Economics at Cornell's Johnson Graduate School of Management. Educated at Georgia Tech and UC Berkeley, he was a Peace Corps Volunteer in Nepal for two years. He is author or co-author of eight books, including *The winner-take all society, Luxury fever, What price the moral high ground?*, and *Passions within reason*.

Chapter 18

Does money buy happiness?

Robert H. Frank

An enduring paradox in the literature on human happiness is that, although the rich are significantly happier than the poor within any country at any moment, average happiness levels change very little as people's incomes rise in tandem over time. Easterlin and others have interpreted these observations to mean that happiness depends on relative income but not on absolute income (Easterlin 1974, 1995). In this chapter I offer a slightly different interpretation of the evidence—namely, that gains in happiness that might have been expected to result from growth in absolute income have not materialized because of systematic changes in the ways in which people spend their incomes.[1]

Measuring subjective well-being

The main method that psychologists have used to measure human well-being has been to conduct surveys in which they ask people whether they are: (1) very happy; (2) fairly happy; or (3) not happy (see Easterlin 1974). Most respondents are willing to answer the question and not all of them respond 'very happy', even in the US, where one might think it advantageous to portray oneself as being very happy. Many people describe themselves as fairly happy, and others confess to being not happy. A given person's responses tend to be consistent from one survey to the next.

Happiness surveys and a variety of other measures of well-being employed by psychologists are strongly correlated with observable behaviors that we associate with well-being (for surveys of this evidence see Frank 1985, chapter 2; Clark and Oswald 1996). If you're happy, for example, you're more likely to initiate social contacts with your friends. You're more likely to respond positively when others ask you for help. You're less likely to suffer from psychosomatic illnesses—digestive disorders; other stress disorders; headaches; vascular stress. You're less likely to be absent from work, or to get involved in disputes at work. And you're less likely to attempt suicide—the

* Robert H. Frank, *How not to buy happiness*, Daedalus, 133:2 (Spring, 2004), pp. 69–79. © 2004 by the Massachusetts Institute of Technology and the American Academy of Arts and Sciences.

[1] This chapter draws heavily on Chapters 5 and 6 of Frank (1999).

ultimate behavioral measure of unhappiness. In sum, it appears that human happiness is a real phenomenon that we can measure (Diener and Lucas 1999; Kahneman 1999).

Much of the interpersonal variation in happiness is hereditary, but environmental factors also matter. People who have many close friends, for example, tend to be significantly happier than others, and also to live longer (Myers 1999).

Income and happiness

How does happiness vary with income? As noted earlier, studies show that when incomes rise for everybody measures of well-being don't change much. Consider the example of Japan, which was a very poor country in 1960. Between then and the late 1980s, its per capita income rose several-fold, placing it among the highest in the industrialized world (see Fig. 18.1). Yet the average happiness levels reported by the Japanese were no higher in 1987 than in 1960. They had many more washing machines, cars, cameras, and other things than they used to, but they did not register significant gains on the happiness scale.

The pattern shown in Fig. 18.1 consistently shows up in other countries as well, and that's a puzzle for economists. If getting more income doesn't make people happier, why do they go to such lengths to get more income? Why, for example, do tobacco company chief executive officers (CEOs) endure the public humiliation of testifying before Congress that there's no evidence that smoking causes serious illnesses?

It turns out that if we measure the income–happiness relationship in another way, we get just what the economists suspected all along. Consider Fig. 18.2, which shows this relationship for the US during a brief period during the 1980s. When we plot average happiness versus average income for clusters of people in a given country at a given time, as in the diagram, rich people are, in fact, much happier than poor people.

The patterns portrayed in Figs 18.1 and 18.2 suggest that if income affects happiness it is relative, not absolute, income that matters. Some social scientists who have pondered the significance of these patterns have concluded that, at least

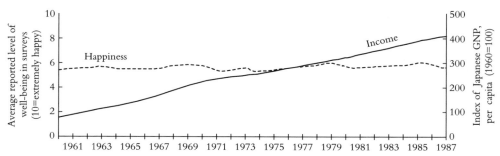

Fig. 18.1 Average happiness versus average income over time in Japan. (Adapted from Veenhoven (1993).)

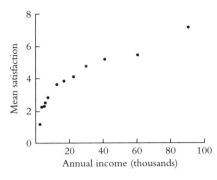

Fig. 18.2 Income versus satisfaction in the US, 1981–4. (Adapted from Diener *et al.* (1993).)

for people in the world's richest countries, no useful purpose is served by further accumulations of wealth (e.g. see Townsend 1979).

On the face of it, this should be a surprising conclusion, since there are so many seemingly useful things that having additional wealth would enable us to do. Would we really not be any happier if, say, the environment were a little cleaner, or if we could take a little more time off, or even just eliminate a few of the hassles of everyday life? In principle at least, people in wealthier countries have these additional options, and it should surprise us that this seems to have no measurable effect on their overall well-being.

There is indeed independent evidence that having more wealth *would* be a good thing, provided it were spent in certain ways. The key insight supported by this evidence is that, even though we appear to adapt quickly to across-the-board increases in our stocks of most material goods, there are specific categories in which our capacity to adapt is more limited. Additional spending in these categories appears to have the greatest capacity to produce significant improvements in well-being.

Adaptation

The human capacity to adapt to dramatic changes in life circumstances is impressive. Asked to choose, most people state confidently that they would rather be killed in an automobile accident than to survive as a quadriplegic. And so we are not surprised to learn that severely disabled people experience a period of devastating depression and disorientation in the wake of their accidents. What we do not expect, however, are the speed and extent to which many of these victims accommodate to their new circumstances. Within a year's time, many quadriplegics report roughly the same mix of moods and emotions as able-bodied people do (Bulman and Wortman 1977). There is also evidence that the blind, the retarded, and the malformed are far better adapted to the limitations imposed by their conditions than most of us might imagine (Cameron 1972; Cameron *et al.* 1976).

We adapt swiftly not just to losses but also to gains. Ads for the New York State Lottery show participants fantasizing about how their lives would change if they won ('I'd buy the company and fire my boss.'). People who actually win the lottery typically report the anticipated rush of euphoria in the weeks after their good fortune. Follow-up studies done after several years, however, indicate that these people are often no happier—and, indeed, are in some ways less happy—than before (Brickman *et al.* 1978).

In short, our extraordinary powers of adaptation appear to help explain why absolute living standards simply may not matter much once we escape the physical deprivations of abject poverty. This interpretation is consistent with the impressions of people who have lived or traveled extensively abroad, who report that the struggle to get ahead seems to play out with much the same psychological effects in rich societies as in those with more modest levels of wealth (Myers 1993, p. 36).

These observations provide grist for the mills of social critics who are offended by the apparent wastefulness of the recent luxury-consumption boom in the US. What many of these critics typically overlook, however, is that the power to adapt is a two-edged sword. It may indeed explain why having bigger houses and faster cars doesn't make us any happier but, if we can also adapt fully to the seemingly unpleasant things we often have to endure to get more money, then what's the problem? Perhaps social critics are simply barking up the wrong tree.

I believe, however, that to conclude that absolute living standards do not matter is a serious misreading of the evidence. What the data seem to say is that as national income grows people do not spend their extra money in ways that yield significant and lasting increases in measured satisfaction. But this still leaves two possible ways in which absolute income might matter. One is that people might have been able to spend their money in other ways that would have made them happier, yet for various reasons did not, or could not, do so. I will describe presently some evidence that strongly supports this possibility.

A second possibility is that, although measures of subjective well-being may do a reasonably good job of tracking our experiences as we are consciously aware of them, that may not be all that matters to us. For example, imagine two parallel universes, one just like the one we live in now and another in which everyone's income is twice what it is now. Suppose that in both cases you would be the median earner, with an annual income of $100 000 in one case and $200 000 in the other. Suppose further that you will feel equally happy in the two universes (an assumption that is consistent with the evidence discussed thus far). And suppose, finally, that you know that people in the richer universe would spend more to protect the environment from toxic waste, and that this would result in healthier and longer, even if not happier, lives for all. Can there be any question that it would be better to live in the richer universe?

My point is that, although the emerging science of subjective well-being has much to tell us about the factors that contribute to human satisfaction, not even its most ardent practitioners would insist that it offers the final word. Whether growth in national income is, or could be, a generally good thing is a question that will have to be settled by the evidence.

And there is in fact a rich body of evidence that bears on this question. One clear message of this evidence is that beyond some point across-the-board increases in spending on *many types* of material goods do not produce any lasting increment in subjective well-being. Sticking with the parallel-universes metaphor, let us imagine people from two societies, identical in every respect save one: In society A, everyone lives in a house with 4000 square feet of floor space, whereas in society B each house has only 3000 square feet. If the two societies were completely isolated from one another, there is no evidence to suggest that psychologists and neuroscientists would be able to discern any significant difference in their respective average levels of subjective well-being. Rather, we would expect each society to have developed its own local norm for what constitutes adequate housing, and that people in each society would therefore be equally satisfied with their houses and other aspects of their lives.

Moreover, we have no reason to suppose that there would be other important respects in which it might be preferable to be a member of society A rather than society B. Thus, the larger houses in society A would not contribute to longer lives, more freedom from illness, or indeed any other significant advantage over the members of society B. Once house size achieves a given threshold, the human capacity to adapt to further across-the-board changes in house size would appear to be virtually complete.

Of course, it takes real resources to build larger houses. A society that built a 4000-square-foot house for everyone could have built 3000-square-foot houses instead, freeing up considerable resources that could be used to produce something else. Hence this central question: Are there alternative ways of spending these resources that could have produced lasting gains in human welfare?

An affirmative answer would be logically impossible if our capacity to adapt to every other possible change were as great as our capacity to adapt to larger houses. As it turns out, however, our capacity to adapt varies considerably across domains. There are some stimuli, such as environmental noise, to which we may adapt relatively quickly at a conscious level, yet to which our bodies continue to respond in measurable ways even after many years of exposure. Indeed, there are even stimuli to which we not only do not adapt over time, but to which we actually become sensitized. Various biochemical allergens are examples, but we also see instances on a more macro scale. Thus, after several months' exposure, the office boor who initially took 2 weeks to annoy you can accomplish the same feat in only seconds.

The observation that we adapt more fully to some stimuli than to others opens the possibility that moving resources from one category to another might yield lasting changes in well-being. Considerable evidence bears on this possibility.

Spending categories that matter

A convenient way to examine this evidence is to consider a sequence of thought experiments in which you must choose between two hypothetical societies. The two societies have equal wealth levels but different spending patterns. In each case, let us again suppose that residents of society A live in 4000-square-foot houses while those in society B live in 3000-square-foot houses.

In each case, the residents of society B use the resources saved by building smaller houses to bring about some other specific change in their living conditions. In the first thought experiment, I review in detail what the evidence says about how that change would affect the quality of their lives. In the succeeding examples, I simply state the relevant conclusions and refer to supporting evidence published elsewhere.

◆ Which would you choose: society A, whose residents have 4000-square-foot houses and a 1-hour automobile commute to work through heavy traffic or society B, whose residents have 3000-square-foot houses and a 15-minute commute by rapid transit?

Let us suppose that the cost savings from building smaller houses are sufficient to fund not only the construction of high-speed public transit, but also to make the added flexibility of the automobile available on an as-needed basis. Thus, as a resident of society B, you need not give up your car. You can even drive it to work on those days when you need extra flexibility, or you can come and go when needed by taxi. The *only* thing you and others must sacrifice to achieve the shorter daily commute of society B is additional floor space in your houses.

A rational person faced with this choice will want to consider the available evidence on the benefits and costs of each alternative. As concerns the psychological cost of living in smaller houses the evidence provides no reason to believe that if you and all others live in 3000-square-foot houses, your subjective well-being will be any lower than if you and all others live in 4000-square-foot houses. Of course, if you moved from society B to society A, you might be pleased, even excited, at first to experience the additional living space. But we can predict that in time you would adapt, and simply consider the larger house the norm.

Someone who moved from society B to society A would also initially experience stress from the extended commute through heavy traffic. Over time, his or her consciousness of this stress might diminish. But there is an important distinction. Unlike the adaptation to the larger house, which will be essentially complete, the adaptation to his new commuting pattern will be only partial. Available evidence clearly shows that, even after long periods of adjustment, most people experience the task of navigating through heavy commuter traffic as stressful (for a survey, see Koslowsky *et al.* 1995).

In this respect, the effect of exposure to heavy traffic is similar to the effect of exposure to noise and other irritants. Thus, even though a large increase in background noise at a constant, steady level is experienced as less intrusive as time passes, prolonged exposure nonetheless produces lasting elevations in blood pressure (Glass *et al.* 1977). If the noise is not only loud but intermittent, people remain conscious of their heightened irritability even after extended periods of adaptation, and their symptoms of central nervous system distress become more pronounced (Glass *et al.* 1977). This pattern was seen, for example, in a study of people living next to a newly opened noisy highway. Twenty-one per cent of residents interviewed 4 months after the highway opened said they were not annoyed by the noise, but that figure dropped to 16% when the same residents were interviewed a year later (Weinstein 1982).

Among the various types of noise exposure, worst of all is exposure to sounds that are not only loud and intermittent, but also unpredictably so. Subjects exposed to such noise in the laboratory experience not only physiological symptoms of stress, but also behavioral symptoms. They become less persistent in their attempts to cope with frustrating tasks, and suffer measurable impairments in performing tasks requiring care and attention (Glass *et al.* 1977).

Unpredictable noise may be particularly stressful because it confronts the subject with a loss of control. David Glass and his collaborators confirmed this hypothesis in an ingenious experiment that exposed two groups of subjects to a recording of loud, unpredictable noises. Whereas subjects in one group had no control over the recording, subjects in the other group could stop the tape at any time by flipping a switch. These subjects were told, however, that the experimenters would prefer that they not stop the tape, and most subjects honored this preference. Following exposure to the noise, subjects with access to the control switch made almost 60% fewer errors than other subjects on a proofreading task and made more than four times as many attempts to solve a difficult puzzle (Glass *et al.* 1977, figs 5 and 6).

Commuting through heavy traffic is in many ways more like exposure to loud, unpredictable noise than to constant background noise. Delays are difficult to predict, much less control, and one never quite gets used to being cut off by others who think their time is more valuable than anyone else's. A large scientific literature documents a multitude of stress symptoms that result from protracted driving through heavy traffic.

One strand in this literature focuses on the experience of urban bus drivers, whose exposure to the stresses of heavy traffic is higher than that of most commuters, but who have also had greater opportunity to adapt to those stresses. Compared to workers in other occupations, a disproportionate share of the absenteeism experienced by urban bus drivers stems from stress-related illnesses such as gastrointestinal problems, headaches, and anxiety (Long and Perry 1985). Many studies have found sharply elevated rates of hypertension among bus drivers relative to a variety of control groups, including bus drivers themselves during their pre-employment physicals (Ragland *et al.* 1987; Pikus and Tarranikova 1975; Evans *et al.* 1987). Additional studies have found elevations of stress hormones such as adrenaline, noradrenaline, and cortisol in urban bus drivers (Evans *et al.* 1987). And one study found elevations of adrenaline and noradrenaline to be strongly positively correlated with the density of the traffic with which urban bus drivers had to contend (Evans and Carrere 1991). More than half of all urban bus drivers retire prematurely with some form of medical disability (Evans 1994).

A 1-hour daily commute through heavy traffic is presumably less stressful than operating a bus all day in an urban area. Yet this difference is one of degree rather than of kind. Studies have shown that the demands of commuting through heavy traffic often result in emotional and behavioral deficits upon arrival at home or at work (Glass and Singer 1972; Sherrod 1974). Compared to drivers who commute through low-density traffic, those who commute through heavy traffic are more likely to report feelings of annoyance (Stokols *et al.* 1978). And higher levels of commuting distance, time, speed, and months of commuting are significantly positively correlated with increased systolic and diastolic blood pressure (Stokols *et al.* 1978, table 3).

The prolonged experience of commuting stress is also known to suppress immune function and shorten longevity (DeLongis *et al.* 1988; Stokols *et al.* 1978). Even spells in traffic as brief as 15 minutes have been linked to significant elevations of blood glucose and cholesterol, and to declines in blood coagulation time—all factors that are positively associated with cardiovascular disease. Commuting by automobile is also positively linked with the incidence of various cancers, especially cancer of the lung, possibly because of heavier exposure to exhaust fumes (Koslowsky *et al.* 1995, chapter 4). Among people who commute to work, the incidence of these and other illnesses rises with the length of commute (Koslowsky *et al.* 1995) and is significantly lower among those who commute by bus or rail (Taylor and Pocock 1972; Koslowsky and Krausz 1993) and lower still among non-commuters (European Foundation for the Improvement of Living and Working Conditions 1984). Finally, the risk of death and injury from accidents varies positively with the length of commute and is higher for those who commute by car than for those who commute by public transport.

In sum, there appear to be persistent and significant costs associated with a long commute through heavy traffic. We can be confident that neurophysiologists would find higher levels of cortisol, norepinephrine, adrenaline, noradrenaline, and other stress hormones in the residents of society A. No one has done the experiment to discover whether people from society A would report lower levels of life satisfaction than people from society B. But, since we know that drivers often report being consciously aware of the frustration and stress they experience during commuting, it is a plausible conjecture that subjective well-being, as conventionally measured, would be lower in society A. Even if the negative effects of commuting stress never broke through into conscious awareness, however, we would still have powerful reasons for wishing to escape them.

Table 18.1 Which would you choose: society A or society B?

Thought experiment	Society A	Society B
1	Everyone lives in 4000 ft² houses and has no free time for exercise each day	Everyone lives in 3000 ft² houses and has 45 minutes each day available for exercise
2	Everyone lives in 4000 ft² houses and has time to get together with friends 1 evening each month	Everyone lives in 3000 ft² houses and has time to get together with friends 4 evenings each month
3	Everyone lives in 4000 ft² houses and has 1 week of vacation each year	Everyone lives in 3000 ft² houses and has 4 weeks of vacation each year
4	Everyone lives in 4000 ft² houses and has a relatively low level of personal autonomy in the workplace	Everyone lives in 3000 ft² houses and has a relatively high level of personal autonomy in the workplace

On the strength of the available evidence, then, it appears that a rational person would have powerful reasons to choose society B, and no reasons to avoid it. And yet, despite this evidence, the US is moving steadily in the direction of society A. Even as our houses continue to grow in size, the average length of our commute to work continues to grow longer. Between 1982 and 2000, for example, the time penalty for peak-period travelers increased from 16 hours per year to 62 hours, the period of time when travelers might experience congestion increased from 4.5 hours to 7 hours, and the volume of roadways where travel is congested has grown from 34% to 58% (Schrank and Lomax 2002). The Federal Highway Administration predicts that the extra time spent driving because of delays will rise from 2.7 billion vehicle hours in 1985 to 11.9 billion in 2005 (Clark 1994, p. 387).

Table 18.1 lists four similar thought experiments that ask you to choose between societies that offer different combinations of material goods and free time to pursue other activities. Each case assumes a specific use of the free time and asks that you imagine it to be one that appeals to you. (If not, feel free to substitute some other activity that does.)

The choice in each of these thought experiments is one between conspicuous consumption (in the form of larger houses) and what, for want of a better term, I shall call 'inconspicuous consumption'—freedom from traffic congestion, time with family and friends, vacation time, and a variety of favorable job characteristics. In each case the evidence suggests that subjective well-being will be higher in the society with a greater balance of inconspicuous consumption (for a detailed survey of the supporting studies, see Frank 1999, chapter 6). And yet in each case the actual trend in US consumption patterns has been in the reverse direction.

The list of inconspicuous consumption items could be extended considerably. Thus we could ask whether all living in slightly smaller houses would be a reasonable price to pay for higher air quality, for more urban parkland, for cleaner drinking water, for a reduction in violent crime, or for medical research that would reduce premature death. And in each case the answer would be the same as in the cases we have considered thus far.

My point in the thought experiments is not that inconspicuous consumption is always preferable to conspicuous consumption. Indeed, in each case we might envision a minority of rational individuals who might choose society A over society B. Some people may simply dislike autonomy on the job, or dislike exercise, or dislike spending time with family and friends.

But if we accept that there is little sacrifice in subjective well-being when all have slightly smaller houses, the real question is whether a rational person could find *some* more productive use for the resources thus saved. Given the absolute sizes of the houses involved in the thought experiments, the answer to this question would seem to be yes.

And this suggests that the answer to the question posed in my title ('Does money buy happiness?') is that it depends. Considerable evidence suggests that, if we all work longer hours to buy bigger houses and more expensive cars, we do not end up any happier than before. As for whether increases in absolute income *could* buy happiness, however, the evidence paints a very different picture. The less we spend on conspicuous consumption goods, the better we can afford to alleviate congestion, the more time we can devote to family and friends, to exercise, sleep, travel, and other restorative activities, and the better we can afford to maintain a clean and safe environment. On the best available evidence, reallocating our time and money in these ways would result in healthier, longer, and more satisfying lives.

So why hasn't subjective well-being been rising?

It might seem natural to suppose that when per capita income rises sharply, as it has in most countries since at least the end of the Second World War, most people would spend more on both conspicuous and inconspicuous consumption. In many instances, this is in fact what seems to have happened. Thus, the cars we buy today are not only faster and more luxuriously equipped, but also safer and more reliable. If both forms of consumption have been rising, however, and if inconspicuous consumption boosts subjective well-being, then why has subjective well-being not increased during the last several decades?

A plausible answer is that, whereas some forms of inconspicuous consumption have been rising, others have been declining, often sharply. There have been increases in the annual number of hours spent at work in the US during the last 2 decades. Traffic has grown considerably more congested, savings rates have fallen precipitously, personal bankruptcy filings are at an all-time high, and there is at least a widespread perception that employment security and autonomy have fallen sharply. Declines in these and other forms of inconspicuous consumption may well have offset the effects of increases in others.

The more troubling question is *why* we have not used our resources more wisely. If we could all live healthier, longer, and more satisfying lives by simply changing our spending patterns, why haven't we done that?

As even the most ardent free-market economists have long recognized, the invisible hand cannot be expected to deliver the greatest good for all in cases in which each individual's well-being depends on the actions taken by others. This qualification was once thought important in only a limited number of arenas—most importantly, activities that generate environmental pollution. We now recognize, however, that the interdependencies among us are considerably more pervasive. For present purposes, chief among them are the ways in which the spending decisions of some individuals affect the frames of reference within which others make important choices.

Many important rewards in life—access to the best schools, to the most desirable mates, and even, in times of famine, to the food needed for survival—depend critically on how the choices we make compare to the choices made by others. In most cases, the person who stays at the office 2 hours longer each day to be able to afford a house in a better school district has no conscious intention to make it more difficult for others to achieve the same goal. Yet that is an inescapable consequence of his action. The best response available to others may be to work longer hours as well, thereby to preserve their current positions. Yet the ineluctable mathematical logic of musical chairs assures that only 10% of all children can occupy top-decile school seats, no matter how many hours their parents work.

That many purchases become more attractive to us when others make them means that consumption spending has much in common with a military arms race. A family can choose how much of its own money to spend, but it cannot choose how much others spend. Buying a smaller-than-average vehicle means greater risk of dying in an accident. Spending less on an interview suit means a greater risk of not landing the best job. Yet, when all spend more on heavier cars or more finely tailored suits, the results tend to be mutually offsetting, just as when all nations spend more on armaments. Spending less—on bombs or on personal consumption—frees up money for other pressing uses, but only if everyone does it.

What, exactly, is the incentive problem that leads nations to spend too much on armaments? It is not sufficient merely that each nation's pay-off from spending on arms depend on how its spending compares with that of rival nations. Suppose, for example, that each nation's pay-off from spending on non-military goods also depended, and to the same extent as for military goods, on the amounts spent on non-military goods by other nations. The tendency of military spending to siphon off resources from other spending categories would then be offset by an equal tendency in the opposite direction. That is, if each nation had a fixed amount of national income to allocate between military and non-military goods, and if the pay-offs in each category were equally context-sensitive, then we would expect no imbalance across the categories.

For an imbalance to occur in favor of armaments, the reward from armaments spending must be *more* context-sensitive than the reward from non-military spending. And since this is precisely the case, the generally assumed imbalance occurs. After all, to be second-best in a military arms race often means a loss of political autonomy, clearly a much higher cost than the discomfort of having toasters with fewer slots.

In brief, we expect an imbalance in the choice between two activities if the individual rewards from one are more context-sensitive than the individual rewards from the other. The evidence described earlier suggests that the satisfaction provided by many conspicuous forms of consumption is more context-dependent than the satisfaction provided by many less conspicuous forms of consumption. If so, this would help explain why the absolute income and consumption increases of recent decades have failed to translate into corresponding increases in measured well-being.

References

Brickman, P., Coates, D., and Janoff-Bulman, R. (1978). Lottery winners and accident victims: is happiness relative? *J. Pers. Soc. Psychol.* **36**, 917–27.

Bulman, R.J. and Wortman, C.B. (1977). Attributes of blame and 'coping' in the 'real world': severe accident victims react to their lot. *J. Pers. Soc. Psychol.* **35**, 351–63.

Cameron, P. (1972). Stereotypes about generational fun and happiness vs. self-appraised fun and happiness. *Gerontologist* **12**, 120–3.

Cameron, P., Titus, D., Kostin, J., and Kostin, M. (1976). *The quality of American life.* Russell Sage, New York.

Clark, A. and Oswald, A. (1996). Satisfaction and comparison income. *J. Public Econ.* **61**, 359–81.

Clark, C.S. (1994). Traffic congestion. *CQ Res.* May 6, 387–404.

DeLongis, A., Folkman, S., and Lazarus, R.S. (1988). The impact of daily stress on health and mood: psychological and social resources as mediators. *J. Pers. Soc. Psychol.* **54**, 486–95.

Diener, E. and Lucas, R.E. (1999). Personality and subjective well-being. In *Well-being: the foundations of hedonic psychology* (ed. D. Kahneman, E. Diener, and N. Schwarz), pp. 213–29. Russell Sage, New York.

Diener, E., Sandvik, E., Seidlitz, L., and Diener, M. (1993). The relationship between income and subjective well-being: relative or absolute? *Soc. Indic. Res.* **28**, 195–223.

Easterlin, R. (1974). Does economic growth improve the human lot? In *Nations and households in economic growth: essays in honor of Moses Abramovitz* (ed. P. David and M. Reder), pp. 89–125, Academic Press, New York.

Easterlin, R. (1995). Will raising the incomes of all increase the happiness of all? *J. Econ. Behav. Org.* **27**, 35–47.

European Foundation for the Improvement of Living and Working Conditions (1984). *The journey from home to the workplace: the impact on the safety and health of the community/workers.* European Foundation for the Improvement of Living and Working Conditions, Dublin.

Evans, G.W. (1994). Working on the hot seat: urban bus drivers. *Accident Anal. Prev.* **26**, 181–93.

Evans, G.W. and Carrere, S. (1991). Traffic congestion, perceived control, and psychophysiological stress among urban bus drivers. *J. Appl. Psychol.* **76**, 658–63.

Evans, G., Palsane, M., and Carrere, S. (1987). Type A behavior and occupational stress: a cross-cultural study of blue-collar workers. *J. Pers. Soc. Psychol.* **52**, 1002–7.

Frank, R.H. (1985). *Choosing the right pond.* Oxford University Press, New York.

Frank, R.H. (1999). *Luxury fever.* The Free Press, New York.

Glass, D.C. and Singer, J. (1972). *Urban stressors: experiments on noise and social stressors.* Academic Press, New York.

Glass, D.C., Singer, J., and Pennebaker, J. (1977). Behavioral and physiological effects of uncontrollable environmental events. In *Psychological perspectives on environment and behavior* (ed. D. Stokols), pp. 131–51. Plenum, New York.

Kahneman, D., Diener, E., and Schwarz, N. (eds.) (1999). *Well-being: the foundations of hedonic psychology.* Russell Sage, New York.

Kahneman, D. (1999). Objective happiness. In *Well-being: the foundations of hedonic psychology* (ed. D. Kahneman, E. Diener, and N. Schwarz), pp. 3–25. Russell Sage, New York.

Koslowsky, M. and Krausz, M. (1993). On the relationship between commuting, stress symptoms, and attitudinal measures. *J. Appl. Behav. Sci.* **29**, 485–92.

Koslowsky, M., Kluger, A.N., and Reich, M. (1995). *Commuting stress.* Plenum, New York.

Long, L. and Perry, J. (1985). Economic and occupational causes of transit operator absenteeism: a review of research, *Transport Rev.* **5**, 247–67.

Myers, D.G. (1993). *The pursuit of happiness: who is happy and why?* Avon, New York.

Myers, D.G. (1999). Close relationships and quality of life. In *Well-being: the foundations of hedonic psychology* (ed. D. Kahneman, E. Diener, and N. Schwarz), pp. 374–91. Russell Sage, New York.

Pikus, W. and Tarranikova, W. (1975). The frequency of hypertensive diseases in public transportation. *Terapevischeskii Arch.* **47**, 135–7.

Ragland, D., Winkleby, M., Schwalbe, J., Holman, B., Morse, L., Syme, L., and Fisher, J. (1987). Prevalence of hypertension in bus drivers. *Int. J. Epidemiol.* **16**, 208–14.

Schrank, D. and Lomax, T. (2002). *The 2002 Urban Mobility Report.* Texas Transportation Institute, June 2002, <http://mobility.tamu.edu/>.

Sherrod, D.R. (1974). Crowding, perceived control, and behavioral aftereffects. *J. Appl. Soc. Psychol.* **4**, 171–86.

Stokols, D., Novaco, R.W., Stokols, J., and Campbell, J. (1978). Traffic congestion, type A behavior, and stress. *J. Appl. Psychol.* **63**, 467–80.

Taylor, P. and Pocock, C. (1972). Commuter travel and sickness: absence of London office workers. *Br. J. Prev. Soc. Med.* **26**, 165–72.

Townsend, P. (1979). The development of research on poverty. In *Social security research: the definition and measurement of poverty*, Department of Health and Social Research, pp. 15–45. HMSO, London.

Veenhoven, R. (1993). *Happiness in nations: subjective appreciation of life in 56 nations 1946–1992.* Erasmus University Press, Rotterdam.

Weinstein, N.D. (1982). Community noise problems: evidence against adaptation. *J. Environ. Psychol.* **2**, 82–97.

Johan Galtung is director of TRANSCEND, an NGO for mediation in protracted conflicts. He has been a professor on all continents, holds 12 honorary doctorates, and is the publisher of more than a hundred books with more than a hundred translations; among them *Transcend and transform* and *Peace by peaceful means*.

Chapter 19*

Meeting basic needs: peace and development

Johan Galtung

The science of well-being: a prologue

To start with, I shall give some rough definitions of the concepts used in this chapter:

- body: the material base, with upper limits on growth;
- mind: the memory of cognitions and emotions, more or less retrievable (verbalizable, articulable), with upper limits on growth;
- spirit: the capacity to reflect on body and mind, including the capacity to identify and change codes/scripts and modify body and mind, with no known upper limits on growth;
- the subconscious: the unreflective part of the mind;
- the collective subconscious: the subconscious shared by a gender, generation, class, nation etc.—also called code, script, deep culture, social cosmology;
- behavior: the physical and verbal movement or non-movement of the body as observed from the outside; largely script-driven;
- action: reflective behavior, as acts of commission or omission;
- interaction: reflective behavior anticipating the behavior of the other; reciprocity, as opposed to autism;
- social system: a set of interaction relations;
- society: a sustainable social system;
- structure: the sets of actors, interaction relations, relations among relations, and so on;
- culture: the sets of mappings of anything on good–bad, right–wrong, true–false, beautiful–ugly, sacred–profane, etc.

★ © The Royal Society 2005

'*Homo mensura*', Protagoras said, 'man is the measure of all things.' But what is the measure of man, and of human well-being? Is it the moral purity/wisdom of the *brahmin*/clergy, the physical prowess/power of the *kshatriya*/aristocrat, the wealth of the *vaisya*/merchant? Or is it what the *sudra*/common human being can strive for—*healthy longevity* or plain *happiness*? Or, to lean even more on the Hindu depository of human wisdom, is it the four overlapping life stages for all castes and classes in society, the *dharma* of morality, the *artha* of competence, the *kama* of enjoyment, or the *moksha* of liberation, fulfillment, always with all four in mind as we move through life? The focus shifts along the *dharma–artha–kama–moksha* axis. But whoever goes in for only one will lose even that one, like a King Midas surrounded by gold, and gold only.

We sense schools of thought, classes, areas, and eras, behind such time-tested formulas. There are many of them. What do we want? None of the above? All of the above? Perhaps the latter.

Does 'well-being' cover all of that? Perhaps, but perhaps not. The term certainly covers an emotion, a 'sense of well-being', 'I now feel well'. This is a little bland, not very adequate for euphoric peaks, for those rare instances when love is experienced as the triple union of body, mind, and spirit at the same time. 'Well-being' is not the word for the best of times, for those peaks. It is more a term reflecting the *altiplano*, the highlands, than the peaks. Helen Fisher (2004) also sees love in terms of three components: sexual desire–lust; attachment; and 'romantic love'. The first is testosterone-driven and corresponds to the 'union of the bodies'. The third is dopamine–norepinephrine-driven and corresponds to the 'union of the minds' with emotions resonating in the sense of suffering the partner's suffering and enjoying the partner's joy. 'Attachment' is a somewhat bland term for the 'union of the spirits', one expression being 'shared life projects'.

It is up to the holder of the emotion to decide whether words like 'happiness' or 'well-being' are adequate. Not only the feelings, but also the words chosen are 'subjective'. But that does not exclude subjective well-being from having objective causes and effects or, more broadly, conditions and consequences. In my autobiography (Galtung 2000, pp. 407–10) there is a chapter on *Lykke*, happiness. In it I suggest two formulations that encapsulate 'well-being' for me:

♦ consciousness of the positive aspects of any situation, including consciousness of the suffering we do not have, that is, enjoying whatever health, good human relations, and other good things, we have;

♦ attaching a conscious, reflective meaning to your life—a life project—and finding meaning in your marriage as a joint life project.

For me these have been sufficient conditions. 'Hormonal cascades' (endomorphin? oxytocin?) may serve as intervening variables between consciousness (even at low levels) and the feeling of well-being. And no doubt much can be obtained by coming closer to the positive of any kind, partners, friends, beloved things, whatever we believe in. Thus, Ruthellen Josselson (1998) points out that friendships have a tendency to be put on the back burner, with work and family up front. But this can also be read positively: we have reserves to draw on!

There are three propositions behind what has been said above.

1 'Well-being' is a subjectively held emotion, but conditions and consequences can be analyzed objectively.

2 Those conditions and consequences vary greatly from person to person, and for the same person, depending on the circumstances.

3 The verbal expression of that emotion also varies from person to person, and for the same person depending on the circumstances.

From (3) it follows that verbal approaches in exploring the presence or absence of well-being are problematic. The semantic habits will vary with the culture, the person, and for the same person over time. The sentence 'I am happy' is frequently heard in the USA, but much less in the UK: does that reflect a lower level of happiness, or different semantic habits? Semantic habits are good indicators of deep culture. The assumption in the deep culture of American society seems to be that, in the best of all societies, the cause of not being happy is entirely your own (laziness, stupidity, inability to be 'a good team-worker', a 'part of the solution, not of the problem'). Thus, for an American to say 'I am not happy' is akin to confessing a personal failure. On the other hand, in a French, let alone a Russian context, to say 'je suis heureux' is interpreted as an inability to exit from a state of infantile naiveté. To say that the US goddess is Polyanna and that of the Russians is Cassandra is not to imply that people in the USA are happy and people in Russia unhappy, even though the frequency of use of the word 'happy' may point in that direction. But that conceals the nagging doubts about Polyanna's wisdom and the delight that may derive from proving Cassandra right, again and again. (Charles de Gaulle is supposed to have said that 'happiness is for idiots'.) A Japanese would even have difficulties with the pronoun 'I'. All of the above applies even more to the phrase, 'I love you'.

Any study using verbal indicators of 'well-being' must include an in-depth dialogue to explore social, even personal semantics. If the use of words like 'happy' and 'love' is correlated with gender, generation, class, nation, country, then what does that mean? Differences in semantic habits? Different levels of feelings of happiness or love? Both? Or neither? Thus, women are often held to be better at verbalizing emotions. Does that imply that women have more of those emotions?

From (2) I draw the conclusion that a science positing *sufficient* conditions for well-being is problematic. The set of conditions is too extensive. Well-being may be felt under a wide variety of circumstances. For example, this author had a high level of well-being in prison as a conscientious objector under circumstances designed to deter law-breaking by inducing a sense of ill-being.[1] However, a science positing conditions for ill-being, the absence of which would be the *necessary* conditions for well-being, may be less problematic, being less extensive. Or so I would assume.

Finally, from proposition (1) I draw the conclusion that in exploring well-being we are approaching the essence of the human condition.

Five approaches to well-being

That essence has been approached through well-known approaches, based on word-pairs other than *ill-being/well-being*.

1 The peace-based approach tries to move human beings from *violence* to *peace* at the micro/meso/macro/mega levels, that is, within or among persons, societies, states/nations, regions/civilizations. (For an exploration of this, see Galtung (2004); for a more general theory of peace see Galtung (1996).)

2 The development approach seeks to move human beings from *misery* to *wellness* and, at the societal level, to cure the disease from which less developed countries are seen as suffering—underdevelopment. The definition of well-being in this chapter is more positive than this focus on deficit relative to what more developed countries have.

3 The health-based approach seeks to move a person from *disease* (ill(ness)) to *health* (ease, well(ness)).

4 The human rights-based approach seeks to move societies from *repression* to *freedom* (for one approach to human rights, see Galtung 1994).

5 The Buddhist approach seeks to move the individual self and collective self from *dukkha* (suffering) to *sukha* (fulfillment).[2]

For human beings–body, mind, and spirit–*ill-being/well-being* might cover all five word-pairs. Thus, five windows have been opened above, five perspectives for exploring the essence of well-being. To be sure, there are more windows, more perspectives.

The Buddhist formula is the most general if violence, misery, disease, *or* repression all imply suffering, and if we assume a human striving away from suffering. But we cannot assume that the negation of suffering—fulfillment, bliss—implies the negation

---

[1] For 6 months in the winter 1954/55 I was in the central prison of Oslo, Norway, for refusing to do the additional (to the 12 months corresponding to military service) alternative service, unless it could be for peace. I have never had such intense experiences of music (over a miserable, blessed radio) and of literature, and of colors as under those conditions of gray, sensory deprivation.

[2] The words *dukkha* and *sukha* are from Pali, which is to Sinhala what Sanskrit is to Hindi (or Latin to French, etc.). For one approach to Buddhism, see Galtung (1993).

of violence, misery, disease, *and* repression, even if that is logically correct (i.e. the counterpositive proposition that $p \rightarrow q$ equals $-q \rightarrow -p$).

These concepts are too 'fuzzy' to be captured by standard Aristotelian/Cartesian logic. A Buddhist answer might run as follows: true, there is suffering to the body and to the mind when any one of the four calamities hits.[3] But the spirit is able to extricate itself from insults to body and mind, finding fulfillment in the midst of calamities. It all depends on who is in command: body/mind, or the spirit. There may be *sukha* in *dukkha*, and vice versa. Emotional logic is *yin/yang* (both–and) rather than the rational Aristotelian/Cartesian either–or, *tertium non datur*. In a later section, this both–and, dukkha–sukha approach will be explored.

An objection to the other four approaches is that, as commonly conceived of, they are only body/mind-oriented. Violence is conceived of in terms of violence to the body/mind, as killed, wounded, displaced or raped, and traumatized; misery is conceived of as hunger or disease; disease is conceived of as somatic or mental but not applying to the spirit; and human rights protect the body and the mind but not the more elusive spirit.

But how about the Abrahamic pact, some would say contract, to stay away from sin (a major sin being non-submission to the Lord) in return for salvation and eternal life? The word-pair sin/salvation fills many with awe and certainly dwarfs such enlightenment ideas as peace, development, health, and human rights. If well-being can be derived from engaging in that project, fine. But a major problem is that consummation of well-being as salvation is 'not of this world', making any hypothesis non-falsifiable and hence inaccessible to science.

The focus here is on testable well-being in this world, and well-being for the total human being—body, mind, and spirit. Well-being applies to the spirit, but, as for the mind, it is unobservable. But theories about well-being of the spirit and mind can be tested for their consequences in this world. Thus, a major part of behavior under conditions of crisis, complexity, and/or striving to attain consensus (called the '3 C conditions') is script-driven, by individual or collective subconscious, not what we call 'rational', brain-processed, and capable of being expressed in words.

This tripartite conceptualization of human beings sees the body as material, the mind as the depository of individual and collective, cognitive, and emotional memory that also programs our physical and verbal behavior, and the spirit as the capacity to reflect, also on body and mind, including on how I/we are programmed and on how to change that program. The growth potential of the body is limited. Sports are pushing those limits a little. The growth potential of the mind is also limited,

---

[3] The Buddhist 'four calamities' are very similar to the four calamities of the *Apocalypse*: conquest (the white horse); war (the red horse); misery (the black horse); hunger and pestilence (the pale horse). The spiritual calamity, the void, is missing, but it may be said to underlie the last book of the New Testament, *Revelation* (the horses of the *Apocalypse* are from 6:1–8).

with gestalts serving as protection against overload. But there is no limit to spiritual growth. This is where the creative capacity to transcend, to *go beyond*, is located. The well-being of the spirit grows with that creativity, or so we assume.

The basic human needs approach

The five approaches mentioned above are supposed to 'move' individuals and collectivities from ill-being to well-being. This presupposes a general human drive in that direction, which should not be confused with acceptance of the social rituals of the five approaches and their practitioners. There is a general gradient from ill-being to well-being. One language for expressing this drive is in terms of *basic human needs* (BHN; where inclusion of the word 'human' underlines the body–mind–spirit totality); this assumes an innate, universal, drive to have those needs satisfied—not neglected or even assaulted. We then assume that:

- neglecting to satisfy basic needs is a sufficient condition for ill-being; and
- satisfying basic needs is a necessary condition for well-being.

What follows is an effort to root well-being in basic human needs—needs in plural in order to have several factors to work with. 'Ill-being' and 'well-being' are useful words, but too holistic. Some subdivision is an analytical necessity.

Of course, this is begging the question: what is a basic need? Are we engaging in *obscurum per obscurius*, referring the obscure 'well-being' to the equally or more obscure 'basic human needs'? Not quite. 'Basic' indicates rock-bottom, non-negotiable, *sine qua non*. Without basic human needs satisfied through adequate satisfiers, body, mind, and/or spirit growth are negated. 'Human' indicates 'rooted inside the human being'. 'Needs' points outside—something is needed from the outside. *Satisfaction depends on the inside–outside interface*; so do neglects and assaults. The 'outside' means the environment, meaning the natural, human-made, and human environment. The needs approach serves as a bridge from the inner human to the environmental worlds. Needs are felt on the inside, and depend on the environment to be satisfied or assaulted.

How do we identify basic human needs? By what methods? Among the approaches rejected are reliance on scriptures or armchair exercises by philosophers, social scientists, etc. There have to be fresh inputs of empirical elements. The world is changing and human beings are changing with the world and basic human needs are changing with them—or vice versa. Two approaches to basic human needs will be accepted here: *observed needs* rooted in human physiology; and *expressed needs* emerging from dialogues (not from interviews or questionnaires; for the use of dialogue as a social science tool see Galtung 1988a, pp. 68–92).

For the *observed needs* the openings of the body, and the quality and quantity of inputs and outputs, serve as a guide. The inputs are food–water–air and impressions through the classical five senses—sights, sounds, smells, tastes, and sensations. The outputs are excretion, ejaculation, ovulation, giving birth, lactation, speech.

Sexuality is both. But throughput quality and quantity are also rooted in human needs. Muscles must be used and sleep is needed to restore balances.

Food–water–air are indispensable for the human physiology and so are housing and clothing to protect against excessive cold, heat, humidity, and wind. Impressions are stored in the mind, processed at low or high levels, and expressed in speech and writing through the spirit. The term *wellness* covers a sense of physiologically based satisfaction.

For the *expressed* needs, dialogues about 'what you cannot live without' serve as a guide. This study was carried out when the author directed the GPID (Goals, Processes and Indicators of Development Project) of the United Nations University in Tokyo, 1976–81. The idea was to root development, goals as well as processes, in basic human needs—conceiving of the goal of development as *well-being for all*. The idea of identifying basic needs by asking people was at first considered somewhat subversive—do people really know what their needs are? The Marxist concept of false consciousness was invoked. As there is something to that objection, we decided to use dialogue rather than interview with pre-set answers, and with the possibility of probing, also in depth, starting with the question 'what can't you live without'. This was done in about 50 countries, but in a non-standardized way with no protocols worth preserving. The findings were as reported, and all I can say is that anyone is free to do studies along this line and be rewarded by the insights 'common people' have into their own life situation and the human condition in general (see Galtung 1980*a, b*, 1988*b*).

In this study wellness was seen in terms of five inputs, *los cinco bienes fundamentales* (a frequently used expression in socialist Cuba, summarizing their political program): food, clothing, housing, health, and education in the broad sense of communicative competence. Other needs came up, such as *survival*, the ultimate *sine qua non*. So did *freedom* as choice—of spouse, job, residence, lifestyle, polity, and economy. So did meaning (*identity*)—to live *for* something, not only *from* something. The identification may be egoistic, i.e. focused on self, or altruistic as in love for spouse, family, friends, colleagues, neighbors, networks. The focus may be on 'big self', such as generation, gender, race, class, nation, country, region, and humanity, or on God. Egoism and altruism may exclude each other. Whoever has only one focus may lose even that one. It is better to have more and, indeed, one wise formula is to disperse the focus of identity over several points on the range from self to God so that if one focus fails there are always the others. In sum, identity is seen as a two-way street: the need is not only to love but to be loved, to be esteemed, to get signs of the family's, network's, nation's, region's, God's love and esteem.

We conclude that four basic needs are necessary for *well-being*.

1 Well-being → survival + wellness + freedom + identity.

If we now accept violence, misery/disease, repression, and alienation as negations of these four, then any one of them should be a sufficient condition for the negation of well-being, *ill-being*. Thus, (1) → (2).

2 Violence or misery/disease or repression or alienation → ill-being.

We have not established sufficient conditions for well-being, but we have postulated four necessary conditions as universally valid, even if the concrete interpretations of the 10 terms in (1) and (2) vary with space, time-regions, down to person P in situation S.

However, to explore the social conditions for well-being, the necessary conditions were what we wanted, as a rock-bottom, *sine qua non* basis for any *politics of well-being*. Individual-level variations are beyond our social perspective, being too casuistic.

Now consider the time dimension. No one knows for sure what the future may carry in its womb. To the suffering we could add *fear*, the premonition that things will get worse; and to the well-being *hope*, the sense that things will get better. Cassandra specializes in the former, seeing well-being as not sustainable whereas ill-being is. Polyanna specializes in the latter, seeing ill-being as not sustainable whereas well-being is.[4] These are subjective personality traits that conceal the objective contexts postulated in (1) and (2). They may also be semantic habits, like a preference for subjunctive or indicative modes.

How do needs relate to the five windows opened as approaches to well-being? They are related to a division of labor among the approaches. Peace-based approaches focus on negating violence, development- and health-based approaches on negating misery and disease, human rights-based approaches on negating repression, and Buddhist approaches on negating all four, including alienation. In this context, one possible definition of 'alienation' would be as 'spiritual void'.

Life is lived, even for a long time, with wellness and in freedom. But it is meaningless. 'What is the project of your life?' draws a blank. The negation of survival may be homicide. And the negation of identity may be suicide. According to WHO (2001, p. 28), unipolar depressive disorder is the number 1 cause of years of lives lived with disability (YLDs), in all ages and in 14–44 year olds, both genders (in the number 2 position there are hearing loss, iron-deficiency anemia, and alcohol use disorders; bipolar affective disorders are in ninth position for both sexes, all ages and schizophrenia in seventh position). A link between unipolar depression and lack of sense of meaning with life does not seem far-fetched, nor does seeing depression as a phase on the way to suicide. We are talking about 12% of the world population, 10% of men and 14% of women.

Identity, any meaning-producing identity, might fill that void. Life can also have meaning even when short and brutish. But we did not get from our dialogue partners any hierarchy or priority. The non-negotiable cannot be put on hold or

---

[4] Cassandra sounds somewhat Russian and Polyanna somewhat American—stereotypes no doubt, but probably quite good as rules of thumb. They may, of course, both be proven wrong.

traded against other needs, like the 'identity against survival + wellness + freedom' formula just indicated. For a concrete person in a concrete situation, the identity with one's country may take priority over anything else. But change the person and/or the situation and it may turn out the other way round. No universal hierarchy, such as 'first attention to the survival and the somatic aspects of wellness, then comes the time for freedom and identity', can be imposed on those four needs.

In such hierarchies we sense the bias of Abrahamic religions, putting soma below psyche. On the contrary, we would rather assume the borders between the classes of basic human needs to be as fuzzy as the borders between body, mind, and spirit. Of course, there is something spiritual in good food and sex, and, of course, there is room for growth in identity because it can unfold to ever deeper identity, etc.

The four basic needs, when met/satisfied, are four overlapping regions in that *altiplano* of well-being mentioned above. There are peaks of higher levels of satisfaction, worth scaling occasionally, but not in order to take up residence. They are too taxing, too exposed in the longer run—like love peaking as the triple union of body, mind, and spirit—to be sustainable forever. Moving around on the *altiplano*, like the Hindu project, is very preferable to life in the crevices with one or more needs neglected.

The four Hindu stages and the four basic human needs are related but not identical. *Dharma* builds meaning, *artha* secures wellness, *kama* takes wellness to higher levels, and *moksha* takes it to the level of the spirit. The stages do not emphasize survival and freedom but can be interpreted as a recipe for their realization.

Well-being, peace, and development: a comparative survey

How do people see the prospects for peace (absence of violence) and development (absence of misery)—two parts of well-being?

At the end of the 1960s the present author directed a comparative interview study of the images 9000 persons in 10 countries around the world held of the year 2000 (see Ornauer *et al.* 1976). We used the year with the three 0s as a projection screen of hopes and fears. We did not ask 'do you feel well', 'are you happy', or any words subject to very elusive variations in semantic habits, but found the 'Cantril ladder' more useful to gauge optimism/pessimism. The 'instrument' gives the respondent a chance to place himself on a ladder with 9 (in the original, 11) rungs, and s/he is simply asked 'where do you think you/your country/the world stand(s)', from the worst possible to the best, 'at present, 5 years ago, in 5 years, and in the year 2000' (Ornauer *et al.* 1976, p. 74). The general methodology was as outlined in Galtung (1967).[5] We assumed that the neutral 'where do you stand?' taps into 'well-being', and found, not surprisingly, that the higher the social position, the higher did they place themselves. But we were also interested in how they placed their own country, particularly over time, and more particularly relative to the level of development.

> . . . most nations show perceptions of amelioration. . . . But only for six nations are the ratios monotonically increasing, with no dips at all. And which are the six nations that show no dips? They are all the less developed nations in the sample [Czechoslovakia, Spain, India, Japan, Poland, and Yugoslavia—this was in 1969]. Britain, the Netherlands and Finland all look back to better pasts—and Norway as the only one also looks forward to a worse future. [Ornauer et al. 1976, p. 74]

The more developed countries saw the future in terms of mental illness and narcotics, more desire for success, more interest in material things, and in terms of more criminality and unemployment. The less developed countries saw only benefits in development.

A methodological point: 'Had we used only the present our conclusion would have been simply that the nations that are best off are also the most satisfied. And the conclusion would easily have been "more development, more satisfaction"; which would then shade into "more development, more optimism" ' (Ornauer et al. 1976, p. 79). Keeping in mind the Cassandra and Polyanna perspectives in the text, a time dimension is recommended in all well-being studies to check for perceived or factual sustainability.

The findings of the survey can be summarized as follows (Ornauer et al. 1976, p.118).

1 As to *domestic perspectives*, the organizing axis was the level of technical–economic development. Countries high on this dimension are pessimistic, bewildered, and uncertain, probably: (a) because they see the negative effects of this type of development; (b) because they feel they have exhausted the program of their societies and that the future is without challenging and clear goals. Countries low in development follow the program defined and developed by countries that are already disillusioned by it.

2 When it comes to *international perspectives*, the organizing axis was not East–West, as opposed to the North–South axis that seemed to prevail for the domestic perspectives. The overwhelming impression was not one of a humanity divided

---

[5] Galtung (1967, chapter 4.2) gives a critique, inspired by the Department of Sociology at Columbia University, of the use of significance tests in the social sciences. Sampling is a minor source of error in the social sciences. Moreover, there is a tendency to use significance level as a measure of correlation. Of course, with sufficiently big samples even minor correlations become 'significant' statistically, but not substantially. As a rule of thumb, decent correlations should at least account for half the variance to serve as a basis for theory and practice.

by the stark Cold War at the time, but a humanity united in a desire for peace and a high consensus when it comes to how it could be obtained. We may rather talk about a government–people contradiction, cutting across the East–West and North–South axes. The *international hopes* for world peace and disarmament, absence of war, and a world government drew more consensus than the *domestic hopes* for population control, better health care, enough food, more equality, and cooperation. They understood by development something social like mental health, no use of drugs, less 'rat race' and divorce, less materialism, less criminality, more employment. But they got 'economic growth'.

The conclusion is that *what people did not ask for they got; what they did ask for (peace) they did not get*. A poor deal indeed!

This is of key importance for well-being. The governments, democratic or not, have their programs; the people have theirs. A development program people do not really believe in alienates. And the respondents did not believe much in the peace program either.

So what did they believe in? They were given 25 'strategies to obtain peace', and were asked to choose. The five most popular were (Ornauer *et al.* 1976, p. 101):

- 'hunger and poverty must be abolished all over the world';
- 'increased trade, exchange, and cooperation also between countries that are not on friendly terms';
- 'improve the United Nations to make it more efficient than today';
- 'the gap between poor and rich countries must disappear';
- 'people all over the world should freely choose their governments'.

However, these are not the peace theories inspiring governments. Hunger and poverty are increasing. When countries 'are not on friendly terms', economic sanctions are used, killing hundreds of thousands of innocent people. The UN is sabotaged, and particularly by one country, the USA, very often in the minority of 1, 2, or 3, breaking international law and in addition not paying UN dues. The gap between poor and rich countries is increasing. True, more people than ever can now freely choose their governments; but electionism is a very narrow approach to democracy. And democracies are not always peaceful as witnessed by recent belligerence (dubbed 'democratic totalitarianism' by Zinoviev), including the US punishment/revenge/incapacitation response to 9/11.

Again semantics comes into play. Governments often use the word 'peace' but pursue national interest, if possible backed by military power policies. That is not among the popular strategies chosen above. And the suggested peace strategy 'to obtain peace countries must be members of military alliances so that no country or group of countries dare attack others' was among the least popular, number 17.5 out of the 25 strategies offered. People can be mobilized for such policies, but do not believe much in them as *peace* policies. If we add fear to suffering, see war as a major source of fear, and governments as not pursuing peace, then a major necessary condition for well-being is not met at the societal level.

Thus, in almost all countries there were more pessimists than optimists about the possibility that the particular proposal they found most likely to lead to peace would in fact lead to peace by the year 2000. And the important question 'is there anything ordinary people can do?' elicited the standard microlevel answers such as 'improve oneself', 'improve interpersonal relations', with a few adding 'protest, demonstrations'. People feel powerless, deprived of ways and means, alienated, with little difference among individual respondents. People come up with ideas for the microlevel because the foreign affairs macrolevel is monopolized higher up.

But, in general, people are simply realistic. Or, as we concluded 30 years ago, 'Mankind has so far made for itself a world where the phenomena seen as most threatening to people in general are also the phenomena most beyond what the common person can directly influence.' People feel helpless, for good reasons. It is easy to explain. The state system is also a war system, and the right to wage wars was an important part of the Westphalia 'peace' of 1648 (denied Japan in the famous Article 9 in its 1947 peace constitution). But the contradiction remains. Not only do people not want wars, but they also think there are ways of avoiding them. Their priorities among the peace philosophies are eminently sensible. But this is not what governments prioritize.

'Peace seen as unobtainable and development seen as counterproductive' may have changed since 1969, but we doubt it. What has changed, however, is the growth of non-governmental organizations (NGOs) and civil society in general. People are taking both development and peace into their own hands, often ahead of governments. In principle, this should reduce alienation and decrease ill-being.

On the politics of well-being and the dialectics of politics

The task of the politics of well-being is not to implement any formula for a single peak experience of well-being for all for the following reasons.

1 Being a single experience, that would be the search for a singular utopia for a plurality of humans, meaning essentially totalitarian.

2 Being a peak experience, it is not sustainable however enjoyable.

3 Being a peak, it will only be for those having at their disposal, at the same place and point in time, all societal and psychological conditions, meaning a small elite.

A multipeak approach is better. A better task is to design the politics of the *altiplano* of necessary conditions. A suggested formula would be: *take worries out of people's lives*.

I grew up in a Norway with an intact welfare state doing that. People worried about disease and premature decease, but they did not worry about being unable to pay the costs of health care. Of course, people worried about not being capable of a high-level professional education, but not about being unable to defray the costs. Both health and education were essentially free. With full employment, salaries covered the other three—food, clothing, and housing. Criminal threats to survival were very low. Freedom as choice in politics, in food, clothing, and housing, was high; in lifestyle, health, and education more limited. The welfare state delivered standardized medical treatment and education, based on the strong position of school medicine and public school education, also aiming at a single class society, low on private alternatives. Identity had a name: *solidarity*, with all members of society. And I came to a USA filled with worries along all these lines, 'how can we get our kids through college; what if cancer strikes?' And I now witness how Norway follows the US lead down from the *altiplano* into the abyss of multiple worries. Why?

According to the basic needs approach, the politics of well-being is the politics of survival, wellness, freedom, and identity *for all*. As mentioned, there can be no trade-offs, such as more freedom for less wellness. Any politics, such as the liberal/ conservative predilection for freedom or the Marxist/radical one for wellness, limits the politics of basic needs. Nothing short of the full package is demanded. Any empirical regularity to the effect that 'you cannot have them all' may be a summary of the past, but is not a death sentence for tomorrow. 'Reality' should stimulate a search to create new social realities where the seemingly incompatible become compatible.

Creativity touches the spiritual dimension of human beings, the capacity to be on top of reality rather than vice versa. Economics as an approach to well-being is filled with trade-offs and curves, avoiding any 'both-and' in favor of 'either-or' and compromises; it is an institutionalization of anti-spirituality. In no way does this imply that transcending apparent incompatibilities is easy even if human history is filled with examples. (For an effort to work this 'negativist' rather than positivist approach seamlessly into social science methodology see the chapter entitled 'Science as invariance-finding and invariance-breaking activity' in Galtung (1977).)

The problem can be stated as 'Political parties are mainly based on economic class, not on all basic human needs for all.'

Table 19.1 takes a bird's-eye view of the needs around the world. All $4 \times 9 = 36$ pluses and minuses in the table are debatable. But what matters is the profile. The USA is high on freedom of choice and on identity with itself; the EU weak point is exactly identity, as is also true in socialist and capitalist Eastern Europe/Russia where people got freedom of choice instead of law and order after the terrors of Stalinism, and instead of wellness in the guise of welfare. A good deal? Most of the Third World is low on all measures of well-being.

Humankind is far from well-being: no region scores +4. And the task is like squaring the circle. 'Law and order' for survival is a part of the conservative package,

Table 19.1 State of world well-being, early 21st century

	Survival	Wellness	Freedom	Identity	No. of +s
USA	−	−	+	+	2
European Union	+	+	+	−	3
Eastern Europe					
Socialist	+	+	−	−	2
Capitalist	−	−	+	−	1
China	+	−	−	−	1
Japan	+	+	+	−	3
Southeast Asia	+	−	+	−	2
West Asia	−	−	−	+	1
Latin America, Africa, and South Asia	−	−	−	−	0
No. +s	5	3	5	2	15

'welfare state' for wellness is a part of the socialist package, and 'human rights' for freedom is a part of the liberal package, found also in the other two. Identity may be cause and effect of the struggle for autonomy by a suppressed group—the women, the young, the non-white, the excluded, the non-dominant nations—but is often rejected by middle-aged, non-excluded White men from the dominant nation, in short, those who tend to run political parties. Thus, we have a double problem: the needs package is badly reflected in the political agendas, and it is not easily implemented.

Identity is particularly problematic. There are many foci for identity. Two of them are lethal combinations. Religion/ideology fueling a state or even region may spell external war and hence is bad for survival. And when it stirs up a nation or even civilization it may spell internal war or wars without borders. Today, this would apply to fundamentalist Judeo-Christianity in USA/Israel and its homologue in fundamentalist Islam in western Asia.

But political parties as we know them are custom-tailored to the nation state with the exception of parties in the Indian Union and the European Union that are also regional. The world has no strong political forces standing for a global, inclusive identity, say, as human beings, for example, through global governance by an improved United Nations.

For people low on identity a polarizing but identity-producing conflict may even be welcome, buying identity at the expense of survival, and often also at the expense of freedom and wellness. And then we are deeply steeped in the anti-human politics of trading one need for the other. Or, pay no attention to the basic need for identity and it may raise a rather ugly head in highly sectarian ways.

How about that major intellectual/political UN break-through, the 'human development index' (HDI), reported in the United Nations Development Program's superb *Human Development Reports*? Among the hundreds of interesting findings let us just focus on one set (United Nations Development Program 1998, p. 20).

- Of the 174 countries, 98 rank higher on HDI than on GDP per capita, suggesting that they have converted economic prosperity into human capabilities very effectively. This achievement is noteworthy for such low-income countries as Lesotho, Madagascar, the United Republic of Tanzania and Viet Nam.

- For 73 countries the ranking on the HDI is lower than on GDP per capita, suggesting that they have failed to translate economic prosperity into correspondingly better lives for their people.

- The link between economic prosperity and human development is thus neither automatic nor obvious.

The HDI is a brilliant index of wellness, but not of well-being. It reflects only the *bienes fundamentales*, and actually only health and education. The underlying idea is that deficits in food, clothing, and housing will show up as disease and particularly in *infant mortality* and *life expectancy*. And that the basic indicator of communicative competence is *literacy*. HDI is based on those three.

All of this is certainly debatable. But the HDI is a human-, not a system-level indicator, focusing on individuals, not on abstractions like economic growth. There is something fascist in the economist's *strong market fetishism*, compatible with the classical fascist *strong state fetishism*, in turn compatible with *state planning fetishism*.

HDI ranks countries and Norway came out as number one years in succession. This has not been much reported in the Anglo-American media, perhaps partly because they themselves did not 'win' and partly because the HDI rules do not reflect their system/market rules. But Norwegians also feel uneasy because of the aspects of well-being not reflected by wellness. Survival is not the problem; Norway is not (yet) a violent society. But Norway may be an example of equality limiting freedom, as indicated above in connection with the welfare state approach to health and education. In Norway this is known as the 'Jante law', i.e. 'you shall not believe that you are better than us, etc.', formulated by the brilliant Danish author Axel Sandemose who emigrated to Norway (and probably knew something about the Jante mentality). The level of pluralism (or lack thereof) in medical and pedagogical orientation should be reflected if the alternatives are accessible to all. But the major shortcoming in the Norwegian well-being situation is along the identity dimension. Identities built around religion, ideology, and nationalism are withering away, leaving a vacuum and much loneliness. Some global identity is taking shape, however.

Again, if people want well-being, and we live in a democratic society where what people want becomes politics, then a politics of well-being for all should be possible. The burden of proof falls on those who fail to endorse well-being for all. If that is not the end, then what is? To maintain power and class privileges? A politics of egoism as opposed to altruism? Inspired by Adam Smith's theorem that a zillion egoisms become one altruism 'as by an invisible hand'? That theorem is obviously false. Given the egoism and lack of solidarity in institutionalized capitalism, there should be well-being for all by now. As this is not the case, Smith's God may be doubted.

The politics of well-being for all is the politics of satisfying basic needs for all. There will be a left wing giving priority to those most in need, and a right wing with the hypothesis that an excess high up will trickle down to those low down with a deficit. The latter hypothesis is obviously false. Trickling down is not automatic; pumping up is, with increasing gaps and even more doubts on capitalism.

Let us now bring in three of the approaches to well-being again.

♦ *Peace* in a narrow sense is a project against direct violence.
♦ *Development* in a narrow sense is a project for wellness; with health.
♦ *Human rights* in a narrow sense is a project for freedom.

But they all can easily take on the other needs and become broader.

♦ *Peace*. Undoing direct violence against survival; structural violence against wellness, freedom, and identity; and cultural violence legitimizing direct and structural violence.
♦ *Development*. Creating societies that guarantee well-being for all.
♦ *Human rights*. Protecting all basic human needs as human rights.

How about a politics of well-being protected by four human rights?

1 the right to live in a social and international order where everything is done to solve conflict by peaceful means;
2 the right to livelihood through gainful employment or otherwise;
3 the right to cultural identities but no right to impose them;
4 the right to options in how to implement the other rights.

This is inspired by the effort to build consensus using the human rights tradition in general, and more particularly the very important article 28 of the *Universal declaration of human rights* (UN 1948):

> Article 28. Everyone is entitled to a social and international order in which the rights and freedoms set forth in this Declaration can be fully realized.

To a rock-bottom, non-negotiable, *sine qua non* individual-level focus and basis should correspond an equally solid social and global-level basis protecting those irreducible rights. This is in no way a utopian project. The *Universal declaration of human rights* is already such a societal basis, not always well implemented, sometimes broken. But it exists as a signed, ratified, and celebrated fact. Of the four new rights proposed to cover better the four classes of basic human needs, the second, about livelihood, another word for wellness, is already relatively well covered in the 16 December 1966 *Covenant on social, economic, and cultural rights* (UN 1966). For the other three rights, political struggle would be needed. But that already serves as an indication that the arena is global, the UN, and that the carriers are enlightened countries. This in no way excludes similar struggles inside countries by NGOs and other parties.

We are searching for means–ends consistency in politics, for peace by peaceful means, development by developmental means, health by healthy means, human rights by respecting human rights, and environment by environmental means. And we are searching for cross-referencing, for peace by developmental means, and so on. We cannot promote the satisfaction of one need at the expense of another, as said so often above. But we are actually quite close; there are many places in the world where the social order of article 28 is quite well implemented. The problem is the international order.

So let us have a second look at politics: first, at politics in general. Politics translates advocacy into agency, using some force F to move inert social conditions. Power splits into *cultural* power to condition, persuade; *economic* power to reward or deprive; *military* power to punish or defend; and *political* power to decide. Class is a structure distributing power from the fully empowered (high on all four rights) to the totally powerless (low on all four). Equal power means power profiles in equilibrium (horizontal, at a middle level) or in disequilibrium (high and low balancing each other). Power profiles in disequilibrium can be socially very dynamic and can also be very disruptive (see Galtung 1978).

Power can be exercised in two ways: as an act of commission, $F > 0$, or as an act of omission, $F = 0$. What is not done is as important as what is done. But acts of commission are events, and are easily seen in a positivist paradigm focusing on the positively occurring. Acts of omission are non-events, in need of a 'negativist' paradigm with an equal focus on the negatively non-occurring. Neglect of the importance of omission is a major bias in Western civilization, already mirrored in the many 'nots' in the 10 commandments and the very high level of omission needed for legal culpability. As long as you do nothing you are safe; people who speak out and demonstrate against obvious wrongs are called 'activists', but people who seek safety and comfort in omissions are not given the appellation 'passivists'. Whereas acts of commission are unambiguous, acts of omission are ambiguous as omissions, non-acts, can be identified in very many directions.

The politics of status quo can come about by everybody doing nothing (Newton's first law), or by a political force bringing about another force, opposite and equal (Newton's third law). If not equal, the resultant force will move the conditions depending on their inertia (Newton's second law). Accumulated, quantitative moves may then lead to a qualitative transformation into a new social reality (Hegel's first law), as a synthesis of the opposed forces, the thesis and antithesis (Hegel's second law), and then a new force emerges and negates the synthesis (Hegel's third law). History spirals on.

Politics is dialectics. Linear means–end relations assume detachability from contexts that produce counterforces, unrealistic in the really existing world.

Backlashes and blowbacks are normal. $F > 0$ implies dynamism. But analogously to Newton's second and third laws, we get Hegel's first, second, and third laws.

Can we get some ideas about politics from Newton and Hegel? We have an international or world system where many things work not too badly and others very badly in terms of people killed and suffering violence and misery, where the citizens still have little say over crucial development parameters and no say at all over peace policies, and where pathological identities wreak havoc with the system. So, Newton says, do nothing and it will continue like that. But Newton also says, apply a force and you will get a counterforce, as in the present terrorism/state terrorism vicious cycle. If you want to move the system, and the international system is very inert, then you have a choice between using tremendous force to get a positive resultant, or the wiser approach of not provoking too much counterforce. Seek consensus. Move together. The policy suggested above comes out of that approach.

Will this produce any change or just move the system to some other place? *Plus ça change, plus c'est la même chose*? That is where Hegel's dialectics is so much more realistic than Newton's mechanics. While Newton's three laws are operating like clockwork, Hegel's three laws are operating inside Newtonian mechanics. Make incremental quantitative changes, Hegel says, and there will be a jump to a new quality, a real change, not *la même chose*. That new reality comes about as a synthesis of the force, the *actio*, the thesis, and the counterforce it has elicited, the *reactio*, the antithesis. To this can be added the distinction between three types of synthesis: compromise; the negative transcendence of neither–nor; and the positive both–and transcendence, negating the past that was (Galtung 2004). And Hegel then adds the third law attributed to him, that the negation will in turn be negated by the same process, but not always back to the past. History moves on, in other words. There is no end.

The Buddhist panetics approach to well-being

Panetics, the science of suffering, was developed by the late Ralph Siu, a remarkable Chinese-American. From his Chinese background he made creative use of neither the Daoist nor the Confucian strand of the immensely rich Chinese culture, but of the Buddhist component. And more particularly, *dukkha*, suffering, and the message of Buddha that, although life is replete with suffering, that suffering can be reduced, even eliminated. You shall live your life so as not to cause suffering in any form of sentient life, 'sentient' meaning the capacity to experience suffering also found in animals and plants. One sentient life is yourself. Your duty is also to alleviate your own suffering. The cause of suffering is greed, craving, excessive attachment. To reduce them is the way out. (For a Christian approach see Downey (2003).)

Siu, trained as a scientist, developed a measure of suffering, a 9-point dukkha scale from '1', 'noticeable', for example, a dentist appointment tomorrow, to '9', the 'unbearable, wanting to die', to leave this life (see Siu 1993, p. 9). The unit, one dukkha, is 'one person suffering level 1 for one day' ('0' means no suffering). This permits us to make comparisons across time and space for any set of humans, from the individual via the couple/family and the country to the world, for any length of time. The suffering of a social unit is computed as the 'number of persons × average intensity of suffering × duration in days'.

Like HDI, this is in the Protagoras *homo mensura* tradition, 'Man is the measure of all things.' In a community, a country, you may have glittering buildings or other material achievements, you may be politically/ideologically correct, based on the purest implementation of the single true teaching. But, if the reality of the experience is human suffering, then what is so great about it?

This measure is *subjective*. The individual person is the judge of his/her situation. My suffering is mine. Nobody is going to tell me I am not suffering when I am, nor that the suffering is good for me. Nor is anybody going to tell me I am suffering when in fact I am not, because according to them the objective circumstances should make me suffer. I decide, upon mature reflection.

The measure is also profoundly *egalitarian*. Each person's suffering is given the same weight in the terse 'number of persons involved'. This sounds trivial, and yet is filled with dramatic implications. We humans suffer from serious fault-lines in our social constructions: between us and other forms of life; between genders, generations, and races; between classes, nations, and states; regions and civilizations. Genocide is massive violence across such fault-lines, direct or structural. Yet we avoid taking in the suffering on the other side, drawing lines between worthy and unworthy sufferers. The suffering of dehumanized, even demonized persons does not count. This measure makes us all equal before the suffering. Like before the law. Or before God. My enemy counts like myself.

Finally, the measure is *non-theoretical*, only reporting that human beings feel and report suffering. The measure does not tell us why. Maybe this is an improvement on Buddha's reductionist insistence on greed and craving as the prime cause. Perhaps it is better to identify the causes in each concrete case. There is a deeper reason hidden behind this formula: the theory we have may itself cause immense suffering. A formula to obtain the good society is usually a formula to reduce suffering. But the formula may not include the suffering to end suffering. Those who resist help are dehumanized as unworthy sufferers. You cannot make omelets without crashing some eggs. For Siu that is only another form of suffering.

In his article 'Panetics—the study of human suffering', Ralph Siu (1988) asks the question, 'what are the types of suffering being inflicted by Americans upon fellow Americans?' And he presents two tables, one for the types of suffering, and one for the agents on the causal side of that suffering. By far most suffering, in his estimate, is linked to unemployment and poverty; much less falls under such headings as the justice system, crime, and smoking (alcohol somewhat more). The agents he finds most responsible for suffering are the business leaders, the 'president' (presumably standing for the government), and the public media persons; much less responsible for suffering are the criminals.

In short, his image, expressed in *megadukkhas*, is of a society where crime may be an irritant but is of little significance relative to the suffering caused by capital and state and expressed in unemployment and poverty. The role of lawyers and church leaders is very minor.

At this point perhaps one should come to the rescue of business leaders and presidents, getting them off the hook shaped by the word 'agent'. Agency is too close to intended action, and those people hardly wish to inflict the suffering of human degradation known as unemployment and poverty on anybody. It is more structure as usual. What they do is the unreflective, recurrent, behavioral pattern of acts of omission, enacting the structures within which they operate. Profit and growth are higher on their priorities than full employment and the elimination of misery. 'Structural violence' is intended to make this distinction lest we fall into the trap of blaming particular people rather than particular structures. True, there are those who justify unemployment because employers can be more choosey, picking from among the unemployed, and may even recommend a higher level of that immense suffering. But such people are often economists blinded to externalities by their misleading theories rather than pragmatic business leaders.

Let us take this reasoning one step further by applying it to the practice of peace and development. How does a peace researcher try to come to grips with the effects of a war? And how is that similar to, and different from, the panetics approach? Table 19.2 takes us beyond the conventional military headquarter's terminology, 'dead, wounded, material damage', into the often forgotten externalities of violence/war (see Galtung 1998). The panetics approach focuses on *humans*, and rightly so. More particularly, the focus is on human beings who are alive, not dead. As Siu (1988) expresses it: 'Although there may be fear, anxiety, and anguish stemming from the anticipation of death, death as such is regarded as being, for our purpose, suffering-free for the deceased. The related sorrows, worries, and economic and emotional deprivations to the bereaved and dependents can, of course, be severe. These need to be taken into account in estimating the total panetic burden on society as a whole.'

But that vision can also be expanded. Table 19.2 makes space for the Buddhist approach to life in general by including the category of damage to nature. And it gives a glimpse into peace theory through the inclusion of the structural damage done to the *social* and *world* spaces, the cultural damage, and the damage over *time*, pointing to an understanding of how violence breeds violence. *Dukkha* captures all of that.

The *dukkha* measure is theory-free. But to act so as to reduce *dukkha* we cannot be theory-free. The task of panetics is:

Table 19.2 Visible and invisible effects of direct violence

Material, visible effects	Non-material, invisible effects
Effects on nature	
Depletion and pollution; damage to diversity and symbiosis	Less respect for non-human nature; reinforcing 'man over nature'
Effects on humans	
Somatic effects:	*Spiritual effects*:
Numbers killed, wounded, raped, displaced, and in misery; widows; orphans; soldiers unemployed	Numbers bereaved and traumatized; general hatred; general depression; general apathy; revenge addiction; victory addiction
Effects on society	
The material damage to buildings; the material damage to infrastructure: road, rail, mail, telecommunication, electricity, water, health, education	The damage to social structure: to institutions, to governance; the damage to social culture: to law and order, to human rights
Effects on the world	
The material damage to infrastructure: breakdown of trade, international exchange	The damage to world structure; the damage to world culture
Effects having a time dimension	
Delayed violence: land-mines, unexploded ordnance; transmitted violence: genetic damage to offspring	Structure transfer to next generation; culture transfer to next generation; *kairos* points of trauma and glory
Effects on culture	
Irreversible damage to human cultural heritage, to sacred points in space	Violence culture of trauma, glory; deterioration of conflict-resolving capacity

1 to identify the suffering and its magnitude, and then

2 to identify theories and practices for its reduction.

My own studies lead me to believe in violent cultures and structures rather than violent people as the major causes for violence in general and war in particular. We may punish the bad actors and derive some satisfaction from that. But, unless we change some aspects of culture and structure, those very same aspects will reproduce bad actors, and those actors will then reproduce the violence. Some aspects are very deeply rooted and, moreover, there is the Pogo principle: 'I have met the enemy and he is us.' Hence the inclusion in the table of structure, culture, and of time in order to get some handles on the dynamics of suffering.

Panetics focuses on the infliction of suffering. If we now assume that the dead suffer no more and that the wounded can be healed, this leaves us with the bereaved and the traumatized. Multiply the military figures about casualties at least by a factor of 10 for the primary bereaved (near family and friends), by a factor of 100 for the secondary bereaved (extended family, neighbors, colleagues), and by the 1000s for the tertiary bereaved, those of the same nation, and panetics shows the madness of war very clearly. Our usual statistics conceal that suffering. In addition, panetics, like peace studies, will register equally the suffering on all sides, not only our own.

Table 19.3 Comprehending genocide defined as killing in the massive category: eight fault-lines and direct versus indirect killing

No.	Fault-lines	Direct violence	Structural violence
1	Nature	Slaughter	Depletion, pollution
2	Gender	Killing women: abortion, infanticide, witch-burning	Patriarchy as prison of women
3	Generation	Abortion Euthanasia	Exclusion of young Exclusion of old

4	Deviance
	  Criminal
	  Mental
	  Somatic
		
		Death penalty
		Euthanasia
		Euthanasia
			
			Exclusion
			Exclusion
			Exclusion

5	Race	Eradication, slavery	Colonialism, imperialism, slavery

6	Class
	  Military
	  Economic
	  Political
	  Cultural
		
		Elimination (all)
		
			Exploitation (body)
			Repression (mind)
			Alienation (spirit)

7	Nation/culture/ideology	'Genocide' narrowly defined	The state as prison of nations
8	State/country/territory	War (for food, sacrifice, winning)	Imperialism

Humans have the unfortunate habit of constructing fault-lines in the human condition. We create categories, the self/other divides, and *massive category violence = genocide* may fly across the divide. A case is the holocaust of European Jews. The Holocaust was truly horrendous, and yet, why is there a Holocaust Museum in Washington DC for *shoa* and not for Native Americans, African-Americans, Hawaiians? In a sense the reason is obvious: it is easier to highlight the massive crimes of others than one's own crimes.

Table 19.3 gives a more complete image of the told and untold massive suffering humans are capable of inflicting, according to the fault-line (adding the line between normal and deviant) and according to the type of violence, direct (intended) and structural (unintended). As the subject will be fully covered in a forthcoming book I will make only some quick comments here. The worst of the 16 cases, in terms of numbers affected, are the massive killings of women, particularly through abortion of unborn girls and infanticide (one hundred million having been reported missing between the UN world census in 1980 and 1990), and the massive deaths everyday of people from misery (one-quarter from starvation and three-quarters from easily preventable and curable diseases, perhaps 100 000 per day). Keep the bereaved in mind.

Genocide certainly goes beyond 'genus' defined as 'nation'. The massive suffering is related not only to Armenia, Nanjing, Gulag/KZ, Shoa, Hiroshima–Nagasaki, Cambodia (both US bombing and the Red Khmer regime), Rwanda. Admittedly, 'genus', narrowly defined, may point in that direction. But if we expand to 'massive category killing' the perspective broadens, bringing in crimes by commission and omission by the world's ruling class: Anglo-American White middle-aged males.

A more adequate discourse would include the whole paradigm—all 16 cases. It would also focus on the structural violence of patriarchy, misery, etc., not only on the acts of commission. It would focus on the violence to nature and to animals also, because it serves as training in taking life, 'soulless life' being destroyed. It certainly would focus on gender and generation, for example, putting children and the aged in ghettos known as kindergartens and old age homes, seeing exclusion as mental preparation for killing. It would focus on race used to define non-Whites as closer to soulless animals, 'feeling less pain and bereavement' with no 'real' families. It would focus on our handling of deviants: destroyed, not killed. And it would focus on class, putting 'the dangerous classes' in city ghettos.

Nation is privileged in 'genocide studies', particularly if the victims are Whites like 'us'. And states cling to the legitimacy of murder in war. Legitimacy, the cultural violence justifying direct and structural violence by means of hard Christianity, hard Darwinism, and hard economism, is key both to diagnosis, prognosis, and therapy.

Peace research has categories and panetics has measures of the suffering. But what does peace research have to offer as peace? Peace includes the *absence of direct violence* engaged in by military and others. But it also includes the *absence of structural violence*, the non-intended slow, but massive suffering caused by economic and political structures. And the *absence of the cultural violence* that legitimizes direct and/or structural violence. All these absences add up to *negative peace*, also obtained in mutual isolation. Better than violence, but not good enough. It is not peace because *positive peace* is missing (for more on this see Galtung 1996, part I). This approach is summarized in Table 19.4.

There are six tasks. First, eliminating the direct violence that causes suffering. Then eliminating the structures that cause suffering through economic inequity, or walls placing Jews or much of Palestine in ghettos. Then eliminating cultural

Table 19.4 Six approaches to peace

	Categories of violence		
	Direct (harming, hurting)	Structural (harming, hurting)	Cultural (justifying violence)
Negative peace	(1) Absence of = ceasefire	(2) Absence of = no exploitation; or structure = *atomie*	(3) Absence of = no justification; or culture = *anomie*
Positive peace	(4) Presence of = cooperation	(5) Presence of = equity, equality	(6) Presence of = culture of peace, and dialogue
Peace	Negative + positive	Negative + positive	Negative + positive

themes that justify one or the other. Then come the three tasks of building direct, structural, and cultural peace. The parties exchange goods, not 'bads', not violence. The structural version of that builds cooperation into the structure as something sustainable under the heading of equity for the economy[6] and equality for the polity, that is, reciprocity, equal rights, benefits, and dignity, 'what you want for yourself you should also be willing to give to the other.' Equality for the polity involves democracy (one person, one vote) and human rights (every one is entitled . . .), but not only within countries, also among them. Finally, a culture of peace, positively confirming and stimulating, must be established.

There is more peace around than meets the naked eye. When solid, as in a good family, neighborhood, organization, most of these six tasks have been very well done. The international order, however, is lagging behind.

Conclusion: on the well-being/ill-being sukha/dukkha dialectic

Siu's measure does not include *sukha*, fulfillment. It is like peace research only focusing on violence and war, development/health research only focusing on misery/disease, and human rights focusing only on infractions and, for that reason, understanding little of peace, development/health, and compliance.

Following Siu, however, we could easily construct a 9-point *sukha* scale from '1', the 'noticeable', for example, a lunch appointment tomorrow, to '9', 'I wish this moment could last forever'. One sukha, then, is 'one person enjoying level 1 for one day' ('0' means no bliss). This permits us to make comparisons across time and space for any set of humans, from the individual via the couple/family and the country to the world, for any length of time, measuring bliss/well-being as the 'number of persons × average intensity of fulfillment × duration in days'. Or simply adding individual sukha levels, not forgetting the dispersion.

The danger would be to fall into the Jeremy Bentham trap of using some people's *sukha* to compensate, through averages, for other people's *dukkha*, like economists constructing a measure of per capita wealth whereby the well-being of the wealthy conceals the suffering of those in misery. Averages bring no comfort to the sufferer. Reducing her/his suffering does. So the first priority in suffering-reduction should be to John Ruskin's *Unto this last*, to those who suffer most.

Building on another strand in Siu's Chinese culture, we would not conceive of the sukha measure as the positive part of a dukkha–sukha scale from −9 via 0 to +9. We would build on the Daoist assumption that there can be dukkha in sukha and sukha in dukkha. The two are orthogonal but not mutually exclusive. There

---

[6] Equity for the economy is a very weak, very undeveloped field, both in economic theory and practice. The *International covenant on economic, social and cultural rights* (UN 1966) is an effort to achieve equity, but it has not yet been ratified by the United States.

can even be moments of unbearable ambiguity (S = D = 9), like the extremes of love; vast time stretches of blandness (S and D both equal or close to 0); relatively pure sukha zones (S >> D); and relatively pure dukkha zones (D >> S).

Combine this with the focus on the bereaved and traumatized and we sense the enormous suffering through the millennia. But we only sense this if we open our hearts and minds to the suffering of everybody, not privileging some suffering at the expense of others. As a matter of fact, probably one of the worst forms of dehumanization is to deny some categories their suffering, invoking theories like 'they suffer less because they are so used to it'. Deprive a human being of suffering and you deprive him/her of subjectivity. Deprive him/her of subjectivity and you have constructed a non-human, as in behaviorist social science.

A focus on violence is a focus, essentially, on dukkha, not denying that in inflicting, even in receiving violence, there may be elements of sukha. Like the sukha derived from being an instrument of God's will against Satan's ilk, and the sukha derived from being God's people as proved by the fact of being the victims of Satan's ilk.

But how about a focus on development? Using the definition above, linking development to well-being, the work for development is sukha-producing work, just like the work for peace in a limited, and limiting, sense is dukkha-reducing work. Let us now think both at the same time.

The ideal is, of course, *sukha high–dukkha low*; the worst is *sukha low–dukkha high*. But how about the other two? Sukha low–dukkha low is very frequent—dull, tepid poverty bordering on misery; not much suffering, but not much happiness either. Or hard work for survival, forgetting about the other three needs. And that opens a perspective on maldevelopment: *it is not so much dukkha as deprived of sukha*—life lived worrying so much about dukkha that there is no space for sukha. And correspondingly for sukha high–dukkha high: not dull and tepid, but too much and too ambiguous. It may be tolerable as a *kairos* (the moment lasting "forever") time capsule but not as *chronos* flowing through time. We gain more insight studying well-being and ill-being, fulfillment and suffering, together.

However, there is the asymmetry between necessary and sufficient conditions. As mentioned there is apprehension about Utopia, not because it defines well-being, but because it defines it for everybody forever. The problem is not utopias, in plural, with lower case u, as hypotheses people develop for good lives well lived. The problem is one single Utopia formula for the social conditions supposed to produce sukha, which invariably, when enforced, will lead to enormous dukkha. A good example of this is what is happening in Afghanistan and Iraq at present.

Hence the focus on the necessary conditions of well-being. Reduce ill-being by satisfying the basic needs, those *sine qua non*. Reduce suffering, dukkha. Take worries out of people's lives. But don't expect them to proclaim happiness, and certainly not to be grateful. And pay attention to the time dimension. Make the satisfaction sustainable.

From this I sense a simple bridge to subjective happiness, a *pons asinorum*, philosophically unsophisticated: learn to appreciate the pain not suffered, the worries not felt. Derive well-being also from the negative absence of ill-being. Derive sukkha from non-dukkha. Enjoy sitting warm and dry with cold rain pouring down outside. Even when suffering some illness, enjoy the health of 99% of the body. Enjoy the love and esteem you feel. Worry less about love lost and esteem missing.

Ralph Siu (1993) expresses this very well in his preface and conclusion.

> We wish to share a vision: The Day when the leaders will have become nearly half as conversant with panetics as they are with economics in their exercise of dominance and stewardship, and the people at large nearly half as clear concerning dukkhas as they are concerning dollars in their expectations, and dispensations of entitlements and rights.

and

> The time when the decline in the average number of dukkhas inflicted on an individual per time unit is universally accepted as a standard and basic measure of social progress, [and] the steady decline in the number of dukkhas sustained by all the residents in a country for the year is accepted as an essential and explicit measure of the excellence of government.

And I could add a second formula for subjective happiness to the joy from non-suffering above: the well-being derived from living a life with meaning. Like working for the reduction of ill-being/dukkha for all. Be careful, however. Hitler's life also had meaning, purpose in life, derived from the exceedingly bad ideas he believed in. Hence the Buddhist admonition to decrease the dukkha and increase the sukha in *all* others, and yourself. For this, as George Bernard Shaw points out, the 'golden rule' of doing unto others as you want them to do unto you is insufficient: *Their tastes may be different.* Respect others. Know their tastes.

References

Downey, J.K. (2003). Suffering as common ground. In *Constructing human rights in the age of globalization* (ed. M. Monshipouri, N. Englehart, A.L. Nathan, and K. Philip), Chapter 12. M.E. Sharpe, London.

Fisher, H. (2004). *Why we love*. Holt, New York.

Galtung, J. (1967). *Theory and methods of social research*. Columbia University Press, New York and Allen and Unwin, London.

Galtung, J. (1977). *Methodology and ideology*. Ejlers, Copenhagen.

Galtung, J. (1978). *Peace and social structure, essays in peace research*, Vol. III. Ejlers, Copenhagen.

Galtung, J. (1980*a*). The basic needs approach. In *Human needs: a contribution to the current debate* (ed. K. Lederer, J. Galtung, and D. Antal), pp. 55–125. Oelgeschlager, Cambridge, Massachusetts.

Galtung, J. (1980*b*). The basic needs approach [in French]. In *Il faut manger pour vivre: controverses sur les besoins fondamentaux et le développement* (ed. P. Spitz, J. Galtung, R. Preiswerk, G. Berthoud, M. Guillaume, G. Rist, A. Allain, J.-P. Bärfuss, G. Etienne, and L. Kanemann), pp. 51–127. Presses Universitaires de France, Paris.

Galtung, J. (1988*a*). *Methodology and development*. Ejlers, Copenhagen.

Galtung, J. (1988*b*). The basic needs approach. In *The power of human needs in world society* (ed. R.A. Coate and J.A. Rosati), pp. 128–59. Lynne Rienner, Boulder, Colorado.

Galtung, J. (1993). *Buddhism: a quest for unity and peace*. Sarvodaya, Colombo, Sri Lanka.

Galtung, J. (1994). *Human rights in another key*. Polity Press, London.

Galtung, J. (1996). *Peace by peaceful means*. Sage, London.

Galtung, J. (1998). *Tras la violencia, 3R: reconstrucción, reconciliación, resolución*. Gernika Gogoratuz, Bilbao; also in English at <www.transcend.org>.

Galtung, J. (2000). *Johan uten land*. Aschehoug, Oslo. [Publications in other languages forthcoming.]

Galtung, J. (2004). *Transcend and transform: an introduction to conflict work*. Pluto, London.

Josselson, R. (1998). *The pleasures of girls' and women's friendships*. Three Rivers Press, Michigan.

Ornauer, H., Sicniski, A., Wiberg, H. and Galtung, J. (Eds.) (1976). *Images of the world in the year 2000*. Humanities Press, Atlantic Highlands, New Jersey.

Siu, R. (1988). Panetics—the study of human suffering. *J. Hum. Psychol.* **28** (3), 6–22.

Siu, R. (1993). *Panetics and dukkha, an integrated study of the infliction of suffering and the reduction of infliction*, Panetics Trilogy, Vol. II. The International Society for Panetics, Washington, DC.

United Nations (UN) (1948). *Universal declaration of human rights*, General Assembly resolution 217A (III), UN Document A/810 at 71. UN, Geneva.

United Nations (UN) (1966). *International covenant on economic, social and cultural rights*, General Assembly resolution 2200A (XXI), 21 UN GAOR Supp. (No. 16) at 49, UN Document A/6316. UN, Geneva.

United Nations (UN) Development Program (1998). *Human development report 1998*. Published for the UN Development Program by Oxford University Press, New York.

World Health Organization (WHO) (2001). *The world health report 2001, mental health: new understanding, new hope*. WHO, Geneva.

Nic Marks and Hetan Shah work closely together on the Well-being Programme at **nef** (the new economics foundation). Nic leads the well-being research, and Hetan is director of new economics. Both are interdisciplinarians, with Hetan having qualifications in philosophy, politics, economics, law, and history and Nic in statistics, economics, management studies, change processes, and psychotherapy. **nef**'s Well-being Programme seeks to address the question: what would society and government look like if people's well-being was one of its ultimate aims? Details of the programme can be found on the **nef** website.

Chapter 20*

A well-being manifesto for a flourishing society[1]

Nic Marks and Hetan Shah

Introduction

The idea that government should be concerned with people's well-being or happiness is no longer considered frivolous. There has been a surge of interest in this area, sparked by the growing research on well-being and not least by the devastating research finding that, whilst economic output has nearly doubled in the last 30 years in the UK and other developed countries, life satisfaction levels have remained resolutely flat. Whilst governments are slowly adjusting to this new reality, there are some signs of change in the UK. For example, the Prime Minister's Strategy Unit has reviewed the policy implications of psychological research on life satisfaction (Donovan and Halpern 2002). At the local level, government has gained new powers to promote economic, social, and environmental well-being. In order to take forward the thinking on well-being and policy, this well-being manifesto was printed in its original form in September 2004 by **nef** (the new economics foundation)—a think tank in London that focuses upon creating an economy based on social justice, environmental sustainability, and well-being. It was printed with the aim of raising the profile of 'well-being politics' in the run up to the general elections in the UK in 2005, and was launched at the time of the party political conferences in the autumn of 2004. The full report was accompanied by a 'mini manifesto'—a credit card–sized fold-out concertina with the key points of the manifesto. Whilst the focus of the piece is on UK policy and the examples are largely drawn from the UK, we believe that directions set out in this piece will apply to many developed country contexts.

The well-being evidence

One of the key aims of a democratic government is to promote the good life: a flourishing society where citizens are happy, healthy, capable, and engaged—in

[1] This chapter is based on *A well-being manifesto for a flourishing society*, originally published in 2004 by The New Economics Foundation, London, who hold the copyright.

other words, with high levels of well-being. For most of human history, trying to understand what led to well-being was the stuff of philosophy or poetry. Recently, however, some psychologists and sociologists have turned away from studying illness and dysfunctionality and begun to study well-being, happiness, and flourishing. The results have profound implications for individuals and for government. This well-being manifesto seeks to answer the question: 'what would politics look like if promoting people's well-being was one of government's main aims?'

One of the main ways in which governments in developed countries try to promote well-being is through increasing economic growth. The logic to this is that, by increasing national and individual incomes, people have more choices about how they should lead their lives. Psychologists, however, have thrown a large spanner in the works. The relationship between economic prosperity and both individual and social well-being in developed countries seems to have broken down. For example, whilst economic output has almost doubled in the UK in the last 30 years, life satisfaction has remained resolutely flat (see Fig. 20.1). Whilst rates of depression are difficult to track over long periods of time, due to changes in diagnosis and in how people perceive depression, they do seem to have risen over the last 50 years in developed countries. For example, Diener and Seligman (2004) cite strong evidence that young people are experiencing more depression, and that 'people born earlier in this century have experienced much less depression in their lifetime than people born later'. In order to make better judgements about what government should be doing in response to this situation, we should consider the lessons from well-being research.

Well-being research

What is well-being? Some academics argue that well-being is best understood in terms of our overall happiness or satisfaction with life. But evidence shows that there is much more to life than satisfaction: people also want to be leading rich and fulfilling lives—developing their capabilities, fulfilling their potential, and leading

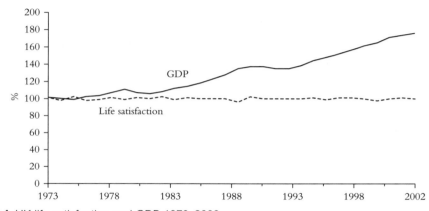

Fig. 20.1 UK life satisfaction and GDP, 1973–2003.

socially useful lives. We would suggest it is important to supplement life satisfaction with other measures, particularly as people's satisfaction with their lives seems particularly prone to adaptation effects, with average life satisfaction being relatively stable at a country level (see Fig. 20.1). For a fuller explanation of, and justification for, a multidimensional model of well-being see Marks *et al.* (2004).

Therefore, **nef**'s model of well-being has two personal dimensions and a social context.

- People's *satisfaction with their life* which is generally measured by an indicator called *life satisfaction*: this captures satisfaction, pleasure, and enjoyment.

- People's *personal development* for which there is not yet one standard psychological indicator—the concept includes being engaged in life, curiosity, 'flow' (a state of absorption where hours pass like minutes), personal development and growth, autonomy, fulfilling potential, having a purpose in life, and the feeling that life has meaning. See Ryan and Deci (2001) for their closely related conception of 'eudaimonic' approaches to happiness.

- People's *social well-being*—a sense of belonging to our communities, a positive attitude towards others, feeling that we are contributing to society and engaging in pro-social behaviour, and believing that society is capable of developing positively. This schema is based on Corey Keyes's classification of social well-being (Keyes and Haidt 2003).

For people to lead truly *flourishing* lives they need to feel they are personally satisfied and developing, as well as functioning positively in regard to society. Unfortunately, too many people are instead *languishing*—living unhappy, unfulfilled lives as well as lacking social and community engagement. Estimates from the US suggest that less than 20% of the population is flourishing and over 25% is languishing, with the rest being somewhere in between. Whilst we do not use them in their strictest sense here, 'flourishing' and 'languishing' as scientific concepts have been developed by Corey Keyes and the data quoted use Keyes's methodology (Keyes 2002).

There has been some recent interest by UK policy-makers in life satisfaction, but this only tells part of the story. Whilst life satisfaction seems strongly (and inversely) related to mental health and depression, personal development seems to be more strongly linked to overall health, longevity, resilience, and the ability to cope with adverse circumstances and 'thrive' in life. For example, older people who score highly on the personal development dimension have a different biological profile and are therefore less likely to develop serious illnesses in later life (Singer and Ryff 2001). This dimension of personal development is also closely related to the kind of individual characteristics that underpin current UK government agendas around active citizenship or enterprise. Well-being is not just about a passive happiness; it is also about an active engagement with life and with others. Promoting well-being is a 'good' in itself as well as a contributor to other ends (see Box 20.1).

Box 20.1 **Well-being promotes a better society**

Well-being is an important end in itself. It also has many benefits and contributes to other important ends (Lyubomirsky, King, and Diener, 2005). Evidence shows that happy people are more:

- ◆ sociable
- ◆ generous
- ◆ creative
- ◆ active
- ◆ tolerant
- ◆ healthy
- ◆ altruistic
- ◆ economically productive
- ◆ long living

Therefore, promoting individual well-being is not just an important end in itself, but also has useful consequences for a flourishing society in all sorts of other ways, including the enhancement of people's social well-being.

The basic findings of well-being research

Where does our well-being come from? There is more research on the satisfaction and happiness elements of well-being than on the personal development and social well-being aspects. Nevertheless, what we know is fascinating. US research suggests that there are three main influences (Lykken 1999; Sheldon *et al.* 2005).

1 *Genetics.* We have a predisposition to a certain level of happiness. Researchers argue that it is the largest influence on our happiness level and it explains about 50% of the variation in people's current happiness. There is, of course, an interaction between genetic predispositions and our upbringing and environment in that genes create a predisposition, but they sometimes require environmental conditions that would activate them. Therefore, there is an important interaction between genes and the environment, and thus the environment can be moulded to facilitate positive predispositions and prevent the activation of negative predispositions.

2 *Life circumstances.* These include factors such as our income, material possessions, and marital status, as well as contextual circumstances such as our neighbour-hood, whether we have just moved jobs or home, and indeed our favourite British obsession: the weather. People tend to adapt to changes in circumstances quite quickly and so these factors have only a small influence on people's

happiness—researchers estimate it explains only 10% of the variation. But, as a society, we tend to spend a disproportionate amount of time focused on this aspect of our lives. In particular, money is often seen as a key to happiness, but it is not necessarily so.

3 *Intentional activities.* These are pursuits that we actively engage in and they account for 40% of the variation in our happiness. They include actual behaviours and aims such as working towards our goals, socializing, exercising, doing engaging and meaningful work; they also include cognitive activities such as appreciating and savouring life or looking at the bright side of things. Adaptation rarely occurs for these kinds of activities, as they are impermanent or can be infinitely varied. It is clear that this is the area where we can make the most difference to our own well-being.

So where does well-being come from? Is it from money? The short answer is 'not after our basic needs are met'. Whilst there is some correlation across nations between wealth and satisfaction, many developing countries have very high levels of happiness (for the international comparison data presented in Fig. 20.2, the correlation is 0.49, which suggests that gross national product (GNP) can explain about 25% of the variation in happiness between nations). People in developed countries don't seem happier than they were 40 or 50 years ago, despite rising living standards (see Fig. 20.1). Within countries, however, very rich people tend to

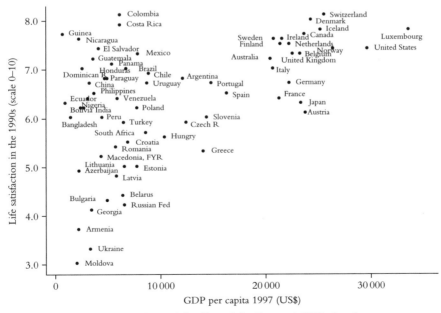

Fig. 20.2 An international comparison of the life satisfaction and GDP of nations (Veenhoven 2003).

Box 20.2 Why our circumstances don't make us happier

One of the great insights of well-being psychology has been to explore why circumstances have such little effect on our happiness. The answer is two-fold: adaptation and comparison. Human beings are phenomenally adaptable, and we tend to learn very quickly to view our current position as normal. Within a year, lottery winners are little happier, or even less happy, than they were before they won the big prize (Brickman *et al.* 1978). As we adapt, our expectations about what will make us happy rise. We are also creatures of comparison. We compare ourselves to where we want to be, and to other people. As we achieve our goals, we change whom we compare ourselves to and find a new source of unhappiness! The goal posts are always moving.

be a little bit happier than the moderately well off, both of whom are happier than the seriously impoverished.

How do we make sense of these seemingly contradictory findings? The evidence suggests that, after our basic needs are met, we adapt very quickly to the material gains that come from increases in income. We also compare ourselves to others and, unless we are at the top of the pile, this can lead to dissatisfaction (see Box 20.2). Whilst it is clear that people gain status from factors other than their income level, the dominance of economic comparisons is probably reflected in the finding that the very wealthy are a little happier than those who are closer to average income levels, i.e. it is the status gained from income rather than the income itself that is important (Diener *et al.* 1985).

Not only does economic growth not bring happiness, but many of the things associated with growth can reduce our well-being. By focusing purely on economic indicators, we have missed the negative side-effects of economic growth and efficiency. These might include the depletion of environmental resources, the stress from working long hours, and the unravelling of local economies and communities when the out-of-town supermarket settles nearby (Simms 2003). Research also shows that people who are materialistic tend to be less happy than those who value other things (Kasser 2002) and it is undeniable that we live in a culture that promotes materialism. Perhaps it is no surprise that, alongside a near doubling of economic output in the last 30 years, we have seen depression and mental illness rise—the UK Office for National Statistics (ONS) estimates that 10% of the population is depressed at any given time (ONS 2001). In addition, levels of trust have fallen in the UK from around 60% in the 1950s to around 30% today (Performance and Innovation Unit 2002). The economically efficient society is not necessarily a society that promotes high well-being.

So what else affects our well-being?

One of the most important factors that promotes well-being is our personal relationships. Marriage and long-term cohabiting relationships have a beneficial effect on well-being. Having intimate friendships and family networks is also an important contributor to well-being (Diener and Seligman 2002) as is belonging to some kind of community or social group. The biggest message of the whole body of well-being research is that, as a society, we now devote too much time to increasing our standard of living and not enough time to fostering our relationships.

While unemployment is one of the biggest sources of unhappiness, good quality work can be an important source of well-being. In particular, it can bring 'flow'— a state of absorption where hours pass like minutes. The most important condition for creating flow at work is for the job to be challenging and for the level of the challenge to be matched to the skills and capabilities of the worker.

In terms of health, how we *perceive* our condition is the crucial factor—our objective health status matters less. Those people who see themselves as healthy are happier than those who do not, even when their objective level of health is the same. So hypochondriacs are right to complain—but it is their unhappiness itself that makes them sicker! And, as your mother told you, exercise brings happiness both in the short and longer term (Sarafino 2002).

Education has been shown to have little effect on life satisfaction in itself but it appears to be an important factor in our personal development; for example, young people's satisfaction with school was shown to be the most important influence on their personal development in **nef**'s recent report (Marks *et al.* 2004). In addition, education is likely to be key in the promotion of social well-being.

Our physical environment appears to have some limited independent effect on our well-being. For example, living close to open green space has been shown to enhance people's well-being (Maller *et al.* 2002), while ugly or toxic environments can diminish well-being. We adjust very quickly to many environmental factors, including our climate. So the dream of moving to sunnier climes is unlikely to deliver permanent well-being enhancement, with surveys showing that people living in sunny California are no happier than those living in the frozen midwestern US (Schkade and Kahneman 1998).

What can government do?

It is obvious that government cannot and should not attempt to *make* us happy but that the impacts of government can have profound effects on the culture within which we live. Are we a materialistic society focused on individual gain? Or are we a companionable, sustainable society that has the time to enjoy the fruits of its economic prosperity?

Government already does a great deal to promote well-being; the fact that we live in a democratic and stable state is an important prerequisite to our well-being.

This is highlighted by the other extreme: the low well-being of unstable ex-Soviet states (see Fig. 20.2) and the fact that the lowest ever national score of life satisfaction was 1.6 on a 10-point scale, which occurred after the government of the Dominican Republic was overthrown (Diener and Seligman 2004). As well as providing a secure base for people to get on and live flourishing lives, the UK government has recognized through recent legislation that there is indeed potential for the further promotion of well-being. In the Local Government Act 2000, all local authorities were given a new *power of well-being* that enables them to do anything that promotes social, economic, and environmental well-being. **nef** has carried out a pilot project with Nottingham City Council building transdepartmental well-being indicators—supporting the council in the twin aims of working towards 'joined-up government' and shedding light on policies that would enhance the well-being of young people in the city (Marks *et al.* 2004; see Box 20.3).

This manifesto provides some ways of thinking about policy to promote well-being. It is by no means the final word, rather a space in which to begin the debate and push forward the directions of the thinking. To do this, we suggest that there are eight interrelated areas where government could take action to promote well-being.

Box 20.3 **The power of well-being and Nottingham City Council**

Nottingham City Council (NCC) and **nef** worked together to innovatively use the local government power of well-being by creating a single set of well-being indicators across several departments. NCC's community strategy partially focuses on young people and, therefore, the well-being indicators were used with 7–19-year-olds. The creation and the use of well-being indicators supported the joining up of government by concentrating on one of the true ends of policy-making—well-being—rather than on the means. They also shed light on the reality of how young people in Nottingham were really faring. The work shows, for example, that 32% of young people were at the very least unhappy and could be at risk of mental health problems. The indicators also provide valuable new information on a range of policy-making questions. For example, they show that victims of crime had lower well-being, but that fear of crime did not affect the well-being of young people. They also highlight the importance of providing local opportunities for participating in sport and other engaging activities. This is valuable information in relation to spending priorities. (There are also important results in relation to education—see Box 20.6.)

1 Measure what matters: produce a set of national well-being accounts.

2 Create a well-being economy: employment; meaningful work; and environmental taxation.

3 Reclaim our time through improving our work–life balance.

4 Create an education system to promote flourishing.

5 Refocus the National Health Service to promote complete health.

6 Invest in early years and parenting.

7 Discourage materialism and promote authentic advertising.

8 Strengthen civil society, social well-being, and active citizenship.

This is not to say that the only aim of government is to maximize personal well-
being: we are not advocating utilitarianism. Other goals, such as promoting social
justice or sustainable development, are also critical and may conflict with individual
well-being. For example, our desire for cheap flights—subsidized by the lack of tax
on aviation fuel (Simms 2000)—is clearly in conflict with environmental sustainability. However, well-being can often go hand in hand with social justice or
environmental sustainability. Improving social conditions can have well-being
effects for the worst off; environmental sustainability can be interpreted partly in
terms of balancing the well-being needs of future generations against those of
today's generation. In practice, well-being research opens up new arguments and
potential solutions for creating an economy that is environmentally sustainable.
Since it shows that well-being is not necessarily linked to growing consumption
and economic growth, it is possible that we can both reduce our environmental
impacts and improve our quality of life.

1 Measure what matters: produce a set of national well-being accounts

Government spends hundreds of millions of pounds on measuring economic and
social indicators, with the ONS alone spending over £142 million in 2002/3
on the collection of social and economic statistics (ONS 2004). One of the big
surprises of well-being research, however, is the disjunction between people's
standard of living and their happiness. As we have seen, indicators of economic
growth such as GDP (gross domestic product) are poor measures of well-being
(Jackson 2004). Therefore, with the exception of surveys regarding the prevalence
of mental illness, we have little systematic knowledge of how people in the UK are
really faring psychologically.

A set of national well-being accounts should be created that covers the main
components of individual well-being—life satisfaction and personal development—
as well as a range of components of well-being including engagement,
meaningfulness, trust, and measures of ill-being such as stress and depression
(Diener and Seligman 2004). The indicators should also include measures of

well-being beyond the personal, what we call social and ecological well-being, in other words, how we feel about and how engaged we are with the society and the environment in which we live. Well-being is not purely an individual phenomenon: it is rooted in our broader communities.

A set of national well-being accounts would help to focus the minds of policymakers on the true end of policy. In particular, it would help identify the worst off in well-being terms. Policy tends to think of the worst off as the income-poor. We need to supplement this with broader definitions of poverty and ill-being and what we might call the opposite of flourishing—*languishing*. Whilst, on average, the income-poor may have lower well-being, there will be many individual cases where those who are languishing are not financially poor. It may be that to tackle their problems the best interventions are, for example, in the field of mental health, education, or finding meaningful work.

A set of national well-being accounts would help us to understand which kinds of economic growth enhance well-being and which reduce it. More well-being research would help individuals make better choices about their own lives. This would rebalance the media coverage focused on economic and financial indicators. At present, for example, many people put too much weight on financial considerations when choosing what sort of work to do. In order to rebalance this, the public needs to have more systematic knowledge about, for example, how pleasurable or engaging or meaningful different kinds of work tend to be.

In addition to national well-being accounts, all local authorities should carry out well-being audits of their areas. This would create a single set of indicators that all departments would connect with, as ultimately most of their purposes concern the well-being of their population. Nottingham City Council, together with **nef**, has already begun to develop some of these indicators for young people with important implications for policy (see Box 20.3).

2 Create a well-being economy: employment, meaningful work, and environmental taxation

One of the biggest messages coming from the whole literature on well-being is that increasing economic growth and efficiency does not necessarily improve well-being. So what new directions should a well-being economy move in?

Work is often characterized by economists as a necessary evil so that we can have income to enjoy our leisure time. The well-being research shows that work is far broader in its effects than this narrow view suggests. Good work can profoundly affect our well-being by providing us with purpose, challenge, and opportunities for social relationships; it can constitute a meaningful part of our identity. Thus the well-being economy needs to be concerned with the quality of work in which we engage.

It should be noted that work is broader than employment. It includes unpaid work, voluntary work, or work at home, such as caring for children and the elderly as well as domestic chores. We tend not to value unpaid work in our society, even though our economy depends on it. The value of unpaid work at home in the UK has been estimated as approximately £150 000 million in 2002—the equivalent of an extra 15% on GDP that year (Jackson 2004). A recent (as yet unpublished) **nef** survey found that people rated their voluntary work as significantly more meaningful and satisfying than their paid work. Thus, as well as considering the quality of work in the economy, we should recognize and support unpaid work through valuing it in the national accounts, increasing opportunities for volunteering, and consider instituting a participation income or a citizen's service.

Well-being literature provides many insights into what makes for good work. It is in the self-interest of business to promote good work as there are linkages between quality work, productivity, and worker retention. Some of the key lessons that are emerging from research about quality work are the following.

- Jobs should be redesigned as far as possible to fit people's particular strengths and interests.
- Workers should be given more autonomy—the ability to make decisions and have control over their work.
- Work should promote 'flow'—it should be designed to challenge, but not be so challenging that it provokes anxiety and stress (see Box 20.4).
- There should be opportunities for interpersonal contact and friendships at work (Harter *et al.* 2002).

Box 20.4 **Flow**

Mihaly Csikszentmihalyi defined the concept of 'flow' to describe experiences where we are completely absorbed in what we are doing, and where time feels like it passes very quickly. He argues that we experience 'flow' when we are engaged in activities that are challenging but for which we have the skills to meet the challenge. Different people find 'flow' in different activities, but the state is the same whether it is derived through mountain biking, having a good conversation, or playing cards. Far more profound and enduring rewards come from engaging our signature strengths (Seligman 2002) rather than from passive consumption of TV, alcohol, or excessive food. Csikszentmihalyi argues that work is one of the most important sources of 'flow' in our lives. His research suggests that around 15% of people have never experienced 'flow', whilst around 20% say they feel it every day, with others somewhere in between (Csikszentmihalyi 1997; for a slightly different more detailed approach to assessing and measuring flow see Delle Fave and Massimini 2004).

- There should be opportunities for workers to develop their skills and capabilities at work.
- In general, workers should not be relocated unless it is absolutely necessary, as it destroys both their own and their families' social relationships and prevents them from forming community ties.

There are many models of good workplaces whose lessons need to be drawn out and disseminated to employers. Government itself is the largest employer in the UK; thus it should also take on board the well-being research and explicitly seek to promote the well-being of its employees.

Focusing just on paid work, research shows that unemployment has terrible effects on well-being (Argyle 2001). This goes far beyond the effect that loss of income has and includes the loss of identity, meaning, and social relationships. Research also shows that even the employed are unhappier in an economy with high unemployment, partly due to the lack of security (Di Tella *et al.* 2003). Thus well-being research reinforces the case for a full-employment economy, even if this is not the most efficient or highly productive macroeconomic scenario. At present, UK unemployment is relatively low, but this is, in part, due to the hidden unemployment of those people who are claiming incapacity (disability) benefits but are still capable of work. Recent research suggests unemployment figures would double if they included people on incapacity benefits who were capable of working (Beatty and Fothergill 2004). Government, therefore, needs to help these often hard-to-reach groups to find meaningful work and to support the creation of appropriate work opportunities. This can be done, for example, through supporting organizations such as intermediate labour markets and social firms.

Research shows that material gain has little impact on well-being once basic needs are met. Hence a pound in the pocket of a poorer person is worth more in well-being terms than a pound to a rich person. In Europe, there is also a relationship between social inequality and lower well-being: in other words, we do not like to live in an unfair society (Alesina *et al.* 2001). This is an instance where well-being research strengthens the existing social justice case for material redistribution. The case is even stronger in relation to promoting overseas development through measures such as debt cancellation, reduction of UK trade barriers, and increased aid.

The well-being of future generations depends on protecting our environment from climate change, resource depletion, and other environmental problems. One of the most important changes that policy-makers can make to move us towards this is to fundamentally change the incentives system. We need to start moving towards a system of taxing environmental *bads*, such as fossil fuels, and reducing the tax burden on *goods*, such as work. Such a shift would have to be implemented with measures to prevent it from having regressive effects and hitting the worst off. Increasing environmental taxes and simultaneously reducing payroll taxes could have a small but positive effect on employment if the payroll tax cuts were aimed at lower-income groups (Turner 2001). Hence a 'double dividend' of reduced pollution and increased employment is possible here.

3 Reclaim our time

Economic prosperity has not necessarily brought us stronger families, better relationships, or more resilient and vibrant communities. In fact, it appears that economic pressures have diverted us from these things which really matter.

The single most important shift we need to make as a society to promote well-being is to improve our work–life balance. Whilst, as discussed above, we need to structure work in such a way that it produces well-being, the evidence shows that overall we are spending too much time at work and not enough time doing other things. Research shows clearly that individuals consistently mistake how much happier an increase in income will make them. In the US, research that has tracked people across time shows that, at any given stage, most people believe that 20% more income would make them happier. But measuring their life satisfaction a few years later when they have achieved that rise in income shows that they are no happier—they have adapted to the new level of income (Easterlin 2003). So, we work too hard to bring in the extra income to consume more, but it makes us no happier. In the process we are left with less and less time to spend with our children, family, and friends, to exercise, to engage in voluntary work, or to pursue recreational interests, which the evidence shows are some of the real sources of well-being. We do not adapt to our relationships and activities in the same way that we adapt to the level of income.

Economists argue that our choices reveal what makes us most happy. The evidence suggests otherwise: our choices—'revealed preferences'—do not always show what makes us happiest. We make systematic mistakes about where to allocate our time. There may be a case, therefore, for action at the collective level.

For some decades, as a society, we have tended to take our productivity gains in the form of income. The former trend of our moving towards a 7-hour day has now swung in the opposite direction back towards the 10-hour day of a century ago. But, as our increasing income isn't making us any happier, well-being data suggest we should start taking our income gains in the form of that really scarce resource: time. But this is a problem that is difficult to tackle at the individual level. We don't want to be left behind our neighbours; we don't want people to think we are not committed to our work; and what is the point of downshifting if everyone else is still working long hours anyway?

One way to restore the adequate balance between work and life in society is for the UK to end individual opt-outs to the European Union Working Time Directive and thus institute a maximum 48-hour working week. (The Directive institutes a maximum average 48-hour working week over a reference period, but the UK allows individuals to opt out of the 48-hour week, so the Directive has had little effect in practice in the UK.) If we wanted to go further, we could then reduce this maximum working week an hour each year until we reach a 35-hour week. This gradual approach would also give business time to make plans, rather than being given little real transition period as in the French case where they introduced a 35-hour week in 2000. If we took our productivity gains in the form of time, assuming annual labour productivity gains of 2%, we could be working an average 30-hour week with an unchanged standard of living in a decade (Hamilton 2004) if accompanied by appropriate pensions reform, re-engagement of those on incapacity benefits into the labour market, and managed migration. This policy is likely to be popular as suggested by a recent European Union survey on attitudes at work with over 50% of workers wanting to reduce their working week to an average of only 34 hours. Indeed, the report also noted that people 'remarkably, would even accept a corresponding drop in income to achieve this' (European Foundation for Improvement of Living and Working Conditions 2003). Economic analysis suggests that this policy should not have a negative effect on employment and that reducing the working week for the majority might also offer more work opportunities for the unemployed and underemployed. As Adair Turner (2001) notes, 'across the board "labour standards"—such as . . . the French thirty-five hour week . . . amount to a diversion . . . of potential post-tax income towards greater leisure but they have no necessary long-term impact on total labour costs per hour . . . [T]he idea that such rights will have any sustained impact on employment levels is wrong.'

Reducing working hours could also reduce commuting, which would have important well-being benefits, as well as lowering environmental impacts. Recent research shows that 'people who commute one hour (one way) would have to earn 40% more [than they do] in order to be fully compensated' for the loss in well-being that they suffer (Stutzer and Frey 2003). In other words, we don't fully factor into our thinking (and pay bargaining) the truth about just how awful the daily commute is. Once more, the research shows that we make systematic errors about predicting our well-being.

A reduction in working hours would need to be accompanied by redistributive measures for the poorest, who are working long hours to survive. This could be done through increasing progressive taxation. Another way to address this and some other wider issues, such as the 'benefits' trap' where the incentive structure of state benefits discourages working, is through instituting a universal citizen's income (see Box 20.5). Such a basic income would enable all citizens to create a life of meaningful paid and unpaid work. We should explore in detail the feasibility of such an income.

Box 20.5 Citizen's income

A 'citizen's income' would be a tax-free income paid to all people (including children) by the state regardless of employment status or social circumstances. It would enable people to make wider choices about how to allocate their time among employment, unpaid work, parenting, and leisure. The principle of a citizen's income is already recognized in the form of child benefit and pensions. It could be financed through a mixture of reduced expenditure on existing benefits and administration, and new taxation. Present benefit administration costs are extremely high and would be significantly reduced as there would be far less means testing required and fewer benefits in existence.

The basic income would have important redistributive effects. It would also promote employment as it would reduce the level at which paid work becomes worthwhile. The present benefits system discourages work as moving from unemployment to paid work may bring very little rise in disposable income after travel costs, childcare, etc. are taken into account.

A citizen's income could be introduced through a transitional system, which could begin as tax neutral, with the basic income replacing the existing system of tax allowances, benefits, and state pension over time. (See <www.citizensincome.org> for more details about the citizen's income.)

Regardless of how far down we bring working hours through regulation, it is clear that we should, in any event, increase flexible working provisions. All employees should be entitled to ask for flexible working patterns including flexibility around working hours, job sharing, taking pay rises in the form of time rather than income, increased parental leave and elder care, compressed working weeks, and teleworking. In addition to flexible working patterns, there is a need for flexibility over the whole of a person's working life. Patterns of breaks from work may emerge as people choose to take time off for education, parenting, or to pursue leisure activities. A redistribution of work may allow a more positive role to emerge for older people so that they can continue to engage with the employment market.

Finally, as part of the move towards a more convivial society, we could also raise the number of public holidays in the UK from eight to closer to the European average of 11. We could even have a vote for when these might be held and what they would celebrate—one option might be a day to spend time in the local community.

4 Create an education system that promotes flourishing

The purpose of our education system is unclear in the eyes of most children, parents, and teachers. Research by **nef** shows that between primary and secondary school there is a huge drop in young people's well-being (Marks *et al.* 2004). This is both in terms of their overall well-being as assessed by life satisfaction and curiosity (a proxy measure for personal development in this context) and specifically their well-being whilst at school (see Fig. 20.3, Box 20.6, and Table 20.1).

This result suggests that our secondary education system is not supporting young people to naturally grow and flourish, which implies that they have lower well-being, both currently and across their lives, than they might have done if the education system was different.

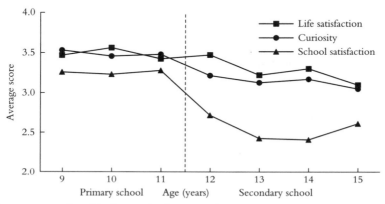

Fig. 20.3 Average overall well-being and school satisfaction by age.

Box 20.6 **Happiness and curiosity in school**

In a pilot study by **nef** and Nottingham City Council to measure well-being in the local area, over 1000 young people completed detailed questionnaires to measure their well-being. The data show that not only did both their satisfaction with life and their curiosity in life (a proxy measure for personal development) both fall as they got older, but also their satisfaction with their school experience plummeted between primary and secondary school, and did not recover. Specifically, their levels of interest in school and the belief that they were learning something dropped a great deal (see Table 20.1).

Table 20.1 Attitudes towards school of primary and secondary school students

Question	Percentage of students strongly agreeing	
	Primary school	Secondary school
I learn a lot at school	71	18
School is interesting	65	12
I enjoy school activities	65	18

The purpose of the education system should be explicitly to promote individual and societal well-being, both now and in the future. It should aim to create capable and emotionally well-rounded young people who are happy and motivated. To this end, all schools should have a strategy to promote emotional, social, and physical well-being. This is not just about rethinking the curriculum—important as that is. This should be a 'whole school' approach that considers a range of matters including the school culture and environment, giving pupils a say, the methods of teaching and assessment, and school governance. Whilst various initiatives are taking forward some aspects of these ideas, they need to be integrated into a comprehensive well-being strategy. There is a briefing paper from the National Healthy School Standard that outlines some of the main ways to take this forward (Health Development Agency (HDA) 2004).

At its heart, education policy must acknowledge that the best way of enabling people to realize their potential is to value them for who they are rather than for their performance against targets. There is evidence to show that focusing heavily on testing can destroy learning, innovation, experimentation, and original thinking (Deci and Ryan 2002).

Creative learning requires creative teaching. The heavy regime of targets and centralized policy prescriptions is destroying the natural curiosity and imagination of teachers. Teachers need more space to explore with each other the challenges they face.

The curriculum also needs to be broadened to promote well-being in order to meet the full range of needs of young people in our society. As well as the obvious need to broaden the vocational elements of the curriculum and to focus on life skills, there are some wider issues. We need to cultivate the variety of 'intelligences' we have—including musical, spatial, physical, interpersonal, and intrapersonal (Gardner 1993). Therefore, we need to refocus the curriculum to include more opportunities around sports, arts, and creativity. There should be a commitment to enhancing social and emotional intelligence as much as academic intelligence. We can, for example, build on the work done in the UK on emotional literacy (there is a campaign to promote emotional literacy called 'Antidote'; see <www.antidote. org.uk>) and the pilots in the US that are looking at how far optimism (a well-established buffer against depression and low well-being) and character strengths can be encouraged (e.g. The Penn Optimism Program).

As discussed earlier, we systematically make errors in relation to choices about our well-being. Early on in their lives, young people should be exposed to the evidence about the kinds of satisfaction derived from different sorts of life choices, perhaps through broader study of the good life: questions around how we should lead our lives, and what well-being is. nef has run two successful summer schools on well-being for 16-year-olds that have enabled them to radically change their views on what the good life is. Broader education of this kind will enable young people to become more autonomous and make better decisions about their lives rather than be led by scripts forced on them by peers, the media, and advertising.

An education system that promotes flourishing is likely to lead to higher productivity, a more entrepreneurial society, and greater active citizenship, as well as less antisocial behaviour and fewer mental health problems.

5 Refocus the National Health Service to promote complete health

There are clear links between promoting health and well-being. It is becoming increasingly clear that psychological factors influence people's physical health to a very large degree. One of the most astonishing findings is the huge positive influence of happiness on longevity (see Box 20.7). Even after allowing for all sorts of other influences—such as income levels, marital status, and even loneliness—researchers on the effects of positive ageing have found that happy people live up to 7.5 years longer. The positive effects from well-being were even larger than the effects for body mass, smoking, and exercise (e.g. Levy *et al.* 2002). Other studies show that unhappy people are more likely to regard themselves as ill, which obviously creates pressures from the 'unhappy-well' on the current health system (Argyle 2001). In addition, risks of cardiovascular illnesses have been estimated to be twice as high for people with depression or mental illnesses, and one-and-a-half times as high for those who are more generally unhappy (Keyes 2004).

The UK National Health Service (a tax-funded service that provides largely free treatment to users) should consider its purpose to be the promotion of complete health, which is defined by the World Health Organization (WHO) as 'a state of complete physical, mental and social well-being and not merely the absence of

---

Box 20.7 **The nun study—happiness and longevity**

In the 1930s, a group of young nuns were asked to write a short autobiography. These papers were recently re-analysed in terms of the amount of positive emotions expressed in the writing. A strong relationship was found to exist between the amount of positive emotion expressed (taken as a proxy for well-being) and the longevity of the nuns (who have had very similar lifestyles with regard to, for example, diet and living standards). Ninety per cent of the quarter who had expressed the most positive emotion in their autobiographies were still alive at the age of 85 compared to just 34% of the quarter who had expressed the least positive emotion. There is discussion about how the relationship between happiness and age actually operates: are happy people less stressed or, for example, do they look after their bodies better? What is clear, however, is that there is a strong relationship between well-being and longevity (Danner *et al.* 2001).

disease or infirmity'. This statement is from the preamble to the Constitution of the WHO as adopted by the International Health Conference, New York, 19–22 June 1946; signed on 22 July 1946 by the representatives of 61 States (Official Records of the World Health Organization, no. 2, p. 100) and entered into force on 7 April 1948. Interestingly, the definition has not been amended since 1948.

More integration between physical health provision and the promotion of personal and social well-being is required to make this happen. An incredible amount is spent on our 'health' service, but most of it focuses on dealing with physical symptoms of sickness. It is acknowledged amongst policy-makers that there is a need to shift the system from being treatment-oriented to being more prevention-oriented. Whilst we are taking some steps towards this, we need to accelerate this process and do a great deal more. We need to invest and commit to disease prevention and public health promotion rather than focusing on technical solutions to ill-health.

The system itself could express the values it promotes by being more personable, socially engaging, and empowering. In part, this could be achieved through putting more emphasis on smaller-scale health enterprises. 'Cottage-hospital'-style approaches are more popular with patients and more successful in developing working partnerships with local communities. Institutions should also have more freedom to set locally agreed health targets in response to local need. Frontline staff could be given discretionary powers and budgets that allow them to short circuit long-term bureaucracy, if necessary. At the local level, government could further encourage and resource community-based institutions such as healthy living centres to take forward a 'complete health' agenda.

The research also suggests that any well-being health policy needs to think of illness in more than just physical terms. In part, this is through recognizing the importance of mental health. The evidence shows that it is our self-perceived health rather than our objective health that matters for our well-being. Mental health is, therefore, by definition, crucial to our well-being and up to one in six people in the UK suffer from depression, anxiety, or phobias at any given time (Social Exclusion Unit 2004). Other estimates suggest that self-reported work-related stress, anxiety, or depression account for approximately 13.5 million reported lost working days per year in Britain (Health and Safety Executive 2003). Depression has risen substantially during a period of growing economic prosperity and, whilst there are distributional effects, the problems are by no means confined to the financially poor. We need to tackle mental health far more systematically, both by dealing with its causes and also through a range of interventions such as strengthening medical care, providing increased support for families and carers and communities, and giving the mentally ill more opportunities to work.

Given the importance of close relationships to our well-being, the health service should explore ways in which it could help deal with relationship issues. Greater provision of relationship and personal counselling, as well as targeted support for groups vulnerable to acute stresses, such as new parents, are likely to be central to this.

Treating people holistically also means that health professionals need to go beyond just treating the biomedical causes of disease to thinking about the social and psychological aspects of how patients are treated. Research shows that, when medical staff are 'patient-centred' and have empathy, involve patients in decision-making, and treat them as real human beings, this has profound consequences for their health. One study notes that, when physicians supported clients to be involved in the decision-making and were empathetic, 'patients showed improved maintenance of healthy behaviour change, greater satisfaction, better adherence to medication, better physical and mental health, fewer healthcare visits . . .' (Williams *et al.* 2000). There is an increasing recognition that patients are 'co-creators' of their health and this needs to be taken on board across the system. To this end, all health institutions should have some system in place to involve patients as partners in the business of delivering health and there needs to be investment in training frontline staff on good practice around this (Boyle *et al.* 2004).

6 Invest in the very early years and parenting

It is increasingly recognized in policy circles that the earliest years of a child's life are very important for their long-term well-being—with cost–benefit analyses showing that investment in the area will pay itself back many times over both financially and non-financially through increased academic achievement, enhanced health outcomes, and reduced social disruption. The majority of attention, however, has been focused on the provision of pre-school childcare for the over-threes—for example through the UK's Sure Start programme. Whilst in no way challenging the importance of this policy, it is vital that an equal amount of attention and money is invested in the first three years of a child's life. The UK is a low spender in this area with only 0.3% of GDP spent on childcare as compared to 2% in Sweden.

There is a need for a balance between the well-being needs of the child and the parents and also the broader issue of future impacts on social well-being. Children need responsive individual attention in their first years, and they need a lot of it. Whilst they often do the best they can, day nurseries can rarely offer this in a consistent manner. They suffer from high turnover of staff and, for both legal and economic reasons, have to have one adult looking after up to three babies—the current UK legal requirement for ratios of adults to children is 1:3 for under-twos, 1:4 for two-year-olds rising to 1:8 for three-year-olds and over (Sure Start 2003). Recent research has begun to suggest that children suffer well-being impacts by spending too much time at day nurseries; with the finding that, if children spend more than 12 hours per week in nursery day care, they are more likely to become insecure and aggressive in later childhood (Melhuish 2004). Whilst it is possible that high-quality nurseries could exist if we were prepared to pay for the quality and quantity of staff, the costs would be very similar to the cost of extending parental leave or supporting child-minders to provide one-to-one care.

A recent report commissioned by the National Audit Office notes, 'in Sweden in the 1980s childcare in the first year of life and later was extremely common due to the extensive availability of state funded provision. However, with the extension of parental leave, parents voted with their feet, and the use of childcare in the first 18 months of life decreased dramatically' (Melhuish 2004). As well as sensing that hands-on parenting is better for their own children, Swedish parents might also be responding to the fact that being with their child is good for their own personal well-being. For, although having children is an enormously stressful time, with many well-being indicators tending to fall (particularly satisfaction with one's relationship with one's partner), in-depth studies show that, of all the activities that parents have to do, it is spending time with their new child that is the most intrinsically rewarding (Delle Fave and Massimini 2004).

It would seem that the best well-being solution for both parents and children is for paid parental leave to be extended to cover at least the whole of a child's first two years of life. This is could be taken by either parent, or potentially shared between them (a desirable option for many couples). The Swedish system provides 13 months of paid leave between a couple, with a further three months available on the payment of a token sum. Whilst such a system might appear difficult to institute from a cost perspective, investment in this area is estimated to pay handsomely for itself in detailed cost–benefit analyses (Melhuish 2004). This solution, however, will clearly not suit all, and, for those parents who need or wish to work, high-quality childcare should be subsidized. This may take unconventional forms such as subsidies for grandparents as well as registered child minders and day nurseries. The role of *au pairs* and nannies, and the qualifications they require, may also need to be examined.

There is also a need to support people to be the best parents they can be. The social environment of the family setting is vital to the future personal well-being of the growing child, and also can affect the social well-being of the whole community. Children brought up in 'risky families'—characterized by conflict, aggression, and relationships that are cold, unsupportive, and neglectful—are far more likely to suffer from future mental illness and also be more socially disruptive (Repetti *et al.* 2002). It is well known that bringing up young children is stressful and increased conflict between parents is a widely observed phenomenon. In well-being terms this is often called the parenting paradox. Adults with young children often report lower satisfaction with their lives and particularly with their relationships with their partners. This may be partially a classic trade-off between more pleasure-orientated satisfaction and more meaning-orientated growth (e.g. MacGregor and Little 1998). There are, however, possible policy interventions that could soften the drop in well-being for parents, thereby enhancing the child's future well-being and also reducing the risk of future socially disruptive behaviour. These might include:

- prenatal parenting (and relationship) skills classes organized through the NHS;

- supportive home visits for new parents from community midwives or nurses whose purpose could be extended to the well-being of the whole family rather just the health of the infant (Mueller 2003);

- more support for community services such as parent and toddler groups where parents self-organize activities and provide support for each other;

- greater support for intergenerational exchanges, so that younger parents could have mentors who were older members of the community—surrogate grand-parents, as it were.

It has also been shown that children suffer adverse effects if their caregiver suffers depression. There needs to be more targeted work tackling depression amongst parents, with a particular focus on postpartum depression. This kind of radical investment in the very early years is likely to yield significant well-being dividends both for the individual and society.

7 Discourage materialism and promote authentic advertising

We live in a highly materialistic society. By 'materialistic' here we mean a value set that believes that material goods will lead to well-being to the exclusion of focusing on other factors. The evidence shows, however, that materialistic people are less happy (Kasser 2002). Material consumption is also the primary driver of many of our environmental problems. It is obviously extremely difficult for policy to intervene to change cultural norms. One area that is open for policy change, however, is advertising. Most advertising rests on the pretence that material goods or services will deliver a variety of non-material benefits, and ultimately happiness. This is patently untrue. Advertising raises the standard of what we consider to be normal, inducing unrealistic comparisons. We need more authentic advertising—products should not pretend to deliver any more than they can. Research shows that young children lack the critical capacity to filter out the subtle selling messages of advertising and, in fact, cannot distinguish between the advertising and the program they are watching. In nef's survey of young people in Nottingham it was clear that materialistic values took an early hold on children—particularly the boys—with over 60% of children thinking it 'is very important to make a lot of money when I grow up' and nearly 70% wanting 'to have a really nice house filled with all kinds of cool stuff'. For these reasons it is worth considering a ban on commercial advertising aimed at the under-eights and a strong code of conduct for commercial advertising for 9–12-year-olds and teenagers. Sweden has a ban on TV advertising aimed at the under-12s and a prohibition against direct marketing aimed at children under the age of 16. These laws are limited, of course; for example, the TV advertising ban cannot extend to cable television channels that broadcast from outside the country; and materialism has more sources than just advertising. Nevertheless, curbing commercial advertising aimed at young people would be an important step in creating a well-being society.

Research shows that a major source of well-being is engaging in challenging activities. The nation's favourite pastime of gardening during the 1970s and 1980s, however, has been replaced by a new, more materialistic, favourite pastime—shopping. We also spend a great deal of our time watching television; research shows that the state induced by passively watching television is similar to that of a mild form of depression! The well-being society is one that spends less time shopping and in front of the television, and more time engaging in active pastimes such as sports, arts, and other hobbies. Research suggests that, whilst we get more well-being from these kinds of activities than 'easy' behaviour like shopping and watching TV, we do not always work up the will to do the more demanding activity. Therefore, society should endeavour to make the well-being choice the easy choice. Policy could intervene to increase the amount of cheap, local leisure provision, such as sports centres and arts venues, as well as informal open spaces and parks. A society more engaged in meaningful pastimes is likely to be less focused on the illusion that material goods will bring it happiness.

As suggested earlier, the education system could also give young people the opportunity to question where their values come from and think about conceptions of the good life. This would also provide an important buffer against materialistic values.

8 Strengthen active citizenship, social well-being, and civil society

Research shows clearly that we derive well-being from engaging with one another and in meaningful projects. In particular, the personal development aspect of well-being is likely to be linked with engaging actively with life and our communities. These findings have implications for policy in relation to civil society, active citizenship, and public-service delivery.

Well-being research comes together with work that has been done on social capital (e.g. Putnam 2000) to show the profound importance of our communities and relationships to our quality of life (see Box 20.8). Community engagement not only improves the well-being of those involved but also improves the well-being of others (Helliwell and Putnam 2004). The relationship is positive in both directions: involvement increases well-being and happy people tend to be more involved in their community. Therefore, interventions in this area should lead to a positive upward spiral.

This evidence bolsters the case for government to support different sorts of community engagement, civil society organizations, and spaces. This could be done

Box 20.8 **Social well-being and social capital**

Robert Putnam describes social capital as 'features of social organisation, such as networks, norms and social trust that facilitate co-ordination and co-operation for mutual benefit'. The major components of social capital are trust, norms, reciprocity, and networks and connections. Social capital has been shown to have positive economic effects. It also has important effects on health and well-being. We are social animals: for example, Putnam cited the extraordinary statistic that if you presently do not belong to any group, joining a club or society of some kind halves the risk that you will die in the next year (from a lecture by Robert Putnam at the Royal Society conference in London in November 2003).

Corey Keyes (1998) defines social well-being as 'the appraisal of one's circumstance and functioning in society'. It is clearly a concept related to social capital but it differs in that it is exclusively based on people's own perceptions ascertained through survey questions. The concept includes how people feel about the society in which they live, their sense of belonging, as well as how much they contribute to society. Social well-being is a relatively new research area, but we would expect there to be important links, as well as differences between social well-being and individual well-being.

partly through setting up a 'citizen's service'—extending the jury service concept, where citizens are obliged to serve as jurors for court trials, to a range of other activities such as volunteering, citizen's panels, environmental juries, etc. There is also a particular need for younger and excluded groups to have more opportunities to engage in the community and in politics. For example, in one study it was noted that 'Young people . . . consistently referred to their feelings of powerlessness and the limited opportunities for them to engage in politics until the age of 18' (Joseph Rowntree Foundation 2000). Time banks and reward card schemes are ways in which marginal groups can be attracted into community engagement through a form of 'mutuality' where they gain clear benefits from what they are doing. (See, for example, Burns and Smith 2004 and, for a discussion about reward cards, see Holdsworth and Boyle 2004.) Government can also intervene where people are unable to engage due to financial circumstances or the regulations of the benefits regime. A means-tested 'participation income' could be paid to those who would otherwise have been unable to carry out voluntary, community, or caring activities (e.g. Atkinson 1996). These kinds of interventions would increase active citizenship throughout society and, in particular, would help to involve people who have traditionally been on the margins of political engagement.

Well-being research has some implications for the way in which citizens are involved in public-service delivery. There is increasing focus on the use of choice in public services, but often this is restricted to choice over service providers. This suffers from a range of well-documented problems. Sometimes individual choice does not lead to optimum outcomes at the collective level. We cannot always choose what we want as individuals but only in the collective (e.g. in relation to a better global environment—see Levett *et al.* 2003). We may not have the information to make the best choices. It is often difficult to know what constitutes a good school or hospital. In the end it is those schools or hospitals with good reputations that end up able to choose the people they take on due to increased demand—the very reverse of what choice was said to deliver. Psychologist Barry Schwartz has shown that too much choice can lower our well-being as we feel overwhelmed (Schwartz 2004).

Choice in public services can be far broader than the choice of public service provider (e.g. Leadbetter 2004). It is being increasingly recognized through concepts such as 'personalization' and 'co-production' that it can be very productive to involve people in the design and delivery of the services that they receive (Burns and Smith 2004). Increasing movement towards this kind of choice in public services will give the public-service user autonomy of the sort that is closely linked to well-being in many other contexts, such as in the workplace, and therefore, is likely to increase well-being. It also builds on Swiss research that shows that there is a strong link between democratic involvement and happiness (see Box 20.9).

In particular, the governance of public services is presently dominated by a plethora of targets and indicators, usually set by central government. As part of a move to put users in the centre of public services, we should move away from this command-and-control model towards a model based on accountability to stakeholders. One way to do this would be to drop the swathes of targets and

Box 20.9 **Democracy and happiness**

Participatory democracy makes us happier. This is the finding of research comparing Swiss cantons (districts) that differ in the extent to which they use referenda for making major decisions (Frey and Stutzer 2002). Most interesting of all, around two-thirds of the well-being effect can be attributed to actual participation itself, and only one-third to the improvement in policy as a result of the participation. This was discovered through looking at the well-being of foreigners resident in Switzerland, who get the well-being benefit from the improved decision-making, but not from the participation itself. This implies that an increased ability to participate—both in politics and in the way public services are delivered—may have positive well-being dividends.

replace them with a process of stakeholder engagement and accountability as recommended by corporate standards such as AA1000 (Institute of Social and Ethical Accountability <www.accountability.org.uk>). This leads to measurement of what matters in the eyes of stakeholders, and the publication of an externally verified report of these factors.

Well-being research has pioneered the use of subjective indicators. To build on the learning that has emerged from this, public-service deliverers could monitor users' satisfaction with services and focus on this as a key indicator to improve in conjunction with more 'objective' indicators. It should be noted, however, that, whilst subjective indicators are important as a supplement to today's focus on objective data, they don't capture everything because people sometimes mitigate negative experiences and therefore express satisfaction even where services are objectively of low quality (Eckersley 2004). The Canadian government has set user satisfaction as the most important indicator across public-service delivery. This method focuses on one of the main ends of public services, but allows flexibility in the means to promote the satisfaction. Our present system has it the wrong way around: the centre tends to specify the means rather than the ends.

Call for a flourishing society

Well-being research points the way to a society that could be profoundly different. If we take the right steps we could move towards a happier, more vibrant society where people are actively engaged in their communities. A flourishing society would be healthier, more productive, entrepreneurial, creative, and engaged.

Well-being research is, however, still fairly new. Whilst striking lessons have already emerged from the research, we need to understand better what well-being is and what brings it about. In particular, there needs to be more research about the second dimension of well-being—that of personal development—and the policy implications that emerge from it, rather than the current interest that focuses solely on life satisfaction. Similarly, we need to go beyond just looking at individual well-being to consider social well-being: how we feel about, and contribute to, our broader society.

Whilst much more funding needs to be provided for high-quality interdisciplinary research on well-being especially regarding the impact of public policy, policy-makers already can, and should, ask themselves: 'What would policy look like if it were seeking to promote well-being?' This should be one of the defining questions of politics in developed countries.

Acknowledgements

We would like to thank the AIM Foundation for the funding that made this work possible. We would also like to thank the people who commented on and inputted into the manifesto including James Baderman, David Boyle, Jessica Bridges-Palmer,

Brendan Burchill, Andrew Clark, Pat Conaty, Richard Easterlin, Richard Eckersley, Felicia Huppert, Tim Jackson, Corey Keyes, Sonja Lyubomirsky, Marie Marks, James Park, James Robertson, Nikul Shah, Andrew Simms, Rajat Sood, Joar Vittersø, and Stewart Wallis. Particular thanks to Andrea Westall, for her support and considerable input, and to Mary Murphy for getting the original version of the document to publication.

References

Alesina, A., Di Tella, R., and MacCulloch, R. (2001). *Inequality and unhappiness: are Europeans and Americans different?*, NBER Working Paper 8198. National Bureau of Economic Research, Cambridge, Massachusetts.

Argyle, M. (2001). *The psychology of happiness*. Routledge, London.

Atkinson, T. (1996). The case for a participation income. *Political Quarterly (Oxford)* **67**, 67–70.

Beatty, C. and Fothergill, S. (2004). *The diversion from "unemployment" to "sickness" across British regions and districts*. Centre for Regional Economic and Social Research, Sheffield Hallam University, Sheffield.

Boyle, D., Conisbee, M., and Burns, S. (2004). *Towards an asset-based NHS*. nef, London.

Brickman, P., Coates, D., and Janoff-Bulman, R. (1978). Lottery winners and accident victims: is happiness relative? *J. Pers. Soc. Psychol.* **36**, 917–27.

Burns, S. and Smith, K. (2004). *Co-production works!* nef, London.

Csikszentmihalyi, M. (1997). *Finding flow, the psychology of engagement in everyday life*. Perseus Books, New York.

Danner, D., Snowdon, D., and Friesen, W. (2001). Positive emotion in early life and longevity: findings from the nun study. *J. Pers. Soc. Psychol.* **80**, 804–13. [quoted in Seligman 2002.]

Deci, E. and Ryan, R. (2002). The paradox of achievement—the harder you push, the worse it gets. In *Improving academic achievement: contributions of social psychology* (ed. J. Aronson), pp. 59–85. Academic Press, New York.

Delle Fave, A. and Massimini, F. (2004). Parenthood and the quality of experience in daily life. *Soc. Indicat. Res.* **67**, 75–106.

Diener, E. and Seligman, M. (2002). Very happy people. *Psychol. Sci.* **13**, 81–4.

Diener, E. and Seligman, M. (2004). Beyond money: toward an economy of well-being. *Psychol. Sci. Public Interest* **5** (1), 1–31.

Diener, E., Horwitz, J., and Emmons, R. (1985). Happiness of the very wealthy. *Soc. Indicat. Res.* **16**, 263–74.

Di Tella, R., MacCulloch, R., and Oswald, J. (2003). The macroeconomics of happiness. *Rev. Econ. Soc. Statist.* **85** (4), 807–27. Available online at <http://mitpress.mit.edu/catalog/item/default.asp?tid=11119andttype=6>.

Donovan, N. and Halpern, D. (2002). *Life satisfaction: the state of knowledge and the implications for government*. Prime Minister's Strategy Unit, London. Available online at <http://www.number-10.gov.uk/su/ls/paper.pdf>.

Easterlin, R. A. (2003) Explaining happiness. *Proc. Natl Acad. Sci. USA* **100** (19), 11176–83.

Eckersley, R. (2004). *Well and good: how we feel and why it matters*. Text Publishing, Melbourne, Australia.

European Foundation for Improvement of Living and Working Conditions (2003). *A new organisation of time over working life*. Available online at <www.eurofound.ie>.

Frey, B. and Stutzer, A. (2002). *Happiness and economics*. Princeton University Press, Princeton, New Jersey.

Gardner, H. (1993). *Frames of mind: the theory of multiple intelligences*. Basic Books, New York.

Hamilton, C. (2004). *Growth fetish*. Pluto Press, London.

Harter, J., Schmidt, F., and Hayes, T. (2002). Business-unit-level relationship between employee satisfaction, employee engagement and business outcomes: a meta-analysis. *J. Appl. Psychol.* **87**, 268–79.

Health Development Agency (HDA) (2004). *Promoting emotional health and wellbeing through the National Healthy Schools Standard*. Available online at <www.had-online.org.uk>.

Health and Safety Executive (2003). *Self-reported work-related illness in 2001/02—results of a household survey*. Available online at <http://www.hse.gov.uk/statistics/causdis/swi0102.pdf >.

Helliwell, J. and Putnam, R. (2004). The social context of well-being. *Phil. Trans. R. Soc. Lond. B*, **359**, 1435–46.

Holdsworth, M. and Boyle, D. (2004). *Carrots not sticks—the possibilities of a sustainable reward card for the UK*. National Consumer Council, London.

Jackson, T. (2004). *Chasing progress*. nef, London.

Joseph Rowntree Foundation (2000). *Young people's politics: political interest and engagement amongst 14- to 24- year-olds*. York Publishing, York.

Kasser, T. (2002). *The high price of materialism*. MIT Press, Cambridge, Massachusetts.

Keyes, C. (1998). Social well-being. *Soc. Psychol. Q.* **61** (2), 121–40.

Keyes, C. (2002). The mental health continuum: from languishing to flourishing in life. *J. Health Soc. Behav.* **43**, 207–22.

Keyes, C. (2004). The nexus of cardiovascular disease and depression revisited: the complete mental health perspective and the moderating role of age and gender. *Aging Ment. Health* **8** (3), 266–74.

Keyes, C. and Haidt, J. (2003). *Flourishing: positive psychology and the life well-lived*. American Psychological Association, Washington, DC.

Leadbetter, C. (2004). *Personalisation through participation*. Demos, London.

Levett, R. with Christie, I., Jacobs, M., and Therivel, R. (2003). *A better choice of choice: quality of life consumption and economic growth*. Fabian Society, London.

Levy, B.R., Slade, M.D., Kunkel, S.R., and Kasl, S.V. (2002). Longevity increased by positive self perceptions of age. *J. Pers. Soc. Psychol.* **83** (2), 261–70.

Lykken, D. (1999). *Happiness: the nature and nurture of joy and contentment*. St Martin's Press, New York.

Lyubomirsky, S., King, L., and Diener, E. (2005). Is happiness a strength? *Psychol. Bull.*, in press.

MacGregor, I. and Little, B. (1998). Personal projects, happiness and meaning—on doing well and being yourself. *J. Pers. Soc. Psychol.* **74** (2), 494–512.

Maller, C., Townsend, M., Brown, P., and St. Leger, L. (2002). *Healthy parks, healthy people*. Deakin University and Parks Victoria, Victoria, Australia. Available online at <http://www.parkweb.vic.gov.au/resources/mhphp/pv1.pdf >.

Marks, N., Shah, H., and Westall, A. (2004). *The power and potential of well-being indicators*. nef, London. Summary available online at <www.neweconomics.org>.

Melhuish, E. (2004). *Literature review of the impact of early years provision on young children*. National Audit Office, London. Available online at <www.nao.org.uk/publications/nao_reports>.

Mueller, G. (2003). *Parental stress and marital satisfaction: some results from a home visiting experiment for maintaining the quality of life of young families*. University of Fribourg, Fribourg, Switzerland.

Office of National Statistics (ONS) (2001). *Psychiatric morbidity among adults living in private households 2000.* Available online at <www.statistics.gov.uk>.

Office of National Statistics (ONS) (2004). *Annual report and accounts 2002–3.* Available online at <www.statistics.gov.uk>.

Performance and Innovation Unit (2002). *Social capital—a discussion paper.* Cabinet Office, London. Available online at <http://www.number-10.gov.uk/su/social%20capital/socialcapital.pdf>

Putnam, R. (2000). *Bowling alone: the collapse and revival of American community.* Simon and Schuster, New York.

Repetti, R., Taylor, S., and Seeman, T. (2002). Risky families: family social environments and the mental and physical health of offspring. *Psychol. Bull.* **128** (2), 330–66.

Ryan, R. and Deci, E. (2001). On happiness and human potentials: a review of research on hedonic and eudaimonic well-being. *Annu. Rev. Psychol.* **52**, 141–66.

Sarafino, E. (2002). *Health psychology.* Wiley, New York.

Schkade, D. and Kahneman, D. (1998). Does living in California make people happy? A focusing illusion in judgments of life satisfaction. *Psychol. Sci.* **9**, 340–6.

Schwartz, B. (2004). *The paradox of choice.* HarperCollins, New York.

Seligman, M. (2002). *Authentic happiness.* Free Press, New York.

Sheldon, K.M., Lyubomirsky, S., and Schkade, D. (2005). Pursuing happiness: the architecture of sustainable change. *Rev. Gen. Psychol.,* in press.

Simms, A. (2000). *Collision course: free trade's free ride on the global climate.* **nef**, London. Available online at <www.neweconomics.org>.

Simms, A. (2003). *Ghost Town Britain.* **nef**, London. Available online at <www.neweconomics.org>.

Singer, B. and Ryff, C. (Eds.) (2001). *New horizons in health: an integrative approach.* National Academies Press, Washington, DC.

Social Exclusion Unit (2004). *Mental health and social exclusion.* Office of the Deputy Prime Minister, London.

Stutzer, A. and Frey, B. (2003). *Stress that doesn't pay: the commuting paradox,* Working paper series 151. Institute for Empirical Research in Economics, University of Zurich, Zurich.

Sure Start (2003). *National standards for under 8s day care and child minding.* Department for Education and Skills (DfES), London.. Available online at <www.childcarelink.gov.uk>.

Turner, A. (2001). *Just capital.* Pan Books, London.

Veenhoven, R. (2003). *World database of happiness. Catalog of happiness queries.* Available at <www.eur.nl/fsw/research/happiness>. Erasmus University, Rotterdam.

Williams, G., Frankell, R., Campbell, T., and Deci, E. (2000). Research on relationship-centred care and healthcare outcomes from the Rochester biopsychosocial program: a self-determination theory integration. *Fam. Syst. Health* **18**, 79–90.

Index